POLARIZED LIGHT IN LIQUID CRYSTALS AND POLYMERS

BICENTENNIAL 1807 WILEY 2007 BICENTENNIAL

The Wiley Bicentennial–Knowledge for Generations

Each generation has its unique needs and aspirations. When Charles Wiley first opened his small printing shop in lower Manhattan in 1807, it was a generation of boundless potential searching for an identity. And we were there, helping to define a new American literary tradition. Over half a century later, in the midst of the Second Industrial Revolution, it was a generation focused on building the future. Once again, we were there, supplying the critical scientific, technical, and engineering knowledge that helped frame the world. Throughout the 20th Century, and into the new millennium, nations began to reach out beyond their own borders and a new international community was born. Wiley was there, expanding its operations around the world to enable a global exchange of ideas, opinions, and know-how.

For 200 years, Wiley has been an integral part of each generation's journey, enabling the flow of information and understanding necessary to meet their needs and fulfill their aspirations. Today, bold new technologies are changing the way we live and learn. Wiley will be there, providing you the must-have knowledge you need to imagine new worlds, new possibilities, and new opportunities.

Generations come and go, but you can always count on Wiley to provide you the knowledge you need, when and where you need it!

William J. Pesce
President and Chief Executive Officer

Peter Booth Wiley
Chairman of the Board

POLARIZED LIGHT IN LIQUID CRYSTALS AND POLYMERS

Toralf Scharf

Institute of Microtechnology
University of Neuchâtel
Switzerland

BICENTENNIAL 1807 WILEY 2007 BICENTENNIAL

WILEY-INTERSCIENCE
A JOHN WILEY & SONS, INC., PUBLICATION

Published by John Wiley & Sons, Inc., Hoboken, New Jersey
Published simultaneously in Canada

Library of Congress Cataloging-in-Publication Data:

Scharf, Toralf, 1967–
Polarized light in liquid crystals and polymers / by Toralf Scharf.
p. cm.
Includes bibliohgraphical references and index.
ISBN-13: 978-0-471-74064-3
ISBN-10: 0-471-74064-0
1. Polarization (Light) 2. Liquid crystals. 3. Liquid crystal polymers. 4. Light-Transmission. I. Title.

QC441.S33 2007
535.5'2–dc22 2006042964

10 9 8 7 6 5 4 3 2 1

CONTENTS

PREFACE

Liquid crystal displays are nowadays the standard for mobile/portable information displays. With the increasing interest in displays, the subject of liquid crystal optical devices has grown to become an exciting and expansive field of research and development. However, a fundamental difference between displays and more generalized optical elements is the resolution. An optical element needs accurate control of the shape of the wavefront over a certain area. To assure high quality, the lateral resolution has to be very high. To achieve that, pixilation is done by a diffractive approach and gradient index optics are used in relatively thick liquid crystal layers. While the first concept is the classical approach for information displays, discussion of gradient index optics with liquid crystals is rarely laid out. The goal of this work is to fill that gap and to enable effective analysis and design of high-resolution optical elements with liquid crystals and polymers.

This book presents the basic principles and provides a systematic treatment of light propagation in liquid crystals and polymers from ray tracing to Bragg reflection. It is intended to be a textbook for engineers and scientists as well as students. In addition, it is structured to serve as a reference book for optical design of photonic systems, including anisotropic materials as well. To best serve this range of requirements, the book covers three main subjects: theoretical analysis, practical characterization, and examples of concrete devices.

The theoretical section gives the basis for further analysis and repeats some well-known concepts for completeness. It adds some less common methods such as ray tracing and finite difference time domain simulations. The former method is used for the design of classical optical elements like prisms and lenses, while the latter is indispensable for the analysis and design of diffractive optics and texture analysis. Practical characterization of liquid crystal elements is often done with the polarization microscope. A chapter is dedicated to standard methods of observation. It is based on a minimum of theoretical understanding. Analysis of birefringent devices is only possible if the internal structure is analyzed. The local orientation of liquid crystals and polymers is given by their texture. A particular effort was made to discuss examples of how the texture translates into optical microscope images. The discussions help to analyze high-resolution optical devices and liquid crystal textures. It is known that technologies advance rapidly, but the underlying principles of operation remain. Therefore the discussion is focused in last parts on example devices and systems to show how the theoretical concepts apply. Owing to the amount of published literature it is almost impossible to cover all of the details, therefore I have selected typical approaches.

The field of applied optics with anisotropic materials is interdisciplinary involving optics, materials science, and electronics. It is therefore difficult to cover all aspects adequately. I hope that this book is useful to understand effects, design new devices, and explain functional principles of the optics of liquid crystal and polymers.

Many of the results presented were produced in collaboration with several students. The author is grateful to Christian Bohley, Gerben Boer, Manuel Bouvier, and Sylvain Jaquet for their excellent work. The author also wishes to thank his colleagues at the University of Neuchâtel and especially at the Institute of Microtechnology for their help, discussions and continued support. Special thanks are given to Joachim Grupp, Naci Basturk, Rolf Klappert and Norbert Koptsis from ASULAB S.A. in Marin, and to Sung-Gook Park, Celestino Padeste, Helmut Schift, and Jens Gobrecht from PSI Villigen. Without their help and technological support, it would be impossible for me to provide such a variety of examples on liquid crystal technology. The author also wishes to thank Christian Bohley and Gerben Boer for proofreading parts of the manuscript and their comments. I am particularly indebted to Ralf Stannarius, for his patient reading of the manuscript and helpful suggestions and corrections.

Toralf Scharf

Neuchâtel, Switzerland
September 2006

1

POLARIZED LIGHT

1.1 INTRODUCTION

A normal beam of light in isotropic material consists of many individual waves, each vibrating in a direction perpendicular to its path. Measurable intensities therefore refer to a superposition of many millions of waves. Normally, the vibrations of each ray have different orientations with no favored direction. In some cases, however, all the waves in a beam vibrate in parallel planes in the same direction. Such light is said to be polarized, that is, to have a directional characteristic. More specifically, it is said to be linearly polarized, to distinguish it from circularly and elliptically polarized light (to be discussed later). Light from familiar sources such as a light bulb, the Sun, or a candle flame is unpolarized, but can easily become polarized as it interacts with matter. Such light is called natural light. Reflection, refraction, transmission, and scattering all can affect the state of polarization of light. The human eye cannot easily distinguish polarized from natural light. This is not true for all animals (Horváth, 2004). In fact, light from the sky is considerably polarized (Minnaert, 1954) as a result of scattering, and some animals, such as bees, are able to sense the polarization and use it as a directional aid. A major industrial use of polarized light is in photoelastic stress analysis (Föppl, 1972; Rohrbach, 1989; Dally, 1991). Models of mechanical parts are made of a transparent plastic, which becomes birefringent when stressed. Normal forces are applied to the model, which is then examined between polarizers. Between crossed polarizers, unstressed regions remain dark; regions under stress change the polarization of light so that light can be transmitted. Figure 1.1 shows an example of such stressed plastic parts

Polarized Light in Liquid Crystals and Polymers. By Toralf Scharf

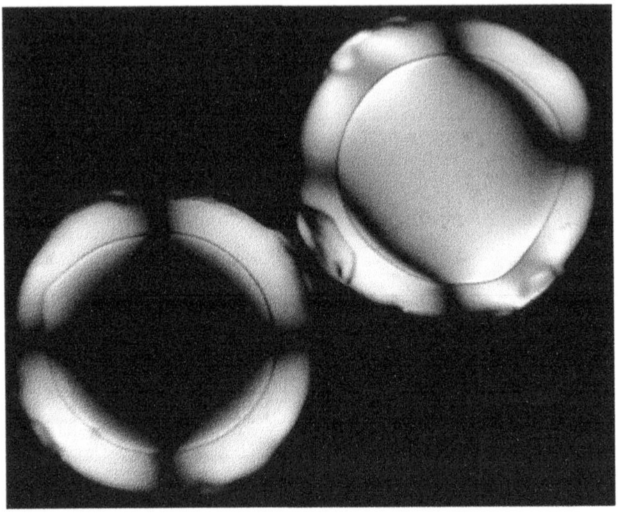

Figure 1.1 Plastic disks of 1 mm diameter arranged between crossed polarizers. The disks are microfabricated by photolithography in a photo-polymerizable isotropic material (SU8). The polarizer and analyzer are horizontally and vertically oriented respectively. The left disk is not stressed, but the right is pressed together from top left and bottom right. The stress produces birefringence that cerates interference colors.

between crossed polarizers. Disks of 1 mm diameter are shown. One is compressed and shows characteristic interference fringes due to stress. When white light is used, each wavelength is affected differently; the result is a highly colored contour map showing the magnitude and direction of the stresses. In Figure 1.1, the points where mechanical force is applied become visible as bright spots on the top-left and bottom-right positions. Another way to influence the polarization of light is optical activity. Many compounds are optically active; that is, they have the ability to rotate the plane of plane-polarized light. This property can be a molecular property and may be used to measure the concentrations of such compounds in solutions. More examples can be found in work by Pye (2001), Shurcliff (1962) and Minnaert (1954). Today, with advanced methods for measuring light fields now being available, polarized light still attracts a lot of attention. Recently, a discussion on singularities of light fields has led to a more intense discussion on polarization states and their propagations (Berry, 1999, 2003).

1.2 CONCEPT OF LIGHT POLARIZATION

Polarization is a property that is common to all types of vector waves. Electromagnetic waves also possess this property (Weller, 1979). For all types of vector waves, polarization refers to the behavior with time of one of the field vectors appropriate to that wave, observed at a fixed point in space. Light waves are electromagnetic in nature and require four basic field vectors for their complete description: the electrical field strength **E**, the electric displacement density **D**, the magnetic field strength **H**, and the magnetic flux density **B**. Of these four vectors the electric field strength **E** is chosen to define the state of polarization of light waves (Born, 1993). This choice is based on the fact that, when light interacts with matter, the force exerted on the electrons by the electric field of the light waves is much greater than the force exerted on these electrons by the magnetic field of the wave. In general, once the polarization of **E** has been determined, the polarization of the three remaining vectors **D**, **H**, and **B** can be found, because the field vectors are interrelated by Maxwell's field equations and the associated constitutive (material) relations. In the following we will focus our attention on the propagation of light as defined by the behavior of the electric field vector $\mathbf{E}(\mathbf{r}, t)$ observed at a fixed point in space, **r**, and at time, t. In general the following statement holds: The change of polarization properties of light is initiated by a symmetry break while light propagates in a certain direction. This symmetry break can be made simply by the geometry of obstacles in the propagation path, by high dimensional order, or by anisotropy on a molecular or atomic level. Imagine a wave traveling in space (vacuum) in a direction described by a wave vector **k** as shown in Figure 1.2. If such a wave hits a surface of an isotropic body of different refractive index at normal incidence, the symmetry with respect to the propagation direction is preserved and the state of polarization is not changed. The situation becomes different when the incidence is no longer normal to the surface. Now, a projection of the different vector components of the electromagnetic field has to be performed. This leads to different equations for reflection and transmission coefficients and finally to the Fresnel equations (see Chapter 3 for details), which allow the calculation of the change of the polarization state. Examples for highly ordered systems with anisotropy are crystals, liquid crystalline phases, and ordered polymers. Here, even for normal incidence, the anisotropy of the material can lead to a different interaction

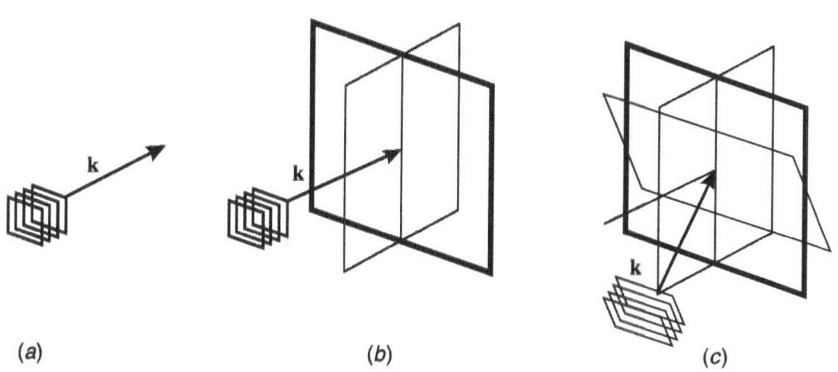

Figure 1.2 A plane wave traveling in direction **k**. (*a*) The planes of constant phases (wavefronts) are indicated with the small layers. If the wave hits an interface normal as in (*b*), the symmetry of the problem is maintained. For isotropic materials no change of the polarization state of light is expected. If the light train has an oblique incidence as in (*c*) a change will be observed that leads to different reflection and transmission properties for differently polarized light trains.

of the vector components of the incident light: Projection of the electric field vectors on the symmetry axes of the system is needed to describe light propagation correctly. Matrix methods are convenient in this case to describe the change of polarization, for example the Jones matrix formalism (Chapter 4). Nanostructured materials with structures smaller than the wavelengths of light also fall into this category. An example is zero-order gratings with their particular polarization properties (Herzig, 1997). Light scattering also leads to polarization effects. For example, according to Rayleigh's classical scattering laws, any initially unpolarized beam becomes polarized when scattered. For this reason, the diffuse scattered light from the Sun in the atmosphere is partially polarized (Minnaert, 1954). On the molecular level, optical activity is an exciting case of interaction of polarized light with anisotropic molecules. Such a phenomenon is observed in sugar in solutions. The description is made by using a particular reference frame that is based on special states of light polarization (circularly polarized light). Optical activity is found naturally in crystals and can also be induced with electrical and magnetic fields and through mechanical stress.

Polarized light interaction happens on every length scale and is therefore responsible for a multitude of different effects. A description is particular difficult if all kinds of mechanisms overlap.

1.3 DESCRIPTION OF THE STATE OF POLARIZATION

To describe a general radiation field, four parameters should be specified: intensity, degree of polarization, plane of polarization, and ellipticity of the radiation at each point and in any given direction. However, it would be difficult to include such diverse quantities as intensity, a ratio, an angle, and a pure number in any symmetrical way in formulating the equation of propagation. A proper parametric representation of polarized light is therefore a matter of some importance. The most convenient representation of polarized light uses a set of four parameters, introduced by Sir George Stokes in 1852. One standard book on polarization optics (Goldstein, 2003) is based on this formulation and offers a deep insight into the formalisms by giving examples. It seems advantageous to use a description of light polarization that is linked to a measurement scheme. That is the case for the Stokes formalism. Assume for a moment that one has tools to separate the linear polarized (of different directions) and circular polarized light (of different sense of polarizations) from an incident light beam. If we know the direction of propagation of the light we are able to determine the properties related to polarization. Four values have to be measured to identify the state of polarization (including the ellipticity), the direction of the ellipse, and the degree of polarization (Gerrard, 1994). The scheme introduced by Stokes is based on the measurement of intensities by using ideal polarization components. Measurement of the total intensity I is performed without any polarizing component. Next, three intensities have to be measured when passing through an ideal polarizer (100% transmission for linear polarized light, 0% for extinction) at 0° 45° and 90° orientation, respectively. The coordinate system is fixed with respect to the direction of propagation. The last measurement uses a circular polarizer. All these measurements, together, allow the determination of the Stokes parameters S_0, S_1, S_2, S_3. The degree of polarization is given by comparing the total intensity with the sum of the ones measured with polarizers. The direction of the polarization ellipse can be found by analyzing the measurement with linear polarizers, carried out under different angles of the polarizer. The sense of rotation and the ellipticity is accessible when all measurements with polarizers are considered. To obtain this in a more quantitative manner, we start with a description of a transversal wave and the polarization states following the description in the work of Chandrasekhar (1960). We assume propagation in an isotropic material. The polarization state description is closely related to the propagation direction. In a ray model the propagation direction is easily defined as the vector of the ray direction. To have easy access to the main parameters of light polarization one assumes that the propagation direction is known and defines a coordinate system. Let z be the direction of propagation and the **k** vector. Then the two components of the electrical field can be assigned with a phase and amplitude such that

$$E_x = E_{x0}\sin(\omega t + \varphi_x) \qquad \text{and} \qquad E_y = E_{y0}\sin(\omega t + \varphi_y), \tag{1.1}$$

where E_x and E_y are the components of the vibration along directions x and y, at right angles to each other, ω is the angular frequency of vibration, and E_{x0}, E_{y0}, φ_x, and φ_y are constants. Figure 1.3 presents the geometrical arrangement. The field components vary in time with a certain phase shift and describe an ellipse. If the principal axes of the ellipse described by (E_x, E_y) are in directions making an angle Θ and $\Theta + \pi/2$ to the direction x, the equations representing the vibration take the simplified forms

$$E_\Theta = E_0 \cos\beta \sin\omega t \qquad \text{and} \qquad E_{\Theta+\pi/2} = E_0 \sin\beta \cos\omega t, \tag{1.2}$$

where β denotes an angle whose tangent is the ratio of the axis of the ellipse traced

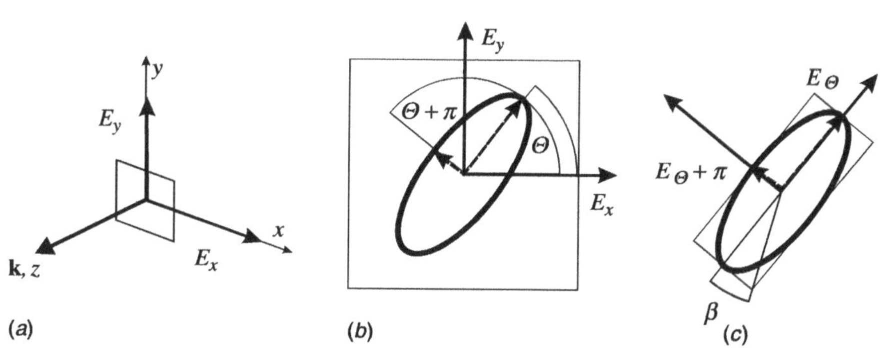

Figure 1.3 To a plane wave traveling in the z-direction with the wave vector **k**, one can assign a plane with **k** as the normal vector, which contains the field components of the electric field E_x and E_y.

by the end point of the electric vector as shown in Figure 1.3. We shall suppose that the numerical value of β lies between 0 and $\pi/2$ and that the sign of β is positive or negative according to whether the polarization is right-handed or left-handed. E_0 denotes a quantity proportional to the mean of the electric field vector, whose square is equal to the intensity $\hat{I}$ of the beam:

$$\hat{I} = E_0^2 = E_{x0}^2 + E_{y0}^2 = \hat{I}_x + \hat{I}_y. \tag{1.3}$$

Note that the intensity $\hat{I}$ introduced here is not the intensity measured by a detector. A detector measures time averaged values of the electric fields that is half of $\hat{I}$ because of the time average of the trigonometric functions in Equation (1.1).

The formulas connecting the representations of polarized light with four intensities S_0, S_1, S_2, and S_3 are important and can be combined in the following manner. The electrical field components of the field amplitude E_Θ and $E_{\Theta+\pi/2}$ in the x- and y-directions are summed (see Fig. 1.3). Starting from the representation in Equation (1.2) we obtain for the vibrations in the x- and y-direction the expressions

$$E_x = E_0(\cos\beta\cos\Theta\sin\omega t - \sin\beta\sin\Theta\cos\omega t) \tag{1.4a}$$

$$E_y = E_0(\cos\beta\sin\Theta\sin\omega t + \sin\beta\cos\Theta\cos\omega t). \tag{1.4b}$$

These equations can be reduced to the form of Equation (1.1) by letting

$$E_{x0} = E_0\sqrt{\cos^2\beta\cos^2\Theta + \sin^2\beta\sin^2\Theta} \tag{1.5a}$$

$$E_{y0} = E_0\sqrt{\cos^2\beta\sin^2\Theta + \sin^2\beta\cos^2\Theta} \tag{1.5b}$$

and

$$\tan\varphi_x = \tan\beta\tan\Theta \qquad \text{and} \qquad \tan\varphi_y = -\tan\beta/\tan\Theta. \tag{1.6}$$

The intensities $\hat{I}_x$ and $\hat{I}_y$ in the directions x and y are therefore given by

$$\hat{I}_x = E_{x0}^2 = I(\cos^2\beta\cos^2\Theta + \sin^2\beta\sin^2\Theta) \tag{1.7a}$$

$$\hat{I}_y = E_{y0}^2 = I(\cos^2\beta\sin^2\Theta + \sin^2\beta\cos^2\Theta). \tag{1.7b}$$

Further, according to Equations (1.5) and (1.6) we may readily verify that

$$2E_{x0}E_{y0}\cos(\varphi_x - \varphi_y) = 2E_0^2(\cos^2\beta - \sin^2\beta)\cos\Theta\sin\Theta = \hat{I}\cos 2\beta\sin 2\Theta. \tag{1.8}$$

Similarly,

$$2E_{x0}E_{y0}\sin(\varphi_x - \varphi_y) = \hat{I}\sin 2\beta. \tag{1.9}$$

From the foregoing equations (1.3–1.7) it follows that whenever the regular vibrations representing an elliptically polarized beam can be expressed in the form of Equation (1.1), we can at once write the relations

$$S_0 = \hat{I} = E_{x0}^2 + E_{y0}^2 = \hat{I}_x + \hat{I}_y \quad (1.10a)$$

$$S_1 = E_{x0}^2 - E_{y0}^2 = I\cos 2\beta \cos 2\Theta = \hat{I}_x - \hat{I}_y \quad (1.10b)$$

$$S_2 = 2E_{x0}E_{y0}\cos(\varphi_x - \varphi_y) = \hat{I}\cos 2\beta \sin 2\Theta = (\hat{I}_x - \hat{I}_y)\tan 2\Theta \quad (1.10c)$$

$$S_3 = 2E_{x0}E_{y0}\sin(\varphi_x - \varphi_y) = \hat{I}\sin 2\beta = (\hat{I}_x - \hat{I}_y)\frac{\tan 2\beta}{\cos 2\Theta}. \quad (1.10d)$$

These are the Stokes parameters representing an elliptically polarized beam. We observe that among the quantities S_0, S_1, S_2, and S_3 defined as in Equations (1.10) there exists the relation

$$S_0^2 = S_1^2 + S_2^2 + S_3^2. \quad (1.11)$$

Further, the plane of polarization and the ellipticity follow from the equations

$$\tan 2\Theta = \frac{S_2}{S_1} \quad \text{and} \quad \sin 2\beta = \frac{S_3}{\sqrt{S_1^2 + S_2^2 + S_3^2}}. \quad (1.12)$$

For the representation given in Equation (1.1) one considers the amplitude and the phases to be constant. This is not realistic. The high angular frequency ω of the electromagnetic oscillation of light allows us to suppose that the phases and amplitudes may be constant for millions of vibrations and yet changes irregularly millions of times a second. In an polarized light beam these irregular variations must be such that the ratio of amplitudes and the difference of phases are absolute constant. One is able to measure the mean intensity in any direction in the transverse plane. The apparent intensities $\hat{I}_x$ and $\hat{I}_y$ will be given by their mean values of electric field components E_x and E_y. Equations (1.10) can be rewritten by using mean intensities and allows to apply the formalism to measurements.

1.4 THE STOKES CONCEPT

Using Equations (1.10), the so-called Stokes column **S** for completely polarized light is defined as

$$\mathbf{S} = \begin{pmatrix} S_0 \\ S_1 \\ S_2 \\ S_3 \end{pmatrix} = \begin{pmatrix} E_{x0}^2 + E_{y0}^2 \\ E_{x0}^2 - E_{y0}^2 \\ 2E_{x0}E_{y0}\cos(\varphi_x - \varphi_y) \\ 2E_{x0}E_{y0}\sin(\varphi_x - \varphi_y) \end{pmatrix}. \quad (1.13)$$

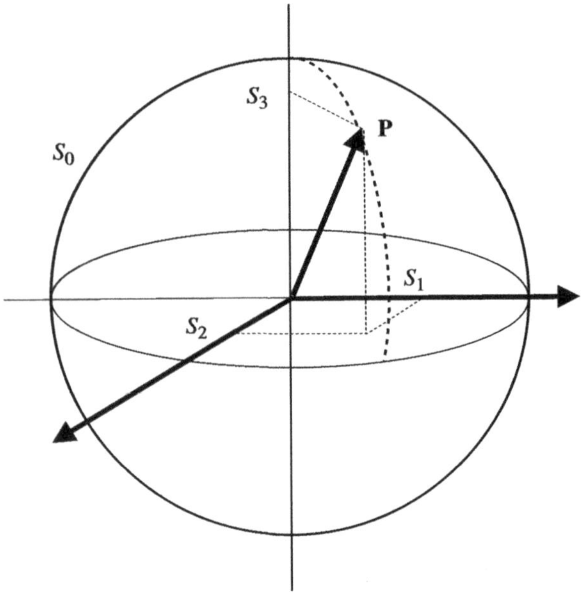

Figure 1.4 Poincaré sphere representation of the polarization states of a monochromatic wave. In most definitions the sphere has a radius of 1. One can introduce the parameter S_0 as the sphere radius to consider partially polarized light or different intensities.

The four elements of **S** are directly obtained from intensity measurements and are related by the expression

$$S_0^2 = S_1^2 + S_2^2 + S_3^2, \tag{1.14}$$

which was shown above. S_0 is the intensity of the beam, S_1, S_2, and S_3 can have any real value between $-S_0$ and $+S_0$. Equation (1.14) allows a representation of the polarization states on the surface of a sphere, which is called a Poincaré sphere as shown in Figure 1.4. Here, S_1, S_2, and S_3 may be regarded as Cartesian coordinates of a point P on a sphere of radius S_0. Every possible state of polarization of a monochromatic plane wave corresponds to a point on the Poincaré sphere and vice versa. The right-handed circular polarization is represented by the north pole, the left-handed polarization by the south pole, the linear polarizations by points in the equatorial plane, and the elliptical states by the points between the poles and the equatorial plane. The right-handed polarization points lie above the equatorial plane and the left-handed lie below. According to a description of the polarization by complex numbers, the ensemble of the polarization states is mapped onto the complex plane; here it is mapped onto the surface of the Poincaré sphere and it exists as a unique projection between them.

Stokes columns can also describe partially polarized light. This can be thought of as combinations of several mutually incoherent beams of different polarizations. The combination can be obtained by addition of the intensities, which are represented by the individual elements of the Stokes column. These are directly related to the so-called coherency matrix (Goodman, 2000). For partially polarized light the condition of Equation (1.14) will no longer be fulfilled. A measure for the validity of this condition can be the parameter p, given by

$$p = \frac{\sqrt{S_1^2 + S_2^2 + S_3^2}}{S_0}, \tag{1.15}$$

which is called the degree of polarization. The degree of polarization p is equal to 1 for fully polarized light (mathematically, this is valid for all Stokes columns that are transformed directly from Maxwell columns) and equal to 0 for nonpolarized (natural) light. Stokes columns with a degree of polarization between 0 and 1 represent partially polarized light beams. Stokes vectors of partially polarized light beams can also be represented with the help of the Poincaré sphere. One can establish a matrices formalism that relates the Stokes vector of a light beam leaving an optical device with the Stokes vectors of the input beam. This matrix is called a Mueller matrix after its inventor. It is a 4 × 4 matrix with real elements. The Stokes vectors $\mathbf{S}$ are then transformed by

$$\mathbf{S}' = \begin{pmatrix} S'_0 \\ S'_1 \\ S'_2 \\ S'_3 \end{pmatrix} = \begin{pmatrix} M_{11} & M_{12} & M_{13} & M_{14} \\ M_{21} & M_{22} & M_{23} & M_{24} \\ M_{31} & M_{32} & M_{33} & M_{34} \\ M_{41} & M_{42} & M_{43} & M_{44} \end{pmatrix} \begin{pmatrix} S_0 \\ S_1 \\ S_2 \\ S_3 \end{pmatrix}. \tag{1.16}$$

Not every real 4 × 4 matrix can be a Mueller matrix $\mathbf{M}$. There exist conditions for testing the consistence of a matrix to be a Mueller matrix (Brosseau, 1998). These conditions can be formulated as inequalities. Mueller matrices can also be specified for nondepolarizing devices and then are called Mueller–Jones matrices.

1.5 THE JONES CONCEPT

In order to characterize the propagation of plane waves of light and the effect of anisotropic optical devices, it is convienent to use matrix methods. Several matrix methods have been developed (Jones, 1941, 1942, 1947, 1948, 1956; McMaster, 1961; Born, 1993). Under the condition of completely polarized beams a 2 × 2 matrix method is sufficient for the description of the state of polarization. The light is represented only by its electric field and we examine a monochromatic plane wave of angular frequency ω, which propagates in direction z, perpendicularly to the x–y-plane. Then, the amplitude and phase behavior of the plane wave is determined by the two electric field vectors

$$E_x = E_{x0}e^{i(\omega t + \varphi_x)} \quad \text{and} \quad E_y = E_{y0}e^{i(\omega t + \varphi_y)}, \tag{1.17}$$

from which we obtain

$$\mathrm{Re}(E_x) = E_{x0}\cos(\omega t + \varphi_x) \quad \text{and} \quad \mathrm{Re}(E_y) = E_{y0}\cos(\omega t + \varphi_y). \tag{1.18}$$

E_{x0} and E_{y0} are the amplitudes of the two electric fields, φ_x and φ_y their phases with a phase difference $(\varphi_y - \varphi_x)$. Note that here the definition of the phase difference is consistent with that in Section 1.3, although the absolute phases are different.

The column with the elements E_x and E_y is called the Maxwell column, and the corresponding time-independent vector

$$\mathbf{E} = \begin{pmatrix} E_x \\ E_y \end{pmatrix} = \begin{pmatrix} E_{x0}\mathrm{e}^{\mathrm{i}\varphi_x} \\ E_{y0}\mathrm{e}^{\mathrm{i}\varphi_y} \end{pmatrix} \tag{1.19}$$

is called the Jones vector. The polarization of the plane wave described by Equation (1.19) can be defined by using a complex number notation. The intensity of $\mathbf{E}$ is $\mathbf{EE}^*$ (* means the complex conjugate). The vector in Equation (1.19) has the same polarization and intensity as

$$\mathbf{E} = \begin{pmatrix} E_x \\ E_y \end{pmatrix} = \begin{pmatrix} E_{x0} \\ E_{y0}\mathrm{e}^{\mathrm{i}(\varphi_y - \varphi_x)} \end{pmatrix}. \tag{1.20}$$

There is a phase shift φ_x relative to Equation (1.19).

A polarization-dependent device can be characterized by a 2×2 matrix $\mathbf{J}$ that connects the incoming and outcoming waves in the Jones vector description by

$$\mathbf{E}_{\mathrm{out}} = \begin{pmatrix} J_{11} & J_{12} \\ J_{21} & J_{22} \end{pmatrix} \begin{pmatrix} E_x \\ E_y \end{pmatrix}. \tag{1.21}$$

Note that the elements of $\mathbf{J}$ are in general complex. The Jones matrix describes the linear transformation of the Jones vector of a plane wave by reflection or transmission. The Jones matrix of a device that is composed of several devices in series can be obtained as the product of the Jones matrices of the individual devices. Jones matrices can be measured with special setups by analyzing the outgoing intensities for incoming beams of different polarizations.

For nondepolarizing devices these exists a mathematical correspondence between the Mueller–Jones and Jones matrices, namely

$$\mathbf{M} = \mathbf{A}(\mathbf{J} \otimes \mathbf{J}^*)\mathbf{A}^{-1}, \tag{1.22}$$

where $\mathbf{J}$ is the corresponding Jones matrix, $\mathbf{M}$ the corresponding Mueller matrix, and $\mathbf{A}$ is defined by

$$\mathbf{A} = \begin{pmatrix} 1 & 0 & 0 & 1 \\ 1 & 0 & 0 & -1 \\ 0 & 1 & 1 & 0 \\ 0 & \mathrm{i} & -\mathrm{i} & 0 \end{pmatrix}, \tag{1.23}$$

where $\otimes$ designates the so-called Kronecker product (or outer product), which transforms two 2×2 matrices into a 4×4 matrix by multiplication of submatrices component for component. The form of **M** in terms of the Jones matrix elements is

$$\begin{pmatrix} \frac{1}{2}\left(|J_{11}|^2 + |J_{12}|^2 + |J_{21}|^2 + |J_{22}|^2\right) & \frac{1}{2}\left(|J_{11}|^2 - |J_{12}|^2 + |J_{21}|^2 - |J_{22}|^2\right) & \mathrm{Re}(J_{11}^* J_{12} + J_{21}^* J_{22}) & -\mathrm{Im}(J_{11}^* J_{12} + J_{21}^* J_{22}) \\ \frac{1}{2}\left(|J_{11}|^2 + |J_{12}|^2 - |J_{21}|^2 - |J_{22}|^2\right) & \frac{1}{2}\left(|J_{11}|^2 - |J_{12}|^2 - |J_{21}|^2 + |J_{22}|^2\right) & \mathrm{Re}(J_{11}^* J_{12} - J_{21}^* J_{22}) & -\mathrm{Im}(J_{11}^* J_{12} - J_{21}^* J_{22}) \\ \mathrm{Re}(J_{11}^* J_{21} + J_{12}^* J_{22}) & \mathrm{Re}(J_{11}^* J_{21} - J_{12}^* J_{22}) & \mathrm{Re}(J_{11}^* J_{22} + J_{12}^* J_{21}) & -\mathrm{Im}(J_{11}^* J_{22} - J_{12}^* J_{21}) \\ \mathrm{Im}(J_{11}^* J_{21} + J_{12}^* J_{22}) & \mathrm{Im}(J_{11}^* J_{21} - J_{12}^* J_{22}) & \mathrm{Im}(J_{11}^* J_{22} + J_{12}^* J_{21}) & \mathrm{Re}(J_{11}^* J_{22} - J_{12}^* J_{21}) \end{pmatrix}. \tag{1.24}$$

Mueller matrices of optical devices can be determined by Stokes vector measurements for different polarizations of the incident light, also for partially polarized light. Examples for Jones and Mueller matrices of particular devices and their measurement schemes can be found in the works of Gerrard (1994) and Brosseau (1998).

1.6 COHERENCE AND POLARIZED LIGHT

One particular beauty of experiments with polarized light is the appearing of colors appearing between crossed polarizers. This is due to interference effects in polarized light. However, interference is only possible under certain circumstances. This chapter will explain conditions when interference occurs and what kind of description is adequate for handling such problems. Three topics will be discussed: the effect of the spectra on interference, the effect of the quality of the source on interference, and the conditions for interference with polarized light.

1.6.1 Spectra and Coherence

A great majority of optical sources emit light by means of spontaneous emission from a collection of excited atoms or molecules, as is the case for the Sun, incandescent bulbs, and gas discharge lamps, for example. Such radiation, consisting of a large number of independent contributions, is referred to as thermal light. This kind of light must ultimately be treated as a random process (Goodman, 2000). In the preceding chapter it was assumed that light is deterministic or "coherent." An exception was the description of partially polarized light with Stokes columns. An example of completely coherent light is a monochromatic wave with a description like $\mathbf{U} = \mathrm{Re}(\mathbf{U}_0 e^{i\omega t})$, for which the complex amplitude $\mathbf{U}_0$ is a deterministic complex valued function, for example, $\mathbf{U}_0 = \mathbf{E}_0 e^{i\mathbf{kr}}$ in the case of a plane wave. The dependence of the wave function on time and space is perfectly periodic and therefore predictable. If the light train is composed of several frequencies forming a continuous spectrum, the situation is more complicated. As a result, the temporal coherence of light influences the appearance of the interference effects. To quantify the influence one considers the spectrum of light as confined to a narrow band centered at a central frequency ν_0 or wavelength λ_0. The spectral width, or linewidth, of light is the width of $\Delta\nu$ of the spectral density $S(\nu)$ (Saleh, 1991). The coherence time is introduced as a parameter of characterization. A light source of broad spectrum has a short coherence time, whereas a light source with narrow linewidth has a long coherence time. For a given coherence time τ_c, the distance that light travels in this time is $l_c = c\ \tau_c$. An experiment that produces interference effects with different path differences around l_c provides information about the coherence of light. The coherence properties of light can be demonstrated by an experiment where a birefringent material of varying thickness is put between crossed polarizers. A possible experimental setup is as follows. A convex lens and a microscope slide are treated with a polyimide and rubbed to form a uniform parallel aligned liquid crystal cell of variable thickness. This cell is filled with a highly birefringent liquid crystal mixture, such as BL006 from Merck ($\Delta n = 0.28$, clearing temperature $T_{NI} = 110°C$). Such a sample is observed between crossed polarizers where the rubbing direction is set to 45° with respect to the analyzer. Figure 1.5

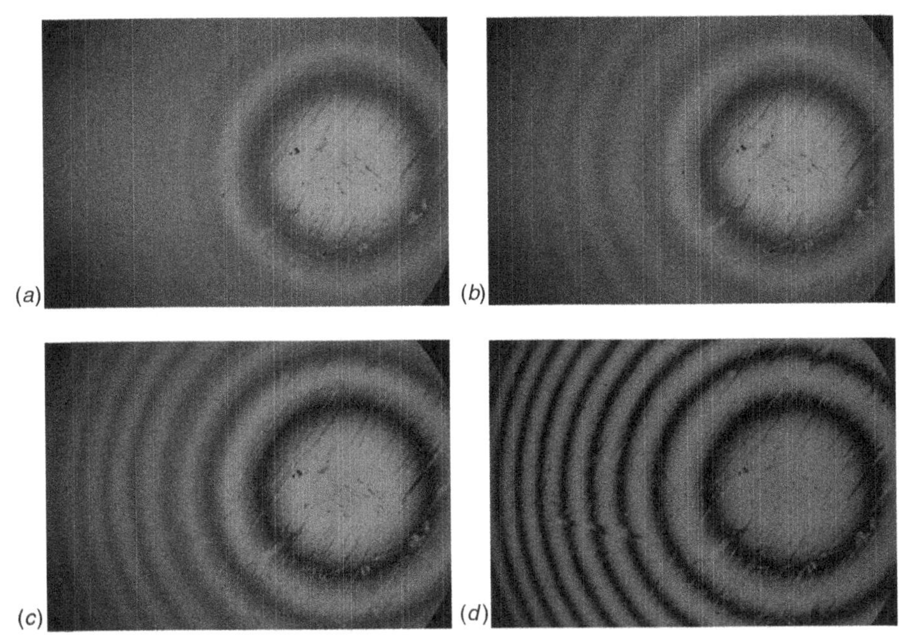

Figure 1.5 Interference effects of a varying thickness liquid crystal cell observed between crossed polarizers. The thickness increases from the center of the round fringe at the right to the left. The width of the spectra of illumination is changed. It can be observed that the contrast of the interference fringes varies for different types of spectra. The smaller the bandwidths of the spectra the higher the contrast. (*a*) White light, $\Delta\lambda = 400$ nm at $\lambda_0 = 550$ nm; (*b*) $\Delta\lambda = 200$ nm at $\lambda_0 = 550$ nm; (*c*) $\Delta\lambda = 100$ nm at $\lambda_0 = 550$ nm; (*d*) $\Delta\lambda = 5$ nm at $\lambda_0 = 550$ nm.

shows the interference fringe system that can be seen for different illumination conditions.

The coherence of a light source can be increased by using optical filters to reduce its spectral width. To illustrate this effect, Figure 1.5 shows the result of increased coherence on the interference fringes. For white light, only the central fringe appears in good contrast. If the spectrum is narrowed to 200 nm, the contrast increases remarkably and additional fringes becomes visible. Decreasing the spectral width to 100 nm and finally to 5 nm leads to pronounced fringes at the highest contrast. The visibility of fringes is a measure of the coherence of light. A plot of the intensity over position for two different spectra is shown in Figure 1.6. The contrast of fringes is highest for the bandwidth of 5 nm and much smaller for the larger spectrum of 200 nm bandwidth. The positions of the peaks of maximum intensity are not the same for the different illuminations.

In the measurements above we have used rectangle-shaped spectral functions. The spectral width is the difference in width of the smallest and highest frequencies in the spectra. But the shape of the spectral density can have a different form and a definition of the spectral width becomes necessary. There are several definitions of spectral widths. The most common is the full width at half maximum (FWHM) of the spectral density $S(\nu)$. The relation between the coherence time and spectral width depends on the spectral profile, as indicated in Table 1.1. A particular situation arises if the light has no continuous spectrum. In particular, in the case of discharge lamps with spectral lines, the definition of the coherence time needs special care. In general, it is possible to define an equivalent coherence time. In such an approach, actual coherence properties can be simulated by using weighted averages.

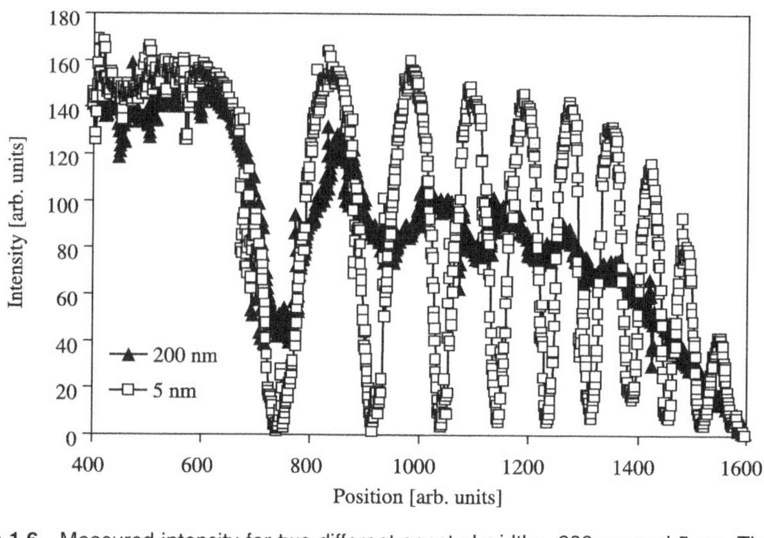

Figure 1.6 Measured intensity for two different spectral widths: 200 nm and 5 nm. The mean intensity is the same. The fringe contrast is reduced for broadband illumination with the 200 nm bandwidth.

Table 1.1 Relation between spectral width and coherence time. Saleh, Teich 1991.

Spectral Density Shape	Spectral Width $\Delta\nu_{\mathrm{FWHM}}$
Rectangular	$\frac{1}{\tau_c}$
Lorentzian	$\frac{1}{\pi\tau_c} \approx \frac{0.32}{\tau_c}$
Gaussian	$\frac{(2\ln 2/\pi)^{1/2}}{\tau_c} \approx \frac{0.66}{\tau_c}$

1.6.2 Spatial Coherence

An additional important parameter, which determines the visibility of fringes in an interference experiment, is the spatial coherence. Light sources have in general a certain dimension. An area of coherent emission related to this is called the coherence area. The coherence area is an important parameter that characterizes random light and must be considered in relation to other dimensions of the optical system. For example, if the area of spatial coherence is greater than the size of the aperture through which light is transmitted, the light may be regarded as spatially coherent. Similarly, if the coherence area is much smaller than the resolution of the optical system, it has to be regarded as incoherent. Light radiated from an extended radiating hot surface has an area of coherence in the order of λ^2, where λ is the central wavelength. In most practical cases it must be regarded as incoherent. Interference occurs when coherent waves are superimposed. Those portions of the light field that produce interference have to be spatially coherent. The spatially coherent parts of light can be identified as spatial modes of the light field. The light field is considered as being composed of a number of such modes. These modes can have any shape. The shape depends on the boundary conditions imposed when light is generated or detected. Independent of the particular mode shape, there exists a relation between the cross-sectional area A_{M} and the solid angle Ω_{M}. For a mode at a given position this relation is always given by

$$A_{\mathrm{M}}\Omega_{\mathrm{M}} = \lambda^2, \tag{1.25}$$

where λ is the wavelength of the light. Equation (1.25) links properties of a spatially coherent emitting area λ^2 to parameters of a coherent light beam. As only light that is spatially coherent and belonging to the same spatial mode can interfere, the coherent power of the light is equal to the power P_{mode} in one spatial mode. P_{mode} is directly related to the radiance B,

$$B = \frac{\mathrm{d}P}{\mathrm{d}A\mathrm{d}\Omega}, \tag{1.26}$$

which is the power dP per unit area dA and unit solid angle $d\Omega$, through

$$P_{\text{mode}} = B\Omega_{\text{M}}A_{\text{M}} = B\lambda^2. \quad (1.27)$$

The radiance B, as defined in Equation (1.26), measures the quality of light in terms of coherence. Liouville's theorem of optics states that the radiance B cannot be increased by passive optical elements such as lenses or mirrors. This means also that the coherent power of the power per mode P_{mode} cannot be increased by passive elements.

1.6.3 Coherence and Polarization

There are additional conditions to be satisfied in order to observe interference effects with polarized light. They are referred to as the rules of Fresnel and Aarago (Goldstein, 2003):

1. Two waves linearly polarized in the same plane can interfere.
2. Two waves linearly polarized with perpendicular polarizations cannot interfere.

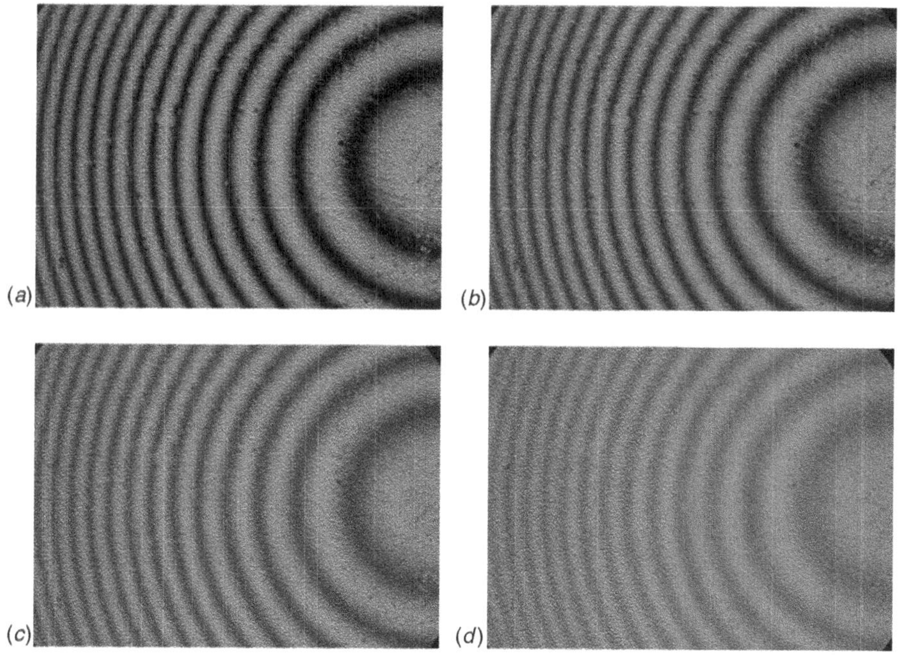

Figure 1.7 Interference fringes of a liquid crystal cell of varying thickness with planar alignment for different voltages: (*a*) 0 V; (*b*) 10 V; (*c*) 15 V; and (*d*) 30 V. The contrast of the fringes varies as a function of the degree of polarization of light. For ideal conditions, in 0 V, maximum contrast is found. Increasing the voltage reduces the degree of polarization of the incoming light and the contrast is reduced.

3. Two waves linearly polarized with perpendicular polarizations, if derived from the perpendicular components of unpolarized light and subsequently brought into the same plane, cannot interfere.
4. Two waves linearly polarized with perpendicular polarizations, if derived from the same linearly polarized wave and subsequently brought into the same plane, can interfere.

These conditions can be worked out with the Stokes formalism and it is interesting to see how this formalism applies (Goldstein, 2003). We will not treat the different cases here. We shall instead show the influence of the degree of polarization on interference with the example discussed above. Using liquid crystal display technology it is possible to build switchable polarizers with dichroic dyes diluted in liquid crystals. The performance, that is, the extinction ratio or pure transmission, of such a polarizer is not very good but still sufficient to perform the experiment. The dichroic liquid crystal cell in planar alignment with a positive dichroic dye absorbs electromagnetic energy for one polarization direction (Bahadur, 1990). As a result, the outgoing beam becomes polarized. If the absorption is not complete only a part of the concerned electrical field component is reduced in amplitude. This results in partial polarization. The degree of polarization can be changed by influencing the strength of absorption of the dichroic dye. With the dichroic liquid crystal cell this may be achieved by changing the orientation of the dichroic dye molecules in the liquid crystal host by applying a voltage (Bahadur, 1990). The result is an electrically switchable polarizer that allows the choice of degree of

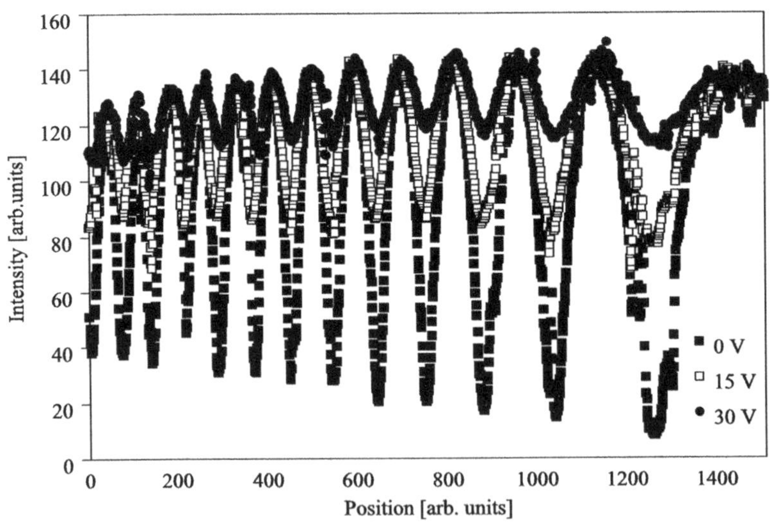

Figure 1.8 Intensity as a function of position for different voltages for the same measurement setup as described in the text. When the degree of polarization is changed, the contrast of fringes is varied but the positions of the maxima and minima remain the same. The contrast reduction from right to left is a result of scattering present in the system.

polarization. A microscope was equipped with such a polarizer to observe the interference fringes of a variable thickness cell. Figure 1.7 shows the case when the cell of variable thickness is illuminated with green light at 550 nm (spectral width $\Delta\lambda = 5$ nm). For increasing voltage, the degree of polarization is reduced. In Figure 1.7 photographs of the interference fringes are shown for different voltages. Increasing the voltage leads to a decrease of contrast for the interference fringes. Only the polarized part can produce interference fringe contrast. The higher the voltage is the viewer the intensity of the polarized light becomes. The fringe contrast decreases but the overall intensity is maintained. To prove this, a plot of intensity as a function of position is given in Figure 1.8.

The positions of the maxima and minima stay the same as seen in Figure 1.8; only the contrast is changed. In comparison to Figure 1.6, the maximum intensity is always the same and the minimum changes with voltage.

REFERENCES

Bahadur, B. (ed.) (1990) *Liquid Crystal – Application and Uses*, Books 1–3, World Scientific Singapore.

Berry, M., R. Bhandari, and S. Klein (1999) Black plastic sandwiches demonstrating biaxial optical anisotropy, *Eur. J. Phys.* **20**, 1–14.

Berry, M.V. and M. Dennis (2003) The optical singularities of birefringent dichroic chiral crystals, *Proc. R. Soc. Lond. A* **459**, 1261–1292.

Born, M. and E. Wolf (1993) *Principles of Optics*, 6th edn, Pergamon, New York.

Brosseau, C. (1998) *Fundamentals of Polarized Light*, Wiley, New York.

Chandrasekhar, S. (1960) *Radiative Transfer*, Dover, New York.

Dally, J.W. and W.F. Riley (1991) *Experimental Stress Analysis*, 3rd edn, Part III, Optical Methods for Stress Analysis, McGraw-Hill, New York.

Föppl, L. and E. Mönch (1972) *Spannungsoptik*, 3rd edn, Springer, Berlin.

Gerrard, A. and J.M. Burch (1994) *Matrix Methods in Optics*, Dover, New York.

Goldstein, D. (2003) *Polarized Light*, Marcel Dekker, New York.

Goodmann, J.W. (2000) *Statistical Optics*, Wiley, New York.

Herzig, H.P (ed.) (1997) *Micro-Optics*, Taylor and Francis, London.

Horváth, G. and D. Varjú (2004) *Polarized Light in Animal Vision*, Springer, Berlin.

Jones, R.C. (1941) A new calculus for the treatmant of optical systems I–III, *J. Opt. Soc. Am.* **31**, 488–503.

Jones, R.C. (1942) A new calculus for the treatmant of optical systems IV, *J. Opt. Soc. Am.* **32**, 486–493.

Jones, R.C. (1947) A new calculus for the treatmant of optical systems V–VI, *J. Opt. Soc. Am.* **37**, 107–112.

Jones, R.C. (1948) A new calculus for the treatmant of optical systems VII, *J. Opt. Soc. Am.* **38**, 671–685.

Jones, R.C. (1956) A new calculus for the treatmant of optical systems VIII, *J. Opt. Soc. Am.* **46**, 126–131.

McMaster, W. (1961) Matrix representation of polarization, *Rev. Mod. Phys.* **33**, 8–28.

Minnaert, M. (1954) *Light and Colour*, Dover, New York.

Pye, D. (2001) *Polarised Light in Science and Nature*, Institute of Physics Publishing, Bristol.

Rohrbach, C. (ed.) (1989) *Handbuch für Experimentelle Spanungsanalyse*, Chapter D2, VDI Verlag, Düsseldorf.

Saleh, B.E.A. and M.C. Teich (1991) *Fundamentals of Photonics*, Wiley, New York.

Shurcliff, W.A. (1962) *Polarized Light*, Harvard University Press, Cambridge.

Weller, W. and H. Winkler (1979) *Elektrodynamik*, Teubner, Leipzig.

2

ELECTROMAGNETIC WAVES IN ANISOTROPIC MATERIALS

2.1 INTRODUCTION

The Maxwell equations are the foundations of modern electromagnetic theory. Maxwell's original formulation comprised 20 equations in 20 variables. The modern mathematical formulation of Maxwell's equations is credited to Oliver Heaviside and Willard Gibbs, who reformulated Maxwell's original system of equations and introduced the vector calculus. The vector notation produced a symmetric mathematical representation. This highly symmetrical formulation inspired later developments in fundamental physics. The electromagnetic field equations have an intimate link with special relativity; the magnetic field equations can be derived from consideration of the transformation of the electric field equations under relativistic transformations at low velocities.

2.2 ANALYTICAL BACKGROUND

Maxwell's equations describe all (classical) electromagnetic phenomena. With the electrical field vector **E**, the magnetic field vector **H**, the electric displacement vector **D**, and the magnetic induction vector **B** they can be expressed as (Born, 1993)

$$\begin{aligned} \nabla \times \mathbf{E} &= -\frac{\partial \mathbf{B}}{\partial t}, \quad & \nabla \times \mathbf{H} &= \mathbf{J} + \frac{\partial \mathbf{D}}{\partial t} \\ \operatorname{div} \mathbf{B} &= 0, & \operatorname{div} \mathbf{D} &= \rho, \end{aligned} \tag{2.1}$$

Polarized Light in Liquid Crystals and Polymers. By Toralf Scharf

where $\mathbf{J}$ is the current and ρ is the electrical charge. The first equation is Faraday's law of induction, the second is Ampere's law as amended by Maxwell to include the displacement current $\partial\mathbf{D}/\partial t$, the third and fourth are Gauss's laws for the electric and magnetic fields. The displacement current term $\partial\mathbf{D}/\partial t$ in Ampere's law is essential in predicting the existence of propagating electromagnetic waves. Equations (2.1) are in SI units. The quantities $\mathbf{E}$ and $\mathbf{H}$ are the electric and magnetic field intensities and are measured in units of [V/m] and [A/m], respectively. The quantities $\mathbf{D}$ and $\mathbf{B}$ are the electric and magnetic flux densities, expressed in units of [C/m^2] and [Weber/m^2], or [Tesla]. $\mathbf{B}$ is also called the magnetic induction. The quantities ρ and $\mathbf{J}$ are the volume charge density and electric current density (charge flux) of any external charges (i.e., not including any induced polarization charges and currents.) They are measured in units of [C/m^3] and [A/m^2]. The right-hand side of the equation div $\mathbf{B} = 0$ is zero because there are no magnetic monopole charges.

The charge and current densities ρ, $\mathbf{J}$ may be thought of as the sources of the electromagnetic fields. For wave propagation problems, these densities are localized in space. The generated electric and magnetic fields are radiated away from these sources and can propagate to large distances compared to the wavelength. Away from the sources, that is, in source-free regions of space, Maxwell's equations take the simpler form:

$$\begin{aligned} \nabla \times \mathbf{E} &= -\frac{\partial \mathbf{B}}{\partial t}, \qquad & \nabla \times \mathbf{H} &= \frac{\partial \mathbf{D}}{\partial t} \\ \text{div}\,\mathbf{B} &= 0, & \text{div}\,\mathbf{D} &= 0. \end{aligned} \tag{2.2}$$

The electric and magnetic flux densities $\mathbf{D}$, $\mathbf{B}$ are related to the field intensities $\mathbf{E}$, $\mathbf{H}$ via the so-called constitutive relations. The precise form of those depends on the material in which the fields exist. In vacuum, they take their simplest form:

$$\mathbf{D} = \varepsilon_0 \mathbf{E}, \qquad \mathbf{B} = \mu_0 \mathbf{H}, \tag{2.3}$$

where ε_0, μ_0 are the permittivity and permeability of vacuum, with numerical values

$$\varepsilon_0 = 8.854 \times 10^{-12}\,\frac{\text{As}}{\text{Vm}}, \qquad \mu_0 = 4\pi \times 10^{-7}\,\frac{\text{Vs}}{\text{Am}}. \tag{2.4}$$

The units for ε_0 and μ_0 are the units of the ratios D/E and B/H, that is, C/m^2 and V/m. From the two quantities ε_0 and μ_0 one can define two other physical constants: the speed of light c_0 and characteristic impedance of vacuum η_0:

$$c_0 = \sqrt{\frac{1}{\varepsilon_0 \mu_0}} = 2.9998 \times 10^8\ \text{m/s}, \qquad \eta_0 = \sqrt{\frac{\mu_0}{\varepsilon_0}} = 377\,\Omega. \tag{2.5}$$

The next simplest form of the constitutive relations is for simple dielectrics and for magnetic materials:

$$\mathbf{D} = \boldsymbol{\varepsilon}_a \mathbf{E}, \qquad \mathbf{B} = \boldsymbol{\mu}_a \mathbf{H}. \tag{2.6}$$

These are typically valid at low frequencies. The permittivity ε_a and permeability μ_a are related to the electric and magnetic susceptibilities of the material as follows:

$$\varepsilon_a = \varepsilon_0(1+\chi), \qquad \mu_a = \mu_0(1+\chi_m). \tag{2.7}$$

The susceptibilities χ and χ_m are measures of the electric and magnetic polarization properties of the material. For example, we have for the electric flux density

$$\mathbf{D} = \varepsilon_a \mathbf{E} = \varepsilon_0(1+\chi)\mathbf{E} = \varepsilon_0 \mathbf{E} + \chi\varepsilon_0 \mathbf{E} = \varepsilon_0 \mathbf{E} + \mathbf{P}, \tag{2.8}$$

where the quantity $\mathbf{P} = \varepsilon_0 \chi \mathbf{E}$ represents the dielectric polarization of the material, that is, the average electric dipole moment per unit volume. The speed of light in the material and the characteristic impedance are given by

$$c = \sqrt{\frac{1}{\varepsilon_a \mu_a}}, \qquad \eta = \sqrt{\frac{\mu_a}{\varepsilon_a}}. \tag{2.9}$$

The relative permittivity and refractive index of the material are defined as

$$\varepsilon = \frac{\varepsilon_a}{\varepsilon_0} = 1 + \chi, \qquad n = \sqrt{\varepsilon} \tag{2.10}$$

so that $\varepsilon = n^2$ and $\varepsilon \hat{a} = \varepsilon_0 \varepsilon_r = \varepsilon_0 n^2$. Using the definition of Equation (2.10) and assuming a nonmagnetic material ($\mu = \mu_0$), we may relate the speed of light and impedance of the material to the corresponding vacuum values:

$$c = \sqrt{\frac{1}{\varepsilon_a \mu_a}} = \sqrt{\frac{1}{\varepsilon_0 \varepsilon \mu_0}} = \sqrt{\frac{1}{\varepsilon_0 \mu_0}}\frac{1}{n} = \frac{c_0}{n}. \tag{2.11}$$

Similarly, in a magnetic material, we have $\mathbf{B} = \mu_0(\mathbf{H} + \mathbf{M})$, where $\mathbf{M} = \chi_m \mathbf{H}$ is the magnetization, that is, the average magnetic moment per unit volume.

More generally, constitutive relations may be inhomogeneous, anisotropic, nonlinear, frequency-dependent (dispersive), or all of the above. In inhomogeneous materials, the permittivity ε depends on the location $\mathbf{r}$ within the material:

$$\mathbf{D}(\mathbf{r}, t) = \varepsilon_0 \varepsilon(\mathbf{r}) \mathbf{E}(\mathbf{r}, t). \tag{2.12}$$

In anisotropic materials, ε depends on direction and becomes a tensor. The constitutive relations may be written component-wise in matrix (or tensor) form as

$$\begin{pmatrix} D_x \\ D_y \\ D_z \end{pmatrix} = \varepsilon_0 \begin{pmatrix} \varepsilon_{xx} & \varepsilon_{xy} & \varepsilon_{xz} \\ \varepsilon_{yx} & \varepsilon_{yy} & \varepsilon_{yz} \\ \varepsilon_{zx} & \varepsilon_{zy} & \varepsilon_{zz} \end{pmatrix} \begin{pmatrix} E_x \\ E_y \\ E_z \end{pmatrix}. \tag{2.13}$$

It is important to note that for a lossless and nonoptically active medium, $\varepsilon_{ij} = \varepsilon_{ji}$. More complicated forms of constitutive relationships arise in chiral and gyroscopic media and are discussed elsewhere (Serdyukov et al., 2001).

The cross-product of the electric field **E** and the magnetic field **H** is often called the Poynting vector, named after its inventor John Henry Poynting. For an electromagnetic wave it points in the direction of energy flow and its magnitude is the power per unit area crossing a surface that is normal to it. It is given the symbol **S** (in bold because it is a vector),

$$\mathbf{S} = \mathbf{E} \times \mathbf{H} = \frac{1}{\mu\mu_0}\mathbf{E} \times \mathbf{B}, \tag{2.14}$$

where **E** is the electric field, **H** and **B** are the magnetic field and magnetic flux density respectively, and μ is the permeability of the surrounding medium. For an electromagnetic wave propagating in free space μ becomes equal to 1, the permeability of free space. As the electric and magnetic fields of an electromagnetic wave oscillate, the magnitude of the Poynting vector changes with time. The average of the magnitude over a long time T (longer than the period of the wave) is called the irradiance, I:

$$I = \langle S \rangle_T. \tag{2.15}$$

2.3 TIME HARMONIC FIELDS AND PLANE WAVES

Maxwell's equations simplify considerably in the case of harmonic time dependence. Through the Fourier transform, general solutions of Maxwell's equation can be built as linear combinations of single-frequency solutions:

$$\mathbf{E}(\mathbf{r}, t) = \frac{1}{2\pi}\int_{-\infty}^{\infty} \mathbf{E}(\mathbf{r}, \omega)e^{i\omega t}\, d\omega. \tag{2.16}$$

Thus, we assume that all fields have a time dependence $e^{i\omega t}$:

$$\mathbf{E}(\mathbf{r}, t) = \mathbf{E}(\mathbf{r})e^{i\omega t}, \qquad \mathbf{H}(\mathbf{r}, t) = \mathbf{H}(\mathbf{r})e^{i\omega t}, \tag{2.17}$$

where the phasor amplitudes $\mathbf{E}(\mathbf{r})$, $\mathbf{H}(\mathbf{r})$ are complex-valued. Replacing time derivatives by $\partial/\partial t \rightarrow i\omega$, we may rewrite Equation (2.1) in the form

$$\begin{aligned} \nabla \times \mathbf{E} &= -i\omega\mathbf{B}, & \nabla \times \mathbf{H} &= \mathbf{J} + i\omega\mathbf{D} \\ \operatorname{div} \mathbf{B} &= 0, & \operatorname{div} \mathbf{D} &= \rho. \end{aligned} \tag{2.18}$$

In this book, we will consider solutions of Maxwell's equations in two different contexts:

1. Uniform plane waves propagating in dielectrics and birefringent media; and
2. Waves propagating in nonuniform dielectric and birefringent media.

It is therefore instructive to see the simplifications provided by the additional assumption of having plane waves. The plane wave propagating in direction **r** is described by an amplitude vector **E** and wave vector **k**. For a monochromatic wave we assume the form

$$\begin{aligned}\mathbf{E}(\mathbf{r},t) &= \mathbf{E}e^{i(\omega t-\mathbf{kr})}\\ \mathbf{H}(\mathbf{r},t) &= \mathbf{H}e^{i(\omega t-\mathbf{kr})}.\end{aligned} \tag{2.19}$$

The wave vector $\mathbf{k} = (\omega/c)n\,\mathbf{s}$ contains **s** as a unit vector. The unit vector **s** is in the direction of propagation. The phase velocities c/n, or equivalently the refractive index n, are to be determined. Substituting **E** and **H** from Equation (2.19) into Maxwell's Equations (2.18) gives

$$\begin{aligned}\mathbf{k}\times\mathbf{E} &= \omega\mu\mu_0\mathbf{H}\\ \mathbf{k}\times\mathbf{H} &= -\omega\mathbf{D} = -\omega\boldsymbol{\varepsilon}\varepsilon_0\mathbf{E}.\end{aligned} \tag{2.20}$$

By eliminating the magnetic field **H** in Equation (2.20) we obtain

$$\mathbf{k}\times(\mathbf{k}\times\mathbf{E}) + \omega^2\varepsilon_0\mu_0\boldsymbol{\varepsilon}\mu\mathbf{E} = 0. \tag{2.21}$$

Using the equality $\mathbf{a}\times(\mathbf{b}\times\mathbf{c}) = \mathbf{b}(\mathbf{a}\cdot\mathbf{c}) - \mathbf{c}(\mathbf{a}\cdot\mathbf{b})$, the first summand in Equation (2.21) takes the form

$$\begin{aligned}\mathbf{k}\times(\mathbf{k}\times\mathbf{E}) &= \mathbf{k}(\mathbf{k}\cdot\mathbf{E}) - \mathbf{E}(\mathbf{k}\cdot\mathbf{k})\\ &= \begin{pmatrix} -k_y^2-k_z^2 & k_xk_y & k_xk_z\\ k_yk_x & -k_x^2-k_z^2 & k_yk_z\\ k_zk_x & k_zk_y & -k_x^2-k_y^2\end{pmatrix}\begin{pmatrix}E_x\\ E_y\\ E_z\end{pmatrix}.\end{aligned} \tag{2.22}$$

This equation will now be used to solve for the eigenvectors of **E** and the corresponding eigenvalues. In the principal coordinate system the dielectric tensor of nonchiral and nongyroscopic media is a diagonal tensor with three different entries. It can be written as

$$\boldsymbol{\varepsilon} = \begin{pmatrix}\varepsilon_x & 0 & 0\\ 0 & \varepsilon_y & 0\\ 0 & 0 & \varepsilon_z\end{pmatrix}. \tag{2.23}$$

When used in Equation (2.21) and with Equation (2.22) this leads to

$$\begin{pmatrix}\omega^2\varepsilon_0\varepsilon_x\mu\mu_0-k_y^2-k_z^2 & k_xk_y & k_xk_z\\ k_yk_x & \omega^2\varepsilon_0\varepsilon_y\mu\mu_0-k_x^2-k_z^2 & k_yk_z\\ k_zk_x & k_zk_y & \omega^2\varepsilon_0\varepsilon_z\mu\mu_0-k_x^2-k_y^2\end{pmatrix}\begin{pmatrix}E_x\\ E_y\\ E_z\end{pmatrix} = 0. \tag{2.24}$$

For nontrivial solutions to exist, the determinant of the matrix must vanish, hence

$$\det\begin{vmatrix} \omega^2\varepsilon_0\varepsilon_x\mu\mu_0-k_y^2-k_z^2 & k_xk_y & k_xk_z \\ k_yk_x & \omega^2\varepsilon_0\varepsilon_y\mu\mu_0-k_x^2-k_z^2 & k_yk_z \\ k_zk_x & k_zk_y & \omega^2\varepsilon_0\varepsilon_z\mu\mu_0-k_x^2-k_y^2 \end{vmatrix} = 0. \quad (2.25)$$

At a given frequency ω, this equation represents a three-dimensional surface in **k** space. This surface consist of two shells and is known as a normal surface. In general, four points are in common for these two shells. The two lines that go through the origin and these points are known as the optical axis.

2.4 MAXWELL'S EQUATIONS IN MATRIX REPRESENTATION

One interesting representation of Maxwell's equations that allows simplification for special cases is the matrix description (Azzam, 1987). To give an overview one starts with Maxwell's equations:

$$\begin{aligned} \nabla\times\mathbf{E} &= -\frac{\partial\mathbf{B}}{\partial t}, & \nabla\times\mathbf{H} &= \frac{\partial\mathbf{D}}{\partial t}, \\ \operatorname{div}\mathbf{B} &= 0, & \operatorname{div}\mathbf{D} &= 0. \end{aligned} \quad (2.26)$$

Using the matrix description of derivations

$$\text{rot} = \begin{pmatrix} 0 & -\frac{\partial}{\partial z} & \frac{\partial}{\partial y} \\ \frac{\partial}{\partial z} & 0 & -\frac{\partial}{\partial x} \\ -\frac{\partial}{\partial y} & \frac{\partial}{\partial x} & 0 \end{pmatrix} \quad \text{and} \quad \text{div} = \begin{pmatrix} \frac{\partial}{\partial x} \\ \frac{\partial}{\partial y} \\ \frac{\partial}{\partial z} \end{pmatrix}, \quad (2.27)$$

one gets for Faraday's law of induction

$$\nabla\times\mathbf{E} = -\frac{\partial\mathbf{B}}{\partial t} \Rightarrow \text{rot}\,\mathbf{E} = \begin{pmatrix} 0 & -\frac{\partial}{\partial z} & \frac{\partial}{\partial y} \\ \frac{\partial}{\partial z} & 0 & -\frac{\partial}{\partial x} \\ -\frac{\partial}{\partial y} & \frac{\partial}{\partial x} & 0 \end{pmatrix} \begin{pmatrix} E_x \\ E_y \\ E_z \end{pmatrix} = -\begin{pmatrix} \dot{B}_x \\ \dot{B}_y \\ \dot{B}_z \end{pmatrix}. \quad (2.28)$$

The set of Maxwell's equations can be expressed in a vector representation using all field components, electrical and magnetic ones, as follows:

$$\begin{pmatrix} 0 & 0 & 0 & 0 & -\frac{\partial}{\partial z} & \frac{\partial}{\partial y} \\ 0 & 0 & 0 & \frac{\partial}{\partial z} & 0 & -\frac{\partial}{\partial x} \\ 0 & 0 & 0 & -\frac{\partial}{\partial y} & \frac{\partial}{\partial x} & 0 \\ 0 & \frac{\partial}{\partial z} & -\frac{\partial}{\partial y} & 0 & 0 & 0 \\ -\frac{\partial}{\partial z} & 0 & \frac{\partial}{\partial x} & 0 & 0 & 0 \\ \frac{\partial}{\partial y} & -\frac{\partial}{\partial x} & 0 & 0 & 0 & 0 \end{pmatrix} \begin{pmatrix} E_x \\ E_y \\ E_z \\ H_x \\ H_y \\ H_z \end{pmatrix} = \begin{pmatrix} \dot{D}_x \\ \dot{D}_y \\ \dot{D}_z \\ \dot{B}_x \\ \dot{B}_y \\ \dot{B}_z \end{pmatrix}. \tag{2.29}$$

We have still neglected the constitutive relations between **E** and **D** and **H** and **B**. Written in a more compact form

$$\begin{pmatrix} \mathbf{0} & \mathbf{rot} \\ -\mathbf{rot} & \mathbf{0} \end{pmatrix} \begin{pmatrix} \mathbf{E} \\ \mathbf{H} \end{pmatrix} = \begin{pmatrix} \dot{\mathbf{D}} \\ \dot{\mathbf{B}} \end{pmatrix}. \tag{2.30}$$

The time derivative is indicated by the point (dot) above quantities. When using a 6×1 column vector **G**, whose elements are the Cartesian components of **E**, followed by that of **H**, and with **C**, that is a 6×1 column vector, whose elements are **D**, followed by those of **B**, the equation reads

$$\begin{pmatrix} \mathbf{0} & \mathbf{rot} \\ -\mathbf{rot} & \mathbf{0} \end{pmatrix} \mathbf{G} = \dot{\mathbf{C}} \quad \text{or alternatively} \quad \mathbf{OG} = \dot{\mathbf{C}}. \tag{2.31}$$

O here is the 6×6 matrix containing the derivations. In the absence of nonlinear optical effects and spatial dispersion, the constitutive relations $\mathbf{D} = \varepsilon_0 \varepsilon \mathbf{E}$ and $\mathbf{B} = \mu_0 \mu \mathbf{H}$ can be translated into **C** and **G**; One finds

$$\mathbf{C} = \mathbf{MG}, \tag{2.32}$$

where **M** is the 6×6 matrix. **M** carries all the information about the anisotropic optical material properties. This matrix is called the optical matrix and can be portioned as

$$\mathbf{M} = \begin{pmatrix} \varepsilon_0 \boldsymbol{\varepsilon} & \boldsymbol{\rho} \\ \boldsymbol{\rho}' & \mu_0 \boldsymbol{\mu} \end{pmatrix}, \tag{2.33}$$

with the following submatrices:

$$\boldsymbol{\varepsilon} = \begin{pmatrix} \varepsilon_{xx} & \varepsilon_{xy} & \varepsilon_{xz} \\ \varepsilon_{yx} & \varepsilon_{yy} & \varepsilon_{yz} \\ \varepsilon_{zx} & \varepsilon_{zy} & \varepsilon_{zz} \end{pmatrix}, \qquad \boldsymbol{\mu} = \begin{pmatrix} \mu_{xx} & \mu_{xy} & \mu_{xz} \\ \mu_{yx} & \mu_{yy} & \mu_{yz} \\ \mu_{zx} & \mu_{zy} & \mu_{zz} \end{pmatrix},$$
$$\boldsymbol{\rho} = \begin{pmatrix} \rho_{xx} & \rho_{xy} & \rho_{xz} \\ \rho_{yx} & \rho_{yy} & \rho_{yz} \\ \rho_{zx} & \rho_{zy} & \rho_{zz} \end{pmatrix}, \qquad \boldsymbol{\rho}' = \begin{pmatrix} \rho'_{xx} & \rho'_{xy} & \rho'_{xz} \\ \rho'_{yx} & \rho'_{yy} & \rho'_{yz} \\ \rho'_{zx} & \rho'_{zy} & \rho'_{zz} \end{pmatrix}. \tag{2.34}$$

The representation contains the dielectric permittivity tensor $\boldsymbol{\varepsilon}$, the magnetic permittivity tensor $\boldsymbol{\mu}$, and two optical rotation tensors $\boldsymbol{\rho}$ and $\boldsymbol{\rho}'$. These follows the very compact form

$$\mathbf{OG} = \mathbf{M}\dot{\mathbf{G}}. \tag{2.35}$$

For time harmonic fields such as

$$\mathbf{G} = \mathrm{e}^{\mathrm{i}\omega t}\boldsymbol{\Gamma}, \tag{2.36}$$

where $\boldsymbol{\Gamma}$ is the spatial part of $\mathbf{G}$, Equation (2.35) becomes

$$\mathbf{O}\boldsymbol{\Gamma} = \mathrm{i}\omega\mathbf{M}\boldsymbol{\Gamma} \qquad \text{or equivalently} \qquad (\mathbf{O} - \mathrm{i}\omega\mathbf{M})\boldsymbol{\Gamma} = 0, \tag{2.37}$$

which is the spatial wave equation for frequencyω. For nonchiral materials, when the optical activities tensors $\boldsymbol{\rho}$ and $\boldsymbol{\rho}'$ are zero, the equations can be simplified to

$$\begin{pmatrix} \mathbf{0} & \mathbf{rot} \\ -\mathbf{rot} & \mathbf{0} \end{pmatrix}\boldsymbol{\Gamma} = \mathrm{i}\omega\begin{pmatrix} \varepsilon_0\boldsymbol{\varepsilon} & 0 \\ 0 & \mu_0\boldsymbol{\mu} \end{pmatrix}\boldsymbol{\Gamma}. \tag{2.38}$$

That excludes only molecular-based optical activity or chirality, following from effective medium calculations. Consequently we find

$$\begin{pmatrix} \mathrm{i}\omega\varepsilon_0\boldsymbol{\varepsilon} & -\mathbf{rot} \\ \mathbf{rot} & \mathrm{i}\omega\mu_0\boldsymbol{\mu} \end{pmatrix}\boldsymbol{\Gamma} = 0. \tag{2.39}$$

This equation can now be transformed to a simpler form by rearranging the field components and taking into account the properties of the permeability's $\boldsymbol{\varepsilon}$ and $\boldsymbol{\mu}$. Expanding Equation (2.39), under the assumption that the field components of $\mathbf{E}$

and **H** are now representing only spatial variation, we obtain

$$\begin{pmatrix} i\omega\varepsilon_0\varepsilon_{xx} & i\omega\varepsilon_0\varepsilon_{xy} & i\omega\varepsilon_0\varepsilon_{xz} & 0 & \frac{\partial}{\partial z} & -\frac{\partial}{\partial y} \\ i\omega\varepsilon_0\varepsilon_{yx} & i\omega\varepsilon_0\varepsilon_{yy} & i\omega\varepsilon_0\varepsilon_{yz} & -\frac{\partial}{\partial z} & 0 & \frac{\partial}{\partial x} \\ i\omega\varepsilon_0\varepsilon_{zx} & i\omega\varepsilon_0\varepsilon_{zy} & i\omega\varepsilon_0\varepsilon_{zz} & \frac{\partial}{\partial y} & -\frac{\partial}{\partial x} & 0 \\ 0 & -\frac{\partial}{\partial z} & \frac{\partial}{\partial y} & i\omega\mu_0\mu_{xx} & i\omega\mu_0\mu_{xy} & i\omega\mu_0\mu_{xz} \\ \frac{\partial}{\partial z} & 0 & -\frac{\partial}{\partial x} & i\omega\mu_0\mu_{yx} & i\omega\mu_0\mu_{yy} & i\omega\mu_0\mu_{yz} \\ -\frac{\partial}{\partial y} & \frac{\partial}{\partial x} & 0 & i\omega\mu_0\mu_{zx} & i\omega\mu_0\mu_{zy} & i\omega\mu_0\mu_{zz} \end{pmatrix} \begin{pmatrix} E_x \\ E_y \\ E_z \\ H_x \\ H_y \\ H_z \end{pmatrix} = 0. \quad (2.40)$$

In a further simplification we can assume that the magnetic permeability μ is isotropic and can be written as

$$\boldsymbol{\mu} = \mu \begin{pmatrix} 1 & 0 & 0 \\ 0 & 1 & 0 \\ 0 & 0 & 1 \end{pmatrix}. \quad (2.41)$$

If, further on, the dielectric permittivity tensor can be transformed to its diagonal form with a convenient rotation of the coordinate system,

$$\boldsymbol{\varepsilon} = \begin{pmatrix} \varepsilon_{xx} & \varepsilon_{xy} & \varepsilon_{xz} \\ \varepsilon_{yx} & \varepsilon_{yy} & \varepsilon_{yz} \\ \varepsilon_{zx} & \varepsilon_{zy} & \varepsilon_{zz} \end{pmatrix} \Longrightarrow \boldsymbol{\varepsilon} = \begin{pmatrix} \varepsilon_1 & 0 & 0 \\ 0 & \varepsilon_2 & 0 \\ 0 & 0 & \varepsilon_3 \end{pmatrix}, \quad (2.42)$$

one obtains for Equation (2.40)

$$\begin{pmatrix} i\omega\varepsilon_0\varepsilon_1 & 0 & 0 & 0 & \frac{\partial}{\partial z} & -\frac{\partial}{\partial y} \\ 0 & i\omega\varepsilon_0\varepsilon_2 & 0 & -\frac{\partial}{\partial z} & 0 & \frac{\partial}{\partial x} \\ 0 & 0 & i\omega\varepsilon_0\varepsilon_3 & \frac{\partial}{\partial y} & -\frac{\partial}{\partial x} & 0 \\ 0 & -\frac{\partial}{\partial z} & \frac{\partial}{\partial y} & i\omega\mu_0\mu & 0 & 0 \\ \frac{\partial}{\partial z} & 0 & -\frac{\partial}{\partial x} & 0 & i\omega\mu_0\mu & 0 \\ -\frac{\partial}{\partial y} & \frac{\partial}{\partial x} & 0 & 0 & 0 & i\omega\mu_0\mu \end{pmatrix} \begin{pmatrix} E_x \\ E_y \\ E_z \\ H_x \\ H_y \\ H_z \end{pmatrix} = 0. \quad (2.43)$$

Equation (2.43) contains 18 elements that are zero out of 36 elements. In principle, that would allow the rearrangement of the elements and separatation of the zeros. Interchanging columns is possible only if the order of field vectors is changed too, and rows can be independently interchanged. If by suitable coordinate transformation a form with zero submatrices, such as

$$\begin{pmatrix} \mathbf{h} & 0 \\ 0 & \mathbf{g} \end{pmatrix} \mathbf{\Gamma}' = 0, \tag{2.44}$$

can be found, the propagation of polarized light for two distinct polarizations can then be treated as independent. The state of polarization for these two eigenpolarizations of the system does not change when propagated through the medium. Note that we made several simplifications that made this possible:

- The material does not show optical activity; and
- The magnetic permeability is a scalar.

We will now discuss two particular examples where polarization separation is possible.

2.5 SEPARATION OF POLARIZATIONS FOR INHOMOGENEOUS PROBLEMS

The differential equation of Equation (2.40) holds for an inhomogeneous anisotropic material in three-dimensional space. Equation (2.40) can only be solved analytically in some special cases. Under certain circumstances it is possible to simplify the system of equations. For the three-dimensional problem the variations of the fields in any direction in space may be nonzero. Assuming isotropic material with a spatially varying dielectric constant $\varepsilon(\mathbf{r})$ in Equation (2.43) we obtain

$$\begin{pmatrix} i\omega\varepsilon_0\varepsilon(\mathbf{r}) & 0 & 0 & 0 & \frac{\partial}{\partial z} & -\frac{\partial}{\partial y} \\ 0 & i\omega\varepsilon_0\varepsilon(\mathbf{r}) & 0 & -\frac{\partial}{\partial z} & 0 & \frac{\partial}{\partial x} \\ 0 & 0 & i\omega\varepsilon_0\varepsilon(\mathbf{r}) & \frac{\partial}{\partial y} & -\frac{\partial}{\partial x} & 0 \\ 0 & -\frac{\partial}{\partial z} & \frac{\partial}{\partial y} & i\omega\mu_0\mu & 0 & 0 \\ \frac{\partial}{\partial z} & 0 & -\frac{\partial}{\partial x} & 0 & i\omega\mu_0\mu & 0 \\ -\frac{\partial}{\partial y} & \frac{\partial}{\partial x} & 0 & 0 & 0 & i\omega\mu_0\mu \end{pmatrix} \begin{pmatrix} E_x \\ E_y \\ E_z \\ H_x \\ H_y \\ H_z \end{pmatrix} = \mathbf{0}. \tag{2.45}$$

The matrix on the left side in Equation (2.45) cannot be rearranged such that the 6×6 matrix separates into nondiagonal 3×3 submatrices, although there are 18 nonzero elements. This can be shown algebraically. It follows that for the general isotropic inhomogeneous three-dimensional problem the decoupling of two polarizations (TE, transverse electric, and TM, transverse magnetic) is not possible.

2.6 SEPARATION OF POLARIZATIONS FOR ANISOTROPIC PROBLEMS

The differential equation in its vector representation of Equation (2.40) holds for an inhomogeneous anisotropic material in three dimensions. We want to find conditions for the dielectric tensor under which it is possible to simplify the equations by separation of two polarizations that can be regarded as independent. Let us consider light propagating in the x–y-plane as shown in Figure 2.1. There is no variation in the z-direction and all derivatives of z will be zero. The incoming propagation vector **k** lies in the shadowed plane. We distinguish two main polarizations: the transverse magnetic (TM) polarization and the transverse electric (TE) polarization as shown in Figure 2.2. For the TM mode only the field components E_x, E_y, and H_z are nonzero, whereas $E_z = H_x = H_y = 0$. For the TE mode only the components E_z, H_x, and H_y are nonzero. Note that in both cases light is propagating in the x–y-plane. We are

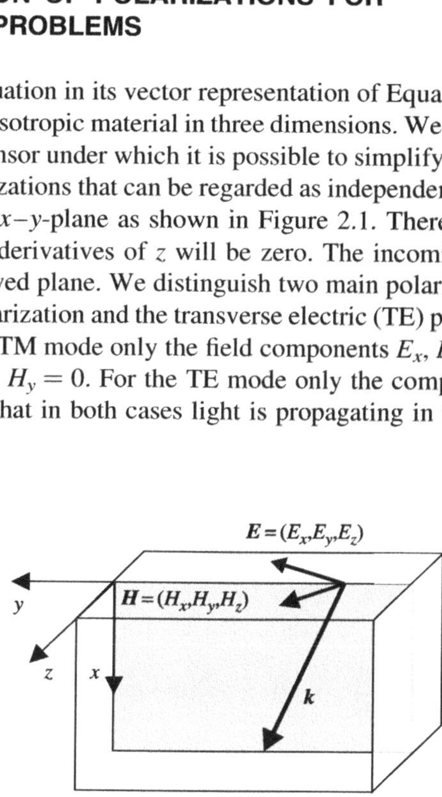

Figure 2.1 Geometry and polarization definitions for the two-dimensional problem. The plane of incidence x–y is shown in gray.

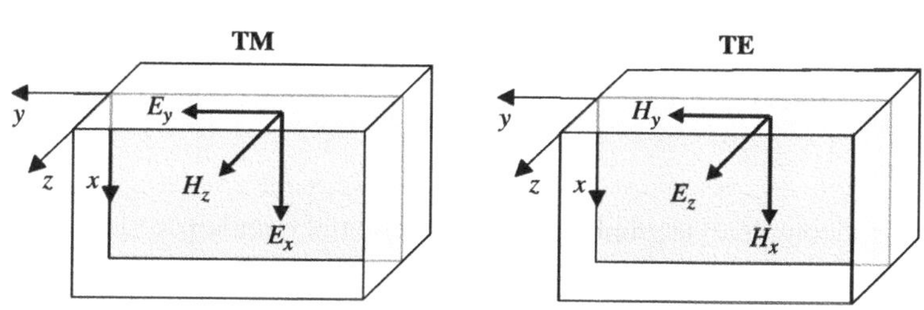

Figure 2.2 Separation of polarization states for anisotropic problems when the dielectric tensor fulfills special conditions.

looking for the conditions that the propagation of these two orthogonal polarizations is completely independent, that is, that an entering TE and TM wave remains TE and TM polarized, respectively. This will be the case if Equation (2.40) is separable into two systems of differential equations as in Equation (2.44), the first with the field components E_x, E_y, H_z (TM) and the second with E_z, H_x, and H_y (TE). In order to examine this, we rearrange the fields in Equation (2.40), taking into account that all derivatives $\partial/\partial z$ are zero. We then obtain

$$\begin{pmatrix} i\omega\varepsilon_0\varepsilon_{xx} & i\omega\varepsilon_0\varepsilon_{xy} & i\omega\varepsilon_0\varepsilon_{xz} & 0 & 0 & -\frac{\partial}{\partial y} \\ i\omega\varepsilon_0\varepsilon_{yx} & i\omega\varepsilon_0\varepsilon_{yy} & i\omega\varepsilon_0\varepsilon_{yz} & 0 & 0 & \frac{\partial}{\partial x} \\ -\frac{\partial}{\partial y} & \frac{\partial}{\partial x} & 0 & 0 & 0 & i\omega\mu_0\mu \\ 0 & 0 & \frac{\partial}{\partial y} & i\omega\mu_0\mu & 0 & 0 \\ 0 & 0 & -\frac{\partial}{\partial x} & 0 & i\omega\mu_0\mu & 0 \\ i\omega\varepsilon_0\varepsilon_{zx} & i\omega\varepsilon_0\varepsilon_{zy} & i\omega\varepsilon_0\varepsilon_{zz} & \frac{\partial}{\partial y} & -\frac{\partial}{\partial x} & 0 \end{pmatrix} \begin{pmatrix} E_x \\ E_y \\ E_z \\ H_x \\ H_y \\ E_z \end{pmatrix} = \mathbf{0}. \quad (2.46)$$

Exchanging columns 3 and 6 of the matrix in the left-hand side of Equation (2.46) gives

$$\begin{pmatrix} i\omega\varepsilon_0\varepsilon_{xx} & i\omega\varepsilon_0\varepsilon_{xy} & -\frac{\partial}{\partial y} & 0 & 0 & i\omega\varepsilon_0\varepsilon_{xz} \\ i\omega\varepsilon_0\varepsilon_{yx} & i\omega\varepsilon_0\varepsilon_{yy} & \frac{\partial}{\partial x} & 0 & 0 & i\omega\varepsilon_0\varepsilon_{yz} \\ -\frac{\partial}{\partial y} & \frac{\partial}{\partial x} & i\omega\mu_0\mu & 0 & 0 & 0 \\ 0 & 0 & 0 & i\omega\mu_0\mu & 0 & \frac{\partial}{\partial y} \\ 0 & 0 & 0 & 0 & i\omega\mu_0\mu & -\frac{\partial}{\partial x} \\ i\omega\varepsilon_0\varepsilon_{zx} & i\omega\varepsilon_0\varepsilon_{zy} & 0 & \frac{\partial}{\partial y} & -\frac{\partial}{\partial x} & i\omega\varepsilon_0\varepsilon_{zz} \end{pmatrix} \begin{pmatrix} E_x \\ E_y \\ E_z \\ H_x \\ H_y \\ E_z \end{pmatrix} = \mathbf{0}. \quad (2.47)$$

This means a complete decoupling of the differential equation systems for TE and TM polarization if the nondiagonal 3×3 submatrices are zero. This condition is fulfilled if $\varepsilon_{xz} = \varepsilon_{yx} = 0$. Taking into account the general symmetry properties of the dielectric tensor for a lossless and nonoptically active medium, $\varepsilon_{ij} = \varepsilon_{ji}$, the dielectric tensor $\boldsymbol{\varepsilon}$ that fulfills this conditions has the form

$$\boldsymbol{\varepsilon} = \begin{pmatrix} \varepsilon_{xx} & \varepsilon_{xy} & 0 \\ \varepsilon_{xy} & \varepsilon_{yy} & 0 \\ 0 & 0 & \varepsilon_{zz} \end{pmatrix}. \tag{2.48}$$

2.7 DIELECTRIC TENSOR AND INDEX ELLIPSOID

In treating birefringent crystals, one considers materials that do not show the same physical properties in all orientations. In this section, we shall address this characteristic further by introducing the tensor concepts and notation that are the language of crystal optics. The starting point of the discussion is the dielectric tensor. The dielectric tensor may be represented by a surface, known as the Fresnel or ray ellipsoid, which is the locus of electric fields **E** that produce the same energy density U. The surface is therefore defined by

$$\frac{1}{2}\mathbf{E}\cdot\mathbf{D} = U = \text{const.}, \tag{2.49}$$

where **D** is given in terms of **E** by Equation (2.13). Writing this explicitly in terms of the components of the dielectric tensor,

$$\varepsilon_{xx}E_x^2 + \varepsilon_{yy}E_y^2 + \varepsilon_{zz}E_z^2 + 2(\varepsilon_{xy}E_xE_y + \varepsilon_{yz}E_yE_z + \varepsilon_{zx}E_zE_x) = 2\frac{U}{\varepsilon_0}, \tag{2.50}$$

where we have used the property $\varepsilon_{ij} = \varepsilon_{ji}$. This reduces the nine components to six, corresponding to the six degrees of freedom that specify the axes and orientation of an ellipsoid. Equation (2.52) is the equation of an ellipsoid, whose axes coincide with the principal axes x, y, and z of the medium, and whose semi-axes, are

$$\frac{1}{\sqrt{\varepsilon_{x'x'}}} = \frac{1}{n_x}, \quad \frac{1}{\sqrt{\varepsilon_{y'y'}}} = \frac{1}{n_y}, \quad \frac{1}{\sqrt{\varepsilon_{z'z'}}} = \frac{1}{n_z}, \tag{2.51}$$

which are proportional to the optical phase velocities. The Fresnel ellipsoid is used to obtain an index ellipse by cutting a cross-section of the three-dimensional ellipsoid perpendicular to the direction of light propagation. For light traveling in any direction, the index ellipse is obtained by making a major cross-section of the ellipsoid perpendicular to that. The index ellipse yields the refractive indices and indicates the vibration direction (of the electric vector) of the light waves. The waves vibrate with the electric vectors along the major and minor axes of the index ellipse. The refractive indices that the two waves experience are given by the major and minor radii of the ellipse. In other words, for each direction of travel in an anisotropic medium, there are two plane-polarized light waves that vibrate perpendicular to each other and travel at different velocities. This is, in fact, the phenomenon of birefringence. These axes, by their orientation and magnitude, therefore define the directions and refractive indices of the fast and slow polarizations.

If the surface of constant energy density is plotted as a function of displacement **D**, rather than electric field **E**, we obtain an alternative representation of the susceptibility tensor known as the index ellipsoid, given, in the principal coordinate frame, by

$$\frac{D_x^2}{\varepsilon_{xx}} + \frac{D_y^2}{\varepsilon_{yy}} + \frac{D_z^2}{\varepsilon_{zz}} = 2\varepsilon_0 U. \tag{2.52}$$

If we normalize the field D via

$$X = \frac{D_x}{\sqrt{2\varepsilon_0 U}}, \qquad Y = \frac{D_y}{\sqrt{2\varepsilon_0 U}}, \qquad Z = \frac{D_z}{\sqrt{2\varepsilon_0 U}}, \tag{2.53}$$

and we use the definition

$$n_x = \sqrt{\varepsilon_{xx}}, \qquad n_y = \sqrt{\varepsilon_{yy}}, \qquad \text{and} \qquad n_z = \sqrt{\varepsilon_{zz}}, \tag{2.54}$$

which assumes $\mu = \mu_0$, then one finds

$$\frac{X^2}{n_x^2} + \frac{Y^2}{n_y^2} + \frac{Z^2}{n_z^2} = 1. \tag{2.55}$$

This is again the equation of an ellipsoid. It is denoted by various names, including the index ellipsoid, index of wave normals, optical indicatrix, and reciprocal ellipsoid. We define the effective index of refraction n_{eff} as

$$n_{\text{eff}} = \sqrt{X^2 + Y^2 + Z^2}. \tag{2.56}$$

This is motivated by the property that

$$U = \frac{1}{2\varepsilon_{\text{eff}}} |D|^2, \tag{2.57}$$

which corresponds to the formula for isotropic media. The value of n_{eff} can be read from the index ellipsoid as follows. Consider a ray emerging from the origin of the coordinate system in the direction of polarization of the field given by **D**. The length of the ray from the origin to the intercepted point on the index ellipsoid is the value of the effective index of refraction. The semi-axes of this ellipsoid are thus given by the principal refractive indices n_x, n_y, and n_z rather than their reciprocals. As the dielectric displacement is perpendicular to the wavevector **k** for plane waves, then

$$\mathbf{D} \cdot \mathbf{k} = 0, \tag{2.58}$$

and the index ellipsoid provides a means of finding the refractive index and polarization axes for any given wavevector **k**. As ray tracing determines the direction of the wavevector rather than the Poynting vector, the index ellipsoid is the more helpful surface for practical calculations. With the help of a geometrical construction it is possible to find two independent plane waves with the orthogonal polarizations $\mathbf{D}_1$ and $\mathbf{D}_2$ that propagate in the medium along a given direction of the wavevector **k** and their corresponding two indices of refraction n_1 and n_2. This is illustrated in Figure 2.3. An ellipse with three unequal principal axes has two circular cross-sections. The plane containing the major and minor axes cuts the ellipsoid into an ellipse with the maximum and minimum possible radii. Somewhere in between is a radius equal to the intermediate axis as illustrated in Figure 2.4. The two circular sections have radii equal to the intermediate axis and intersect along the intermediate axis. Transparent crystalline materials are generally classified into two categories defined by the number of optical axes present in the lattices. Uniaxial crystals have a single optical axis and comprise the largest family of common birefringent specimens, including calcite, quartz, and ordered synthetic or biological structures. The other major class of birefringent materials is biaxial, which have two independent optical axes. The ordinary and extraordinary wavefronts in uniaxial crystals coincide at different axes of the ellipsoid, depending upon the distribution of refractive indices within the crystal. In uniaxial crystals, two of the index ellipsoid's semi-axes are equal; the intersection perpendicular to the third axis is circular, and the refractive index does not vary. The third axis is known as the optical axis.

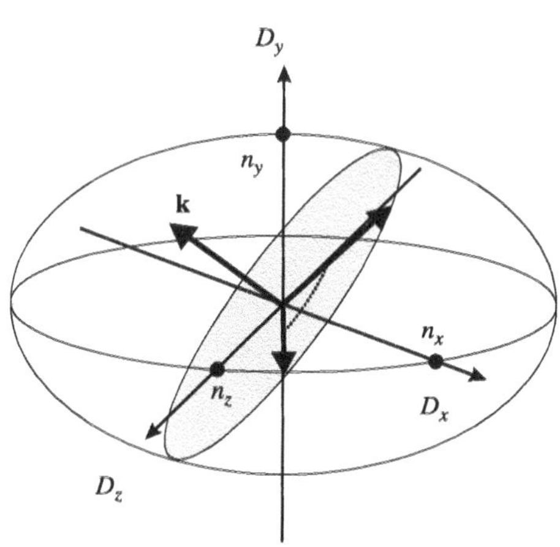

Figure 2.3 The index ellipsoid is the locus of the electric displacement **D** for a constant energy density $\mathbf{E} \cdot \mathbf{D} = U$. The axes of an elliptical section normal to the wavevector **k** indicate the fast and slow field directions. Their lengths corresponds to the refractive indices.

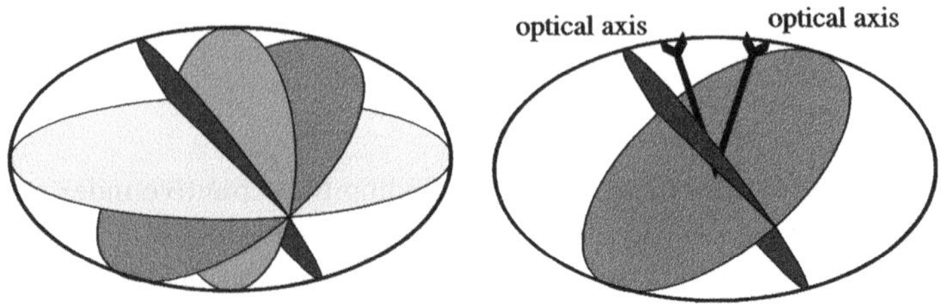

Figure 2.4 An ellipsoid can be cut by different planes and the intersection will give ellipses. For distinct directions the intersections will be circles. In this case all electrical displacement field components lying in this plane see the same refractive index. In these directions the material behaves in an isotropic manner. These directions are called optical axes.

In general, the refractive index for a given propagation direction can take two values, corresponding to the fast and slow polarizations of the beam. In uniaxial crystals, one polarization is inevitably perpendicular to the optical axis and is known as the ordinary polarization; the other, termed extraordinary, will be orthogonal to both the ordinary polarization and the wavevector. Depending upon the relative magnitudes of the ordinary refractive index n_o and extraordinary refractive index n_e, a uniaxial crystal is known to be positive uniaxial ($n_e > n_o$) or negative uniaxial ($n_e < n_o$). As the crystal orientation is changed, the ordinary refractive index remains constant, and the extraordinary refractive index varies. Of particular interest is the effective refractive index defined in Equation (2.56) if uniaxial crystals are considered. The index ellipsoid is a rotational ellipsoid in this case, as shown in Figure 2.5.

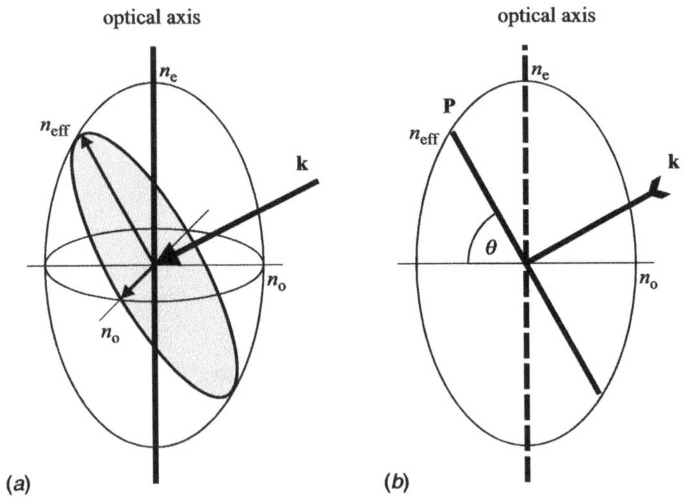

Figure 2.5 Index ellipsoid for the case of uniaxial materials. (*a*) A plane normal to the wavevector defines the plane of the electrical field components that experience n_o and n_{eff}. (*b*) The effective refractive index n_{eff} can be calculated with a geometrical construction in the plane of incidence.

The ordinary vibration direction lies in the circular section of the indicatrix (i.e., perpendicular to the optic axis) with refractive index n_o. Light traveling along the optical axis experiences just this refractive index – the ordinary refractive index. The extraordinary vibration direction lies in the plane of the wavefront and perpendicular to the ordinary vibration direction, and has refractive index n_{eff}. The value of n_{eff} is determined from the ordinary refractive index and the principal extraordinary refractive index n_e, as follows. Consider a cross-section of the indicatrix as shown in Figure 2.5*b*, containing the optical axis and the extraordinary vibration direction. The equation for this ellipse will be

$$\frac{X^2}{n_o^2} + \frac{Y^2}{n_e^2} = 1. \tag{2.59}$$

For the point P it can be seen that $X = n_{eff} \cos\theta$, and $Y = n_{eff} \sin\theta$. Therefore

$$\frac{(n_{eff}\cos\theta)^2}{n_o^2} + \frac{(n_{eff}\sin\theta)^2}{n_e^2} = 1, \tag{2.60}$$

which leads directly to the equation for n_{eff},

$$n_{eff} = \frac{n_e n_o}{\sqrt{n_e^2\cos^2\theta + n_o^2\sin^2\theta}}, \tag{2.61}$$

which fulfils the definition of the effective refractive index given in Equation (2.56).

A third surface of interest is the ray surface or ray velocity surface, which is the locus of the ray velocity. The phase velocity is the speed of propagation along the wavevector and the component of the ray velocity normal to the wavefront. The ray surface has the same shape as the wavefront emerging from a point source.

Table 2.1 Surface of wave propagation and corresponding quantities.

Surface	Property	Coordinates
Fresnel ellipsoid Ray ellipsoid	Constant energy	Electric field
Index ellipsoid Optical indicatrix Ellipsoid of wave normals Reciprocal ellipsoid	Constant energy	Electrical displacement
Normal surfaces Index surfaces	Refractive indices	Wavevector
Ray (velocity) surfaces Wave surfaces	Ray velocities	Poynting vector
(Wave) Normal surfaces	Phase velocities	Wavevector

Freegarde (2006)

As the electric displacement and wavevector are perpendicular ($\mathbf{D} \cdot \mathbf{k} = 0$), the electric displacement **D** is tangential to the ray surface. The ray velocity surface is a simple ellipsoid only for uniaxial crystals.

The linear susceptibility tensor may be represented by a multitude of physically significant three-dimensional surfaces, each of which assumes a variety of names. The refractive index, susceptibility, phase velocity, and ray velocity may each be plotted as functions of wavevector **k**, Poynting vector **S**, electric field **E**, or displacement **D**. Each surface has its own uses, and it is naturally important not to confuse the different representations. As there are two polarizations for any given wavevector, functions of these vectors must be plotted as a pair of surfaces, with one for the ordinary and one for the extraordinary ray.

If the physical properties of a crystalline medium can be described, as in our analysis of birefringence, by a material tensor such as the susceptibility, then the tensor may be invariant under certain symmetry operations, such as rotation, reflection, and inversion. The exact symmetries that apply for any given material reflect the symmetries of the crystal structure itself. The symmetry elements of any physical property of a crystal must include – but are not limited to – the symmetry elements of the point group of the crystal. All crystals fall into one of 32 point groups in crystallography, which can be grouped into systems (cubic, hexagonal, monoclinic, triclinic, orthorhombic, trigonal, tetragonal, isotropic) according to the more important symmetry properties. Cubic crystals, for example, have completely isotropic dielectric properties; tetragonal, trigonal, and hexagonal crystals are uniaxial; and triclinic, monoclinic, and orthorhombic crystals are biaxial. The exact symmetry properties of the various systems and point groups are detailed in the more specialist textbooks, and are useful when specific systems need to be analyzed (Wahlstrom, 1969; Yariv, 2003).

REFERENCES

Azzam, R.M.A. and N.M. Bashara (1987) *Ellipsometry and Polarized Light*, North-Holland, Amsterdam.

Born, M. and E. Wolf (1980) *Principles of Optics*, 6th edn, Pergamon, New York.

Freegarde, T. (2006) *Lecture Notes*, www.phys.soton.ac.mk/quantum/phys3003.html.

Serdyukov, A., I. Semchenko, S. Tretyakov, and A. Sihvola (2001) *Electromagnetics of Bi-anisotropic Materials: Theory and Applications*, Gordon and Breach Science, Harwood Academic, Amsterdam.

Wahlstrom, E. (1969) *Optical Crystallography*, Wiley, New York.

Yariv, A. and P. Yeh (2003) *Optical Waves in Crystals*, Wiley, New York.

3

DESCRIPTION OF LIGHT PROPAGATION WITH RAYS

3.1 INTRODUCTION

Ray tracing is a method of studying light propagation, in which the paths of individual rays of light are followed from their points of origin to a detector plane. The main idea is to use the refractive index as a key parameter to describe the ray path through optical systems. Traditionally, lens design has been carried out using geometric ray tracing to model the propagation of light through optical components. For coherent laser beams, a physical optics method that uses diffraction propagation of electric fields is more appropriate. Nowadays, hybrid approaches that use diffraction methods for modeling beam propagation and geometrical optics to compute effects of arbitrary optical elements are common. For homogeneous, isotropic media, such as air or glass, the rays are straight lines. Refraction, or bending of the rays at optical surfaces, is governed by Snell's law, which may be extended to surfaces with gratings or holograms. Methods for computing the intersection of the rays with optical surfaces of arbitrary shape, as well as reflection, refraction, and bending of rays by gratings are well known. For nonhomogeneous or nonisotropic materials, the ray model has to be extended. For the particular case of high spatial resolution inhomogeneous optical systems ray tracing is not suitable.

Interference of beams and diffraction during beam propagation is often called physical optics propagation. Physical optics propagation, like geometrical optics, has two basic components: a propagation of the beam between surfaces, and a transfer from one side of an optical surface to the other. Several commercial software packages use physical optics and treat that with Gaussian beam algorithms

Polarized Light in Liquid Crystals and Polymers. By Toralf Scharf

(Breault Research Organization, 2005; Photon Engineering, 2005). The most convenient representation of arbitrary light fields and beams is a two-dimensional amplitude array of points. At each point in the array, the complex amplitude of the electric field is stored, which includes amplitude and phase, which is important for interference effects. It is possible to store two arrays, one for each polarization, if desired. In the most general case, the array may have a different number of points in the two orthogonal directions, and the point-to-point spacing may also be different. The primary limitations on the number of points in beam arrays are computer memory and computation time.

In this chapter we shall present the basic principles of ray propagation through anisotropic materials. Only uniaxial materials are considered.

3.2 LIGHT RAYS AND WAVE OPTICS

As pointed out in Section 1.2, time harmonic solution of wave equations leads to considerable simplifications. It is therefore convenient to introduce a real valued wavefunction A($\mathbf{r}$, t) in terms of a complex function

$$\mathrm{A}(\mathbf{r}, t) = \mathrm{A}(\mathbf{r})\mathrm{e}^{\mathrm{i}2\pi\nu t}, \tag{3.1}$$

where the time-independent factor A($\mathbf{r}$) is referred to as the complex amplitude. The function A($\mathbf{r}$, t) is known as the spatial dependent complex wavefunction. This wavefunction has to satisfy the wave equation

$$\nabla^2\mathrm{A}(\mathbf{r}, t) - \frac{1}{c^2}\frac{\partial^2\mathrm{A}(\mathbf{r}, t)}{\partial t^2} = 0 \tag{3.2}$$

and have the same boundary conditions. The differential equation can be obtained by substituting Equation (3.1) into the wave equation (3.2). The result is the so-called Helmholtz equation

$$(\nabla^2 + k^2)\mathrm{A}(\mathbf{r}) = 0, \tag{3.3}$$

where $k = 2\pi\nu/c = \omega/c$ is the wavenumber. The Helmholtz equation will now be used to develop the eikonal equation, the basis of ray tracing and geometrical optics. Ray optics is the limit of wave optics when the wavelength of light is assumed to vanish. Consider a monochromatic wave of free-space wavelength λ_0, in a medium with refractive index $n(\mathbf{r})$ that varies sufficiently slowly with position so that the medium may be regarded as locally homogeneous. We write the complex amplitude in the form

$$\mathrm{A}(\mathbf{r}) = a(\mathbf{r})\mathrm{e}^{\mathrm{i}k_0 S(\mathbf{r})}, \tag{3.4}$$

where $a(\mathbf{r})$ is its magnitude, $k_0S(\mathbf{r})$ its phase, and $k_0 = 2\pi/\lambda$ is the wavenumber. By assuming that $a(\mathbf{r})$ varies sufficiently slowly with position $\mathbf{r}$, it may be regarded as constant within the distance of a wavelength λ_0. If $S(\mathbf{r})$ is held constant, surfaces with constant phase are created that are the wavefronts. The wavefront normals point in the direction of the gradient grad($S(\mathbf{r})$). At a given position r_0, the wave can be regarded as a plane wave with amplitude $a(\mathbf{r}_0)$ and wavevector $\mathbf{k}$ with magnitude $k = n(\mathbf{r}_0)k_0$. The direction will be parallel to the vector given by the gradient of S at $\mathbf{r}_0$. Optical rays are normal to the equilevel surfaces of the function $S(\mathbf{r})$. The function $S(\mathbf{r})$ is called the eikonal. The idea is to associate the local wavevectors with the rays of ray optics. This is illustrated in Figure 3.1. With this analogy, ray optics can serve to determine the approximate effects of optical components on the wavefront normals. Standard examples of a plane wavefront and a circular wavefront are given in Figure 3.2. Substituting Equation (3.4) into the Helmholtz equation, (3.3), provides

$$k_0^2\left(n^2 - |\nabla S(r)|^2\right)a(r) + \nabla^2 a(r) - ik_0\left(2\nabla S(r)\cdot\nabla a(r) + a(r)\nabla^2 S(r)\right) = 0. \quad (3.5)$$

The real and imaginary parts of the left-hand side must both vanish. Equating the real part to zero and using $k_0 = 2\pi/\lambda_0$, we obtain

$$|\nabla S(r)|^2 = n^2 + \left(\frac{\lambda_0}{2\pi}\right)^2 \frac{\nabla^2 a(r)}{a(r)}, \quad (3.6)$$

where $|\nabla S|^2 = \nabla S \cdot \nabla S$. With slow variations over the distance λ_0 one can neglect the second term on the right-hand side because $\lambda_0^2 \frac{\nabla^2 a(r)}{a(r)}$ and one is

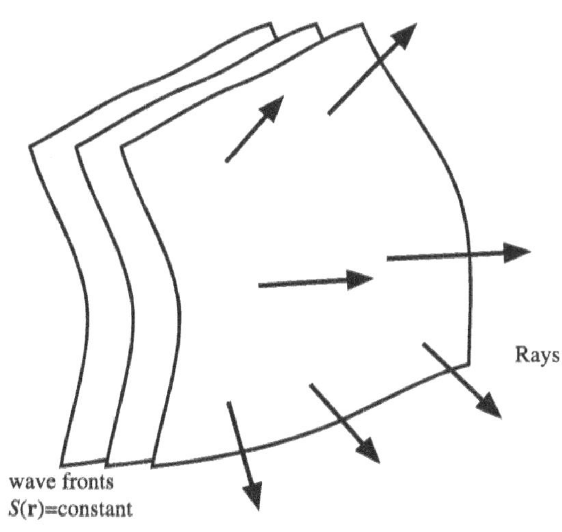

Figure 3.1 The wavefronts are surfaces of constant $S(\mathbf{r})$ and are shown as deformed surfaces. Ray trajectories are normal to that surface.

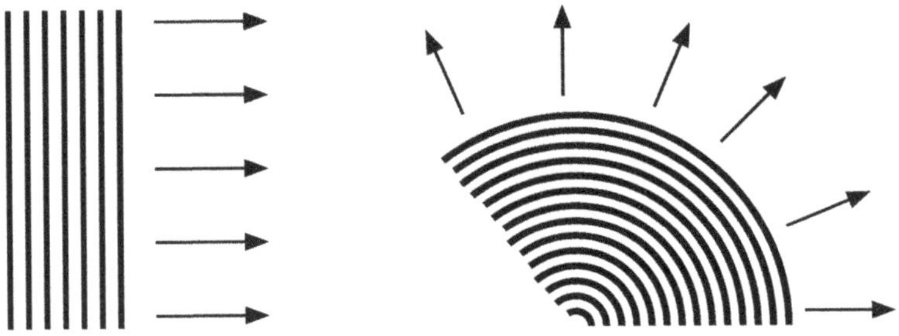

Figure 3.2 The rays of ray optics are orthogonal to the wavefronts of wave optics. Two examples are shown: (*a*) a plane wavefront and (*b*) a spherical wavefront.

left with

$$|\nabla S|^2 \approx n^2 \qquad \text{or} \qquad \left(\frac{\partial S}{\partial x}\right)^2 + \left(\frac{\partial S}{\partial y}\right)^2 + \left(\frac{\partial S}{\partial z}\right)^2 = n^2. \tag{3.7}$$

This is the eikonal equation, which may be regarded as the main postulate of ray optics. Fermat's principle (and the ray equation) can also be shown to follow from the eikonal equation. Therefore, either the eikonal equation or Fermat's principle may be regarded as the principal postulate of ray optics. For anisotropic media similar equations can be developed (Reyes, 1999; Bellver-Cebreros and Rodriguez-Danta, 2001). This leads usually to difficult vector equations.

3.3 LIGHT PROPAGATION THROUGH INTERFACES (FRESNEL FORMULA)

To become familiar with the concepts of light refraction we will discuss Fresnel formulas. Although this is not new, some of the results will be used later in this text when examples of anisotropic–isotropic interfaces are considered. We will first give a short derivation of the basic laws of reflection and refraction and then develop the formulas for the reflection and transmission coefficients for interfaces of isotropic materials. Cases for anisotropic surfaces are treated in Section 3.6.

If a plane wave with wavevector $\mathbf{k}_1$ is incident onto a planar mirror located in free space at the $z = 0$ plane, they will be reflected and a reflected plane wave of wavevector $\mathbf{k}_2$ is created. The angles of incidence and reflection are θ_1 and θ_2 as shown in Figure 3.3. In Figure 3.3 the plane of incidence is shown. The sum of the two waves satisfies the complex form of the wave equation (Helmholtz equation) if the lengths of the wavevectors are not changed, $k_1 = k_2 = k_0$ (Saleh, 1991). Certain boundary conditions must be satisfied at the surface of the mirror. As these conditions are the same at all points (x, y), it is necessary that the wavefronts of the two waves match; that is, the phases must be equal, which

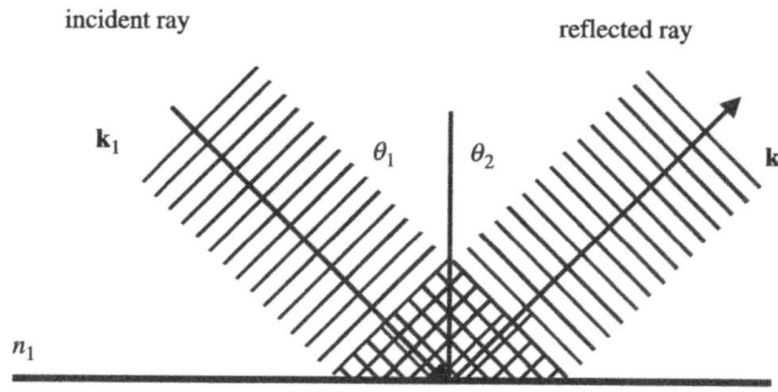

Figure 3.3 Reflection of a plane wave from a planar mirror. Phase matching at the surface of the mirror requires that the angles of incidence and reflection are equal.

immediately results in the condition

$$\mathbf{k}_1 \cdot \mathbf{r} = \mathbf{k}_2 \cdot \mathbf{r} \qquad \text{for all } \mathbf{r} = \begin{pmatrix} x & y & 0 \end{pmatrix} \tag{3.8}$$

or they differ by a constant. Substituting the vector $\mathbf{r} = (x\ y\ 0)$, the wavevectors for the incident wave $\mathbf{k}_1 = (k_0 \sin\theta_1\ 0\ k_0 \cos\theta_1)$ and for the reflected wave $\mathbf{k}_2 = (k_0 \sin\theta_2\ 0\ -k_0 \cos\theta_2)$ into Equation (3.8), we obtain $k_0 \sin\theta_1 x = k_0 \sin\theta_2 x$, from which we see that $\theta_1 = \theta_2$. The angles of incidence and reflection must be equal. Thus the law of reflection of optical rays is applicable to the wavevectors of plane waves too. For a planar boundary between two homogeneous media having refractive indices n_1 and n_2, with the boundary at the $z = 0$ plane as in Figure 3.4, refracted and reflected plane waves of wavevectors $\mathbf{k}_2$ and $\mathbf{k}_3$ emerge. The combination of the three waves satisfies the wave equation for plane waves everywhere. One has to respect the appropriate wavenumber of the medium in which it propagates. The lengths of the wavevector are given by $k_1 = k_3 = n_1 k_0$ and $k_2 = n_2 k_0$. The boundary conditions are invariant to x and y as in the case of the mirror above. The wavefronts of the three waves have to match. These conditions are such that

$$\mathbf{k}_1 \cdot \mathbf{r} = \mathbf{k}_2 \cdot \mathbf{r} = \mathbf{k}_3 \cdot \mathbf{r} \qquad \text{for all } \mathbf{r} = \begin{pmatrix} x & y & 0 \end{pmatrix} \tag{3.9}$$

The wavevectors are now $\mathbf{k}_1 = (n_1 k_0 \sin\theta_1\ 0\ n_1 k_0 \cos\theta_1)$, $\mathbf{k}_3 = (n_1 k_0 \sin\theta_3\ 0 - n_1 k_0 \cos\theta_3)$, and $\mathbf{k}_2 = (n_2 k_0 \sin\theta_2\ 0\ n_2 k_0 \cos\theta_2)$, where θ_1, θ_2, and θ_3 are the angles of incidence, refraction, and reflection, respectively. It follows from Equation (3.9) that the law of reflection holds and $\theta_1 = \theta_3$, and in addition the condition $n_1 \sin\theta_1 = n_2 \sin\theta_2$, the law of refraction also called Snell's law. The boundary conditions are not completely specified in this theory. It is therefore not possible to determine the amplitudes of the reflected and refracted waves using the scalar wave theory. To achieve this one uses the electromagnetic theory of

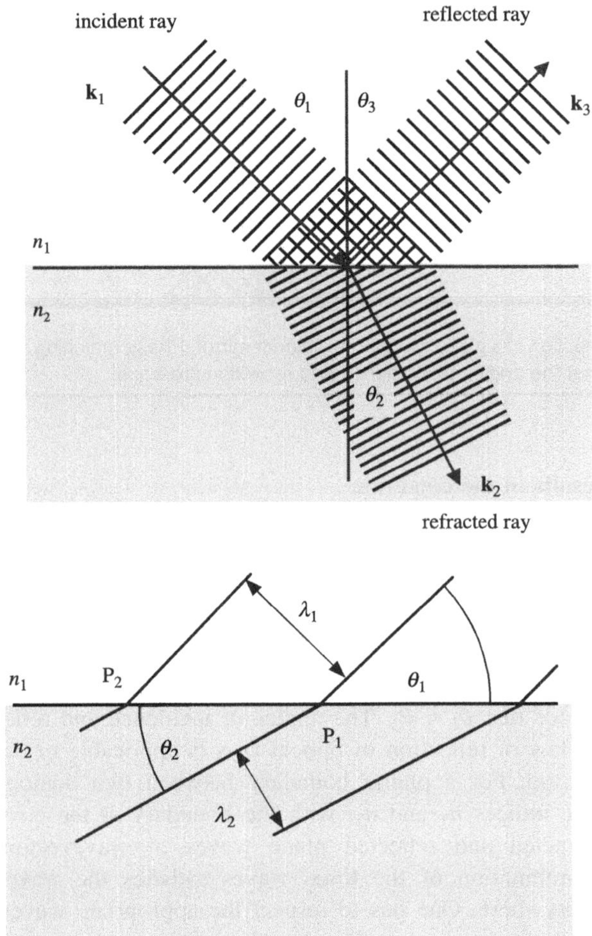

Figure 3.4 Reflection and refraction of a plane wave at a boundary of dielectric material. Matching the wavefront at the boundary, the distance P_1P_2 for the incident wave, $\lambda_1/\sin\theta_1 = \lambda_0/(n_1 \sin\theta_1)$, should be the same as for the refracted wave $\lambda_2/\sin\theta_2 = \lambda_0/(n_2 \sin\theta_2)$. This leads to Snell's law of refraction.

light. To obtain the coefficients of reflection and transmission we examine the reflection and refraction of a monochromatic plane wave incident at a planar boundary between two dielectric media. The media are assumed to be linear, homogeneous, isotropic, nondispersive, and nonmagnetic; the refractive indices are n_1 and n_2. The incident, refracted, and reflected waves are labeled with the subscripts 1, 2, and 3, respectively, as illustrated in Figure 3.5. As shown already, the wavefronts of these waves are matched at the boundary if the angles of reflection and incidence are equal,

$$\theta_1 = \theta_3, \tag{3.10}$$

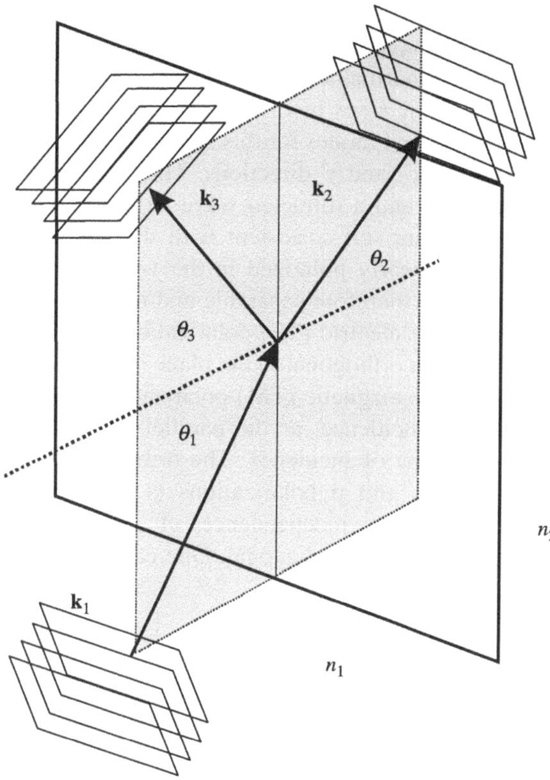

Figure 3.5 Reflection and refraction at the boundary between two dielectric media with refractive indices n_1 and n_2. The incident wave travels with $\mathbf{k}_1$, the reflected and refracted waves with $\mathbf{k}_3$ and $\mathbf{k}_2$, respectively. The wavenormals are perpendicular to the wavevectors and are indicated by the ensemble of planes. The coordinate system allows the calculation of the transmission and reflection coefficients.

and the angles of refraction and incidence satisfy Snell's law,

$$n_1 \sin\theta_1 = n_2 \sin\theta_2. \tag{3.11}$$

To relate the amplitudes and polarizations of the three waves we associate with each wave an x'–y' coordinate system in a plane normal to the direction of propagation. This is indicated in Figure 3.5 by the small planes associated with each vector $\mathbf{k}_i$. The electric field of these waves is conveniently described by vectors

$$\mathbf{J}_1 = \begin{bmatrix} E_{1x} \\ E_{1y} \end{bmatrix}; \quad \mathbf{J}_2 = \begin{bmatrix} E_{2x} \\ E_{2y} \end{bmatrix}; \quad \mathbf{J}_3 = \begin{bmatrix} E_{3x} \\ E_{3y} \end{bmatrix}. \tag{3.12}$$

Transmission and reflection coefficients are obtained by determining the relations between $\mathbf{J}_2$ and $\mathbf{J}_1$ and between $\mathbf{J}_3$ and $\mathbf{J}_1$. These relations are written in the matrix form $\mathbf{J}_2 = \mathbf{t}\mathbf{J}_1$, and $\mathbf{J}_3 = \mathbf{r}\mathbf{J}_1$, where $\mathbf{t}$ and $\mathbf{r}$ are 2×2 matrices describing the transmission and reflection of the wave, respectively. Elements of the transmission and reflection matrices are determined by using the boundary conditions required by electromagnetic theory. That theory imposes that tangential components of **E** and **H** and normal components of **D** and **B** have to be continuous at the boundary. The algebraic steps involved are reduced substantially if we take into account that the two normal modes for this system are linearly polarized waves with polarization along the x'- and y'-directions. This may be proved if we show that an incident, a reflected, and a refracted wave with their electric field vectors pointing in the x-direction are self-consistent with the boundary conditions, and similarly for three waves linearly polarized in the y-direction (Born, 1993). The x- and y-polarized waves are therefore separable and independent. The x-polarized mode is called the transverse electric (TE) polarization or the orthogonal polarization, as the electric fields are orthogonal to the plane of incidence. The y-polarized mode is called the transverse magnetic (TM) polarization, as the magnetic field is orthogonal to the plane of incidence, or the parallel polarization, as the electric fields are parallel to the plane of incidence. The orthogonal and parallel polarizations are also called the s and p polarizations (s for the German senkrecht, meaning "perpendicular"). The independence of the x- and y-polarizations implies that the matrices **t** and **r** are diagonal. One can write

$$\mathbf{t} = \begin{bmatrix} t_x & 0 \\ 0 & t_y \end{bmatrix} \quad \text{and} \quad \mathbf{r} = \begin{bmatrix} r_x & 0 \\ 0 & r_y \end{bmatrix} \tag{3.13}$$

so that

$$E_{2x} = t_x E_{1x}, \qquad E_{2y} = t_y E_{1y}, \qquad E_{3x} = r_x E_{1x}, \qquad E_{3y} = r_y E_{1y}. \tag{3.14}$$

The coefficients t_x and t_y are the complex amplitude transmittances for the TE and TM polarizations, respectively, and similarly the complex amplitude reflectances r_x and r_y. First we examine the case where the polarization is transverse electric (TE) and the electric field is perpendicular to the plane of incidence. Figure 3.6 shows the plane of incidence. At the surface the tangential electrical field, that is, the x-component E_{ix}, has to match which means that

$$E_{1x} + E_{3x} = E_{2x}. \tag{3.15}$$

Using Equation (3.14) one obtains

$$E_{1x} + r_x E_{1x} = t_x E_{1x} \qquad \text{and, further} \qquad t_x = 1 + r_x. \tag{3.16}$$

The situation is more complicated for the magnetic field components as they are not parallel to the surface. To have the tangential components of the field steady their projections (compare to Fig. 3.7) have to fullfil the relation

$$H_1 \cos\theta_1 - H_3 \cos\theta_1 = H_2 \cos\theta_2. \tag{3.17}$$

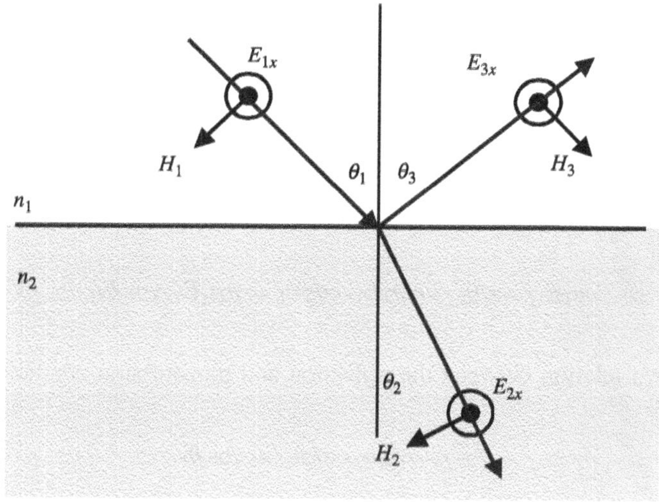

Figure 3.6 Plane of incidence for the TE polarization. The electric field components are perpendicular to the paper plane.

The magnetic field strength **H**, in the case where **E** is perpendicular to **k** and in isotropic materials, is directly linked to the electrical field strength via the material constants such that

$$E/H = \eta = \eta_0/n, \tag{3.18}$$

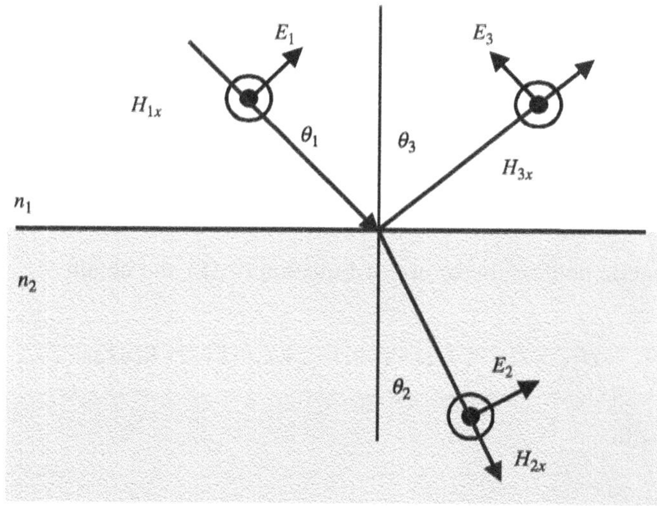

Figure 3.7 Plane of incidence for the TM polarization. This time the magnetic field components are perpendicular to the plane of the paper.

where $\eta = \sqrt{\mu_a/\varepsilon_a}$ (Eq. 2.9) and $\eta_0 = \sqrt{\mu_0/\varepsilon_0}$. Substituting for the magnetic field strengths in Equation (3.17) gives

$$n_1 E_{1x} \cos\theta_1/\eta_0 - n_1 E_{3x} \cos\theta_1/\eta_0 = n_2 E_{2x} \cos\theta_2/\eta_0. \tag{3.19}$$

After simplification, one obtains

$$n_1 E_{1x} \cos\theta_1 - n_1 r_x E_{1x} \cos\theta_1 = n_2 t_x E_{1x} \cos\theta_2, \tag{3.20}$$

which gives a relation between the reflection and transmission coefficient as

$$1 - r_x = t_x n_2 \cos\theta_2/n_1 \cos\theta_1. \tag{3.21}$$

In the case of the transverse magnetic polarization, the magnetic field is perpendicular to the plane of incidence and the projection has to be carried out instead for the electric field.

The boundary conditions for the steady transition of the electrical fields lead to

$$E_{1y} \cos\theta_1 - E_{3y} \cos\theta_1 = E_{2y} \cos\theta_2. \tag{3.22}$$

From Equation (3.14), $E_{3y} = r_y E_{1y}$ and $E_{2y} = t_y E_{1y}$ may be used to replace the unknown fields, giving

$$E_{1y} \cos\theta_1 - r_y E_{1y} \cos\theta_1 = t_y E_{1y} \cos\theta_2, \tag{3.23}$$

which leads directly to

$$1 - r_y = t_y \cos\theta_2/\cos\theta_1. \tag{3.24}$$

For the magnetic field we make use of Equation (3.18) and obtain

$$H_{1x} + H_{3x} = H_{2x} \Longrightarrow n_1 E_{1y} + r_y n_1 E_{1y} = t_y n_2 E_{1y}, \tag{3.25}$$

which results in

$$n_1(1 + r_y) = n_2 t_y \qquad 1 + r_y = t_y n_2/n_1. \tag{3.26}$$

Four equations, Equations (3.16), (3.21), (3.24), and (3.26), with four variables result. The following expressions for the reflection and transmission coefficients, known as the Fresnel equations, can be calculated from these:

$$t_x = \frac{2n_1 \cos\theta_1}{n_1 \cos\theta_1 + n_2 \cos\theta_2}; \qquad r_x = \frac{n_1 \cos\theta_1 - n_2 \cos\theta_2}{n_1 \cos\theta_1 + n_2 \cos\theta_2} \tag{3.27}$$

for TE polarization, and

$$t_y = \frac{2n_1 \cos\theta_1}{n_2 \cos\theta_1 + n_1 \cos\theta_2}; \qquad r_y = \frac{n_2 \cos\theta_1 - n_1 \cos\theta_2}{n_2 \cos\theta_1 + n_1 \cos\theta_2} \tag{3.28}$$

for TM polarization. We have therefore determined the coefficients in the reflection and transmission matrix of Equation (3.13), which can now be written as

$$\begin{aligned} \mathbf{t} &= \begin{bmatrix} \dfrac{2n_1 \cos\theta_1}{n_1 \cos\theta_1 + n_2 \cos\theta_2} & 0 \\ 0 & \dfrac{2n_1 \cos\theta_1}{n_2 \cos\theta_1 + n_1 \cos\theta_2} \end{bmatrix}, \\ \mathbf{r} &= \begin{bmatrix} \dfrac{n_1 \cos\theta_1 - n_2 \cos\theta_2}{n_1 \cos\theta_1 + n_2 \cos\theta_2} & 0 \\ 0 & \dfrac{n_2 \cos\theta_1 - n_1 \cos\theta_2}{n_2 \cos\theta_1 + n_1 \cos\theta_2} \end{bmatrix}. \end{aligned} \tag{3.29}$$

It would be interesting to analyze this behavior in detail, but it has already been covered excellently in other textbooks (Hecht, 1998; Saleh and Teich, 1991).

3.4 PROPAGATION DIRECTION OF RAYS IN CRYSTALS

When ray tracing is considered it is necessary to know how rays propagate in anisotropic materials. As most optical applications are realized using uniaxial optical material, we will discuss only the case of light propagation in uniaxial crystals with plane waves. We start by writing the Maxwell equations for plane waves. In Equation (3.30) we give Maxwell's equations for plane waves as

$$\begin{aligned} \mathbf{k} \times \mathbf{E} &= \omega\mu\mu_0 \mathbf{H}, \\ \mathbf{k} \times \mathbf{H} &= -\omega \mathbf{D} = -\omega\varepsilon\varepsilon_0 \mathbf{E}. \end{aligned} \tag{3.30}$$

Equation (3.30) specifies that the dielectric displacement **D** is perpendicular to the magnetic field vector **H**, as is the electric field vector **E**. Thus, both **E** and **D** have to be perpendicular to **H** and linked via the dielectric constant tensor ε. As shown in Figure 3.8, both **E** and **D** have to be in the same plane. This does not necessarily

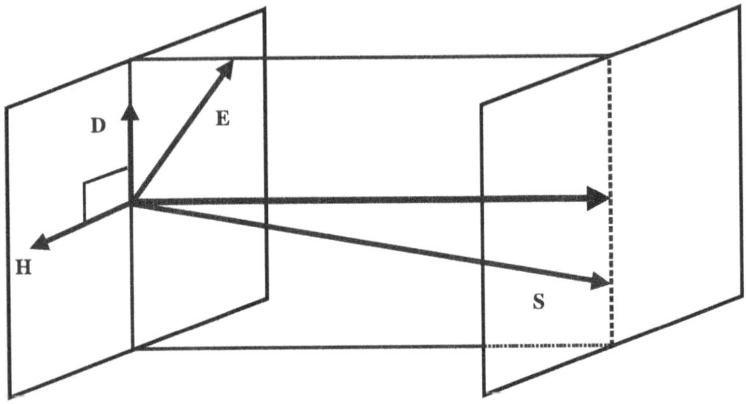

Figure 3.8 Geometry of the basic quantities for light propagation in anisotropic materials. The Poynting vector **S** describes the direction of energy flow and is not parallel to the wavevector **k**.

mean that **E** and **D** are parallel. In general, the algebraic equation that links **D** and **E** through the dielectric tensor ε is a coordinate transformation. The tensor ε transforms the electric field into the electric displacement and represents a change in direction and value. The relation between **E** and **D** for a uniaxial material leads to an expression between **D** and **E**

$$\begin{pmatrix} E_x \\ E_y \\ E_z \end{pmatrix} = \frac{1}{\varepsilon_0}\begin{pmatrix} \varepsilon_{xx} & \varepsilon_{xy} & \varepsilon_{xz} \\ \varepsilon_{yx} & \varepsilon_{yy} & \varepsilon_{yz} \\ \varepsilon_{zx} & \varepsilon_{zy} & \varepsilon_{zz} \end{pmatrix}^{-1}\begin{pmatrix} D_x \\ D_y \\ D_z \end{pmatrix} = \frac{1}{\varepsilon_0}\begin{pmatrix} \varepsilon_{xx}^{-1} & \varepsilon_{xy}^{-1} & \varepsilon_{xz}^{-1} \\ \varepsilon_{yx}^{-1} & \varepsilon_{yy}^{-1} & \varepsilon_{yz}^{-1} \\ \varepsilon_{zx}^{-1} & \varepsilon_{zy}^{-1} & \varepsilon_{zz}^{-1} \end{pmatrix}\begin{pmatrix} D_x \\ D_y \\ D_z \end{pmatrix}. \quad (3.31)$$

The electric field vector can be calculated with the help of the inverse dielectric tensor $\boldsymbol{\varepsilon}^{-1}$. The elements of the inverse dielectric tensor $\boldsymbol{\varepsilon}^{-1}$ are labeled ε_{ij}^{-1}. Equation (3.30) also specifies that the wave vector **k** is perpendicular to **H**; thus, **E**, **D**, and **k** are all in the same plane. Replacing the magnetic field **H** in Equation (3.30) leads to

$$\begin{aligned} &\mathbf{k}\times(\mathbf{k}\times\mathbf{E}) = -\omega^2\mu_0\mu\boldsymbol{\varepsilon}\varepsilon_0\mathbf{E} \Longrightarrow \mathbf{k}\times(\mathbf{k}\times\mathbf{E}) + \omega^2\mu_0\mu\boldsymbol{\varepsilon}\varepsilon_0\mathbf{E} = 0, \\ &-\mathbf{k}\times(\mathbf{k}\times\mathbf{E}) = \omega^2\mu\mu_0\mathbf{D}. \end{aligned} \quad (3.32)$$

In short, we can make the following observations: **D**, **E**, and **k** are in the same plane, which is perpendicular to **H**, and **D** is perpendicular to **k**. Multiplying both sides of Equation (3.32) by ε_0 and using the relation between **E** and **D** in Equation (3.31), one obtains

$$-\mathbf{k}\times\left(\mathbf{k}\times\boldsymbol{\varepsilon}^{-1}\mathbf{D}\right) = \omega^2\mu_0\mu\varepsilon_0\mathbf{D} \quad (3.33)$$

This equation is called the generalized wave equation. Note that **D** was chosen here as the main variable because it is perpendicular to **k**.

Light propagation of plane waves in an anisotropic medium supports two modes of polarization, with different phase velocities represented by two shells in the **k** space. For each incident wave there exist two refracted waves with two different directions and different polarizations. The aim is now to identify, for ray tracing, these waves (rays) and calculate their propagation through optical systems. The nature of waves in anisotropic media is best explained by examining the normal surface (**k** surface) obtained using the determinant of the matrix in Equation (2.25). As well as using the normal surface, there are other ways of graphically representing different aspects of the propagation of an electromagnetic plane wave through an anisotropic medium. As described in Section 2.7, there are several surfaces and representations that are suitable for describing different issues. Assume that the three principal values ε_x, ε_y, and ε_z have been measured and are known. As the refractive index in the different main directions is given by $n_i^2 = \varepsilon_i$, the main refractive indices n_x^2, n_y^2, and n_z^2 are known too. The equation

$$\frac{x^2}{n_x^2} + \frac{y^2}{n_y^2} + \frac{z^2}{n_z^2} = 1 \tag{3.34}$$

defines an ellipsoid called the index ellipsoid (Section 2.7). The length of the principal axis of this ellipsoid is $2n_i^2$, $i = x, y, z$. With the aid of this index ellipsoid we can find the two phase velocities $\mathbf{v}_\mathrm{p}$ and the two directions of **D** that belong to a given wave normal direction **k**. One proceeds as follows:

1. Draw the index ellipsoid as redrawn in Figure 3.9.
2. Draw a vector from the origin pointing in the direction of **k**.
3. Draw a plane through the origin at right angles to **k**. The intersection of that plane with the index ellipsoid is an ellipse. The principal semi-axes of this ellipse define the directions of $\mathbf{D}_1$ and $\mathbf{D}_2$.
4. The lengths of those semi-axes are equal to n_1^2 and n_2^2, respectively; the phase velocities are $(c/n_1)\mathbf{k}$ and $(c/n_2)\mathbf{k}$.

To obtain a complete picture of the ray propagation we need to discuss the normal surface. Up to now, we know only how to determine the normal modes $D_{1,2}$. In uniaxial crystals, two axes have the same refractive index, $n_x = n_y = n_\mathrm{o}$, and $n_z = n_\mathrm{e}$, and with $k_0 = \omega/c$, the equation that determines the normal surface, Equation (2.25), simplifies to

$$\det\begin{vmatrix} n_\mathrm{o}^2 k_0^2 - k_y^2 - k_z^2 & k_x k_y & k_x k_z \\ k_y k_x & n_\mathrm{o}^2 k_0^2 - k_x^2 - k_z^2 & k_y k_z \\ k_z k_x & k_z k_y & n_\mathrm{e}^2 k_0^2 - k_x^2 - k_y^2 \end{vmatrix} = 0 \tag{3.35}$$

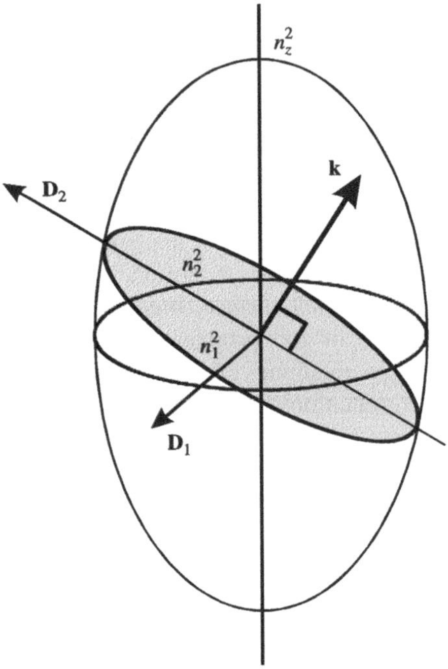

Figure 3.9 Index ellipsoid under consideration for evaluating the normal modes in an anisotropic material for the propagation direction **k**. Semi-axes are given be the square of the principal refractive indices n_x, n_y, n_z. The cut normal to the propagation direction **k** results in an ellipse with the main axes indicating the effective refractive indices n_1 and n_2.

and

$$\left(k_x^2 + k_y^2 + k_z^2 - n_0^2 k_0^2\right)\left(\frac{k_x^2 + k_y^2}{n_e^2} + \frac{k_z^2}{n_o^2} - k_0^2\right) = 0, \tag{3.36}$$

which has two solutions: a sphere

$$k_x^2 + k_y^2 + k_z^2 = n_o^2 k_0^2 \tag{3.37}$$

and an ellipsoid of revolution

$$\frac{k_x^2 + k_y^2}{n_e^2} + \frac{k_z^2}{n_o^2} = k_0^2. \tag{3.38}$$

The normal surface is constructed by first finding the two indices n_1 and n_2 for each direction of wavefront propagation **k**. Then, starting at the origin, two points are located at a distance n_1 and n_2 from the origin in the direction of **k**. The loci of all such points form the two normal surfaces and are represented in Figure 3.10

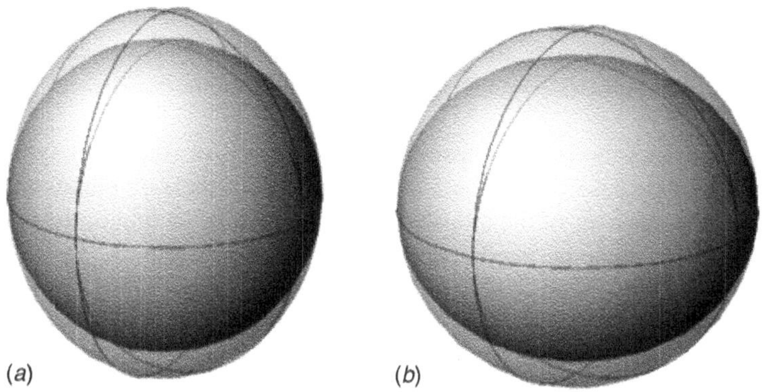

Figure 3.10 Normal surface for (*a*) positive and (*b*) negative uniaxial material. In the positive uniaxial case, the normal surface for the extraordinary propagation is an ellipse and completely surrounds the spherical **k** surface for ordinary propagation. For the negative uniaxial material the sphere for ordinary propagation surrounds the ellipse for extraordinary propagation.

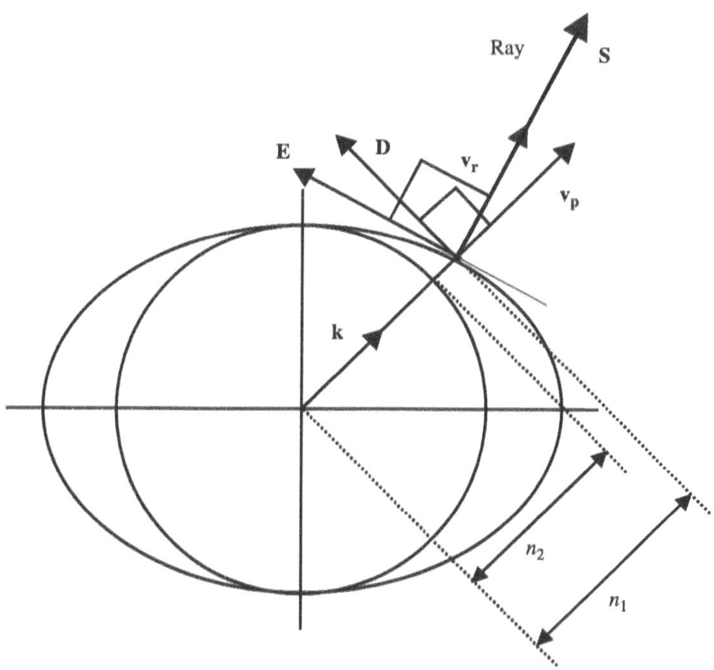

Figure 3.11 With the help of the normal surface the different quantities describing electromagnetic wave propagation are visualized. **D** is perpendicular to the propagation direction **k** and the phase velocity $\mathbf{v}_p$. The electric field **E** is perpendicular to the direction of **S** (energy flow) and the group velocity $\mathbf{v}_r$.

for a positive uniaxial crystal $n_o < n_e$, and for a negative uniaxial crystal $n_o > n_e$. If the index ellipsoid is known, the normal surface can be constructed by using the relation $n_i^2 = \varepsilon_i$ as follows. We consider a uniaxial system.

1. First construct the index ellipsoid. It has one axis of rotational symmetry; in our particular case this is the z-axis. Let **k** lie in the x–z-plane and find the principal semi-axes of the ellipse formed by the intersection of the index ellipsoid and the plane perpendicular to **k** through the origin.
2. Mark off two distances equal to the square root of the lengths of these semi-axes along the direction of **k**. Do this for all possible directions of **k** in the x–z-plane. The points marked off will connect to form a circle and an ellipse.
3. An entire normal surface is obtained by rotating the circle or the ellipse about the z-axis.

The normal surface $n(\mathbf{k})$ may now be used to find the directions of **E**, **H**, **D**, and **S** for the electromagnetic wave propagating with **k**. To summarize, Figure 3.11 shows all quantities drawn in a single viewgraph so as to observe their relationships. It is clear that the geometrical description given here enables a translation into mathematical formulas and then an implementation in ray tracing programs. Often only uniaxial materials can be treated in ray tracing programs. For most applications this is sufficient.

3.5 PROPAGATION ALONG A PRINCIPAL AXIS

The rules that govern the propagation of light in crystals under general conditions are rather complicated (Lekner, 1991; Hodgkinson and Wu, 1997). However, they become relatively simple if the light is a plane wave traveling along one of the principal axes of the crystal. Let us consider therefore a plane wave traveling in the z-direction with the wavevector **k** parallel to z. If we propagate the light along one principle direction, that is, along one of the optical axes, the dielectric tensor has the general form

$$\boldsymbol{\varepsilon} = \begin{pmatrix} \varepsilon_{xx} & \varepsilon_{xy} & 0 \\ \varepsilon_{xy} & \varepsilon_{yy} & 0 \\ 0 & 0 & \varepsilon_{zz} \end{pmatrix}. \tag{3.39}$$

Here no assumptions are made for the crystal properties. The wavevector is given as

$$\mathbf{k} = \begin{pmatrix} 0 \\ 0 \\ nk_0 \end{pmatrix} \qquad (k_0 = \omega/c). \tag{3.40}$$

where n is the refractive index to determine.

The dielectric displacement **D** is perpendicular to **k** and

$$\mathbf{D} = \begin{pmatrix} D_{0x} \\ D_{0y} \\ 0 \end{pmatrix}. \tag{3.41}$$

The electric field can be calculated by using the inverse dielectric tensor. From Equation (3.39) the inverse dielectric tensor is found to be

$$\varepsilon^{-1} = \begin{pmatrix} \dfrac{\varepsilon_{yy}}{\varepsilon_{xx}\varepsilon_{yy} - \varepsilon_{xy}\varepsilon_{yx}} & \dfrac{-\varepsilon_{xy}}{\varepsilon_{xx}\varepsilon_{yy} - \varepsilon_{xy}\varepsilon_{yx}} & 0 \\ \dfrac{-\varepsilon_{yx}}{\varepsilon_{xx}\varepsilon_{yy} - \varepsilon_{xy}\varepsilon_{yx}} & \dfrac{\varepsilon_{xx}}{\varepsilon_{xx}\varepsilon_{yy} - \varepsilon_{xy}\varepsilon_{yx}} & 0 \\ 0 & 0 & \dfrac{1}{\varepsilon_{zz}} \end{pmatrix} = \begin{pmatrix} \varepsilon_{xx}^{-1} & \varepsilon_{xy}^{-1} & 0 \\ \varepsilon_{yx}^{-1} & \varepsilon_{yy}^{-1} & 0 \\ 0 & 0 & \varepsilon_{zz}^{-1} \end{pmatrix}. \tag{3.42}$$

The corresponding equation that links the electric field to the dielectric displacement becomes

$$\varepsilon_0 \mathbf{E} = \boldsymbol{\varepsilon}^{-1}\mathbf{D} = \begin{pmatrix} \varepsilon_{xx}^{-1} & \varepsilon_{xy}^{-1} & 0 \\ \varepsilon_{yx}^{-1} & \varepsilon_{yy}^{-1} & 0 \\ 0 & 0 & \varepsilon_{zz}^{-1} \end{pmatrix} \begin{pmatrix} D_{0x} \\ D_{0y} \\ 0 \end{pmatrix} = \begin{pmatrix} \varepsilon_{xx}^{-1} D_{0x} + \varepsilon_{xy}^{-1} D_{0y} \\ \varepsilon_{yx}^{-1} D_{0x} + \varepsilon_{yy}^{-1} D_{0y} \\ 0 \end{pmatrix}. \tag{3.43}$$

Using the wave equation in the form given by Equation (3.33) leads to

$$-\begin{pmatrix} 0 \\ 0 \\ nk_0 \end{pmatrix} \times \left(\begin{pmatrix} 0 \\ 0 \\ nk_0 \end{pmatrix} \times \begin{pmatrix} \varepsilon_{xx}^{-1} & \varepsilon_{xy}^{-1} & 0 \\ \varepsilon_{yx}^{-1} & \varepsilon_{yy}^{-1} & 0 \\ 0 & 0 & \varepsilon_{zz}^{-1} \end{pmatrix} \begin{pmatrix} D_{0x} \\ D_{0y} \\ 0 \end{pmatrix} \right) = \omega^2 \mu_0 \mu \varepsilon_0 \begin{pmatrix} D_{0x} \\ D_{0y} \\ 0 \end{pmatrix}, \tag{3.44}$$

and further equating one finds in the following the final form, which can be explicitly written as

$$n^2 k_0^2 \begin{pmatrix} \varepsilon_{xx}^{-1} D_{0x} + \varepsilon_{xy}^{-1} D_{0y} \\ \varepsilon_{yx}^{-1} D_{0x} + \varepsilon_{yy}^{-1} D_{0y} \\ 0 \end{pmatrix} = \omega^2 \mu_0 \mu \varepsilon_0 \begin{pmatrix} D_{0x} \\ D_{0y} \\ 0 \end{pmatrix}. \tag{3.45}$$

Two equations result, and with $k_0 = \omega / c$ they can be written, conveniently, in the form

$$\begin{pmatrix} \varepsilon_{xx}^{-1} - \dfrac{1}{n^2} & \varepsilon_{xy}^{-1} \\ \varepsilon_{yx}^{-1} & \varepsilon_{xx}^{yy} - \dfrac{1}{n^2} \end{pmatrix} \begin{pmatrix} D_{0x} \\ D_{0y} \end{pmatrix} = 0. \tag{3.46}$$

This wave equation can be transformed to a diagonal form and delivers the eigenpolarizations $\mathbf{D}_1$ and $\mathbf{D}_2$, as well as two refractive indices corresponding to them. Searching the eigenvalues of the matrix in Equation (3.46) gives for n_1 and n_2 the deterministic equation

$$\frac{1}{n_{1,2}^2} = \frac{1}{2}\left(\left(\varepsilon_{xx}^{-1} + \varepsilon_{yy}^{-1}\right) \pm \sqrt{\left(\varepsilon_{xx}^{-1} + \varepsilon_{yy}^{-1}\right)^2 + 4\varepsilon_{xy}^{-1}\varepsilon_{yx}^{-1}}\right). \tag{3.47}$$

Replacing the values of the inverse dielectric tensor by using Equation (3.42) one obtains for the values n_1 and n_2

$$\frac{1}{n_{1,2}^2} = \frac{1}{2\left(\varepsilon_{xx}\varepsilon_{yy} - \varepsilon_{xy}\varepsilon_{yx}\right)}\left(\left(\varepsilon_{yy} + \varepsilon_{xx}\right) \pm \sqrt{\left(\varepsilon_{yy} + \varepsilon_{xx}\right)^2 + 4\varepsilon_{xy}\varepsilon_{yx}}\right). \tag{3.48}$$

Two eigenpolarizations are found. Please remember that we have used a plane wave traveling in the z-direction for our calculations. The eigenpolarizations are linear polarized because there is no phase delay between the x and y components of each vector D_1 and D_2. Their orthogonal arrangement is shown in the Figure 3.12. The position of the ellipse is given by the angle ψ as

$$\tan 2\psi = \frac{2\varepsilon_{xy}^{-1}}{\varepsilon_{xx}^{-1} - \varepsilon_{yy}^{-1}} \tag{3.49}$$

and with our particular model

$$\tan 2\psi = \frac{2\varepsilon_{xy}}{\varepsilon_{xx} - \varepsilon_{yy}}. \tag{3.50}$$

If the wave travels along one principal axis (the z-axis in our case) and is linearly polarized along an arbitrary direction in the x–y-plane, the propagation can be treated by superposition. A wave traveling along a principal axis and polarized along another principal axis has a phase velocity c_0/n_1, c_0/n_2. As these two components travel with different velocities, they undergo different phase shifts, $\varphi_1 = n_1 k_0 d$ and $\varphi_2 = n_2 k_0 d$ after propagating a distance d. Their phase retardation is

$$\varphi = \varphi_1 - \varphi_2 = (n_1 - n_2)k_0 d. \tag{3.51}$$

When the two components are combined, they form an elliptically polarized wave. In this particular geometry the crystal can therefore be used as a wave retarder device in which two orthogonal polarizations travel at different phase velocities, so that one is retarded with respect to the other.

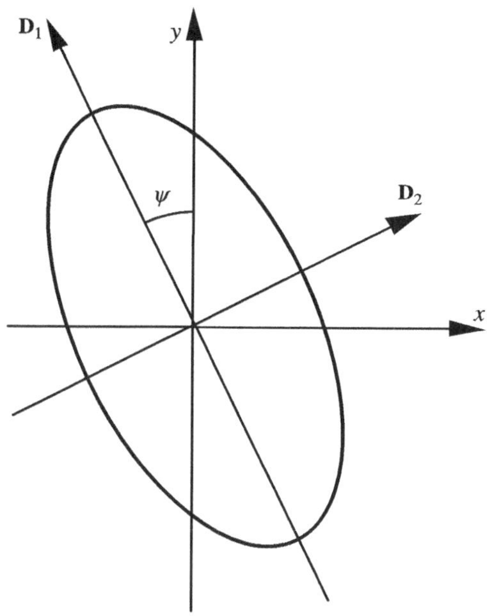

Figure 3.12 Eigenpolarization for crystals where the propagation direction z is parallel to one of the principle axes of the dielectric tensor.

3.6 RAYS AT ISOTROPIC–ANISOTROPIC INTERFACES

If an optical system contains no gradients in the refractive index, the only complicated part is light propagation through interfaces with different refractive index with anisotropy. A general description of the problem includes cases where light propagates from biaxial material into biaxial material. Discussion of this subject is far beyond the scope of this book. In most practical cases it is sufficient to understand the properties when light propagates through an interface of isotropic and uniaxial material with a particular orientation of the optical axis. The main interest lies in the direction of propagation and the change of intensity when traversing such an interface. Optical surfaces are generally curved, and are often curved quite steeply relative to the curvature of the wavefront of the optical beam. Rigorous modeling of propagation of a beam through an arbitrary surface shape would require application of complex boundary conditions on the electric field. Although it may be practical to carryout these computations for planar surfaces, or even spherical surfaces where symmetry may be exploited, almost all practical applications involve skew (nonaxial) beams with arbitrary phase and amplitude incident upon nonspherical surfaces. To prevent unnecessary complications we discuss only the case of a planar surface, but for arbitrary angles of incidence. That can than be expanded to a ray tracing algorithm for free-form surfaces. The key principle is that the wavefronts of the incident wave and the refracted wave must be matched at the boundary. As already discussed, the anisotropic medium supports two modes of different phase velocities. One finds for each incident wave two refracted waves with two different directions, transmission coefficients, and different polarizations. Birefringent ray tracing therefore involves splitting of rays. Let us first examine the refraction of a plane wave at the boundary of an anisotropic uniaxial medium (a crystal). Consider a plane wave incident on the surface. Two refracted waves appear: the ordinary wave and the extraordinary wave. The boundary conditions require that all the wavevectors lie in the plane of incidence and that their tangential components along the boundary be the same. This stays true also for anisotropic surfaces. The incident wavevector is $\mathbf{k}_0$ and we term the refracted ones $\mathbf{k}_1$ and $\mathbf{k}_2$. The projection of $\mathbf{k}_0$ on the boundary has then to match the projection of the two propagation vectors describing light propagation in the birefringent media. Figure 3.13 shows the geometrical situation. The condition requires that the following equations hold for the refracted waves:

$$k_i \sin\theta_i = k_1(\theta_1)\sin\theta_1 = k_2(\theta_2)\sin\theta_2. \tag{3.52}$$

Equation (3.52) is similar to Snell's law, but it is important to note that $\mathbf{k}_1(\theta_1)$ and $\mathbf{k}_2(\theta_2)$ are in general not constant. They vary with the directions of the vectors $\mathbf{k}_1$ and $\mathbf{k}_2$ because of the nonspherical normal surfaces of the anisotropic media. The algebraic problem of determining θ_1 and θ_2 can involve solving a quadratic equation (Francon and Mallick, 1971). We stick to the graphical method for the sake of

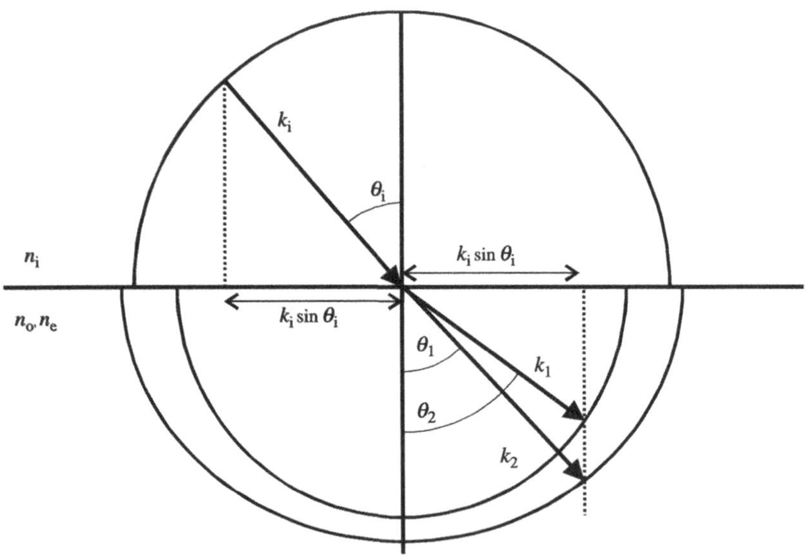

Figure 3.13 Double refraction at an anisotropic boundary. The graphical method allows the determination of the angles of the refracted wavevectors $\mathbf{k}_1$ and $\mathbf{k}_2$.

simplicity. For uniaxial crystals one shell of the normal surface is a sphere and therefore one refracted wavevector is independent of the angle. That is the case for the ordinary ray. Snell's law holds and we find with the refraction index of the incident medium n_i and the ordinary refractive index n_o, that

$$n_i \sin\theta_i = n_o \sin\theta_1 . \tag{3.53}$$

To find the direction of the rays one has to proceed with the construction discussed in Section 3.4. Starting from the normal surface one has to look for the tangential plane to these surfaces and draw the normal to this plane. This normal gives the direction of energy flow. For general configurations the ray thus escapes the plane of incidence because the normal of the tangential plane on the index ellipsoid is not in the plane of incidence anymore. Therefore, the construction becomes particularly complicated, especially if one looks for algebraic descriptions. Let us treat a simple case to get a deeper insight into the change of direction of light propagation. We will limit ourselves to positive uniaxial crystals as long as those are the most important materials for applications. Three cases of particular geometries and positions of the optical axis will be discussed:

1. The optical axis is perpendicular to the interface.
2. The optical axis is parallel to the interface.
3. The optical axis is inclined to the interface.

The three cases are represented in Figure 3.14. The index ellipsoid is shown to aid in visualizing the different situations. It is instructive to construct for a given index ellipsoid the corresponding normal surfaces (**k** surfaces). As we have seen already, these consist of two shells, one for the ordinary and one for the extraordinary propagation. In the uniaxial case, one shell is always a sphere. Because of the analytical relation between the wavevector and the refractive index, the index ellipsoid and the normal surface (**k** surface) look quite different. The optical axis is defined as that direction of propagation where only one wavevector exists, hence where the two shells of the normal surface are touching each other. In comparison to that, the index ellipsoid has its symmetry axis, which coincides with the optical axis (for uniaxial systems), along the largest refractive index (or dielectric constant). As a result, the normal surfaces look like those drawn in Figure 3.15. They have the same symmetry as the index ellipsoid but show a different shape. One should therefore be very cautious with the nomenclature and clearly distinguish between the index ellipsoid and the normal surface. The simplest case is the perpendicular orientation as all incident planes cut the index ellipsoid containing the optical axis in the same way. Therefore all planes behave the same and the ray direction can be directly deduced from the discussion in Section 3.4 and what is shown in Figure 3.11. In this situation the refracted rays stay always in the plane of incidence. More complicated is the situation where the optical axis is in the plane of the interface of the isotropic and anisotropic materials. Even more demanding would be a general discussion of

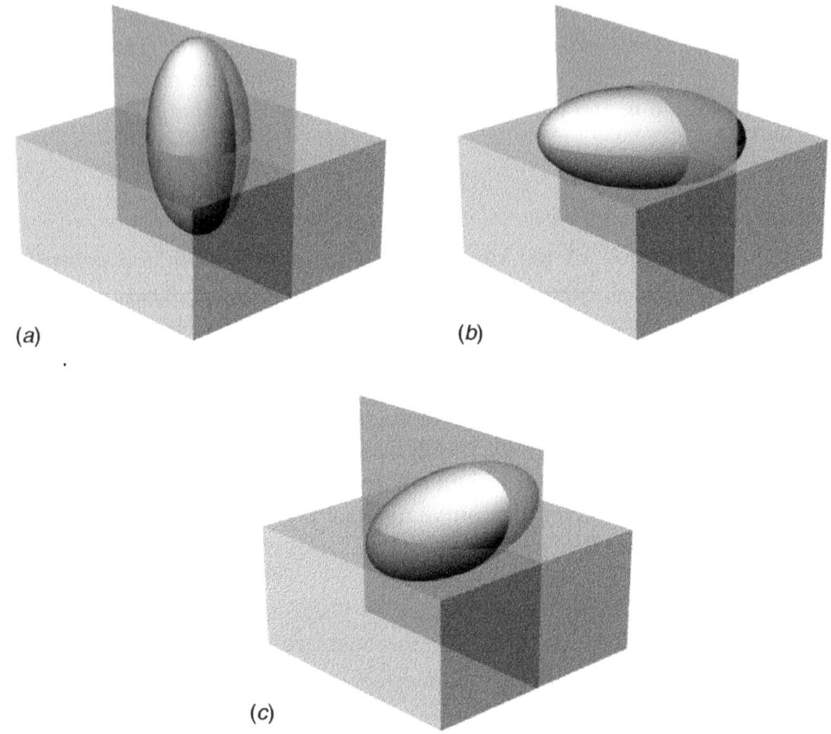

Figure 3.14 (*a*) Perpendicular (homeotropic), (*b*) in plane (planar), and (*c*) oblique orientations of the index ellipsoid with respect to the interface. The plane of incidence is shown for illustration. The index ellipsoid is chosen here because it is linked directly to the anisotropy of the material.

the situation for oblique incidence. We treat only the case where the plane of incidence contains the optical axis and the optical axis is oriented obliquely with respect to the interface, as shown in Figure 3.16. To find the normal modes one has to use the index ellipsoid. The index ellipsoid is shown in Figure 3.17*a*. The plane of incidence is the $w-u$ plane and the normal modes D_1 and D_2 are out of plane in the v-direction and in the plane along the u-direction. The normal mode D_1 sees the refractive index n_o and represents the polarization of the ordinary ray. The normal mode D_2 sees an effective refractive index n_{eff}. The basic constructions to determine ray directions are given in Figures 3.17*b* and *c*. The plane of incidence is shown. Light comes from above and enters the crystal. The normal surfaces (**k** surfaces) for the ordinary and extraordinary propagation are illustrated as a circle and an ellipse, respectively. Ray directions are constructed as normals to tangential planes for corresponding normal surfaces. For normal incidence, the ordinary ray is therefore always parallel to **k**. However, the extraordinary ray points in the direction of the normal to the extraordinary normal surface. As illustrated in Figure 3.17, for normal incidence the extraordinary ray is off the interface normal. Thus, normal incidence creates oblique refraction. Note, however, that the principle of phase matching is still

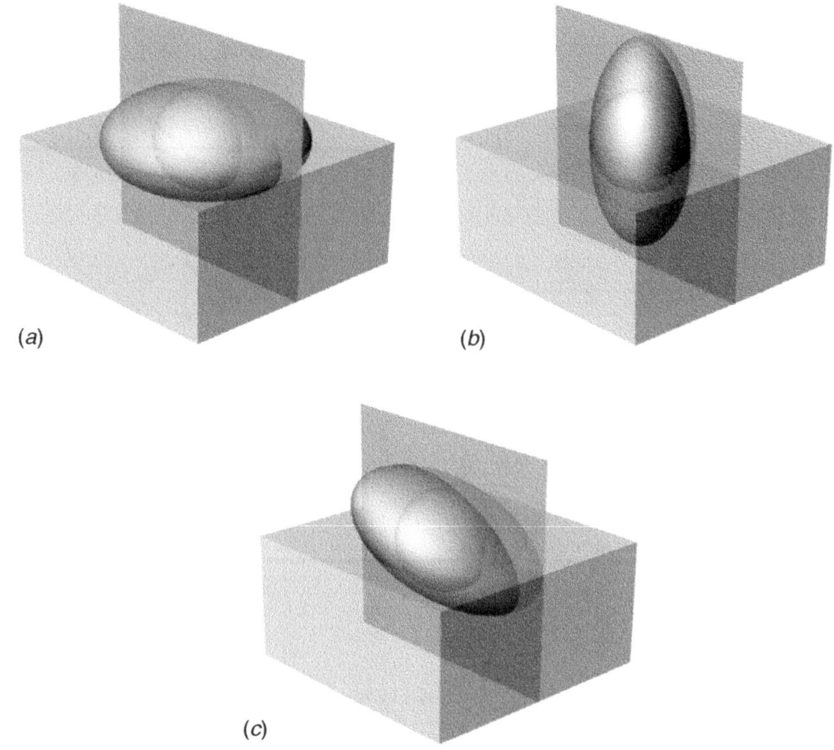

Figure 3.15 Normal surfaces (**k** surfaces) for (*a*) perpendicular (homeotropic), (*b*) in plane (planar), and (*c*) oblique orientations of the optical axis of uniaxial material. The normal surface looks different from the index ellipsoid and has two shells.

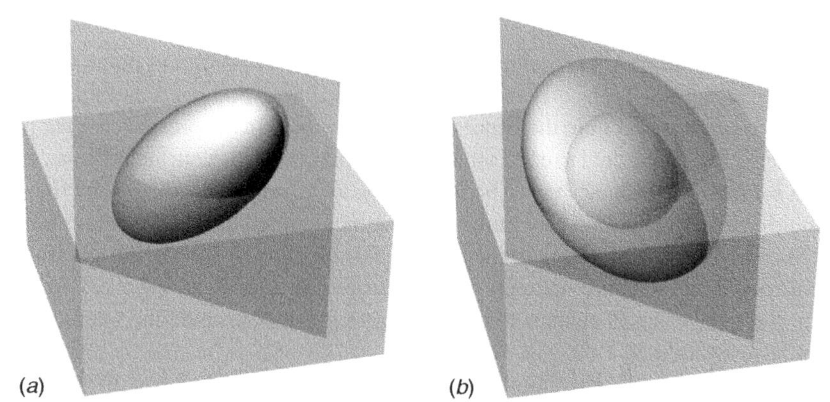

Figure 3.16 Index ellipsoid (*a*) normal (or **k**) surface (*b*) and plane of incidence for oblique orientation of the optical axis. The optical axis is recognized by the axis of revolution of the index ellipsoid and the normal surface. The ordinary and extraordinary rays lie in the plane of incidence when this plane contains the optical axis.

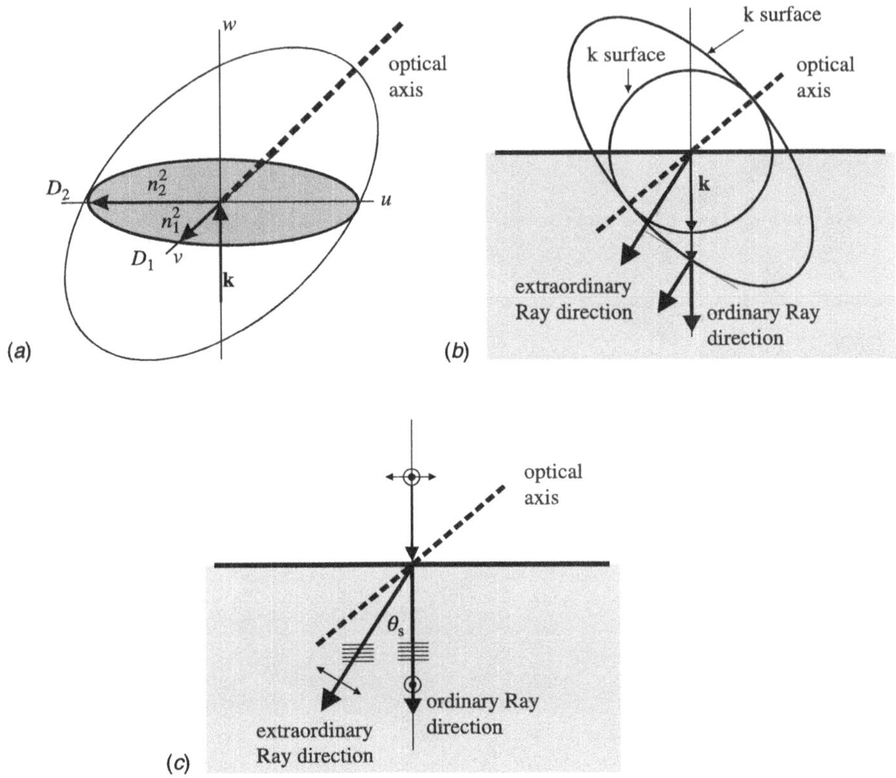

Figure 3.17 Geometrical construction of rays propagating through an interface between isotropic–anisotropic media for normal incidence from above. (*a*) The index ellipsoid is necessary to find the normal modes. (*b*) The normal surface is used to determine the ray directions. (*c*) The ray directions, shown with their polarization and wavefront.

maintained. The ordinary ray and the extraordinary ray have different polarizations. In uniaxial crystals, the eigenstates of polarization are linear polarized and the ordinary and the extraordinary rays are therefore polarized linearly. In the present case the polarization states are orthogonal to each other. The normal surface (**k** surface) allows the off angle θ_s for the extraordinary ray to be found. The wave normals indicated by the small ensemble of lines stay parallel for both rays. The incident polarization contains both TE and TM components and is split at the surface. For oblique incidence, the situation can be constructed in the same manner. More examples are given in the work of Iizuka (2002) where the cases of an anisotropic–anisotropic interface and total internal reflection are also examined.

After having determined the direction of rays, one needs to know the amplitude splitting at the surface for correct ray tracing simulations. Amplitude splitting is represented by the reflection and transmission coefficients at the interface. These coefficients are different for the ordinary and extraordinary ray. For isotropic interfaces they are given by the Fresnel equations (Section 3.3), but for anisotropic materials one has to extend the Fresnel formulas. This can give rather complicated equations as, in the general case, rays do not stay in the plane of incidence. Calculations can be carried out using the following general concept of projecting the incoming wavevector onto the local coordinate system. The local coordinate system is defined by directions of normal modes of the anisotropic material. If one has identified these normal modes, one can calculate the refractive indices in these directions. The Fresnel equations can than be applied by using the normal modes' refractive indices, the directional angles of **k**, and the angle of incidence. Note that the refractive indices of the normal modes are themselves angle-dependent. To illustrate the extension of Fresnel equations we will discuss one simple case of an isotropic–anisotropic interface with uniaxial material. We put the optical axis parallel to the interface. Two situations are considered, one where the optical axis is perpendicular to the plane of incidence and the other where it is parallel to it. Figure 3.18 shows the index ellipsoid for the case where the optical axis is perpendicular to the plane of

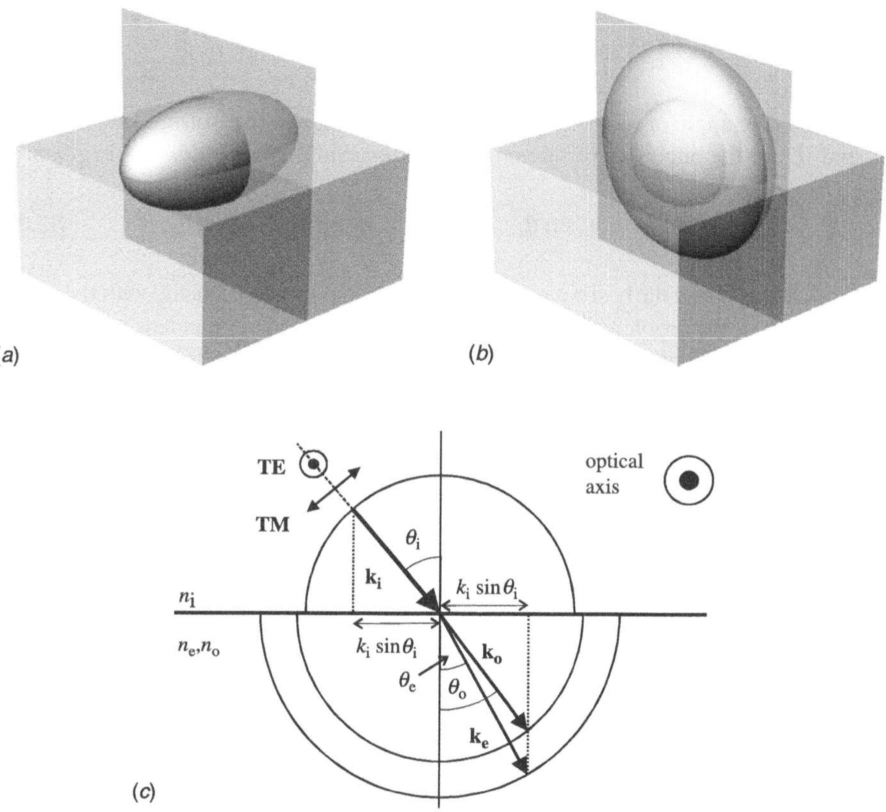

Figure 3.18 Illustration of using the sample geometry to derive the modified Fresnel formulas including anisotropy: (*a*) The index ellipsoid shows the position of the optical axis perpendicular to the plane of incidence; (*b*) the normal surface for that case; (*c*) the plane of incidence with wavevectors and polarizations.

incidence. With the arguments given above, one can directly recognize that the normal modes given by the index ellipsoid coincide with TE and TM polarization as discussed in Section 3.3. The refractive indices for the normal modes TE (x) and TM (y) are n_e and n_o, respectively. Thus the reflection and transmission coefficients can be directly calculated by using the Fresnel formulas and using for each polarization the correct refractive index. The result is

$$\mathbf{t} = \begin{bmatrix} \dfrac{2n_i \cos\theta_i}{n_i \cos\theta_i + n_e \cos\theta_e} & 0 \\ 0 & \dfrac{2n_i \cos\theta_i}{n_o \cos\theta_i + n_i \cos\theta_o} \end{bmatrix} \tag{3.54}$$

and

$$\mathbf{r} = \begin{bmatrix} \dfrac{n_i \cos\theta_i - n_e \cos\theta_e}{n_i \cos\theta_i + n_e \cos\theta_e} & 0 \\ 0 & \dfrac{n_o \cos\theta_i - n_i \cos\theta_o}{n_o \cos\theta_i + n_i \cos\theta_o} \end{bmatrix},$$

where θ_i is the angle of incidence, n_i the refractive index of the isotropic material, and θ_e and θ_o are the angles of refraction. In addition, Snell's law holds and is given by

$$n_i \sin\theta_i = n_e \sin\theta_e \qquad \text{and} \qquad n_i \sin\theta_i = n_o \sin\theta_o. \tag{3.55}$$

The case is particularly simple, because the normal modes coincide with the eigenpolarization of the problem. If the optical axis is in the plane of incidence, the geometry is as given in Figure 3.19. From the position of the index ellipsoid and direction of polarization of TE, we see that TE coincides with one normal mode. The refractive index for this mode D_1 is n_o and it is the ordinary propagation. For TM polarization one has to search the effective refractive index seen by that polarization with the help of the direction of propagation $\mathbf{k}_e$ and the formalism discussed in Section 2.7. With the help of Figure 2.5 we found that for a given direction of propagation $\mathbf{k}$, the effective refractive index was given in Equation (2.61) as

$$n_{\text{eff}} = \frac{n_e n_o}{\sqrt{n_e^2 \cos^2\theta_{\text{eff}} + n_o^2 \sin^2\theta_{\text{eff}}}}. \tag{3.56}$$

The Fresnel formulas can be modified as follows:

$$\mathbf{t} = \begin{bmatrix} \dfrac{2n_i \cos\theta_i}{n_i \cos\theta_i + n_o \cos\theta_o} & 0 \\ 0 & \dfrac{2n_i \cos\theta_i}{n_{\text{eff}} \cos\theta_i + n_i \cos\theta_{\text{eff}}} \end{bmatrix} \tag{3.57}$$

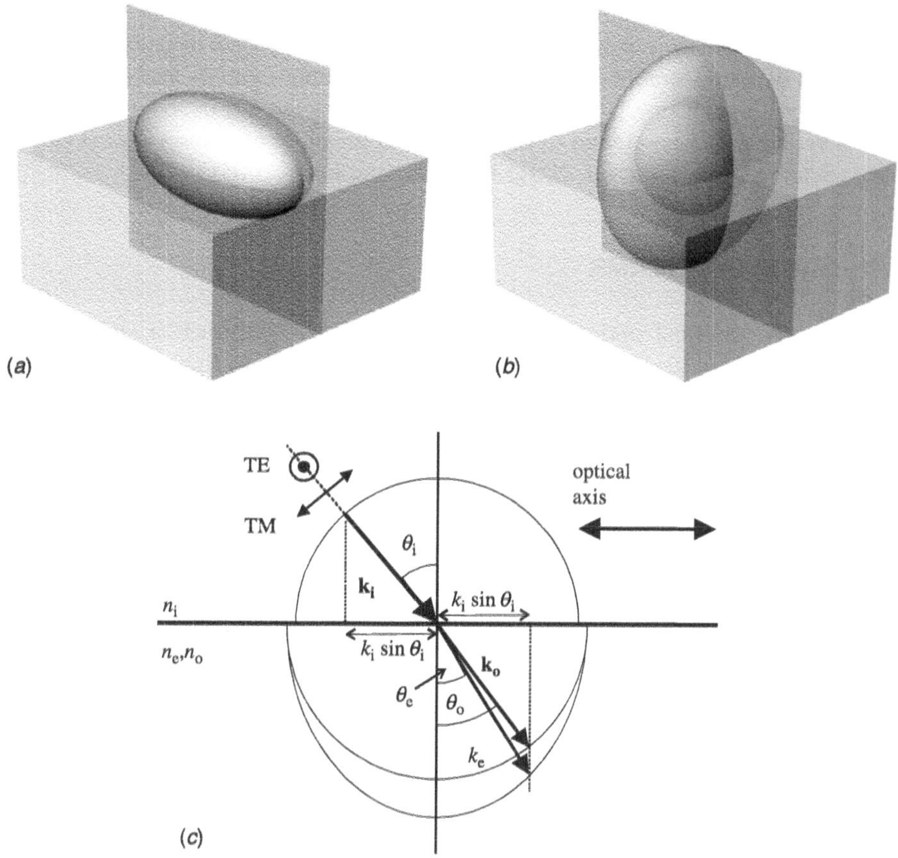

Figure 3.19 Illustration of the sample geometry for deriving the modified Fresnel formulas including anisotropy: (*a*) The index ellipsoid shows the position of the optical axis parallel to the plane of incidence; (*b*) the normal surface for that case; (*c*) the plane of incidence with wavevectors and polarizations.

and

$$\mathbf{r} = \begin{bmatrix} \dfrac{n_i \cos\theta_i - n_o \cos\theta_o}{n_i \cos\theta_i + n_o \cos\theta_o} & 0 \\ 0 & \dfrac{n_{\text{eff}} \cos\theta_i - n_i \cos\theta_{\text{eff}}}{n_{\text{eff}} \cos\theta_i + n_i \cos\theta_{\text{eff}}} \end{bmatrix}.$$

For the relations between the angles, Snell's law stays valid and one has

$$n_i \sin\theta_i = n_{\text{eff}} \sin\theta_{\text{eff}} \qquad \text{and} \qquad n_i \sin\theta_i = n_o \sin\theta_o. \tag{3.58}$$

Note that the θ_{eff} is determined with the help of Figure 3.19*c*. These examples show already that the calculations lead to complex formulas of the angle of incidence for the simple cases where the optical axis is in the plane of incidence. Stratified media are often studied, and their general reflection and transmission coefficients are better calculated by using stratified media approaches including rigorous analysis of light propagation through interfaces. Details are given Chapter 4. For the general case the interested reader is referred to the specialized literature of ellipsometry (Lekner, 1991; Schubert and Herzinger, 2001; Tompkins and Irene, 2005).

3.7 GAUSSIAN BEAMS

Traditionally, physical optics calculations are limited to diffraction simulations, which are modeled using two very different methods. Both of these methods have limitations. In the first method one can derive the pupil function of the optical system if enough rays are traced through the system and the optical path differences between the rays and a reference are calculated to produce an optical path difference map. A Fourier transform of this function yields the point spread function, which in turn may be Fourier transformed to obtain the modulation transfer function of the optical system. This method is probably the most common technique implemented in current optical design software. This type of software is specifically designed to simulate imaging optical systems. One of its major limitations is that one must have well-defined pupils. Furthermore, it is difficult to perform the diffraction calculations in arbitrary skew directions and even out of focus. The second method simulates optical fields with a complex matrix and propagates the optical field by Fourier and Fresnel transform techniques of scalar wave diffraction theory. It requires different propagation techniques depending upon whether one wants to compute the diffraction pattern in the near or far field.

A fast and powerful technique used in optical ray tracing to simulate diffraction and other physical optics phenomena like interference is Gaussian beam superposition (Arnaud, 1985). Such a superposition principle, in combination with ray tracing, is a relatively new concept and is realized in several commercially available software packages such as ASAP (Breault Research Organization, 2005) and FRED (Photon Engineering, 2005). The physical optics technique is substantially enhanced to include the effects of coherence, polarization, amplitude, and phase. All the fundamental characteristics needed to properly simulate physical optics phenomena are considered.

We do not want to examine in detail Gaussian beam propagation in anisotropic media. We therefore explain only the general concept. For applications to concrete problems the reader is referred to the manual of the simulation tool and to specialized literature (Bhawalkar et al., 1967; Landry and Maldonado, 1996; Zomer, 2003). To start with, one has to have an idea of what a Gaussian beam is. The concept of Gaussian beams was originally developed for lasers (Yariv, 1989; Yariv and Yeh, 2003; Saleh and Teich, 1991). In order to gain an appreciation of the principles of Gaussian beam optics, it is necessary to understand the nature of the laser output beam. A common description starts with a beam emitted from a laser that begins as a perfect plane wave with a Gaussian transverse irradiance profile as shown in

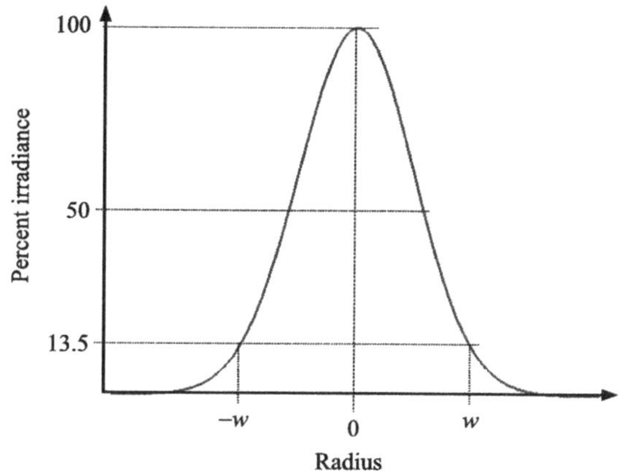

Figure 3.20 Gaussian beam intensity profile. At a distance w the intensity is down to 13.5%. The width of the beam is usually given as $2w$.

Figure 3.20. The Gaussian shape is truncated at some diameter either by the internal dimensions or by some limiting aperture in the optical train. To specify and discuss the propagation characteristics of a beam, we must define its diameter in some way. The commonly adopted definition is the diameter at which the beam irradiance (intensity) has fallen to $1/e^2$ (13.5%) of its peak, or axial, value. Diffraction causes light waves to spread transversely as they propagate, and it is therefore impossible to have a perfectly collimated beam. The spreading of a laser beam is in precise accord with the predictions of pure diffraction theory. The following formulas accurately describe beam spreading for the Gaussian beam formalism. Even if a Gaussian beam wavefront were made perfectly flat at some plane, it would quickly acquire curvature $R(z)$ and begin spreading in accordance with

$$R(z) = z\left(1 + \left(\frac{\pi w_0^2}{\lambda z}\right)^2\right) \tag{3.59}$$

and

$$w(z) = w_0\sqrt{1 + \left(\frac{\lambda z}{\pi w_0^2}\right)^2}, \tag{3.60}$$

where z is the distance propagated from the plane where the wavefront is flat, λ is the wavelength of light, w_0 is the radius of the $1/e^2$ irradiance contour at the plane where the wavefront is flat, $w(z)$ is the radius of the $1/e^2$ contour after the wave has propagated a distance z, and $R(z)$ is the wavefront radius of curvature after propagating a distance z. $R(z)$ is infinite at $z = 0$, passes through a minimum at some finite z, and rises again toward infinity as z is further increased, asymptotically approaching the value of z itself. The plane $z = 0$ marks the location of the Gaussian waist. It is the place where the wavefront is flat; w_0 is called the beam waist radius. The irradiance distribution of the Gaussian beam can be expressed as

$$I(r) = I_0 e^{-2r^2/w^2} = \frac{2P}{\pi w^2} e^{-2r^2/w^2}, \tag{3.61}$$

where $w = w(z)$ and depends on z, and P denotes the total power in the beam. The total power is conserved, as it is at all cross-sections of the beam. The invariance of the form of the distribution is a consequence of the presumed Gaussian distribution at $z = 0$. Note that if a uniform irradiance distribution had been presumed at $z = 0$, the pattern at $z = \infty$ would have been the familiar Airy disc pattern caused by diffraction. For large z, $w(z)$ asymptotically approaches the value

$$w(z) = \frac{\lambda z}{\pi w_0}, \tag{3.62}$$

where z is presumed to be much larger than $\pi w_0/\lambda$. The irradiance contours asymptotically approach a cone of angular radius

$$\theta \cong \frac{w(z)}{z} = \frac{\lambda}{\pi w_0}. \tag{3.63}$$

This value is the far-field angular radius of the Gaussian beam. The vertex of the cone lies at the center of the waist, as shown in Figure 3.21. It is important to note that, for a given value of λ, variations of beam diameter and divergence with distance z are functions of a single parameter, w_0, the beam waist radius. Unlike conventional light beams, Gaussian beams do not diverge linearly. Near the beam waist, the divergence angle is extremely small; far from the beam waist, the divergence angle approaches the asymptotic limit described above. The Raleigh range z_R,

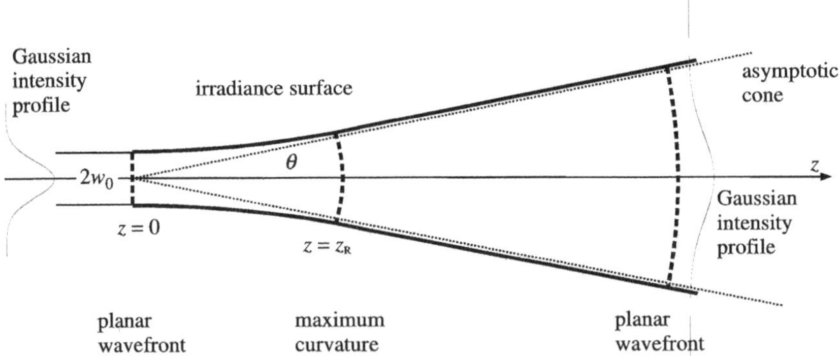

Figure 3.21 Growth in beam diameter as a function of distance from the beam waist.

defines the distance over which the beam radius spreads by a factor of the square root of 2. It is given by

$$z_R = \frac{\pi w_0^2}{\lambda}. \tag{3.64}$$

At the beam waist ($z = 0$), the wavefront is planar ($R(0) = \infty$). For very large distances at $z = \infty$, the wavefront is again planar ($R(\infty) = \infty$). As the beam propagates from the waist, the wavefront curvature, therefore, must increase to a maximum and then begin to decrease. The Raleigh range is the distance from the waist at which the wavefront curvature is a maximum, and is considered to be the dividing line between near-field divergence and mid-range divergence. Far-field divergence must be measured at a distance much greater than z_R (usually $>10z_R$ is sufficient). This is a very important distinction because calculations for spot size and other parameters in an optical train will be inaccurate if near- or mid-field divergence values are used. For a tightly focused beam, the distance from the waist (the focal point) to the far field can be a few millimeters or less. The Gaussian beam superposition algorithm models arbitrary optical fields as a coherent or incoherent sum of fundamental Gaussian beams. The superposition algorithm is similar to Babinet's principle, except that Gaussian beams are used as the basis functions. Gaussian beams can each be represented by a set of real rays describing the coherence, polarization, amplitude, and phase, which may be traced through the opto-mechanical system (Arnaud, 1985). Such a set of rays can characterize the near-field and far-field properties of a Gaussian beam. It is possible to superpose the ray trace information from the individual Gaussian beams to recreate an arbitrary optical field. Figure 3.22 illustrates the concept of beam superposition. Gaussian beams retain their shape as they propagate through free space, and, unlike plane waves, the Gaussian profile falls off rapidly with radial distance. It can be seen as an apodization. If a Gaussian beam interacts with the local curvature of an optical element, it can be assumed that the Gaussian will retain its shape after reflection or refraction. The coherence, polarization, amplitude, and phase of each beam are known, and the information from the propagation can be used to recreate the Gaussian beam at any position in the optical

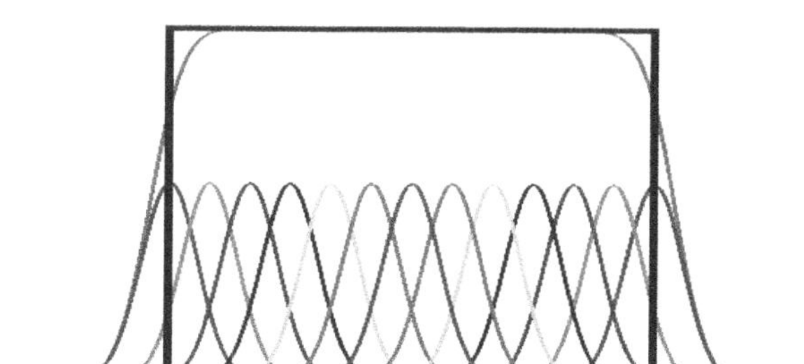

Figure 3.22 Principle of decomposing an optical field into a set of Gaussian "fundamental beams."

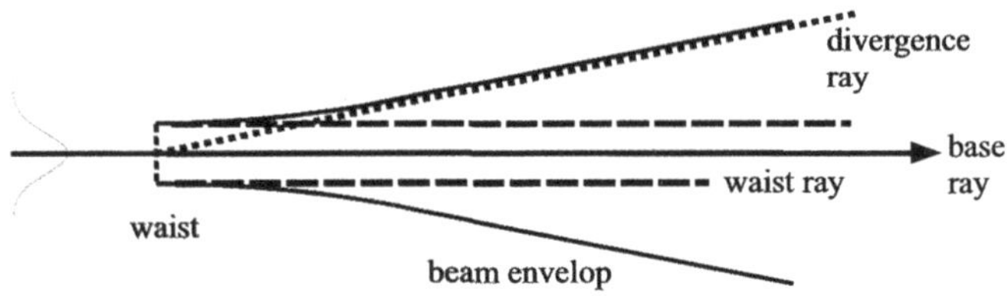

Figure 3.23 Possible ray definition for Gaussian beam ray tracing (ASAP). The base ray defines the direction of propagation. The waist ray and divergence ray allow the simulation of large fields.

system. This reconstruction and summation allows the recreation of the optical field at any position in the optical system. The advantage of the concept is that propagation of an individual Gaussian beam can be done with geometrical ray tracing. Gaussian beams can be represented by a set of geometrical rays as illustrated in Figure 3.23. The most common concept uses several rays. The location and direction of a Gaussian beam are determined by a single ray called the base ray. The base ray is just like the ray used in standard ray tracing codes to set up point sources from grids of rays. The base ray is accompanied by two or more additional rays, which are arranged around the base ray. These additional rays define the width and divergence of the Gaussian beam. They are traced like the base ray. Their size and divergences change and so does the size and divergence of the Gaussian beam. If this is carried out on two orthogonal planes, four rays are traced, which allows the Gaussian to acquire a different width and divergence in each local paraxial axis. Such tracing concepts cerate Gaussian beams that become astigmatic. The mathematical relationships between the widths, locations, and directions of the base and parabasal rays allow the amplitude and phase of the Gaussian beam to be reconstructed.

The way in which the Gaussian beams interact to recreate the optical field is a function of the coherence of the source and the calculation method used to superpose the beams. For example, the Gaussian beams simulating a monochromatic point source are coherent with respect to each other. This means that their electric field amplitudes will be added linearly to calculate the composite optical field. If the individual beams are incoherent with respect to each other, a linear superposition of the squared modulus of the electric fields, which is the energy density, has to be performed to calculate the optical field. For example, a polychromatic source is represented by a series of spectrally apodized monochromatic sources. The Gaussian beams constituting a single wavelength are coherent with respect to each other, but are incoherent with respect to different wavelengths. In the simulation, electric field amplitudes of the Gaussian beams of each wavelength would linearly add (coherent), while linearly adding their energy densities at different wavelengths (incoherent). Such a concept allows the creation of sources with specific coherence properties. The polarization of a source can be included by using a vector description of the electric field. Such a simulation concept allows the creation and simulation of polarized light and its interaction with polarizing components that polarize by reflection/transmission (diattenuation), dichroism (selective absorption), and birefringence (phase retardance). Sources can be polarized in any handedness, orientation, and ellipticity. Variables may be interpreted as Jones vector coefficients if the source model is set up in a specific orientation with respect to the global coordinate system. Doing so will allow the use of the Jones matrix (Chapter 4) or similar calculus for polarized sources and interpretation of results.

REFERENCES

Arnaud, J. (1985) Representation of Gaussian beams by complex rays, *Appl. Opt.* **24**, 538–543.

Bellver-Cebreros, C. and M. Rodriguez-Danta (2001) Eikonal equation, alternative expression of Fresnel's equation and Möhr's construction in optical anisotropic media, *Opt. Comm.* **189**, 193–209.

Bhawalkar, D.D., A.M. Goncharenko, and R.C. Smith (1967) Propagation of Gaussian beams in anisotropic media, *Br. J. Appl. Phys.* **18**, 1431–1441.

Born, M. and E. Wolf (1980) *Principles of Optics*, 6th edn, Pergamon, New York.

Breault Research Organization (2000) Advanced Systems Analysis Program ASAP^{TM}, Breault Research Organization, Inc., 6400 East Grant Road, Suite 350, Tucson, Arizona 85715 USA.

Francon, M. and S. Mallick (1971) *Polarization Interferometers*, Wiley, London.

Hecht, E. (1998) *Optics*, 3rd edn, Addison Wesley, Reading.

Hodgkinson, I.J. and Q.H. Wu (1997) *Birefringent Thin Films and Polarizing Elements*, World Scientific, Singapore.

Iizuka, K. (2002) *Elements of Photonics*, Vols 1 & 2, Wiley, New York.

Landry, G.D. and T.A. Maldonado (1996) Gaussian beam transmission and reflection from a general anisotropic multilayer structure, *Appl. Opt.* **35**, 5870–5879.

Lekner, J. (1991) Reflection and refraction by uniaxial crystals, *J. Phys.: Condens. Matter* **3**, 6121–6133.

Photon Engineering (2005) *Advanced Optical Engineering Software Program FRED*, Photon Engineering Headquarters, 440 South Williams Blvd., Suite # 106, Tucson, Arizona 85711, USA.

Reyes, J.A. (1999) Ray propagation in antisotropic inhomogeneous media, *J. Phys. A: Math. Gen.* **32**, 3409–3418.

Saleh, B.E.A. and M.C. Teich (1991) *Fundamentals of Photonics*, Wiley, New York.

Schubert, M. and C.M. Herzinger (2001) Ellipsometry on anisotropic materials: bragg conditions and phonons in dielectric helical thin films, *Phys. Stat. Sol. A* **188**, 1563–1575.

Tompkins, H.G. and E.A. Irene (2005) *Handbook of Ellipsometry*, William Andrew Publishing, Norwich, New York.

Yariv, A. (1989) *Quantum Electronics*, 3rd edn, Wiley, New York.

Yariv, A. and P. Yeh (2003) *Optical Waves in Crystals*, Wiley, New York.

Zomer, F. (2003) Transmission and reflection of Gaussian beams by anisotropic parallel plates, *J. Opt. Soc. Am. A* **20**, 171–183.

4

STRATIFIED BIREFRINGENT MEDIA

4.1 MAXWELL EQUATIONS FOR STRATIFIED MEDIA

Maxwell's equations, described in Chapter 2, can only be handled analytically in special cases, such as plane waves in free space or in homogeneous anisotropic media, or media of special geometries. Therefore, many numerical methods have been developed, based on different approximations or geometrical assumptions. In this chapter we introduce matrix methods and apply them to particular problems. The main idea of approximation methods for monochromatic light propagation is to reduce the number of electromagnetic field variables. This is possible if the medium is assumed to be stratified. For stratified media plane waves are assumed. These two assumptions allow the elimination of field components, and the first-order Maxwell differential equations can be replaced by second-order differential equations. The aim is then to solve these equations taking into account the dielectric properties of the medium. Every layer of the stratified medium is characterized by a matrix. The 2×2 matrix methods for isotropic stratified media were first introduced by Jones (1941, 1942, 1947, 1948a,b, 1956a,b) and Abelès (1950). We learned also from Chapter 2 that the separation of orthogonal polarizations is generally not possible for the two- and three-dimensional anisotropic problem. However, in assuming a stratified structure, the three-dimensional problem can be reduced to the treatment of four electromagnetic field variables. Matrix methods based on these equations were developed by Berreman (1972) and Yeh (1979). Both methods introduce eigenmodes of the fields. We will first examine the classical Jones matrix formalism valid for normal incidence and then discuss approximations for oblique incidence based on that. The 4 × 4 method of Berreman, which can be considered as a generalization of the Abelès 2 × 2 matrix method (Abelès, 1950) is introduced at the end.

4.2 JONES FORMALISM IN EXAMPLES

The propagation of polarized light in inhomogeneous anisotropic media has been studied by using Berreman's 4 × 4 formulation (Chen et al., 1997). Among recent efforts, a faster method was developed in the work of Wöhler et al. (1988) to increase computing speed. However, it is still demanding and for a lot of problems not necessary to use the 4 × 4 matrix method. A 2 × 2 matrix method that can be directly deduced from the 4 × 4 matrix methods (Lu and Saleh, 1992, Ong, 1991) has been developed. Berreman's formulation of stratified anisotropic optics is transformed into a bidirectional vector formulation, described by differential equations. Two forward waves and two backward waves are considered. A perturbation expansion can be used to obtain a generalized 2 × 2 vector description that is equivalent to Berreman's description for normal incidence. The first-order term represents the transmission, and the higher-order terms represent multiple reflections with the medium. If one governs only the forward propagating terms, one calls that method the Jones matrix formalism. The Jones matrix is the conventional technique for calculating the optical properties of birefringent layered media for normal incidence. It was introduced by Jones in the 1940s in a series of papers (Jones, 1941, 1942, 1947, 1948a,b, 1956a,b). We would like to discuss the basic properties by examining several examples to enlighten the formalism of calculations. Because there is already a lot of literature available on this subject we will be brief. Let us start by recalling the concept. For light traveling along a certain direction and perpendicular to that direction a plane is spanned. In this plane we establish a coordinate system with x- and y-directions. This coordinate is used to describe the state of polarization of light by introducing a vector system with complex numbers that represent the amplitude of phase of polarization components. The Jones vector is given as

$$\mathbf{V} = \begin{pmatrix} V_x \\ V_y \end{pmatrix} = \begin{pmatrix} V_{x0}e^{i\varphi_x} \\ V_{y0}e^{i\varphi_y} \end{pmatrix}. \tag{4.1}$$

Note, compared to the original introduction in Chapter 1, we have changed the notation to be consistent with the literature (Yeh, 1999). The intensity of $\mathbf{V}$ is $\mathbf{VV}^*$ (* indicates the complex conjugate). The vector in Equation (4.1) has the same polarization and intensity as

$$\mathbf{V} = \begin{pmatrix} V_x \\ V_y \end{pmatrix} = e^{i\varphi_x}\begin{pmatrix} V_{x0} \\ V_{y0}e^{i(\varphi_y-\varphi_x)} \end{pmatrix}. \tag{4.2}$$

The intensity is given as

$$I = \mathbf{VV}^* = |V_x|^2 + |V_y|^2. \tag{4.3}$$

Table 4.1 Selected polarization Jones vectors that have normalized intensity I = 1.

State of Polarization	Jones Vector	Jones Vector of the Orthogonal State
Linearly by horizontal polarized (x-axis)	$\begin{pmatrix} 1 \\ 0 \end{pmatrix}$	$\begin{pmatrix} 0 \\ 1 \end{pmatrix}$
Linearly by vertical polarized (y-axis)	$\begin{pmatrix} 0 \\ 1 \end{pmatrix}$	$\begin{pmatrix} 1 \\ 0 \end{pmatrix}$
Linearly by polarizer at 45° to the x-axis	$\frac{1}{\sqrt{2}}\begin{pmatrix} 1 \\ 1 \end{pmatrix}$	$\frac{1}{\sqrt{2}}\begin{pmatrix} 1 \\ -1 \end{pmatrix}$
Linearly by polarized light at an angle α with the axis	$\begin{pmatrix} \cos\alpha \\ \sin\alpha \end{pmatrix}$	$\begin{pmatrix} -\sin\alpha \\ \cos\alpha \end{pmatrix}$
Left-handed circularly by polarized light	$\frac{1}{\sqrt{2}}\begin{pmatrix} 1 \\ \mathrm{i} \end{pmatrix}$	$\frac{1}{\sqrt{2}}\begin{pmatrix} 1 \\ -\mathrm{i} \end{pmatrix}$
Right-handed circularly by polarized light	$\frac{1}{\sqrt{2}}\begin{pmatrix} 1 \\ -\mathrm{i} \end{pmatrix}$	$\frac{1}{\sqrt{2}}\begin{pmatrix} 1 \\ \mathrm{i} \end{pmatrix}$
Elliptically by polarized light (A, B, δ are constants)	$\frac{1}{\sqrt{A^2+B^2}}\begin{pmatrix} A \\ Be^{i\delta} \end{pmatrix}$	$\frac{1}{\sqrt{A^2+B^2}}\begin{pmatrix} Be^{-i\delta} \\ -A \end{pmatrix}$

For each polarization vector $\mathbf{V}_1$ an orthogonal vector $\mathbf{V}_2$ can be defined for which $\mathbf{V}_1{}^*\mathbf{V}_2 = 0$. Table 4.1 gives examples of polarization Jones vectors with the corresponding orthogonal polarization state. A polarization-dependent device can be characterized by a 2×2 matrix $\mathbf{J}$. This is the Jones matrix, which connects the incoming and the outgoing wave in a vector description by

$$\mathbf{V}_{\text{out}} = \begin{pmatrix} J_{11} & J_{12} \\ J_{21} & J_{22} \end{pmatrix} \mathbf{V}_{\text{in}}. \tag{4.4}$$

The elements of $\mathbf{J}$ are complex. The Jones matrix describes the linear transformation of the Jones vector of a plane wave by reflection, retardation, rotation, absorption, or transmission. The Jones matrix of a device that is composed of several devices in series can be obtained as the product of the Jones matrices of the individual devices. Table 4.2 gives a choice of Jones matrices for different optical elements. There are some fundamental properties that Jones matrices possess. There are particular polarization states that are transmitted unaltered by a polarization element except for a change of amplitude and phase. These polarization states are called eigenpolarizations. Every polarization element has two eigenpolarizations. Any incident light not in an eigenpolarization state is transmitted in a polarization state different from the incident one. Eigenpolarizations are the eigenvectors of the corresponding Jones matrix.

One distinguishes homogeneous and inhomogeneous matrices and elements. An inhomogeneous polarization element is an element whose eigenpolarizations are not orthogonal (Brosseau, 1998). Such an element will display different polarization

Table 4.2 Jones matrices for different optical elements in the local coordinate system of the element.

Optical Element	Jones Matrix
Linear horizontal polarizer (x-axis)	$\begin{pmatrix} 1 & 0 \\ 0 & 0 \end{pmatrix}$
Linear vertical polarizer (y-axis)	$\begin{pmatrix} 0 & 0 \\ 0 & 1 \end{pmatrix}$
Retarder with retardation in x- and y-directions	$\begin{pmatrix} e^{-\frac{2\pi i}{\lambda} n_1 d} & 0 \\ 0 & e^{-\frac{2\pi i}{\lambda} n_2 d} \end{pmatrix}$
Rotator by an angle Θ (equivalent to optical activity)	$\begin{pmatrix} \cos\Theta & -\sin\Theta \\ \sin\Theta & \cos\Theta \end{pmatrix}$

characteristics for forward and backward propagating beams. For inhomogeneous Jones matrices, the eigenpolarizations are not the states of maximum and minimum transmittance. On the other hand, a homogeneous polarization element has eigenpolarizations that are orthogonal. The eigenpolarizations are the states of maximum and minimum transmittance and also of maximum and minimum optical path length (Azzam and Bashara, 1987). A homogeneous element is classified as linear, circular, or elliptical depending on the form of the eigenpolarizations. Often, inhomogeneous elements cannot be simply classified as linear, circular, or elliptical.

An optical system might be described in the Jones matrix formalism and the direction of incidence can be from left or right; we name the Jones matrix for light incidence from left M and from right N. If the hypothetical situation where the time is reversed is considered, the output beam will retrace the path and propagate through the birefringent system. It will become the phase conjugate of the input beam. As a result of the time reversal symmetry, it can be shown that the following relation between the Jones matrices N and M holds:

$$\mathrm{NM}^* = 1. \tag{4.5}$$

An additional problem is the transmission of light forward and backward through an optical system with a change in the propagation direction. The problem lies in the fact that the reference frame changes when the propagation direction is inversed. To characterize the propagation the z-axis is used. When the propagation direction is inversed one has to inverse the coordinate system in the y-direction to maintain the right-handed coordinate frame. To go from forward to backward propagation the matrix T might be introduced as

$$\mathrm{T} = \begin{pmatrix} -1 & 0 \\ 0 & 1 \end{pmatrix}. \tag{4.6}$$

For a simple mirror and staying in the same coordinate frame, the Jones matrix is a unitary matrix because the mirror does not change the state of polarization. To deal with problems of bidirectional propagation it is strongly recommended to stay in the same coordinate system and at each step make the transformation in the local coordinate system.

When all the considered perturbations are assumed to be loss free, the resulting Jones matrices are unitary and take the form

$$\mathbf{J} = \begin{pmatrix} J_{11} & J_{12} \\ J_{21} & J_{22} \end{pmatrix} = \begin{pmatrix} A & -B^* \\ B & A^* \end{pmatrix} \qquad \text{with } |A|^2 + |B|^2 = 1. \tag{4.7}$$

With these few elements we can already start to discuss basic examples. When calculating light propagation through elements, the position of the optical axis has to be known and rotation matrices have to be used to adapt the local coordinate system to the laboratory frame. Let us discuss this in an example. The most important formula is for the intensity transmitted by a birefringent slab between crossed polarizers. Given is the angle of orientation of the optical axis of the retarder, φ. We know that the retarder has different retardation in the x- and y-directions of its local coordinate system. The Jones matrix is given in Table 4.2. We then need to make a coordinate transformation to the laboratory frame. In our special case this is a rotation about an angle φ. Polarizer and analyzer are in the laboratory coordinate system set in the x- and y-directions, respectively. Light propagates in the z-direction. In a two-dimensional coordinate system a rotation about an angle φ is performed by the matrix $R(\varphi)$

$$R(\varphi) = \begin{pmatrix} \cos\varphi & \sin\varphi \\ -\sin\varphi & \cos\varphi \end{pmatrix}. \tag{4.8}$$

One has to rotate into the local coordinate system and back to the laboratory frame. Therefore two rotational matrices will appear. The expression for light propagation through the system is

$$\mathbf{V}_{\text{out}} = P_{\text{vertical}} R(-\varphi) J_{\text{retarder}} R(\varphi) P_{\text{horizontal}} \mathbf{V}_{\text{in}}. \tag{4.9}$$

The Jones vector of the incoming light is successively multiplied by a polarizer Jones matrix in the horizontal direction, a rotation matrix, the Jones matrix for the retarder, a rotation matrix to come back to the laboratory frame, and a polarizer Jones matrix in the vertical direction. Filling in the different matrices gives

$$\begin{pmatrix} V_x^{\text{out}} \\ V_y^{\text{out}} \end{pmatrix} = \begin{pmatrix} 0 & 0 \\ 0 & 1 \end{pmatrix} \begin{pmatrix} \cos\varphi & -\sin\varphi \\ \sin\varphi & \cos\varphi \end{pmatrix} \begin{pmatrix} e^{-\frac{2\pi i}{\lambda} n_1 d} & 0 \\ 0 & e^{-\frac{2\pi i}{\lambda} n_2 d} \end{pmatrix}$$
$$\times \begin{pmatrix} \cos\varphi & \sin\varphi \\ -\sin\varphi & \cos\varphi \end{pmatrix} \begin{pmatrix} 1 & 0 \\ 0 & 0 \end{pmatrix} \begin{pmatrix} V_x^{\text{in}} \\ V_y^{\text{in}} \end{pmatrix}. \tag{4.10}$$

We need to calculate the complete series of matrices to finally obtain the transmitted intensity. After some algebra one finds for the output Jones vector

$$\begin{pmatrix} V_x^{\mathrm{out}} \\ V_y^{\mathrm{out}} \end{pmatrix} = \begin{pmatrix} 0 \\ V_x^{\mathrm{in}} \cos(\varphi)\sin(\varphi) e^{-\frac{2\pi i}{\lambda} n_1 d} - V_x^{\mathrm{in}} \sin(\varphi)\cos(\varphi) e^{-\frac{2\pi i}{\lambda} n_2 d} \end{pmatrix}, \tag{4.11}$$

which can be simplified to

$$\begin{pmatrix} V_x^{\mathrm{out}} \\ V_y^{\mathrm{out}} \end{pmatrix} = \begin{pmatrix} 0 \\ V_x^{\mathrm{in}} \cos(\varphi)\sin(\varphi)\left(e^{-\frac{2\pi i}{\lambda} n_1 d} - e^{-\frac{2\pi i}{\lambda} n_2 d}\right) \end{pmatrix}$$

$$= \begin{pmatrix} 0 \\ V_x^{\mathrm{in}} \sin(2\varphi)\left(\dfrac{e^{-\frac{2\pi i}{\lambda} n_1 d} - e^{-\frac{2\pi i}{\lambda} n_2 d}}{2}\right) \end{pmatrix},$$

$$\begin{pmatrix} V_x^{\mathrm{out}} \\ V_y^{\mathrm{out}} \end{pmatrix} = \begin{pmatrix} 0 \\ V_x^{\mathrm{in}} \sin(2\varphi) e^{-\frac{\pi i}{\lambda} n_1 d} e^{-\frac{\pi i}{\lambda} n_2 d}\left(\dfrac{e^{-\frac{\pi i}{\lambda}(n_1 - n_2)d} - e^{\frac{\pi i}{\lambda}(n_1 - n_2)d}}{2}\right) \end{pmatrix},$$

and

$$\begin{pmatrix} V_x^{\mathrm{out}} \\ V_y^{\mathrm{out}} \end{pmatrix} = \begin{pmatrix} 0 \\ i V_x^{\mathrm{in}} e^{-\frac{\pi i}{\lambda}(n_1 + n_2)d} \sin(2\varphi)\sin\left(\frac{\pi(n_1 - n_2)d}{\lambda}\right) \end{pmatrix}. \tag{4.12}$$

The intensity follows as the sum of quadrates of absolute values of each component:

$$\begin{aligned} I = |V_x|^2 + |V_y|^2 &= 0 + \left| i V_x^{\mathrm{in}} e^{-\frac{\pi i}{\lambda}(n_1 + n_2)d} \sin(2\varphi)\sin\left(\frac{\pi(n_1 - n_2)d}{\lambda}\right)\right|^2 \\ &= \left(V_x^{\mathrm{in}}\right)^2 \sin^2(2\varphi)\sin^2\left(\frac{\pi(n_1 - n_2)d}{\lambda}\right). \end{aligned} \tag{4.13}$$

If one identifies the incoming nonpolarized light intensity with I_0 one finds

$$I = \frac{I_0}{2}\sin^2(2\varphi)\sin^2\left(\frac{\pi(n_1 - n_2)d}{\lambda}\right) \tag{4.14}$$

The transmitted intensity is proportional to the incoming intensity I_0 and depends on the angle φ. This angle is measured relative to the polarizer and analyzer direction, which are oriented in the x- and y-directions, respectively. It can be seen that the transmitted intensity is a sinusoidal function of the angle and retardation difference. For the situation where the angle φ is 45°, the transmitted intensity only depends on the retardation difference. Often, observation in a polarizing microscope is carried

Table 4.3 Jones matrices for uniform liquid crystal textures (normal incidence).

Homeotropic liquid crystal layer	$J^{\text{homeotrop}} = \begin{pmatrix} e^{-\frac{2\pi i}{\lambda} n_o d} & 0 \\ 0 & e^{-\frac{2\pi i}{\lambda} n_o d} \end{pmatrix} = e^{-\frac{2\pi i}{\lambda} n_o d} \begin{pmatrix} 1 & 0 \\ 0 & 1 \end{pmatrix}$
Planar liquid crystal layer	$J^{\text{planar}} = \begin{pmatrix} e^{-\frac{2\pi i}{\lambda} n_e d} & 0 \\ 0 & e^{-\frac{2\pi i}{\lambda} n_o d} \end{pmatrix} = e^{-i\frac{\pi}{\lambda}(n_e+n_o)d} \begin{pmatrix} e^{-i\frac{\pi}{\lambda}(n_e-n_o)d} & 0 \\ 0 & e^{i\frac{\pi}{\lambda}(n_e-n_o)d} \end{pmatrix}$
Hybrid liquid crystal layer (assumption of linear tilt angle over the cell thickness)	$J^{\text{hybrid}} = \begin{pmatrix} e^{-\frac{2\pi i}{\lambda} n_{\text{eff}} d} & 0 \\ 0 & e^{-\frac{2\pi i}{\lambda} n_o d} \end{pmatrix} = e^{-i\frac{\pi}{\lambda}(n_e+n_o)d} \begin{pmatrix} e^{-i\frac{\pi}{\lambda}(n_{\text{eff}}-n_o)d} & 0 \\ 0 & e^{i\frac{\pi}{\lambda}(n_{\text{eff}}-n_o)d} \end{pmatrix}$ with $n_{\text{eff}} = n_o \left(\frac{2}{\pi} \int_0^{\pi/2} \frac{1}{\sqrt{1 - R\sin^2(z)}} dz \right)$ and $R = \frac{n_e^2 - n_o^2}{n_e^2}$
Twisted liquid crystal layer with twist angle Φ, and entrance director in the x-direction	$J^{\text{twisted}} = e^{\frac{i\pi d(n_e+n_o)}{\lambda}} \begin{pmatrix} a & b \\ -b^* & a^* \end{pmatrix}$, with $a = \frac{\Phi}{\gamma}\sin\Phi\sin\gamma + \cos\Phi\cos\gamma - i\frac{\beta}{\gamma}\cos\Phi\sin\gamma$ $b = \frac{\Phi}{\gamma}\cos\Phi\sin\gamma - \sin\Phi\cos\gamma - i\frac{\beta}{\gamma}\sin\Phi\sin\gamma$ $\gamma = \sqrt{\beta^2 + \Phi^2}$ and $\beta = \frac{\pi d(n_e - n_o)}{\lambda}$

out between crossed polarizers. The formula has therefore tremendous importance because it allows the calculation of retardations by using intensity measurements. More elaborate examples can be found in the work of Yeh (1999) and Wu and Yang (2001).

For liquid-crystal thin films, several particular Jones matrices are of interest that are linked to the textures in liquid crystal cell. Because the Jones matrix methods deal only with planar geometries, uniform textures are of interest. We write down in Table 4.3 the Jones matrices for four different liquid crystal textures: homeotropic, planar, hybrid, and twisted. Usually the terms before the brackets are neglected in calculations because they only represent phase terms. However, for an interferometer, such phase delays are important and cannot be neglected anymore. In calculations with beam splitting, the phase term has to be taken into account. The detailed derivation of the Jones matrices is done with the discussion of the optical properties of the textures of liquid crystals in Chapter 8.3.

4.3 EXTENDED JONES MATRIX METHOD

The Jones matrix method is a very powerful technique for the calculation of stacked systems at normal incidence. To a certain extent it can also be used for oblique angles of incidence. However, in particular for the calculation of twisted media under higher angles of incidence, the calculations fail. This is mainly because the method neglects interference effects, where become important at high angles due to Fresnel reflection at the surfaces. To correct this, several extended Jones matrix methods have been introduced (Yeh, 1982; Gu, 1993; Lien, 1997). They provide fast calculation and often prevent unwanted interference effects in rigorous modeling. We describe here the method introduced by Lien (1990, 1997). The basic idea is to calculate for each layer the forward and backward propagating eigenwaves and describe the transmission of light with the forward propagating waves only. This has to be done for oblique angles of incidence. It is worth making the calculation to see how the mathematics is applied to reduce the dimensionality of the problem to a 2×2 matrix calculation. We start with the description of the uniaxial dielectric tensor.

Assume a uniaxial index ellipsoid positioned in a Cartesian coordinate system (x, y, z) with an azimuthal angle (called twist) φ with respect to the x-axis and a polar angle (called tilt) θ with respect to the x–y-plane (as drawn in Fig. 4.1). The local dielectric tensor is then

$$\boldsymbol{\varepsilon} = \boldsymbol{R}_z(\varphi)\boldsymbol{R}_y(\theta)\begin{pmatrix} n_e^2 & 0 & 0 \\ 0 & n_o^2 & 0 \\ 0 & 0 & n_o^2 \end{pmatrix}\boldsymbol{R}_y^{-1}(\theta)\boldsymbol{R}_z^{-1}(\varphi), \tag{4.15}$$

where $\boldsymbol{R}_z(\varphi)$ and $\boldsymbol{R}_y(\theta)$ are rotation matrices around the z-axis and the y-axis, respectively. For a liquid crystal layer the angle θ is identified as the tilt angle and φ represents the angle between the projection of the liquid crystal director and the x–y-plane. It represents a twist angle. The diagonal matrix in Equation (4.15) is the dielectric tensor for the uniaxial medium with the major semi-axis

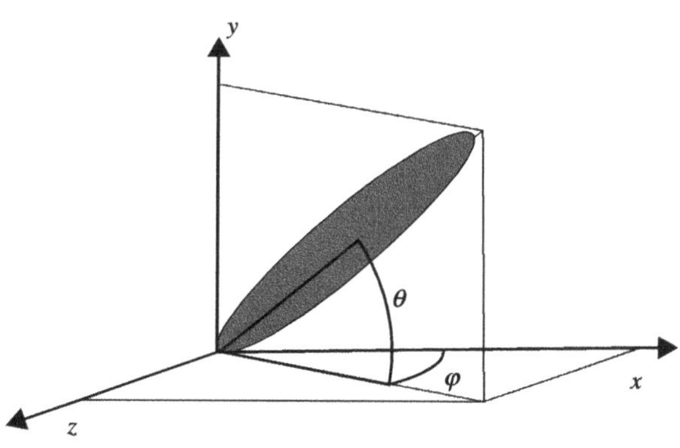

Figure 4.1 Orientation of the uniaxial index ellipsoid in the coordinate system.

lengths n_e, n_o, n_o. From Equation (4.15) we obtain explicitly

$$\boldsymbol{\varepsilon} = \begin{pmatrix} \varepsilon_{xx} & \varepsilon_{xy} & \varepsilon_{xz} \\ \varepsilon_{yx} & \varepsilon_{yy} & \varepsilon_{yz} \\ \varepsilon_{zx} & \varepsilon_{zy} & \varepsilon_{zz} \end{pmatrix}$$
$$= \begin{pmatrix} n_o^2 + \Delta\varepsilon \cos^2\theta \cos^2\varphi & \Delta\varepsilon \cos^2\theta \sin\varphi \cos\varphi & \Delta\varepsilon \sin\theta \cos\theta \cos\varphi \\ \Delta\varepsilon \cos^2\theta \sin\varphi \cos\varphi & n_o^2 + \Delta\varepsilon \cos^2\theta \sin^2\varphi & \Delta\varepsilon \sin\theta \cos\theta \sin\varphi \\ \Delta\varepsilon \sin\theta \cos\theta \cos\varphi & \Delta\varepsilon \sin\theta \cos\theta \sin\varphi & n_o^2 + \Delta\varepsilon \sin^2\theta \end{pmatrix}, \tag{4.16}$$

where $\Delta\varepsilon = n_e^2 \quad n_o^2$. With out-loss of generality we assume a form of the incident wavevector as

$$\mathbf{k} = k_0 \begin{pmatrix} \sin\varphi_0 \\ 0 \\ \cos\varphi_0 \end{pmatrix} = \begin{pmatrix} k_x \\ 0 \\ k_z \end{pmatrix} \tag{4.17}$$

with the polar angle φ_0 and the wavevector in vacuum $k_0 = \omega/c_0$. For a plane wave we know already that Maxwell's equations can be formulated in a vector form (see Eq. (2.21)) as

$$\mathbf{k} \times (\mathbf{k} \times \mathbf{E}) + \mu\boldsymbol{\varepsilon} k_o^2 \mathbf{E} = 0 \quad \text{or alternatively} \quad \frac{\mathbf{k}}{k_o} \times \left(\frac{\mathbf{k}}{k_o} \times \mathbf{E}\right) + \mu\boldsymbol{\varepsilon}\mathbf{E} = 0. \tag{4.18}$$

Furthermore there is a relation given by Eq. (2.20) between the magnetic and the electric fields such that

$$c\mu_0\mu\mathbf{H} = \frac{\mathbf{k}}{k_o} \times \mathbf{E}. \tag{4.19}$$

Writing down Equation (4.18) components leads to

$$\begin{pmatrix} \varepsilon_{xx} - \dfrac{k_z^2}{k_0^2} & \varepsilon_{xy} & \dfrac{k_x k_z}{k_0^2} + \varepsilon_{xz} \\ \varepsilon_{yx} & \varepsilon_{yy} - \dfrac{k_x^2}{k_0^2} - \dfrac{k_z^2}{k_0^2} & \varepsilon_{yz} \\ \dfrac{k_z k_x}{k_0^2} + \varepsilon_{zx} & \varepsilon_{zy} & \varepsilon_{zz} - \dfrac{k_x^2}{k_0^2} \end{pmatrix} \begin{pmatrix} E_x \\ E_y \\ E_z \end{pmatrix} = 0. \tag{4.20}$$

Here the components of $\mathbf{k}$ are given by Equation (4.17) and φ_0 is the angle of incidence. The nontrivial solution is given when the determinant of Equation (4.20) vanishes. We rewrite the determinant for the dielectric tensor specified in Equation (4.16) with its symmetry properties by using the abbreviations

$\kappa_z = k_x/k_0$, $\kappa_z = k_z/k_0$ and obtain

$$\begin{vmatrix} \varepsilon_{xx} - \kappa_z^2 & \varepsilon_{xy} & \kappa_x\kappa_z + \varepsilon_{xz} \\ \varepsilon_{yx} & \varepsilon_{yy} - \kappa_x^2 - \kappa_z^2 & \varepsilon_{yz} \\ \kappa_x\kappa_z + \varepsilon_{zx} & \varepsilon_{zy} & \varepsilon_{zz} - \kappa_x^2 \end{vmatrix} = 0. \tag{4.21}$$

Equating the determinants leads to

$$\begin{aligned} &\kappa_z^4\varepsilon_{zz} + 2\kappa_z^3\kappa_x\varepsilon_{xz} + \kappa_z^2\left(\varepsilon_{yz}^2 + \varepsilon_{xx}\kappa_x^2 - \varepsilon_{xx}\varepsilon_{zz} - \varepsilon_{yy}\varepsilon_{zz} + \varepsilon_{zz}\kappa_x^2 + \varepsilon_{xz}^2\right) \\ &\quad + \kappa_z\left(-2\kappa_x\varepsilon_{xz}\varepsilon_{yy} + 2\kappa_x^3\varepsilon_{xz} + 2\kappa_x\varepsilon_{xy}\varepsilon_{yz}\right) - \varepsilon_{xx}\varepsilon_{zz}\kappa_x^2 - \varepsilon_{xy}^2\varepsilon_{zz} + \varepsilon_{xy}^2\kappa_x^2 \\ &\quad + 2\varepsilon_{xy}\varepsilon_{yz}\varepsilon_{xz} - \varepsilon_{xz}^2\varepsilon_{yy} - \varepsilon_{xx}\varepsilon_{yy}\kappa_x^2 - \varepsilon_{yz}^2\varepsilon_{xx} + \varepsilon_{xx}\varepsilon_{yy}\varepsilon_{zz} + \varepsilon_{xx}\kappa_x^4 + \varepsilon_{xz}^2\kappa_x^2 = 0, \end{aligned} \tag{4.22}$$

where we have used the symmetry properties of the particular dielectric tensor in Equation (4.16): $\varepsilon_{xy} = \varepsilon_{yx}$, $\varepsilon_{xz} = \varepsilon_{zx}$, and $\varepsilon_{yz} = \varepsilon_{zy}$. Next, one introduces the dielectric tensor under consideration in Eq. 4.16. The solutions are:

$$\kappa_{z1} = \frac{k_{z1}}{k_0} = \sqrt{n_o^2 - \frac{k_x^2}{k_0^2}}, \tag{4.23}$$

$$\kappa_{z2} = \frac{k_{z2}}{k_0} = -\frac{\varepsilon_{xz}}{\varepsilon_{zz}}\frac{k_x}{k_0} + \frac{n_o n_e}{\varepsilon_{zz}}\sqrt{\varepsilon_{zz} - \left(1 - \frac{n_e^2 - n_o^2}{n_e^2}\cos^2\theta\sin^2\varphi\right)\frac{k_x^2}{k_0^2}}, \tag{4.24}$$

and

$$\kappa_{z3} = \frac{k_{z3}}{k_0} = -\sqrt{n_o^2 - \frac{k_x^2}{k_0^2}}, \tag{4.25}$$

$$\kappa_{z4} = \frac{k_{z4}}{k_0} = -\frac{\varepsilon_{xz}}{\varepsilon_{zz}}\frac{k_x}{k_0} - \frac{n_o n_e}{\varepsilon_{zz}}\sqrt{\varepsilon_{zz} - \left(1 - \frac{n_e^2 - n_o^2}{n_e^2}\cos^2\theta\sin^2\varphi\right)\frac{k_x^2}{k_0^2}}. \tag{4.26}$$

We have found the wavevectors for the four eigenwaves traveling in the medium. Two of them, k_{z1} and k_{z2}, propagate in the $+z$-direction and are transmitted waves. Vectors k_{z3} and k_{z4} propagate in the $-z$-direction and correspond to reflected waves. The wavevector k_x contains the incidence angle φ_0. Assuming that the reflection is small, one takes only the forward propagating eigenmodes. The electric field vectors can be calculated for each eigenvalue k_{zi}. The electric field of the eigenwave, for example, is

$$\mathbf{E}_1(\mathbf{r},t)=\mathbf{E}_1\mathrm{e}^{\mathrm{i}(\omega t-(k_x x+k_{z1}z))}=\begin{pmatrix}E_{x1}\\E_{y1}\\E_{z1}\end{pmatrix}\mathrm{e}^{\mathrm{i}(\omega t-(k_x x+k_{z1}z))},$$

and

$$\mathbf{E}_2(\mathbf{r},t)=\mathbf{E}_2\mathrm{e}^{\mathrm{i}(\omega t-(k_x x+k_{z2}z))}=\begin{pmatrix}E_{x2}\\E_{y2}\\E_{z2}\end{pmatrix}\mathrm{e}^{\mathrm{i}(\omega t-(k_x x+k_{z2}z))}. \tag{4.27}$$

The electric field components have to satisfy Equation (4.20) and, because of the vanishing determinant condition, only two of the three components of the electric field are independent. This allows the expression of two components of the electric field in terms of the third one in a parametric representation. As a result, only two electric field components have to be considered. The concept is then the following: In a matrix representation one identifies each eigenmode with one matrix component in a mode vector **M**. The link between the mode vector **M** and the electric field components is given by the matrix **S**:

$$\begin{pmatrix}E_x\\E_y\end{pmatrix}=\begin{pmatrix}E_{x1}\\E_{y1}\end{pmatrix}+\begin{pmatrix}E_{x2}\\E_{y2}\end{pmatrix}=\mathbf{S}\begin{pmatrix}M_{x1}\\M_{y2}\end{pmatrix}\quad\text{with }\mathbf{S}=\begin{pmatrix}e_{x1}&e_{x2}\\e_{y1}&e_{y2}\end{pmatrix}. \tag{4.28}$$

Eigenmodes travel without change in the state of polarization through the system but they obey a certain retardation. These retardation terms can be expressed in the matrix **G** that is composed of the phase shift given by the eigenvalues k_{zi}:

$$\begin{pmatrix}M_{x1}\\M_{y2}\end{pmatrix}_d=\mathbf{G}\begin{pmatrix}M_{x1}\\M_{y2}\end{pmatrix}_0 \tag{4.29}$$

with

$$\mathbf{G}=\begin{pmatrix}e^{\mathrm{i}k_{z1}d}&0\\0&e^{\mathrm{i}k_{z2}d}\end{pmatrix}, \tag{4.30}$$

where d is the thickness of the birefringent film traveled through. Using these equations the **E** vector at the bottom of the layer is related to the **E** vector at the top of the layer by

$$\begin{pmatrix}E_x\\E_y\end{pmatrix}_d=\mathbf{J}\begin{pmatrix}E_x\\E_y\end{pmatrix}_0, \tag{4.31}$$

where the extended Jones matrix **J** is

$$\mathbf{J}=\mathbf{S}\mathbf{G}\mathbf{S}^{-1}. \tag{4.32}$$

One has to be aware that the boundary conditions demands for the tangential component of the electric field to be continues at each interface of the layer. Equation (4.32) is the extended Jones matrix representation at an oblique incidence. The equations can be understood as follows. On the left-hand side of Equation (4.32) with the help of the matrix $\mathbf{S}^{-1}$, the electric field vector is transformed at the bottom of the birefringent film in the mode vector. The matrix **G** describes then the propagation of the mode vector through the film and **S** transforms the result back into the electric field vector. The boundary conditions are automatically fulfilled because the transformation back to the electric field vector is performed each time. An important point is that the coordinate system for the field is conveniently chosen to fulfil the boundary conditions. In our case the electrical field components are parallel to the layer system, which is ideal for steady tangential boundary conditions of the electric field. Note that due to the special form of **G**, the matrix **J** can equivalently be rewritten as

$$\mathbf{J}=\begin{pmatrix} 1 & c_2 \\ c_1 & 1 \end{pmatrix}\mathbf{G}\begin{pmatrix} 1 & c_2 \\ c_1 & 1 \end{pmatrix}^{-1}, \tag{4.33}$$

where the components are $c_1 = e_{y1}/e_{x1}$ and $c_2 = e_{y2}/e_{x2}$. The main task now is to find the matrix **S** and the components e_{x1}, e_{y1}, e_{x2}, and e_{y2}. To identify the components of the **S** matrix we go back and use Equations (4.23) and (4.20). Remember that only two electric field components are independent. We obtain for the eigenmode one the expressions

$$\begin{gathered}
\left(\varepsilon_{xx}-n_o^2+\frac{k_x^2}{k_0^2}\right)E_{x1}+\varepsilon_{xy}E_{y1}+\left(\frac{k_x}{k_0}\sqrt{\left(n_o^2-\frac{k_x^2}{k_0^2}\right)}+\varepsilon_{xz}\right)E_{z1}=0,\\
\varepsilon_{xy}E_{x1}+(\varepsilon_{yy}-n_o^2)E_{y1}+\varepsilon_{yz}E_{z1}=0,\\
\left(\sqrt{\left(n_o^2-\frac{k_x^2}{k_0^2}\right)}\frac{k_x}{k_0}+\varepsilon_{xz}\right)E_{x1}+\varepsilon_{yz}E_{y1}+\left(\varepsilon_{zz}-\frac{k_x^2}{k_0^2}\right)E_{z1}=0.
\end{gathered} \tag{4.34}$$

We can now express E_{x1} as a function of E_{y1} by using the last two equations in Equation (4.34). First we substitute E_{z1} and obtain

$$\begin{gathered}
\left(\sqrt{\left(n_o^2-\frac{k_x^2}{k_0^2}\right)}\frac{k_x}{k_0}+\varepsilon_{xz}\right)E_{x1}+\varepsilon_{yz}E_{y1}-\left(\varepsilon_{zz}-\frac{k_x^2}{k_0^2}\right)\frac{\varepsilon_{xy}}{\varepsilon_{yz}}E_{x1}\\
+\left(\varepsilon_{zz}-\frac{k_x^2}{k_0^2}\right)\frac{(\varepsilon_{yy}-n_o^2)}{\varepsilon_{yz}}E_{y1}=0,
\end{gathered} \tag{4.35}$$

which immediately leads to the relation between the x- and y-components of the electric field

$$E_{y1} = \frac{-\sqrt{\left(n_o^2 - \frac{k_x^2}{k_0^2}\right)}\frac{k_x}{k_0} - \varepsilon_{xz} + \left(\varepsilon_{zz} - \frac{k_x^2}{k_0^2}\right)\frac{\varepsilon_{xy}}{\varepsilon_{yz}}}{\left(\varepsilon_{zz} - \frac{k_x^2}{k_0^2}\right)\frac{(\varepsilon_{yy} - n_o^2)}{\varepsilon_{yz}} + \varepsilon_{yz}} E_{x1}. \tag{4.36}$$

From the definition in Equation (4.28) written in the form

$$\begin{aligned}\begin{pmatrix} E_{x1} \\ E_{y1} \end{pmatrix} + \begin{pmatrix} E_{x2} \\ E_{y2} \end{pmatrix} &= \begin{pmatrix} e_{x1} & e_{x2} \\ e_{y1} & e_{y2} \end{pmatrix}\begin{pmatrix} M_{x1} \\ M_{y2} \end{pmatrix} = \begin{pmatrix} e_{x1}M_{x1} + e_{x2}M_{y2} \\ e_{y1}M_{x1} + e_{y2}M_{y2} \end{pmatrix} \\ &= \begin{pmatrix} e_{x1} \\ e_{y1} \end{pmatrix} M_{x1} + \begin{pmatrix} e_{x2} \\ e_{y2} \end{pmatrix} M_{y2},\end{aligned} \tag{4.37}$$

we see that the product $e_{y1}M_{x1}$ can be identified as the electric field component E_{y1}. Hence the constant c_1 defined in Equation (4.33) is given as

$$\frac{E_{y1}}{E_{x1}} = \frac{e_{y1}}{e_{x1}} = c_1 = \frac{-\sqrt{\left(n_o^2 - \frac{k_x^2}{k_0^2}\right)}\frac{k_x}{k_0} - \varepsilon_{xz} + \left(\varepsilon_{zz} - \frac{k_x^2}{k_0^2}\right)\frac{\varepsilon_{xy}}{\varepsilon_{yz}}}{\left(\varepsilon_{zz} - \frac{k_x^2}{k_0^2}\right)\frac{(\varepsilon_{yy} - n_o^2)}{\varepsilon_{yz}} + \varepsilon_{yz}}. \tag{4.38}$$

In the same way, c_2 can be found. Because the eigenvector k_{z2} has a rather inconvenient form, we perform the calculation as in the original paper by Lien (1997). If eigenmode two is considered, Equation (4.20) can be written as

$$\begin{aligned}\left(\varepsilon_{xx} - \frac{k_{z2}^2}{k_0^2}\right)E_{x2} + \varepsilon_{xy}E_{y2} + \left(\frac{k_x k_{z2}}{k_0^2} + \varepsilon_{xz}\right)E_{z2} &= 0, \\ \varepsilon_{xy}E_{x2} + \left(\varepsilon_{yy} - \frac{k_x^2}{k_0^2} - \frac{k_{z2}^2}{k_0^2}\right)E_{y2} + \varepsilon_{yz}E_{z2} &= 0,\end{aligned} \tag{4.39}$$

and

$$\left(\frac{k_z k_x}{k_0^2} + \varepsilon_{xz}\right)E_{x2} + \varepsilon_{yz}E_{y2} + \left(\varepsilon_{zz} - \frac{k_x^2}{k_0^2}\right)E_{z2} = 0.$$

Using the first and the last equation to replace E_{z2}, one finds the relation

$$\left(-\left(\varepsilon_{xx}-\frac{k_{z2}^2}{k_0^2}\right)\left(\varepsilon_{zz}-\frac{k_x^2}{k_0^2}\right)+\left(\frac{k_{z2}k_x}{k_0^2}+\varepsilon_{xz}\right)^2\right)E_{x2}$$
$$=\left(\varepsilon_{xy}\left(\varepsilon_{zz}-\frac{k_x^2}{k_0^2}\right)-\left(\frac{k_xk_{z2}}{k_0^2}+\varepsilon_{xz}\right)\right)\varepsilon_{yz}E_{y2}. \tag{4.40}$$

The final result for the ratio E_{x2}/E_{y2} becomes

$$\frac{E_{x2}}{E_{y2}}=\frac{e_{x2}}{e_{y2}}=c_2=\frac{\left(\varepsilon_{xy}\left(\varepsilon_{zz}-\frac{k_x^2}{k_0^2}\right)-\left(\frac{k_xk_{z2}}{k_0^2}+\varepsilon_{xz}\right)\right)\varepsilon_{yz}}{-\left(\varepsilon_{xx}-\frac{k_{z2}^2}{k_0^2}\right)\left(\varepsilon_{zz}-\frac{k_x^2}{k_0^2}\right)+\left(\frac{k_{z2}k_x}{k_0^2}+\varepsilon_{xz}\right)^2}. \tag{4.41}$$

For a layered system with N layers the equations have to be applied for every layer with the matrix equation

$$\begin{pmatrix}E_x\\E_y\end{pmatrix}_{N+1}=\mathbf{J}\begin{pmatrix}E_x\\E_y\end{pmatrix}_0, \tag{4.42}$$

where

$$\mathbf{J}=\mathbf{J}_N\mathbf{J}_{N-1}\mathbf{J}_{N-2}\mathbf{J}_{N-3}\mathbf{J}_{N-4},\ldots,\mathbf{J}_4\mathbf{J}_3\mathbf{J}_2\mathbf{J}_1 \tag{4.43}$$

is the extended Jones matrix for the entire system. The total optical transmittance has to consider the reflection losses at the entrance of the optical system. We need to respect the Fresnel formulas for correction of the transition between air and a media, for example. The corresponding formulas were developed in Chapter 3. The beauty of the extend Jones matrix method is that one can directly use the Fresnel formulas in this form because the coordinate system and electric field definitions match.

The extended Jones matrix method was introduced for uniaxial material and makes use of the special form of the dielectric tensor. For biaxial media, a similar representation can be found if the geometry is limited to certain orientations of the optical axis. For flat systems, where the dielectric tensor of the biaxial material is rotated only around the z-axis, it was shown that the same eigenmode approach holds (Ong, 1991b; Zhu, 1994).

4.4 THE 4 × 4 BERREMAN METHOD

The propagation of polarized light in stratified anisotropic media can be rigorously described by using a 4 × 4 formulation (Berreman, 1972). The method is based on the assumption of plane waves. We will develop the basic formulas to show the approximations involved and the application of the method. We consider first a

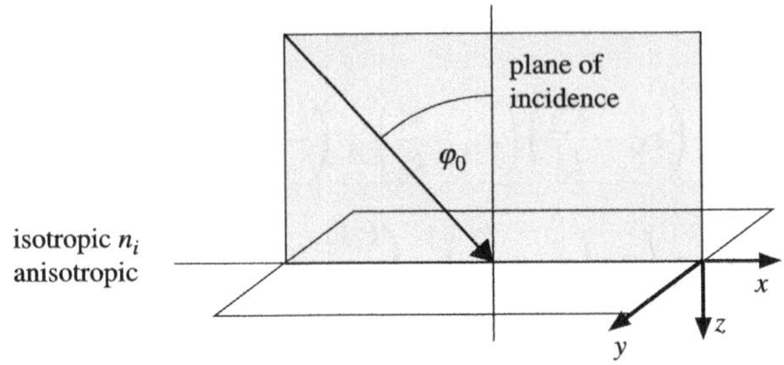

Figure 4.2 Scheme of the geometrical arrangement for a single homogeneous anisotropic layer. Light enters at an incident angle φ_0 in the arbitrarily chosen plane of incidence, x–z.

monochromatic plane wave obliquely incident from an isotropic ambient medium onto a homogeneous anisotropic layer perpendicular to the z-axis as shown in Figure 4.2. The plane of incidence is the x–z-plane. We examine here a three-dimensional problem. However, the dielectric tensor is constant in the x–y-plane and varies only as a function of z. If the x–z-plane is chosen as the plane of incidence, further simplification is possible. The problem is invariant in the y-direction, so all derivatives along y can be set to zero. This can be written as

$$\frac{\partial}{\partial y} \Rightarrow 0. \tag{4.44}$$

The incident plane wave $E = E_0 e^{i(\omega t - \mathbf{kr})}$ must have the same spatial dependence in the x-direction as the incident wave, which is the x-component of the wavevector $\mathbf{k} = [k_x, 0, k_z]$ in the ambient media with index n_o

$$\xi = -k_0 n_i \sin\varphi_0, \tag{4.45}$$

where φ_0 is the angle of incidence and $k_o = \omega/c_o$. Therefore, the variation of all fields in the x-direction is proportional to $\exp(-i\xi x)$ and we can set

$$\frac{\partial}{\partial x} \Rightarrow -i\xi. \tag{4.46}$$

The curl operator becomes, using Equations (4.44) and (4.46),

$$\nabla\times = \begin{pmatrix} 0 & -\dfrac{\partial}{\partial z} & \dfrac{\partial}{\partial y} \\ \dfrac{\partial}{\partial z} & 0 & -\dfrac{\partial}{\partial x} \\ -\dfrac{\partial}{\partial y} & \dfrac{\partial}{\partial x} & 0 \end{pmatrix} = \begin{pmatrix} 0 & -\dfrac{\partial}{\partial z} & 0 \\ \dfrac{\partial}{\partial z} & 0 & i\xi \\ 0 & -i\xi & 0 \end{pmatrix}. \tag{4.47}$$

For Maxwell's equations we obtain

$$\begin{pmatrix} -\dfrac{\partial}{\partial z}H_y \\ \dfrac{\partial}{\partial z}H_x + \mathrm{i}\xi H_z \\ -\mathrm{i}\xi H_y \\ \dfrac{\partial}{\partial z}E_y \\ -\dfrac{\partial}{\partial z}E_x - \mathrm{i}\xi E_z \\ \mathrm{i}\xi E_y \end{pmatrix} = \mathrm{i}\omega \begin{pmatrix} \varepsilon_0\varepsilon_{xx}E_x + \varepsilon_0\varepsilon_{xy}E_y + \varepsilon_0\varepsilon_{xz}E_z \\ \varepsilon_0\varepsilon_{yx}E_x + \varepsilon_0\varepsilon_{yy}E_y + \varepsilon_0\varepsilon_{yz}E_z \\ \varepsilon_0\varepsilon_{zx}E_x + \varepsilon_0\varepsilon_{zy}E_y + \varepsilon_0\varepsilon_{zz}E_z \\ \mu_o\mu H_x \\ \mu_o\mu H_y \\ \mu_o\mu H_z \end{pmatrix}. \tag{4.48}$$

The field components E_z and H_z are linearly dependent on the other field variables. We can write explicitly the third equation and the last equation in the matrix of Equation (4.48) to obtain

$$E_z = -\frac{\xi}{\varepsilon_0\varepsilon_{zz}\omega}H_y - \frac{\varepsilon_{zx}}{\varepsilon_{zz}}E_x - \frac{\varepsilon_{zy}}{\varepsilon_{zz}}E_y \qquad \text{and} \qquad H_z = \frac{\xi}{\omega\mu\mu_0}E_y. \tag{4.49}$$

With this linear dependence we can replace E_z and H_z in Equation (4.8). This leads to four linear, homogeneous, first-order differential equations for the field variables E_x, E_y, H_x, and H_y:

$$\frac{\partial}{\partial z}\begin{pmatrix} E_x \\ H_y \\ E_y \\ -H_x \end{pmatrix} = \begin{pmatrix} -\mathrm{i}\omega\left(\dfrac{-\xi}{\omega}\dfrac{\varepsilon_{zx}}{\varepsilon_{zz}}E_x - \dfrac{1}{\varepsilon_0\varepsilon_{zz}}\dfrac{\xi^2}{\omega^2}H_y + \mu_0\mu H_y - \dfrac{\xi}{\omega}\dfrac{\varepsilon_{zy}}{\varepsilon_{zz}}E_y\right) \\ -\mathrm{i}\omega\left(\varepsilon_0\varepsilon_{xx}E_x - \dfrac{\varepsilon_0\varepsilon_{xz}\varepsilon_{zx}}{\varepsilon_{zz}}E_x - \dfrac{\xi}{\omega}\dfrac{\varepsilon_{xz}}{\varepsilon_{zz}}H_y - \dfrac{\varepsilon_0\varepsilon_{xz}\varepsilon_{zy}}{\varepsilon_{zz}}E_y + \varepsilon_0\varepsilon_{xy}E_y\right) \\ \mathrm{i}\omega\mu_0\mu H_x \\ -\mathrm{i}\omega\begin{pmatrix} \varepsilon_0\varepsilon_{yx}E_x - \dfrac{\varepsilon_0\varepsilon_{yz}\varepsilon_{zx}}{\varepsilon_{zz}}E_x - \dfrac{\xi}{\omega}\dfrac{\varepsilon_{yz}}{\varepsilon_{zz}}H_y \\ +\varepsilon_0\varepsilon_{yy}E_y - \dfrac{\xi^2}{\omega^2}\dfrac{1}{\mu_0\mu}E_y - \dfrac{\varepsilon_0\varepsilon_{yz}\varepsilon_{zy}}{\varepsilon_{zz}}E_y \end{pmatrix} \end{pmatrix}. \tag{4.50}$$

This contains only field components in the x–y-plane. The stratified birefringent media changes in the z-direction. The plane of incidence was set as the x–z-plane. A generalized field vector can be defined that contains these field components as $\psi = [E_x, H_y, E_y, -H_x]$ and we obtain the wave equation for the generalized field vector, which can be written in matrix form as

$$\frac{\partial}{\partial z}\begin{pmatrix} E_x \\ H_y \\ E_y \\ -H_x \end{pmatrix} = -i\omega \begin{pmatrix} -\frac{\xi}{\omega}\frac{\varepsilon_{zx}}{\varepsilon_{zz}} & \mu_o\mu - \frac{1}{\varepsilon_0\varepsilon_{zz}}\frac{\xi^2}{\omega^2} & -\frac{\xi}{\omega}\frac{\varepsilon_{zy}}{\varepsilon_{zz}} & 0 \\ \varepsilon_0\varepsilon_{xx} - \frac{\varepsilon_0\varepsilon_{xz}\varepsilon_{zx}}{\varepsilon_{zz}} & -\frac{\xi}{\omega}\frac{\varepsilon_{xz}}{\varepsilon_{zz}} & \varepsilon_0\varepsilon_{xy} - \frac{\varepsilon_0\varepsilon_{xz}\varepsilon_{zy}}{\varepsilon_{zz}} & 0 \\ 0 & 0 & 0 & \mu_0\mu \\ \varepsilon_0\varepsilon_{yx} - \frac{\varepsilon_0\varepsilon_{yz}\varepsilon_{zx}}{\varepsilon_{zz}} & -\frac{\xi}{\omega}\frac{\varepsilon_{yz}}{\varepsilon_{zz}} & \varepsilon_0\varepsilon_{yy} - \frac{\varepsilon_0\varepsilon_{yz}\varepsilon_{zy}}{\varepsilon_{zz}} - \frac{\xi^2}{\omega^2}\frac{1}{\mu_0\mu} & 0 \end{pmatrix} \begin{pmatrix} E_x \\ H_y \\ E_y \\ -H_x \end{pmatrix} \tag{4.51}$$

or as

$$\frac{\partial}{\partial z}\mathbf{\Psi} = -i\omega\mathbf{\Pi}\mathbf{\Psi}, \tag{4.52}$$

where $\mathbf{\Pi}$ is the differential propagation matrix, which depends on the dielectric tensor of the anisotropic medium, the refractive index of the isotropic medium, the angle of incidence and the frequency of the light. Explicitly it is given by

$$\mathbf{\Pi} = \begin{pmatrix} -\frac{\xi}{\omega}\frac{\varepsilon_{zx}}{\varepsilon_{zz}} & \mu_o\mu - \frac{1}{\varepsilon_0\varepsilon_{zz}}\frac{\xi^2}{\omega^2} & -\frac{\xi}{\omega}\frac{\varepsilon_{zy}}{\varepsilon_{zz}} & 0 \\ \varepsilon_0\varepsilon_{xx} - \frac{\varepsilon_{xz}\varepsilon_{zx}}{\varepsilon_{zz}} & -\frac{\xi}{\omega}\frac{\varepsilon_{xz}}{\varepsilon_{zz}} & \varepsilon_{xy} - \frac{\varepsilon_{xz}\varepsilon_{zy}}{\varepsilon_{zz}} & 0 \\ 0 & 0 & 0 & \mu_0\mu \\ \varepsilon_0\varepsilon_{yx} - \frac{\varepsilon_0\varepsilon_{yz}\varepsilon_{zx}}{\varepsilon_{zz}} & -\frac{\xi c}{\omega}\frac{\varepsilon_{yz}}{\varepsilon_{zz}} & \varepsilon_0\varepsilon_{yy} - \frac{\varepsilon_0\varepsilon_{yz}\varepsilon_{zy}}{\varepsilon_{zz}} - \frac{\xi^2}{\omega^2}\frac{1}{\mu_0\mu} & 0 \end{pmatrix}. \tag{4.53}$$

We consider now a layer between z and $z+h$. Equation (4.52) can be solved numerically by developing a Taylor series. This delivers a generalized field vector ψ in $z+h$, if ψ is known in z. As the formal solution of Equation (4.52) is an exponential function of the field vector ψ, the development is the Taylor series:

$$\mathbf{\Psi}(z+h) = \left(I - i\omega h\mathbf{\Pi} - \frac{\omega^2 h^2}{2!}\mathbf{\Pi}^2 + i\frac{\omega^3 h^3}{3!}\mathbf{\Pi}^3 \pm \cdots\right)\mathbf{\Psi}(z). \tag{4.54}$$

Equation (4.54) describes the linear matrix relationship between the generalized field vectors at both sides of an anisotropic layer of thickness h. Approximations are made by limiting the number of terms in the series of Equation (4.54). Usually, three or four terms are taken into account. If the matrix $\mathbf{\Pi}$ is independent of z over some finite distance h in the direction of the z-axis then we can write

$$\mathbf{\Psi}(z+h)=\mathbf{P}(h)\mathbf{\Psi}(z). \tag{4.55}$$

In some simple cases a general expression for $\mathbf{\Pi}^n$ can be written down in which it may be possible to recognize a closed form for each of the terms in $\mathbf{P}(h)$. An alternative method of expressing $\mathbf{P}(h)$ is to integrate Equation (4.52) to obtain

$$\mathbf{\Psi}(z+h)=\mathbf{P}(h)\mathbf{\Psi}(z)=\mathrm{e}^{-\mathrm{i}\omega h\mathbf{\Pi}}\mathbf{\Psi}(z). \tag{4.56}$$

However, the series solution is useful mainly for numerical problems in which h is small enough such that the first terms of the series give sufficient accuracy. Certain useful symmetry properties of $\mathbf{P}(h)$ are evident from Equation (4.54). In particular

$$\mathbf{P}(mh)=(\mathbf{P}(h))^m, \tag{4.57}$$

where m is a positive or negative integer.

4.5 ANALYTICAL SOLUTION FOR A BIREFRINGENT SLAB

Analytical examples of the Berreman 4 × 4 matrix method are rare. Only very few problems can be solved by writing down the equations. The treatment of a uniaxial crystal slab is possible if the crystal shows particular orientation in the reference frame. Orthorhombic crystals with the principal axis parallel to the xyz-coordinates have a dielectric tensor like

$$\boldsymbol{\varepsilon}=\begin{pmatrix}\varepsilon_x & 0 & 0\\ 0 & \varepsilon_x & 0\\ 0 & 0 & \varepsilon_z\end{pmatrix}=\begin{pmatrix}n_0^2 & 0 & 0\\ 0 & n_o^2 & 0\\ 0 & 0 & n_\mathrm{e}^2\end{pmatrix}. \tag{4.58}$$

If this is put into the differential propagation matrix $\mathbf{\Pi}$ given in Equation (4.53), one finds, with $\mu = 1$,

$$\mathbf{\Pi}=\begin{pmatrix}0 & \left(\mu_o-\dfrac{1}{\varepsilon_0 n_\mathrm{e}^2}\dfrac{\xi^2}{\omega^2}\right) & 0 & 0\\ \varepsilon_0 n_o^2 & 0 & 0 & 0\\ 0 & 0 & 0 & \mu_o\\ 0 & 0 & \varepsilon_0 n_o^2-\dfrac{\xi^2}{\omega^2}\dfrac{1}{\mu_0} & 0\end{pmatrix}. \tag{4.59}$$

To calculate the optical properties we need to determine the matrix $\mathbf{P}(h)$ of Equation (4.55), which relates the generalized field vectors at the entrance and exit of the birefringent layers to each other. As seen from Equation (4.54), **P**(h) is given by

$$\mathbf{P}(h) = I - \mathrm{i}\omega h\,\mathbf{\Pi} - \omega^2 h^2 \frac{\mathbf{\Pi}^2}{2!} + \mathrm{i}\omega^3 h^3 \frac{\mathbf{\Pi}^3}{3!} \pm \cdots, \tag{4.60}$$

and one has to calculate a series of powers of **Π**. Because the **Π** matrix in Equation (4.59) has a block structure, powers of the matrix are found as follows:

$$\begin{aligned}
&\mathbf{\Pi} = \begin{pmatrix} 0 & a^2 & 0 & 0 \\ b^2 & 0 & 0 & 0 \\ 0 & 0 & 0 & u^2 \\ 0 & 0 & v^2 & 0 \end{pmatrix}, \quad \mathbf{\Pi}^2 = \begin{pmatrix} a^2b^2 & 0 & 0 & 0 \\ 0 & a^2b^2 & 0 & 0 \\ 0 & 0 & u^2v^2 & 0 \\ 0 & 0 & 0 & u^2v^2 \end{pmatrix}, \\
&\mathbf{\Pi}^3 = \begin{pmatrix} 0 & a^4b^2 & 0 & 0 \\ a^2b^4 & 0 & 0 & 0 \\ 0 & 0 & 0 & u^4v^2 \\ 0 & 0 & u^2v^4 & 0 \end{pmatrix}, \ldots.
\end{aligned} \tag{4.61}$$

The coefficients a, b, u, and v are identified by comparing Equations (4.59) and (4.61). It is easy to rewrite a general expression for $\mathbf{\Pi}^n$ in this case. For example,

$$\begin{aligned}
(\mathbf{\Pi}^n)_{11} = (\mathbf{\Pi}^n)_{22} &= \begin{cases} 0 & \text{for } n \text{ odd} \\ (ab)^n & \text{for } n \text{ even} \end{cases} \\
(\mathbf{\Pi}^n)_{33} = (\mathbf{\Pi}^n)_{44} &= \begin{cases} 0 & \text{for } n \text{ odd} \\ (uv)^n & \text{for } n \text{ even} \end{cases} \\
(\mathbf{\Pi}^n)_{12} &= \begin{cases} \dfrac{(ab)^n a}{b} & \text{for } n \text{ odd} \\ 0 & \text{for } n \text{ even} \end{cases} \\
(\mathbf{\Pi}^n)_{21} &= \begin{cases} \dfrac{(ab)^n b}{a} & \text{for } n \text{ odd} \\ 0 & \text{for } n \text{ even} \end{cases}
\end{aligned} \tag{4.62}$$

and

$$\begin{aligned}
(\mathbf{\Pi}^n)_{34} &= \begin{cases} \dfrac{(uv)^n u}{v} & \text{for } n \text{ odd} \\ 0 & \text{for } n \text{ even} \end{cases} \\
(\mathbf{\Pi}^n)_{43} &= \begin{cases} \dfrac{(uv)^n v}{u} & \text{for } n \text{ odd} \\ 0 & \text{for } n \text{ even} \end{cases}.
\end{aligned}$$

This allows us to find an analytical solution in the sense that one can equate the general field vector equation. With the help of Equation (4.56) we can express

$\mathbf{P}(h)$ in trigonometric functions:

$$\mathbf{P}(h) = e^{i\omega h \boldsymbol{\Pi}} = \cos(\omega h \boldsymbol{\Pi}) - i \sin(\omega h \boldsymbol{\Pi}). \tag{4.63}$$

The trigonometric functions sin(x) and cos(x) can be developed in a Taylor series as

$$\begin{aligned} \cos x &= 1 - \frac{x^2}{2!} + \frac{x^4}{4!} - \frac{x^6}{6!} \cdots (-1)^n \frac{x^{2n}}{(2n)!} \\ \sin x &= x - \frac{x^3}{3!} + \frac{x^5}{5!} - \frac{x^7}{7!} \cdots (-1)^n \frac{x^{2n+1}}{(2n+1)!}. \end{aligned} \tag{4.64}$$

If we choose the argument of the development correctly, we obtain, for example,

$$\cos(ab\omega h) = 1 - \frac{1}{2!}\omega^2 h^2 a^2 b^2 + \frac{1}{4!}\omega^4 h^4 a^4 b^4 \cdots$$

and

$$\sin(ab\omega h) = \omega h a b - \frac{1}{3!}\omega^3 h^3 a^3 b^3 + \frac{1}{5!}\omega^5 h^5 a^5 b^5. \tag{4.65}$$

The matrix element $(\boldsymbol{\Pi}^n)_{21}$ takes than the form

$$\begin{aligned} (\boldsymbol{\Pi}^n)_{21} &= -i\frac{b}{a}\sin(ab\omega h) \\ &= -i\omega h b^2 + i\frac{1}{3!}\omega^3 h^3 a^2 b^4 - i\frac{1}{5!}\omega^5 h^5 a^4 b^6 \cdots. \end{aligned} \tag{4.66}$$

The same procedure for the other matrix elements leads to $\mathbf{P}(h)$, expressed as

$$\mathbf{P}(h) = \begin{pmatrix} \cos(ab\omega h) & -i\frac{a}{b}\sin(ab\omega h) & 0 & 0 \\ -i\frac{b}{a}\sin(ab\omega h) & \cos(ab\omega h) & 0 & 0 \\ 0 & 0 & \cos(uv\omega h) & -i\frac{u}{v}\sin(uv\omega h) \\ 0 & 0 & -i\frac{v}{u}\sin(uv\omega h) & \cos(uv\omega h) \end{pmatrix}, \tag{4.67}$$

which is a closed analytical expression for the propagation through an birefringent orthoscopic crystal slab of thickness h. One can identify the eigenvalues of $\boldsymbol{\Pi}$ and eigenvectors of the generalized fields.

4.6 REFLECTION AND TRANSMISSION

To use the 4×4 matrix methods one conveniently links them to a 2×2 matrix method that is used to describe the light propagation in isotropic media. One can then directly calculate the reflection and transmission coefficients related to the Fresnel equations. Figure 4.3 shows an anisotropic stratified medium between two ambient isotropic media with indices of refraction n_0 and n_2, where the entering plane wave comes from the ambient medium with index n_0. Let (E_{ip}, E_{is}), (E_{rp}, E_{rs}), and (E_{tp}, E_{ts}) be the components of the incident, reflected, and transmitted electric field vectors, parallel (p) or perpendicular (s) to the plane of incidence x–z. A Medium with $\mu = n$ is assumed using the fact that the orthogonal electric and magnetic fields are related by n_0 and n_2, respectively, in the ambient isotropic media, with the equations

$$\frac{H_{ip}}{E_{is}} = \frac{H_{is}}{E_{ip}} = \sqrt{\frac{\varepsilon_0}{\mu_0}} n_0, \qquad \frac{H_{rp}}{E_{rs}} = \frac{H_{rs}}{E_{rp}} = \sqrt{\frac{\varepsilon_0}{\mu_0}} n_0, \qquad \frac{H_{tp}}{E_{ts}} = \frac{H_{ts}}{E_{tp}} = \sqrt{\frac{\varepsilon_0}{\mu_0}} n_2, \tag{4.68}$$

the generalized field vectors of incidence, reflection, and transmission can be written as

$$\boldsymbol{\Psi}_i = \begin{pmatrix} E_{ip}\cos\varphi_0 \\ \sqrt{\frac{\varepsilon_0}{\mu_0}} n_0 E_{ip} \\ E_{is} \\ \sqrt{\frac{\varepsilon_0}{\mu_0}} n_0 E_{is}\cos\varphi_0 \end{pmatrix}, \qquad \boldsymbol{\Psi}_r = \begin{pmatrix} -E_{rp}\cos\varphi_0 \\ \sqrt{\frac{\varepsilon_0}{\mu_0}} n_0 E_{rp} \\ E_{rs} \\ -\sqrt{\frac{\varepsilon_0}{\mu_0}} n_0 E_{rs}\cos\varphi_0 \end{pmatrix},$$
$$\boldsymbol{\Psi}_t = \begin{pmatrix} E_{tp}\cos\varphi_2 \\ \sqrt{\frac{\varepsilon_0}{\mu_0}} n_2 E_{tp} \\ E_{ts} \\ \sqrt{\frac{\varepsilon_0}{\mu_0}} n_2 E_{ts}\cos\varphi_2 \end{pmatrix}. \tag{4.69}$$

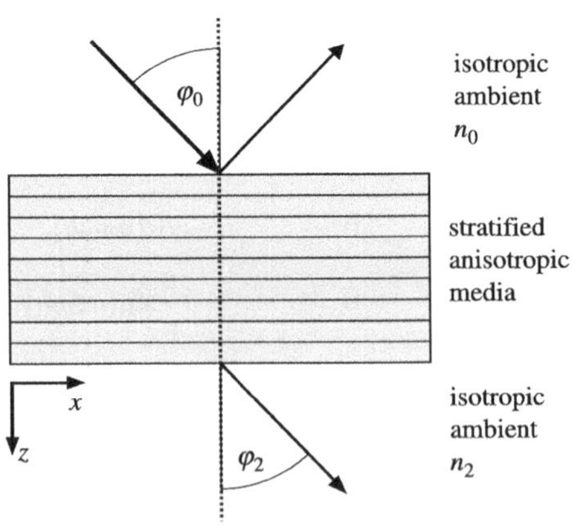

Figure 4.3 Stratified anisotropic slab between isotropic ambient media. If the media on both sides of the slab are the same, the angles φ_0 and φ_2 are identical.

We will now consider a plane wave, which propagates in the stratified medium from one homogeneous anisotropic layer to the next. This plane wave can be described by a generalized field vector $\boldsymbol{\psi}$. Equating $\mathbf{P}(h)$ using Equation (4.54) with the matrix $\boldsymbol{\Pi}$ defined in Equation (4.53) for every single layer relates the generalized field vectors $\psi_i + \psi_r$ and ψ_t. We obtain

$$\begin{pmatrix} E_{\text{tp}}\cos\varphi_2 \\ \sqrt{\frac{\varepsilon_0}{\mu_0}}n_2 E_{\text{tp}} \\ E_{\text{ts}} \\ \sqrt{\frac{\varepsilon_0}{\mu_0}}n_2 E_{\text{ts}}\cos\varphi_2 \end{pmatrix} = \begin{pmatrix} \ell_{11} & \ell_{12} & \ell_{13} & \ell_{14} \\ \ell_{21} & \ell_{22} & \ell_{23} & \ell_{24} \\ \ell_{31} & \ell_{32} & \ell_{33} & \ell_{34} \\ \ell_{41} & \ell_{42} & \ell_{43} & \ell_{44} \end{pmatrix} \begin{pmatrix} (E_{\text{ip}} - E_{\text{rp}})\cos\varphi_0 \\ \sqrt{\frac{\varepsilon_0}{\mu_0}}n_o(E_{\text{ip}} + E_{\text{rp}}) \\ E_{\text{is}} + E_{\text{rs}} \\ \sqrt{\frac{\varepsilon_0}{\mu_0}}n_2(E_{\text{is}} - E_{\text{rs}})\cos\varphi_2 \end{pmatrix}. \tag{4.70}$$

Note that the material inside one layer is assumed to be homogeneous. Therefore, materials with continuously changing dielectric properties have to be divided into sufficiently thin layers. Equation (4.70) results in a system of four linear equations for the four unknowns E_{rp}, E_{rs}, E_{tp}, and E_{ts} if E_{ip} and E_{is} are assumed to be known. This system can be solved numerically. The equations for the reflected and transmitted electric fields can be separated into

$$\begin{pmatrix} E_{\text{rp}} \\ E_{\text{rs}} \end{pmatrix} = \begin{pmatrix} R_{\text{pp}} & R_{\text{ps}} \\ R_{\text{sp}} & R_{\text{ss}} \end{pmatrix}\begin{pmatrix} E_{\text{ip}} \\ E_{\text{is}} \end{pmatrix} \quad \text{and} \quad \begin{pmatrix} E_{\text{tp}} \\ E_{\text{ts}} \end{pmatrix} = \begin{pmatrix} T_{\text{pp}} & T_{\text{ps}} \\ T_{\text{sp}} & T_{\text{ss}} \end{pmatrix}\begin{pmatrix} E_{\text{ip}} \\ E_{\text{is}} \end{pmatrix}, \tag{4.71}$$

where $\mathbf{R}$ and $\mathbf{T}$ are the Jones matrices for reflection and transmission of the system. A more detailed description of the method can be found in Azzam (1987).

Recapping, with the 4×4 Berreman method the reflection and transmission Jones matrices of anisotropic stratified media can be calculated numerically. The ambient media are assumed to be isotropic and semi-infinite. This is a plane wave method; that is, no lateral scattering is taken into account.

REFERENCES

Abeles, F. (1950) Recherches sur la propagation des ondes electromagnetiques sinusoïdales dans les milieux stratifies. Application aux couches minces, *Ann. Phys., Paris* **5**, 596–640.

Azzam, R.M.A. and N.M. Bashara (1987) *Ellipsometry and Polarized Light*, North-Holland, Amsterdam.

Berreman, D.W. (1972) Optics in stratified and anisotropic media: 4 × 4 matrix formulation, *J. Opt. Soc. Am.* **62**(4), 502–510.

Chen, C., A. Lien, and M. Nathan (1997) 4 × 4 and 2 × 2 matrix formulations for the optics in stratified and biaxial media, *J. Opt. Soc. Am. A* **14**, 3125–3134.

Gu, C. and P. Yeh (1993) Extended Jones matrix method. II, *J. Opt. Soc. Am. A* **10**, 966–973.

Jones, R.C. (1941) A new calculus for the treatmant of optical systems I–III, *J. Opt. Soc. Am.* **31**, 488–503.

Jones, R.C. (1942) A new calculus for the treatmant of optical systems IV, *J. Opt. Soc. Am.* **32**, 486–493.

Jones, R.C. (1947) A new calculus for the treatmant of optical systems V–VI, *J. Opt. Soc. Am.* **37**, 107–112.

Jones, R.C. (1948a) A new calculus for the treatmant of optical systems VII, *J. Opt. Soc. Am.* **38**, 671–685.

Jones, R.C. (1948b) A new calculus for the treatmant of optical systems VII, *J. Opt. Soc. Am.* **38**, 671–685.

Jones, R.C. (1956a) A new calculus for the treatmant of optical systems VIII, *J. Opt. Soc. Am.* **46**, 126–131.

Jones, R.C. (1956b) A new calculus for the treatmant of optical systems VIII, *J. Opt. Soc. Am.* **46**, 126–131.

Lien, A. (1990) Extended Jones matrix representation for the twisted nematic liquid-crystal display at oblique incidence, *Appl. Phys. Lett.* **57**, 2767–2769.

Lien, A. (1997) A detailed derivation of extended Jones matrix representation for twisted nematic liquid crystal displays, *Liq. Cryst.* **22**, 171–175.

Lu, K. and B. Saleh (1992) Reducing Berreman's 4 × 4 formulation of liquid-crystal-display optics to 2 × 2 Jones vector equations, *Opt. Lett.* **17**, 1557–1559.

Ong, H. (1991) Reducing the Berreman 4 × 4 propagation matrix method for layered inhomogeneous anisotropic media to the Abeles 2 × 2 matrix method for isotropic media, *J. Opt. Soc. Am. A* **8**, 303–305.

Ong, H.L. (1991) Electro-optics of electrically controlled birefringence liquid-crystal displays by 2 × 2 propagation matrix and analytic expression at oblique angle, *Appl. Phys. Lett.* **59**, 155–157.

Wohler, H., G. Haas, M. Fritsch, and D. Mlynski (1988) Faster 4 × 4 matrix method for uniaxial inhomogeneous media, *J. Opt. Soc. Am. A* **5**, 1554–1557.

Wu, S.-T. and D.-K. Yang (2001) *Reflective Liquid Crystal Displays*, Wiley, New York.

Yeh, P. (1979) Electromagnetic propagation in birefringent layered media, *J. Opt. Soc. Am.* **69**, 742–756.

Yeh, P. (1980) Optics of anisotropic layered media: a new 4 × 4 matrix algebra, *Surface Sci.* **96**, 41–53.

Yeh, P. (1982) Extended Jones matrix method, *J. Opt. Soc. Am.* **72**, 507–513.

Zhu, X. (1994) Explicit Jones transformation matrix for a tilted birefringent plate with its optic axis parallel to the plate surfaces, *Appl. Opt.* **33**, 3502–3506.

5

SPACE-GRID TIME-DOMAIN TECHNIQUES

5.1 INTRODUCTION

Assuming an arbitrary anisotropic, inhomogeneous medium with a known dielectric tensor, matrix methods (like the Berreman 4 × 4 matrix method; see Chapter 4) are no longer applicable to the general case. Because no simplification of Maxwell's partial differential equation system is feasible, it is necessary to revert to Maxwell's equations in their original form. This was done the first time before 1960 in order to solve antenna and radar problems. For the simulation of wave propagation in such electromagnetic engineering systems, principal approaches involving closed-form and infinite-series analytical solutions were developed. These techniques solve the steady-state Maxwell equations in a simplified way. After about 1960, numerical methods, the so-called "frequency domain approaches," became more sophisticated due to the rise in computational speed, storage capacity, and high-level programming languages. The initial computational approaches included high-frequency asymptotic methods (Keller, 1962) and the discretization of integral equations (Harrington, 1968). The application of these methods is limited to simply structured materials. For the special case of periodic structures, the so-called Fourier model method is convenient and can be extended to anisotropic materials and two-dimensional gratings (Glytsis and Gaylord, 1987).

Because of these limitations, an alternative was sought, and was found in the 1970s: a space-grid, time-domain technique. The principal idea of this technique is to solve directly the Maxwell equations over a time domain on spatial lattices. The basic algorithm of a space-grid, time-domain technique was first introduced by Yee in 1966, who presented the basics of the finite-difference time-domain (FDTD) method (Yee, 1966). Once the correct numerical stability criterion for the Yee algorithm was provided (Taflove and Brodwin, 1975a), the FDTD method was applied to electromagnetic pulse problems (Kunz and Lee, 1978); also sinusoidal, steady-state solutions of two- and three-dimensional electromagnetic wave interactions with structured materials (Taflove and Brodwin, 1975b) were found. An important step for completing the availability of the FDTD method was the introduction of numerically stable absorbing boundary conditions (ABC) for the Yee grid. In their first form they were published by Mur to second-order accuracy (Mur, 1981). Kriegsmann et al. worked out an ABC theory (Kriegsmann et al., 1987) and Berenger introduced the highly effective perfectly matched layer (PML) for two- and three-dimensional FDTD grids (Berenger, 1994, 1996). These were enhanced for physically realizable, uniaxial anisotropic PMLs (Sacks et al., 1995; Gedney, 1996). In the meantime, the FDTD method was applied to a multitude of electromagnetic problems like microwave amplifiers, cellular telephones, optical microdisk resonators and photonic bandgap microcavity lasers (Taflove, 2005). Space-grid, time-domain techniques are based on a volumetric sampling of the electric and magnetic fields and on a sampling over a time period. Because the space sampling has to have subwavelength resolution (typically 10–20 samples per wavelength), the discretization will be very extensive for small wavelengths like the optical ones. Furthermore, if field values at some distance are desired, it is likely that this distance will force the computational domain to be excessively large. Far-field extensions are available for the FDTD method, but require a large degree of post-processing. This great disadvantage is gradually being overcome by the technical developments in computer processors and storage capacities. To date, acceptable (relating to the effort) FDTD calculations for wave propagation in large spatial domains are limited to two-dimensional problems. The FDTD method has a very high accuracy in convergence, depending on the sampling resolution. Moreover, the FDTD method is very instructive, because the electric and magnetic field arrays are calculated at every time step and can be simultaneously visualized to illustrate the field dynamics.

5.2 DESCRIPTION OF THE FDTD METHOD

The FDTD method is a direct method for the solution of the Maxwell curl equations

$$\nabla \times \mathbf{E} = -\mu_0 \mu \frac{\partial \mathbf{H}}{\partial t}, \qquad \nabla \times \mathbf{H} = \varepsilon_0 \boldsymbol{\varepsilon} \frac{\partial \mathbf{E}}{\partial t}. \tag{5.1}$$

First, the simulation domain will be discretized at subwavelength resolution. The time evolution of the electromagnetic fields is then computed by means of an explicit iterative updating scheme. So, the discretization happens in time *and* space dimensions. The general geometric situation is described as follows. In a model system the anisotropic material is confined between a substrate and a superstrate as shown in Figure 5.1.

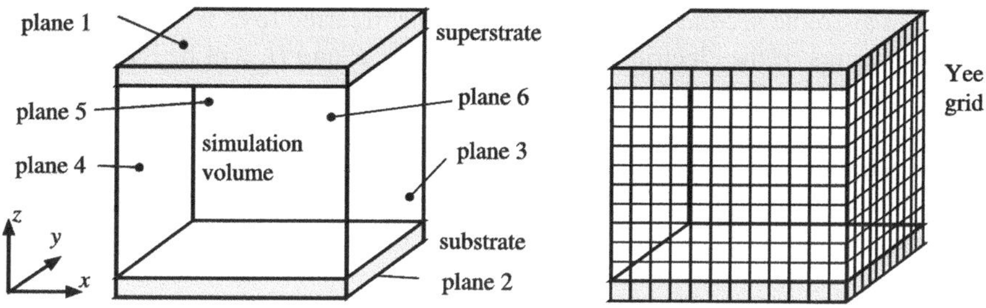

Figure 5.1 General geometry for the FDTD simulation. A simulation volume is defined with different kinds of boundary conditions for the planes 1, 2, 3, 4, 5, and 6. The observation direction is defined by z in this case. The simulation volume might contain the substrate and the superstrate and the volumes to set up the PML boundary conditions.

A simulation volume (box) is defined and within that a grid is established. For two-dimensional problems the implementation and geometry become much easier, as the boundaries are only lines. The simulation volume has to be surrounded by convenient boundary conditions. For problems that are periodic in the x- and y-directions in Figure 5.1, periodic walls might be set up in plane 3, 4, 5, or 6. In the z-direction, the usual observation direction, special care has to be taken to match the external interface. It is convenient to transfer the field values to matrix systems that can describe reflection and transmission for further calculations. To prevent artificial reflections from the z-direction at planes 1 and 2, perfectly matched layer boundaries or similar conditions are used. As an alternative to periodic boundary conditions for planes 3, 4, 5, and 6, perfectly matched layers or perfect reflectors might by used. Most FDTD formulations that are used are based on the Yee grid (Yee, 1966). This is represented in Figure 5.2. It describes two ranked Cartesian space meshes on whose points the electric and magnetic field components, respectively, are alternatively

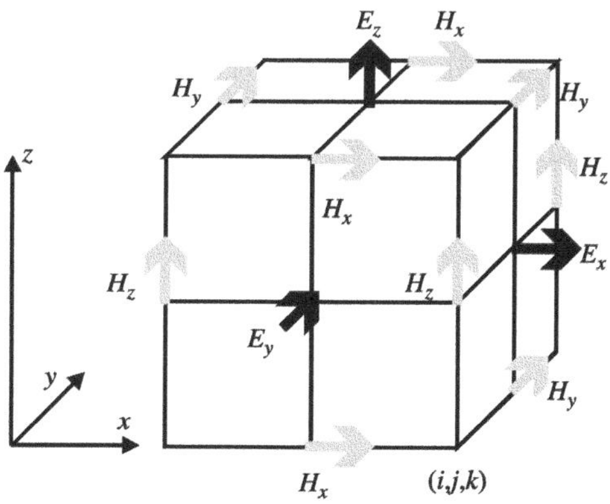

Figure 5.2 Primary and secondary grid in the Yee scheme concept. The electric field components are symbolized with black arrows and the magnetic field components are shown in gray.

updated in a simple second-order accurate scheme, referred to as a leapfrog scheme. The components of the electric and magnetic fields, E and H, will be interpolated at the time points $t_n = n\Delta t$ and $t_{n+1/2} = (n + 1/2)\Delta t$ (n is a positive integer), respectively, where the constant time step Δt has to be chosen depending on the node distances and the refraction index. In the discretized formulation of the partial differential Maxwell equations (Equation (5.2)), the time derivations of the electric and magnetic fields on a grid node can be approximated in the second order with the central difference equations

$$\frac{\delta}{\delta t} E_{i,n+1/2} = \frac{E_{i,n+1} - E_{i,n}}{\Delta t}, \qquad \frac{\delta}{\delta t} H_{i,n} = \frac{H_{i,n+1/2} - H_{i,n-1/2}}{\Delta t}, \tag{5.2}$$

where i stands for the indices x, y, and z. Note that the electric fields are updated at the "integer" time points and the magnetic fields at the "half" time points.

The spatial grid system in Figure 5.2 can be thought of as consisting of a primary integer grid for the actualization of the electric field components and a secondary grid for the actualization of the magnetic fields. The nodes of the secondary grid are at the centers of the primary cells formed by the primary grid vertices. This construction of the two staggered grids is adapted to the spatial central difference equations of Equation (5.2). Imagine the grid as a three-dimensional chessboard on whose black areas lie the nodes for the computation of the electric fields and on whose white areas are the nodes for which the magnetic field will be computed. The discretization of the equation that yields the time derivative of the electric field in the Maxwell equations of Equation (5.2) will be made with

$$\begin{pmatrix} E_{x,n+1} \\ E_{y,n+1} \\ E_{z,n+1} \end{pmatrix} = \begin{pmatrix} E_{x,n} \\ E_{y,n} \\ E_{z,n} \end{pmatrix} + \frac{\boldsymbol{\varepsilon}^{-1}}{\varepsilon_0} \Delta t$$
$$\times \begin{pmatrix} \dfrac{H_{z,n+1/2}(y_0 + \Delta y/2) - H_{z,n+1/2}}{\Delta y} - \dfrac{H_{y,n+1/2}(z_0 + \Delta z/2) - H_{y,n+1/2}}{\Delta z} \\ \dfrac{H_{x,n+1/2}(z_0 + \Delta z/2) - H_{x,n+1/2}}{\Delta z} - \dfrac{H_{z,n+1/2}(x_0 + \Delta x/2) - H_{z,n+1/2}}{\Delta x} \\ \dfrac{H_{y,n+1/2}(x_0 + \Delta x/2) - H_{y,n+1/2}}{\Delta x} - \dfrac{H_{x,n+1/2}(y_0 + \Delta y/2) - H_{x,n+1/2}}{\Delta y} \end{pmatrix}, \tag{5.3}$$

where $\boldsymbol{\varepsilon}^{-1}$ is the local inverse dielectric tensor, and x_0, y_0, and z_0 indicate the current central grid point. The dielectric tensor $\boldsymbol{\varepsilon}$ may be locally different, representing a structured material. The discretization steps in the directions x, y, and z are Δx, Δy, and Δz, which are at the same time the side lengths of the lattice primary cells. All field vectors depend on the time step number and the three spatial coordinates x, y, and z. In our notation, the spatial arguments are not explicitly mentioned

for reasons of simplicity. The electric fields lie on the primary grid nodes, by which we mean $E_{x,n} = E_{x,n}(x_0,y_0,z_0)$ and, for example, $E_{x,n}(y_0 + \Delta y) = E_{x,n}(x_0,y_0 + \Delta y,z_0)$. The magnetic fields lie on the secondary grid nodes. Therefore, we will understand $H_{x,n+1/2}$ as $H_{x,n+1/2}(x_0 - \Delta x/2,\ y_0 - \Delta y/2,\ z_0 - \Delta z/2)$ and, for example, $H_{x,n+1/2}$ $(y_0 + \Delta y/2)$ as $H_{x,n+1/2}(x_0 - \Delta x/2,\ y_0 + \Delta y/2,\ z_0 - \Delta z/2)$.
The magnetic fields will be updated by

$$\begin{pmatrix} H_{x,n+1/2} \\ H_{y,n+1/2} \\ H_{z,n+1/2} \end{pmatrix} = \begin{pmatrix} H_{x,n-1/2} \\ H_{y,n-1/2} \\ H_{z,n-1/2} \end{pmatrix} + \frac{\Delta t}{\mu_0} \begin{pmatrix} \dfrac{E_{y,n}(z_0 + \Delta z) - E_{y,n}}{\Delta z} - \dfrac{E_{z,n}(y_0 + \Delta y) - E_{z,n}}{\Delta y} \\ \dfrac{E_{z,n}(x_0 + \Delta x) - E_{z,n}}{\Delta x} - \dfrac{E_{x,n}(z_0 + \Delta z) - E_{x,n}}{\Delta z} \\ \dfrac{E_{x,n}(y_0 + \Delta y) - E_{x,n}}{\Delta y} - \dfrac{E_{y,n}(x_0 + \Delta x) - E_{y,n}}{\Delta x} \end{pmatrix}, \tag{5.4}$$

where we use the same notation as in Equation (5.3). Here, the material is assumed to be nonmagnetic ($\mu = 1$).

The algorithm, which associates the updating of the electric (Equation (5.3)) and the magnetic (Equation (5.4)) fields, is commonly called the leapfrog time-marching scheme. Starting from initial values of the electric and magnetic fields on all nodes of the lattice, the electric field values will be updated by Equation (5.3), corresponding to a time step of $\Delta t/2$. The next time step (again $\Delta t/2$) will update the magnetic fields using Equation (5.4), using the previous magnetic fields and the electric fields that were computed directly before. After that, Equation (5.3) will be used again for the updating of the electric fields, and so on. Note that all updates of the electric (Equation (5.3)) and magnetic (Equation (5.4)) fields are made at the same time points for the whole lattice.

Applying this time-updating scheme from the Maxwell equations, the local material parameters have to be known. This means that at every grid point in the primary and secondary grids the dielectric tensor has to be defined. The FDTD can also be generalized to materials of arbitrary conductivity and magnetic permeability, but here we restrict the problem to nonconductive and nonmagnetic media.

Usually the spatial steps Δx, Δy, and Δz are equal (the primary cell is then a cube), but in special cases different discretization steps can be advantageous, for example, if a material has a rapidly changing anisotropy in only one direction. Then the space sampling in this direction has to have a higher resolution than in the other directions. Generally, anisotropic materials require a sampling resolution of about 20 per wavelength, although a sampling of 10 per wavelength is sufficient for isotropic materials.

The time advancing updating is based on the condition that the discretization scheme is conditionally stable. In order to ensure that the time step Δt fits the spatial grid, we have to choose $\Delta t < \Delta t_{\max}$, where $\Delta t_{\max}$ is the propagation time of the wave with velocity c_0/n along the spatial step Δx, n is the minimal refraction index of the medium, and c_0 is the velocity of light in a vacuum. In this way we obtain $\Delta t_{\max} = n\Delta x/c_0$ as a necessary condition. Taflove gave, for a three-dimensional isotropic cubic grid, the condition

$$\Delta t_{\max} = \frac{n\Delta x}{c_0\sqrt{3}} \tag{5.5}$$

for numerical stability (Taflove and Hagness, 2005).

5.3 IMPLEMENTATION AND BOUNDARY CONDITIONS

As the surrounding components are required for the computation of the fields, the described scheme cannot be applied at the outermost lattice planes of the simulation grid. Consequently, special updating schemes were developed for the outer grid boundaries. For periodic structures the computation domain can be truncated by applying periodic boundary conditions. This means in the discretization the field values required in Equations (5.3) and (5.4) outside a border will be taken not from the spatially antipodal border, but from the corresponding nodes inside the lattice. Nonperiodic finite structures have to be simulated by the use of absorbing boundary conditions and for this reason the border domains are parts with a sufficiently low level of reflection.

The idea of the Mur boundary condition is the modification of the discretizated Maxwell equations in the truncated domains. The central difference equations Equations (5.3) and (5.4) are replaced by so-called one-way wave equations. The normal derivatives in the update equation are here replaced by derivatives with respect to tangential directions (Mur, 1981).

Nowadays, absorbing boundary conditions are used most often, which are based on the construction of artificial absorbing media in the truncated border domains. The advantage of this idea is that the truncation can be made with orders of magnitude lower errors than when employing one-way wave equations. The truncated border domains are filled virtually with a special anisotropic material for which it can be shown that a plane wave incident from vacuum with an arbitrary wavevector is transmitted without reflection. This phenomenon is referred to as perfect matching. For sufficiently thick layers, the transmitted wave will be exponentially decaying within the truncating medium if the uniaxial material has major axis elements in the diagonal of the dielectric tensor that have an imaginary part larger than zero (Ziolkowski, 1997a,b). The so-called perfectly matched layers (PMLs) were introduced by Berenger for isotropic materials (Berenger, 1994) and later extended to domains of anisotropic materials (Peréz et al., 1998).

A last word on the wave sources: The FDTD is free concerning the definition of source conditions. This means, for example, that an arbitrary light source can be initiated by a simple addition of a constant electric or magnetic field value in the right-hand side of Equations (5.3) or (5.4) respectively, at each time step at the desired source region. Usually, point, line, or Gaussian beam sources are used depending on the simulation requirements. A more sophisticated route towards a plane wave source definition is to use a domain boundary as the source region. This can be done by the total-field/scattered-field formulation, which separates the domain into a scattering and a total field region. The scattering field region contains the material (or scattering) domain and the total field region is free of scatterers. If the electric and magnetic field values are updated at the total-field/scattering-field boundary, both total field and scattered field quantities appear in the second term on the right-hand sides of Equations (5.3) and (5.4). In this formulation it is the virtual interface between the total and the scattered field regions, which serves as the generator of the incident wave. For details and discussion of this method the reader is referred to the work of Mur (1981) and Umashankar and Taflove (1982).

5.4 RIGOROUS OPTICS FOR LIQUID CRYSTALS

For a long period of time liquid crystal optics has been pursued under the assumption that the liquid crystal material can be considered a stratified medium, with the direction of stratification coinciding with the cell normal. Matrix-type solvers have been developed, introducing different levels of complexity within the stratified medium framework. The simplest of all is the Jones method, restricted to forward-only propagation at normal incidence; this was later followed by the extended Jones method, which still traces the forward waves, but now allows for oblique incidence. Forward and backward waves are considered in the Berreman method, which is the most elaborate of the above matrix methods for optics. Matrix methods for optics can accurately describe light-wave propagation in liquid crystal devices when uniform transverse material orientation is maintained over a scale far exceeding the optical wavelength. For example, typical pixels found in flat-panel displays with transverse dimensions of hundreds of micrometers fall well within the stratified medium approximation. When the matrix methods are applied to devices with transverse variation, some approximations are necessarily introduced. A decomposition of the director profile into stratified columns, together with some averaging of the optical response associated with each individual column, is one approach. Alternatively, some type of averaging has to be applied to the original director profile to yield an average optical response. It is obvious that these approximations can be justified only in the limit of transverse variation of device characteristics and director orientation, which are extremely slow in comparison with the optical wavelength. Otherwise, the error introduced is difficult to estimate *a priori* and can certainly be substantial in many practical small-sized liquid crystal devices, which involve significant liquid crystal reorientation along a transverse direction. A step forward in the optics of liquid crystal devices is to consider rigorously the liquid crystal variation both along the normal to the cell surfaces and along a single transverse direction, leading to a two-dimensional treatment of light propagation. This approach has proven to be successful, and it has been implemented with the finite-difference time-domain method (Witzigmann et al., 1998) by solving time-harmonic maxwell equations (Amarasinghe et al., 2004) and the vector beam propagation method (Kriezis and Elston, 1999, 2000b). The wide-angle beam propagation method is suitable for analyzing anisotropic devices involving liquid crystals. The mathematical formulation is based on a system of coupled differential equations involving an electric and a magnetic field component. The contribution of all dielectric tensor elements is included. Usually, the method is numerically implemented, for instance with finite differences.

The above methods were applied to various structures with rapid liquid crystal reorientation on the optical wavelength scale, such as reverse tilt disinclinations in twisted nematic pixel edges (Titus et al., 1999; Kriezis, 2000a,b), ferroelectric liquid crystal domain walls (Kriezis et al., 2000), zenithal bistable nematic devices with surface-relief monogratings (Kriezis and Elston, 2001), optics of liquid crystal defects (Hwang and Rey, 2005a), diffractive optical elements (Scharf and Bohley, 2002; Wang et al., 2005), polymer liquid crystal dispersions and filled microcavities (Kriezis, 2002; Wang et al., 2003), and instabilities in nematic phases (Bohley et al., 2005). It was demonstrated that both methods were consistent within a small margin of error. It was also evident that in many cases the more commonly used matrix methods led to highly erroneous results because the rapid liquid crystal variation caused strong scattering and diffractive effects that were simply ignored.

A more detailed look at the different subjects is given in Chapters 8 to 10, where examples are discussed and different simulation techniques compared.

REFERENCES

Amarasinghe, N., E. Gartland, Jr, and J. Kelly (2004) Modeling optical properties of liquid-crystal devices by numerical solution of time-harmonic Maxwell equations, *J. Opt. Soc. Am. A* **21**, 1344–1361.

Berenger, J.P. (1994) A perfectly matched layer for the absorption of electromagnetic waves, *J. Comput. Phys.* **114**, 185–200.

Berenger, J.-P. (1996) Three-dimensional perfectly matched layer for the absorption of electromagnetic waves, *J. Comput. Phys.* **127**, 363–379.

Bohley, C., J. Heuer, and R. Stannarius (2005) Optical properties of electrohydrodynamic convection patterns: rigorous and approximate methods, *J. Opt. Soc. Am. A* **22**, 2818–2826.

Gedney, S.D. (1996) An anisotropic perfectly matched layer absorbing media for the truncation of FDTD Lattices, *IEEE Trans. Antennas and Propagation* **44**, 1630–1639.

Glytsis, E. and T. Gaylord (1987) Rigorous three-dimensional coupled-wave diffraction analysis of single and cascaded anisotropic gratings, *J. Opt. Soc. Am. A* **4**, 2061–2080.

Harrington, R.F. (1968) *Field Computation by Moment Methods*, Macmillan, New York.

Hwang, D.K. and A.D. Rey (2005a) Computational modeling of light propagation in textured liquid crystals based on the finite-difference time-domain (FDTD) method, *Liq. Cryst.* **32**, 483–497.

Hwang, D. and A. Rey (2005b) Computational modeling of the propagation of light through liquid crystals containing twist disclinations based on the finite-difference time-domain method, *Appl. Opt.* **44**, 4513–4522.

Keller, J. (1962) Geometrical theory of diffraction, *J. Opt. Soc. Am.* **52**, 116–130.

Kriegsmann, G.A., A. Taflove, and K.R. Umashankar (1987). A new formulation of electromagnetic wave scattering using an on-surface radiation boundary condition approach, *IEEE Trans. Antennas and Propagation* **35**, 153–161.

Kriezis, E.E. and S.J. Elston (1999) A wide angle beam propagation method for the analysis of tilted nematic liquid crystal structures, *J. Mod. Opt.* **46**, 1201–1212.

Kriezis, E.E. and S.J. Elston (2000a) Light wave propagation in liquid crystal displays by the finite-difference time-domain method, *Opt. Commun.* **177**, 69–77.

Kriezis, E.E. and S.J. Elston (2000b) A wide-angle beam propagation method for liquid-crystal device calculations, *Appl. Opt.* **39**, 5707–5714.

Kriezis, E.E., S.K. Filippov, and S.J. Elston (2000) Light propagation in domain walls in ferroelectric liquid crystal devices by the finite-difference time-domain method, *J. Opt. A: Pure Appl. Opt.* **2**, 27–33.

Kriezis, E.E. and S.J. Elston (2001), Beam propagation method modelling of zenithal bistable nematic devices: analysis and assessment, *Mol. Cryst. Liq. Cryst.* **359**, 597–608.

Kriezis, E.E. (2002) Comparative study of light scattering from liquid crystal droplets, *Microwave Opt. Tech.* **35**, 437–441.

Kunz, K.S. and K.M. Lee (1978) A three-dimensional finite-difference solution of the external response of an aircraft to a complex transient EM environment. I. The method and its implementation, *IEEE Trans. Electromagn. Compat.* **20**, 328–333.

Mur, G. (1981) Absorbing boundary conditions for the finite-difference approximation of the time-domain electromagnetic-field equations, *IEEE Trans. Electromagn. Compat.* **23**, 377–382.

Peréz, I.V., S.G. García, R.G. Martín, and B.G. Olmedo (1997) Extension of Berenger's absorbing boundary conditions to match dielectric anisotropic media, *IEEE Microwave Guided Wave Lett.* **7**, 302–304.

Sacks, Z.S., D.M. Kingsland, R. Lee, and J.F. Lee (1995) A perfectly matched anisotropic absorber for use as an absorbing boundary condition, *IEEE Trans. Antennas and Propagation* **AP-43**, 1460–1463.

Scharf, T. and C. Bohley (2002) Light propagation through alignment patterned liquid crystal gratings, *Mol. Cryst. Liq. Cryst.* **375**, 491–500.

Taflove, A. and M.E. Brodwin (1975a) Numerical solution of steady-state electromagnetic scattering problems using the time-dependent maxwell's equations, *IEEE Trans. Microwave Theory and Techniques* **23**, 623–630.

Taflove, A. and M.E. Brodwin (1975b), Computation of the electromagnetic fields and induced temperatures within a model of the microwave irradiated human eye, *IEEE Trans. Microwave Theory and Techniques* **23**, 888–896.

Taflove, A. and S.C. Hagness (2005) *Computational Electrodynamics*, Artech House, Boston.

Titus, C.M., P.J. Bos, and J.R. Kelly (1999) Two-dimensional optical simulation tool for use in microdisplays, *International Symposium Society for Information Display*, San Jose, California, 624–627.

Umashankar, K.R. and A. Taflove (1982) A novel method to analyze electromagnetic scattering of complex objects, *IEEE Trans. Electromagn. Compat.* **24**, 397–405.

Wang, B., P. Bos, and C. Hoke (2003) Light propagation in variable-refractive-index materials with liquid-crystal-infiltrated microcavities, *J. Opt. Soc. Am. A* **20**, 2123–2130.

Wang, X., B. Wang, P.J. Bos, J.E. Anderson, J.J. Pouch, and F.A. Miranda (2005) Finite-difference time-domain simulation of a liquid-crystal optical phased array, *J. Opt. Soc. Am. A* **22**, 346–354.

Witzigmann, B., P. Regli, and W. Fichtner (1998) Rigorous electromagnetic simulation of liquid crystal displays, *J. Opt. Soc. Am. A* **15**, 753–757.

Yee, K.S. (1966) Numerical solution of inital boundary value problems involving Maxwell's equations in isotropic media, *IEEE Trans. Antennas and Propagation* **14**, 302–307.

Ziolkowski, R.W. (1997) The design of Maxwellian absorbers for numerical boundary conditions and for practical applications using engineered artificial materials, *IEEE Trans. Antennas and Propagation* **375**, 565–671.

Ziolkowski, R.W. (1997) The incorporation of microscopic material models into the FDTD approach for ultrafast optical pulse simulations, *IEEE Trans. Antennas and Propagation* **45**, 375–391.

6

ORGANIC OPTICAL MATERIALS

6.1 INTRODUCTION

Historically, the benefit of plastic optics has been reduction in cost. Plastic optical elements are much less expensive than those made from glass. However, from a material point of view, completely different properties must be considered when macromolecules are used. Polymer molecules, or more generally macromolecules, are normally anisotropic in their material, electrical, magnetic, and optical properties. As a consequence of their structure (chains, rods, disks, and so on), their moduli and susceptibilities are tensor quantities leading to anisotropic behavior when the molecules are oriented. Orientation may be brought about through application of an external field, which may be rheological, electrical, or magnetic. For alignment of liquid crystals, surface treatments are common. As a result, one may have systems that demonstrate anisotropic refractive index (birefringence), anisotropic absorption of radiation (dichroism), polarized fluorescence, scattering, and/or diffraction.

6.2 POLYMERS FOR OPTICS

Polymers can be easily processed using methods such as injection molding, embossing, stamping, sawing, wet etching, and dry etching. They appear to be an ideal material for optics. Unfortunately, they also have some disadvantages, such as

Polarized Light in Liquid Crystals and Polymers. By Toralf Scharf low melting point and high linear expansion coefficient. Some of the most common optical polymers and copolymers are given in the following list.

- *Polymethyl methacrylate* (*PMMA*): Acrylic is the most commonly used optical plastic. Because its refractive index and dispersion values are similar to those of common crown glasses (particularly BK 7), acrylic is referred to as the crown of optical plastics. Acrylic is easily molded, scratch resistant, and not very water absorptive. It also has a relatively high transmission. Additives to acrylic (as well as to several other plastics) considerably improve its ultraviolet transmittance and stability.
- *Styrene*: Because styrene has a higher index and a lower numerical dispersion value than other plastics, it is often used as the flint element in color-corrected plastic optical systems. Polystyrene is a low-cost material with excellent molding properties. Compared with acrylic, styrene has lower transmission in the UV region of the spectrum and is softer. Because its surface is less durable, styrene is more typically used in nonexposed areas.
- *Methyl-methacrylate styrene* (*NAS*): This copolymer material is composed of 70% acrylic and 30% styrene. The specific blend ratio affects the material's refractive index, which ranges from 1.533 to 1.567. These resins have excellent alcohol resistance, better scratch resistance, and higher impact strength than styrene.
- *Styrene acrylonitrile* (*SAN*): This copolymer is also a member of the styrenic family. It is clear and stiff. These resins have more chemical resistance, better scratch resistance, and higher impact strength than styrene.
- *Polycarbonate* (*PC*): This plastic is very similar to styrene in terms of optical properties such as transmission, refractive index, and dispersion. Polycarbonate, however, has a much broader operating temperature band of −137 to 120°C. The high impact resistance of polycarbonate is its strongest advantage.
- *Cyclic olefin copolymer* (*COC*): Cyclic olefin copolymer provides a high-temperature alternative to acrylic. The material has a similar transmittance (92% through a 3-mm sample) and differential coefficient of linear thermal expansion to that of acrylic.
- *Methylpentene copolymer* (*TPX*): TPX is a 4-methylpentene-1 based polyolefin with a unique combination of transparency and heat- and chemical-resistant qualities. TPX is manufactured and marketed by Mitsui Chemicals.
- *Acrylonitrile butadiene styrene* (*ABS*): ABS is a member of the styrenic resin family. ABS resins are typically known for their toughness, excellent surface gloss, and very good chemical resistance over polystyrene, and also their good heat resistance.

Table 6.1 provides a summary of values that are found in different publications for the different polymers. The following datasheets were used:

Amco plastics: Nova NAS 90 technical datasheet (NAS)
Arla plast: ABS standard datasheet (ABS)

Table 6.1 Properties of selected optical plastics.

	Units	Acrylic	Styrene	NAS®	SAN	PC	TPX	ABS	COC	NOA 61 (cured)	Glass
		Polymethyl methacrylate (Lucite; Plexiglass)	Polystyrene (Dylene; Styron; Lustrex)	Methyl methacrylate styrene copolymer	Styrene acrylonitrile (Lustran; Tyril)	Polycarbonate (Lexan; Merlon)	Methylpentene (TPX)	Acrylonitrile butadiene styrene	Cyclic olefine copolymer	Epoxy UV curable glue	BK 7
Refractive index, n											
n_F (486.1 nm)		1.498	1.604	1.575	1.578	1.599	1.473		1.540	1.5691	1.522
n_d (587.6 nm)		1.492	1.590	1.533–1.567	1.567–1.571	1.585	1.467	1.538	1.530	1.5594	1.517
n_C (656.3 nm)		1.489	1.585	1.558	1.563	1.580	1.464		1.526	1.5562	1.514
Abbe value, V_d		57.4	31.1	35	37.8	29.9	51.9		58	43.3	64.17
Rate of change in index with temperature	$dn/dt \times 10^{-5}/°C$	−10.5	−14.0	−14.0	−11.0	−10.7 to −14.3			−10.1	−18.3	−1.28
Haze	%	2	3	3	3	3	1.5	12	1–2		0–0.17
Uncoated transmittance	%, $d = 3.175$ mm	92	88	91	88	89	92	79–90.6	92		92
Coefficient of linear expansion	$cm/cm \times 10^{-5}/°C$	6.74 at 70°C	6.0–8.0		6.5–6.7	6.6–7.0	11.7	0.83	6.0–7.0		0.71
Deflection temperature											
3.6°F/min 264 psi	°C	92	82		99–104	142		90	123		
3.6°F/min 66 psi	°C	101	110		100	146	90	84	130		
Recommended max. cont. service temp.	°C	92	82	93	79–88	124			130	125	250
Thermal conductivity	cal/sec-cm °C × 10^{-4}	4.96	2.4–3.3	4.5	2.9	4.65	4.0				0.92
Water absorption	%, 24 h at 23°C	0.3	0.2	0.15	0.2–0.35	0.15	0.01		<0.01	0.16	
Advantages		Transmission	High index	Good index range	Stable	Impact strength	Chemical resistance	Durable		Durable, UV curable	

Uncoated luminous transmittance 79%, thickness 6.35 mm; 90.6%, thickness 0.381 mm.
n_F, n_d, n_C are refractive indices measured at different wavelengths.

GE plastics: optical properties of Lexan datasheet (PC)
GS optics: plastic optics charts (PMMA, PC, PS, COC, COP)
Hoechst: Polymere Werkstoffe, sonderdruck B175 (COC)
Norland Adhesives: NOA 61 data sheet
Schott glass: technical datasheet BK 7 (glass)
Rohm: ZK20 technical datasheet (PMMA)
Tycona (Topas) technical datasheet (COC)
Zeonex technical datasheet (COP)
[Ampiatec, 2003; Bass, 1994; Bentec, 2003; GSoptics, 2002; Norland Datasheet, 2005]

6.3 PHYSICAL PROPERTIES OF POLYMERS

A polymer is a substance composed of molecules that have long sequences of one or more species of atoms or groups of atoms linked to each other by primary, usually covalent, bonds. A polymer is composed of macromolecules (Young and Lovell, 1991). Linking together monomer molecules through chemical reactions forms macromolecules – a process known as polymerization. It is this long-chain nature that sets polymers apart from other materials and gives rise to their characteristic properties. Many molecules have a linear skeletal structure, which may be represented by a chain with two ends. Although this is true for many macromolecules, there are also many with nonlinear skeletal structures. Branched polymers have side chains, or branches, of significant length, which are bonded to the main chain at branch points and are characterized in terms of the number and size of the branches. Network polymers have three-dimensional structures in which each chain is connected to all others by a sequence of junction points and other chains. Such polymers are said to be crosslinked and are characterized by their crosslink density, or degree of crosslinking, which is related directly to the number of junction points per unit volume. With low crosslink densities the product is a flexible elastomer; it is a rigid material when crosslink density is high. The most common way to separate the polymers is into three groups: thermoplastics, elastomers, and thermosets.

Thermoplastic polymers are further separated into those that are crystalline and those that are amorphous. Many thermoplastics are completely amorphous and incapable of crystallization, even upon annealing. Amorphous polymers are characterized by their glass transition temperature (T_g). The transition corresponds to the onset of chain motion – below glass transition the polymer chains are unable to move and are frozen in position. Thermoplastics are polymers that become soft and fluid upon heating and can then be transformed into any desired shape, which can be stabilized by subsequent cooling. This behavior is a consequence of the absence of chemical crosslinks in these polymers. This cycle can be repeated at will.

Elastomers are crosslinked rubbery polymers that can be stretched easily to high extensions. They rapidly recover their original dimensions when the applied stress is released. This property is a reflection of their molecular structure, which is a network with low crosslinking density.

Thermoset resins consist of a mixture of reactive, low molar mass compounds, which react with each other upon heating and then transform from a fluid into a solid material as a result of crosslinking or curing. The polymer is formed in a mold, and the product cannot be further melted or dissolved as it consists of one giant network molecule. The crosslinking can also be established by UV radiation.

When a polymer melt is gradually cooled its volume decreases, which can be followed by dilatometry (Challa, 1993). The polymer melt first behaves like a liquid, the viscosity of which increases with decreasing temperature. As a consequence, the mobility of the polymer chains decreases strongly and the normal thermal shrinkage starts to lag behind the temperature. This means that the so-called free volume between the chain segments starts to decrease less quickly. This is the beginning of the glass transition, in which the coordinated motion of chain segments becomes more and more hindered. Finally, the mobility is so low that further shrinkage can only occur by somewhat closer packing of chain segments that are already neighbouring, and the polymer has reached its glassy state. Extrapolation of the volume–temperature lines from the liquid and the glassy state yields the glass transition temperature T_g at the intersection. Obviously, the rate of cooling will determine the location of the glass transition region. If a polymer is stored at a temperature not too far below T_g, the strongly reduced chain mobility will still enable the polymer to achieve closer packing. This phenomenon is called volume retardation and is of great importance for the long-term mechanical behavior of structural polymers. It is understandable that the magnitude of T_g is correlated with chain stiffness and polarity, as these factors determine the mobility of the chains in the polymer and thus the onset of the glass transition. Both transition temperatures T_g and melting temperature T_m are influenced similarly by the chemical structure, and it seems impossible to control T_g and T_m separately. Important also is the effect of plasticizers on T_g. These are generally liquids, which are soluble in the polymer chains, thus diminishing their mutual interactions and promoting their mobility.

In the technical datasheets the deflection temperature is given rather then the glass temperature. For pure (unfilled) polymers the difference between these two values is only a few degrees and can thus be used to give an estimate of the glass transition temperature. The deflection temperature is a measure of a polymer's resistance to distortion under a given load at elevated temperatures. The deflection temperature is also known as the "heat deflection temperature", "deflection temperature under load" (DTUL), or "heat distortion temperature" (HDT) (ASTM, http://www.astm.org).

Plastic materials expand when heated and contract when cooled. The coefficient of linear thermal expansion (CLTE) is the ratio of the change of a linear dimension to the original dimension, for unit change of temperature. When compared to glass, the CLTE for polymers is a factor of 10 greater. This effect will change the radii and thickness of the optics and thus shift the point of focus if the device is used at different temperatures.

Thermal conductivity is the rate at which a plastic conducts heat along its length or through its thickness. It is important for the design engineer to understand this factor if the final product is used above a heat source.

The water absorption property is expressed as the percentage increase in weight of a material due to absorption of water. This property affects mechanical and electrical properties of a specific plastic. Some polymers are hygroscopic. The most common of these are nylon, ABS, PMMA, and PC. Water absorption also influences the optical properties. The polymer materials used for optics may absorb from about 0.003 to about 2% water by weight.

6.4 OPTICAL PROPERTIES OF POLYMERS

Optical polymers are clear plastics that provide excellent light transmission. In photonic applications, they offer advantages over optical glass: They weigh less and can be molded into spherical, aspherical, and nonrotationally symmetric shapes. Typical uses for these optical polymers include lenses for video and still cameras; projection televisions; compact disk drives; light emitting diodes; printers and barcode readers; lightguides; optical films; high-density optical storage media; diffractive optics; flat panel displays; metallized reflectors; and optical fibers, fiber couplers, and connectors. Electronic, vibrational, and magnetic states of the macromolecules of the polymer are all relevant in the spectroscopic analysis of polymers. Most of the optical polymers have a transmission window between 400 and 1600 nm. The small difference in the transmission window between the different polymers can be traced back to their molecular composition. The spectrum for polymers can be divided into three regions: ultraviolet (UV), visible, and infrared (IR). Polymers, in common with all organic materials, absorb light in the UV region of the spectrum. The mechanism for this absorption depends on transitions between electronic energy levels of the bonds within the material; The absorption of a photon causes an upward transition, leading to excitation of the electronic state of the solid. δ-Electrons, involved in covalent bonds, absorb high-energy photons in the UV region, whereas π-electrons absorb photons at longer wavelengths, often in the visible region. The interaction between polymer molecules and the UV components in sunlight can lead to dissociation of δ-bonds and is the cause of photodegradation of polymers (Campbell and White, 1989). For instance, in the case of PMMA, the most significant absorption is caused by the transition of the $n-\pi^*$ orbital of the double bond within the ester group (Borrelli, 1999). Pigments and UV stabilizers are two classes of additives that depend on their response to electromagnetic radiation within this range to provide the very property for which they are included. For most of the polymers, the attenuation is higher than in the visual spectrum and it is different for every polymer. This effect is also used to characterize polymers and thus a lot of information can be found about this subject. The technique is called near infrared spectroscopy (NIR). NIR spectra arise from an overtone of the fundamental vibrations and hence the technique is complementary to IR and Raman spectroscopy. Most of the spectral lines detectable in the NIR arise from vibrations of hydrogen-containing bonds (Campbell and White, 1989). Absorption may be caused by electronic transitions within contaminants. For instance, metal ions have the most influence in the visible wavelength region. Water absorbed in polymers is another factor responsible for increasing loss. The water absorption coefficient of PMMA is higher by one order than that of PS. This attenuation in the visible wavelength region is due to OH vibrations. It is also possible to substitute hydrogen atoms with fluorine atoms. To suppress water vapor absorption, substitution for hydrogen in the core polymer is considered to be effective. Fluorine compounds are more effective at preventing the penetration of moisture into the polymer compared with deuterium substitution. If the CH atom pairs within a polymer are replaced by carbonfluorine (CF), it would cause the fundamental absorption to shift to the longer wavelength region.

Haze is the percentage of transmitted light that deviates from the incident beam by forward scattering when passing through a specimen. Lower haze values imply greater transparency. Haze is caused by the scattering of light within a material, and can be affected by molecular structure, degree of crystallinity, or impurities at the surface or in this interior of the polymer. The haze test method actually measures absorption, transmittance, and deviation of a direct beam by a translucent material. A specimen is placed in the path of a narrow beam of bright light so that some of the light passes through the specimen and some continues unimpeded. Both parts of the beam pass into a sphere equipped with a photodetector. Two quantities can be determined: the total strength of the light beam and the amount of light deviated by more than 2.5° from the original beam. From these two quantities, two values are calculated: haze, or the percentage of incident light scattered more than 2.5°, and the luminous transmittance, or the percentage of incident light that is transmitted through the specimen.

There exist several methods to solidify liquid monomers by in situ polymerization. Three important ones are thermal polymerization (as for thermosets), polymerization by mixing several components, and polymerization by UV radiation. Very often, UV polymerization is used because the degree of polymerization and the moment when the polymerization should happen can be exactly defined. Two classes of materials are important, epoxy-based and acrylic-based polymerizable resins. UV-curable epoxy-resin-based polymer systems are suitable for bonding, sealing, and coating, and post-curing enhances its chemical resistance to acids and solvents. Optical adhesives are liquid photopolymers that are clear and colorless. They will cure when exposed to UV light. As it can be a one-part system and be composed of 100% solids, it offers many advantages in bonding. The only condition is that the adhesive has to be exposed to UV light. Curing time can be fast, and is dependent upon the thickness and the amount of UV light energy available. Sometimes curing agents are used to accelerate the process. Curing is carried out at ambient temperatures and is even possible in the presence of air. Shrinkage upon cure differs for different polymers. The operating temperature extends to 150°C. Polymers have viscosities between 100 mPas and thixotropic and cures in thicknesses of up to several millimeters (NOA, 2005). The materials have excellent clarity but undergo a certain shrinkage. The polymers are cured by UV light with maximum absorption within the near UV range, usually 350–380 nm, which is easy accessible using mercury pressure lamps. It is clear that such materials cannot be used for UV applications as they absorb in the UV region. As an example we show the details for the UV-curable adhesive NOA 61 in the Table 6.1.

6.5 LIQUID CRYSTAL PHASES

Most substances exist in three phases: solid (often crystalline), liquid, and vapor. The difference between these states of matter is the degree of order in the material, which is directly related to the surrounding temperature and pressure. Certain organic substances possess more condensed phases than the basic two. They are referred to as liquid crystals and their constituent molecules are called mesogens. Their different extra phases, found between the solid and liquid states, are called liquid crystalline phases or mesophases (Baron, 2001). The explanation for these intermediate phases lies in the fact that liquid crystal molecules are always shape anisotropic, normally having a more or less rodlike shape. They may also be disc-, or bowl-shaped. The latter two groups are referred to as discotic and bowlic liquid crystals, respectively. Very recently, banana-shaped liquid crystals found particular interest for their extraordinary phase properties (Pelzl et al., 1999). We will limit our discussion to the basic properties of liquid crystalline phases. In the above description of the solid, liquid, and vapor phases, we only took the degree of positional order into account. Because of the anisotropic shape of the liquid crystal molecules, we must now also consider the orientational order in the material. The liquid crystalline phases all exhibit high orientational order, but the order in the molecule positions is limited. An example of a nematic phase is shown in Figure 6.1

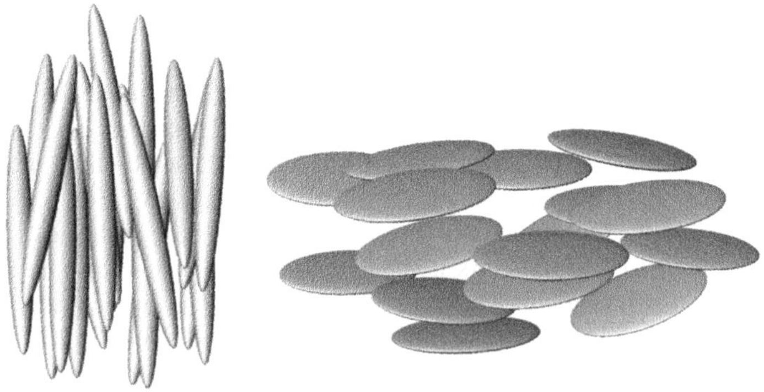

(*a*) (*b*)

Figure 6.1 The nematic phase, the simplest liquid crystalline phase, is characterized by orientational order. They do not have any positional order. Calamitic (*a*) and discotic (*b*) molecules are shown.

for rod-like (calamitic) and disc-like (discotic) molecules. We will concentrate on phases with calamitic molecules, which constitute the major class. The positional arrangement in the liquid crystalline phases varies from complete disorder to order in one dimension (the molecules tend to form layers) or slightly more. A substance in a liquid crystalline phase is therefore a fluid in that it assumes the shape of its container. At the same time the high orientational order gives this fluid a highly anisotropic property, which distinguishes it from normal liquids, which are always isotropic – the liquid phase without order of a mesogenic compound is normally denoted the isotropic phase. The orientational order in the liquid crystal phases is of course not perfect. In fact, if we were to take a snapshot of a liquid crystalline sample we would find quite large differences in orientation of different molecules. This leads to light scattering effects, which are visible under the microscope if nematic phases are observed under particular conditions. There is, however, a locally distinct preferred direction around which the molecules fluctuate. This average molecule orientation is described by a unit vector called the director, denoted by $\mathbf{n}$. It is a special vector as its sign is generally of no importance: $\mathbf{n} = -\mathbf{n}$. This reflects the fact that turning the director 180° is a symmetry operation, which conserves all physical properties of the liquid crystal. When classifying liquid crystal phases we first distinguish between two main types: one with nematic order and one with smectic order. In the nematic phase (of which the cholesteric phase is a special case), the molecules are free to move in all directions. There is no positional order of the centers of mass, but on average they keep their long axes locally parallel. In a smectic state one also has a positional order along one dimension. Some smectic phases have positional order in more than one dimension. A smectic is a layered structure with the molecules oriented parallel or tilted relative to the layer normal. Two examples of such a layer arrangement are shown in Figure 6.2. Two smectic phases, called smectic A and smectic C, have acquired a special importance and are relatively well understood. They are characterized by absence of positional order within the layers. The molecules have some freedom to move within the layers, but they are much less free to move between layers. These smectics can therefore be said to be stacks of two-dimensional fluids, but they behave as crystalline across the layers. There are several smectic phases differing from one another in the tilt angle that the director makes with the layer normal, and also in the arrangement of molecules within each layer. The simplest is the smectic A phase, as illustrated in Figure 6.2*a*. It is characterized by a director parallel to the layer normal and random positional order within the plane. Substances featuring the A phase also often exhibit the smectic C phase (Fig. 6.2*b*) at a lower temperature. In this phase the molecules have the same random order within the layer, but tilt relative to the layer normal. The tilt angle normally increases with decreasing temperature. The other smectic phases are even more crystalline in that they also feature some positional order within the layers (Stegemeyer, 1994; Demus et al., 1998; Dierking, 2003). They may, for instance, exhibit hexagonal packing of the molecules.

In addition to the mentioned phases, the concept of chirality is important with liquid crystalline phases. This is a strictly geometric concept and should therefore

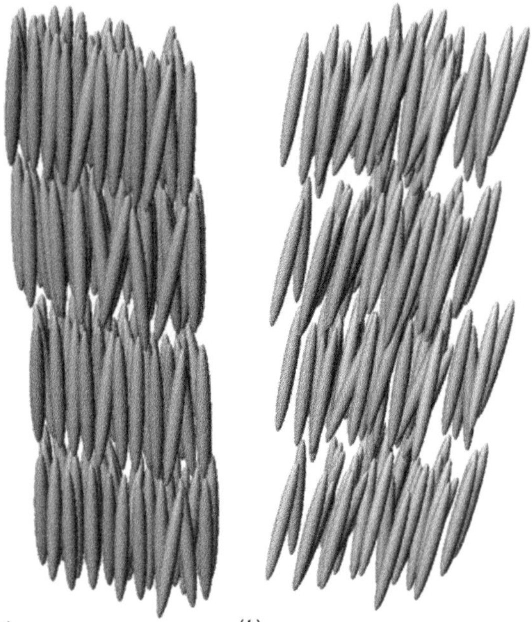

Figure 6.2 Layered phases with their director perpendicular to the layer normal are called smectic A (*a*). If the director is tilted to the layer normal the phase is called smectic C (*b*). The tilt angle varies with temperature.

not be confused with something dependent of the different phases of the liquid crystal. A chiral object has such a shape that it cannot be superposed on its mirror image. An object is either chiral or achiral, meaning that it either lacks or has mirror symmetry. It cannot be both. If a liquid crystalline substance consists of chiral molecules one denotes this by putting a star after the phase labels, for instance N* or C*. Chirality is not a state of matter and there can thus be no phase transition from chiral to achiral, or vice versa. For instance, a nematic phase made up of chiral molecules, which is called a cholesteric, cannot turn into an achiral phase at a certain temperature. Chirality has important consequences for the macroscopic arrangement of the liquid crystal molecules. It therefore affects the optical behavior of the substance.

A typical member of that class formed from chiral nematic mesogens is the cholesteric phase. The cholesteric phase features a helical arrangement of the molecules, as in Figure 6.3. In the cholesteric phase molecules twist around a helix axis. That axis lies perpendicular to the local director. As the twist has a constant strength throughout the sample, we obtain a periodic structure along the helix axis. It is not a positional periodicity, but directional. Every half turn of the helix the molecules are oriented the same way. Remember that the director has no sign – turning it 180° or 360° gives the same appearance. The full helical turn is called the pitch, p, of the cholesteric. The pitch varies with temperature. When a chiral

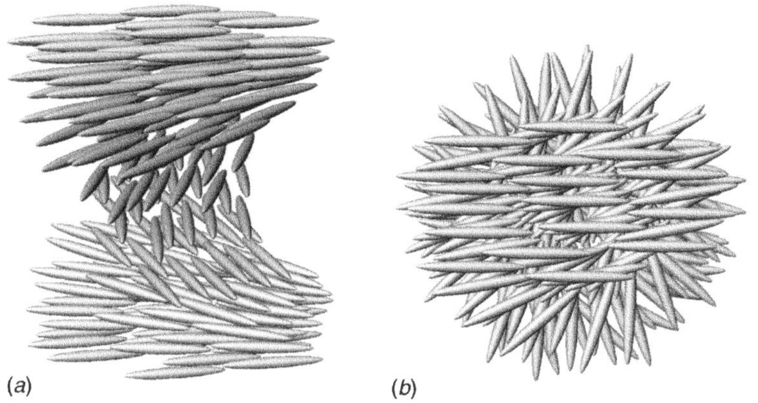

(a) (b)

Figure 6.3 The chiral nematic, or cholesteric, phase forms a helical structure. (*a*) The molecules twist in a plane perpendicular to the paper. (*b*) View along the helical axis. The molecules show a left-hand twist.

substance exhibits smectic phases, it can also lead to a defect-free twisted structure. However, due to the layered structure, no twist can be admitted within the layer. Without breaking the layers, the orientation of the director can vary continuously only between the layers, when going from one layer to the next. The helix axis must now lie perpendicular to the layer planes. In the untilted A* phase this leads to no visible difference compared with the achiral A phase. However, when the director starts to tilt, as in the C* phase, we do obtain a helicoidal macroscopic structure, in that the azimuthal angle of the director changes with a constant value from layer to layer. This gives the structure shown in Figure 6.4. Normally it takes some hundred layers to complete one revolution in the azimuthal angle. In some cases, when the intrinsic twisting power of the material is very high, the continuous layer structure of the smectic A* may break down, permitting a twist with twist axis perpendicular to the director, as in a cholesteric. This means that the smectic structure breaks down into periodic stacks of layers, with a finite twist in between, mediated by a regular array of defects. This kind of defect structure is called the TGB (twist–grain–boundary) phase (Collings and Hird, 1997). It is the smectic correspondence to the blue phase, which may appear out of a cholesteric as a result of a very high twisting power (Chadrasekhar, 1992). A corresponding TGB phase can also form from a smectic C* phase. In both cases the twist axis is along the smectic layers. The TGB phases exist only in a very narrow temperature range.

The liquid crystals described so far are all of the so-called thermotropic type. The name reflects the fact that they change phase depending on temperature. However, liquid crystals may also be lyotropic (Hamley, 2000). Such liquid crystals, although sharing some basic features with the thermotropic liquid crystals, normally differ widely from these in structure, properties, and applications. In many cases, however, the liquid crystal must be considered to belong to both classes. Their organization is normally not based on mesogenic properties of a certain molecule

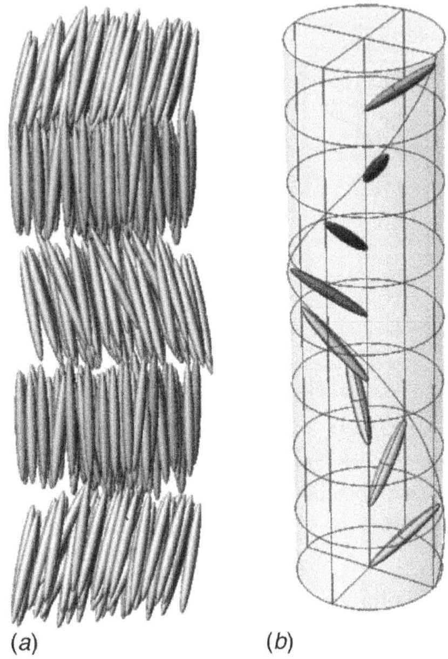

Figure 6.4 Helical structure of the chiral smectic C* phase. The molecules are tilted and rotate around the helical axis from layer to layer as shown in (*a*). The azimuthal angle changes and makes a complete turn within the lengths of the helical pitch, as better seen in (*b*) where lines guide the eye.

but rather on the interaction between two, or even more, molecular units in solution. Therefore their sequence of phases depends more on the concentration of the different components added to the mixture than on temperature. These types of liquid crystal are very common in the biological world, for instance in cell membranes, where we find both thermotropic and lyotropic forms of liquid crystals. The interested reader can get more information about that subject in textbooks on soft matter (Stegemeyer, 1999; Hamley, 2000; Daoud and Williams, 2005). Liquid crystal molecules may also be chemically tied together by polymerization. These concept are explained in more detail in Section 6.6.

Microscopic observations of liquid crystal textures can reveal a great deal of information about their ordering. To understand liquid crystal textures, one needs to understand the basic principles of light propagation in anisotropic media, which we will discussed in detail in Chapter 8. There are a multitude of phases with complex internal structures, which are difficult to identify. The order of these phases is very useful in identifying mesophases when polymorphism (more than one mesophase) occurs. There are several books (Demus and Richter, 1987; Dierking, 2003) and numerous articles that contain excellent photographs of textures of mesophases. To discover the beauty of liquid crystalline textures it is often sufficient to simply place a sample on a slide, add a cover slip, and place this into a polarizing microscope. If it is not in the liquid crystalline phase, heating with an adapted heating stage is required. This sometimes yields beautiful textures, but often it does not. This is because the alignment of the molecules determines what texture will be observed. It is important to remember that a good texture is needed to identify the characteristic features of a particular mesophase. If the texture is not characteristic, this method cannot be used to identify a phase. A totally mixed, aligned texture is useless for identifying mesophases. Texture depends on alignment and there are three principal types that can occur relative to the slide: parallel (homogeneous) perpendicular (homeotropic), and a mixture of these two. Because liquid crystals are often studied in thin films, the alignment is strongly coupled to the substrate surfaces.

Special treatment allows the establishment of particular orientations on substrate surfaces. Figure 6.5 illustrates the basic orientations. The liquid crystal director can be oriented mainly parallel to the surface or mainly perpendicular to the surface. The first case is called planar orientation and the second homeotropic orientation. For smectic phases these terms are often defined with respect to the layer arrangement. If the smectic layers are parallel to the substrate surfaces, the alignment is called homeotropic, and if it is perpendicular it is called planar or bookshelf texture. These are ideal cases when the orientation is not disturbed by defects. Two major factors affect this alignment: the structure of the molecule itself, and the surfaces of the substrate. Many mesogens will readily align in either direction, sometimes on the same substrate. Homogenous alignment is usually possible by treating the surface substrate. The thickness of the sample also has an influence. Obtaining a uniform texture usually requires some time for relaxation of defects, and the viscosity of the material should be low. Polymers usually have larger viscosities than smaller molecules. Consequently, good textures are often difficult to achieve in these materials.

Comparing the subtle differences often found in liquid crystalline textures is only possible by presenting a sequence of texture photos that shows exactly the same sample area. Often a sample will show various textures in different areas of a sample. Good textures obtained on cooling may convert to poor textures on reheating the crystals formed on cooling. Such textures are useless for identification. Homogenous alignment produces textures having many characteristics useful in identifying mesophases. However, the homeotropic texture can be extremely

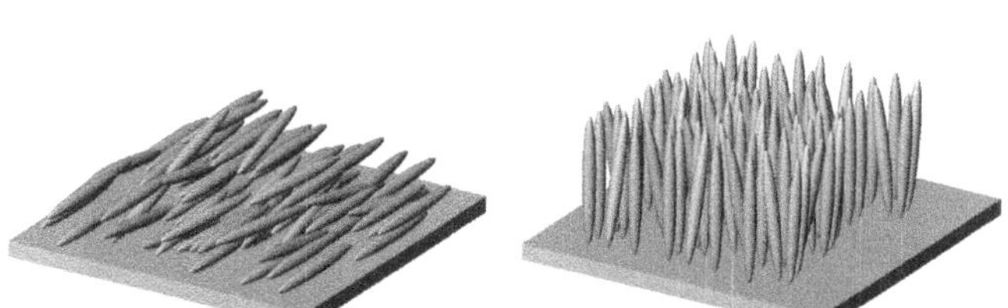

(*a*) (*b*)

Figure 6.5 Uniform planar (*a*) and homeotropic (*b*) textures of liquid crystal directors at a surface. The molecules are oriented parallel in the planar case and perpendicular in the homeotropic case.

useful in providing confirmation of these identifications. A perfect homeotropic texture would be totally black in the polarizing microscope between crossed polarizers, but different from that for the isotropic liquid.

Many effective aligning agents, such as polymers and lecithins, are useful for achieving good homeotropic alignment. Rubbed polyimides and polyvinyl alcohol (PVA) are useful for obtaining homogenous alignment. However, there exist more sophisticated methods that also allow patterning of alignment on a single surface substrate. Figure 6.6 shows a typical homeotropic Schlieren texture having imperfect alignment. Texture photographs are taken with a polarizing microscope and crossed polarizers. The nematic Schlieren texture is often highly colored and consists of disclination points from which two or four brushes or more radiate. It is this texture that is most commonly associated with the nematic phase, but in reality the marbled texture is seen much more frequently. The marbled texture occurs only in the nematic phase. It shows threads (lines) in some form or another. A plain, black texture for homeotropic alignment can be observed only in uniaxial mesophases. If the mesophase is not a uniaxial one, it will show either a Schlieren texture, which is usually gray, or a mosaic platelet texture. Smectic C phases will show a gray Schlieren texture.

Focal conic fan textures are particularly useful in identifying mesophases, as they have specific characteristics for each type of mesophase. For a smectic A phase a typical appearance is shown in Figure 6.7. The fans shown can be much more pronounced. These fans are not observed for nematic phases, but they can occur in chiral nematic (cholesteric) phases. They occur in homogenously aligned samples. Sharp, focal conic fans occur primarily in smectic A and B phases but can also occur in some more highly ordered smectic phases due to paramorphism. In the smectic C phase, these fans convert to broken fans, which often are less sharp. Several discussions of the focal conic texture can be found in the literature (Kléman, 1983).

Figure 6.6 Schlieren texture of the liquid crystal mixture E7 (Merck) between polycarbonate substrates spaced by 10 μm. The brushes indicate defects of different order. The texture is not stable because of the solubility of polycarbonate in the liquid crystal.

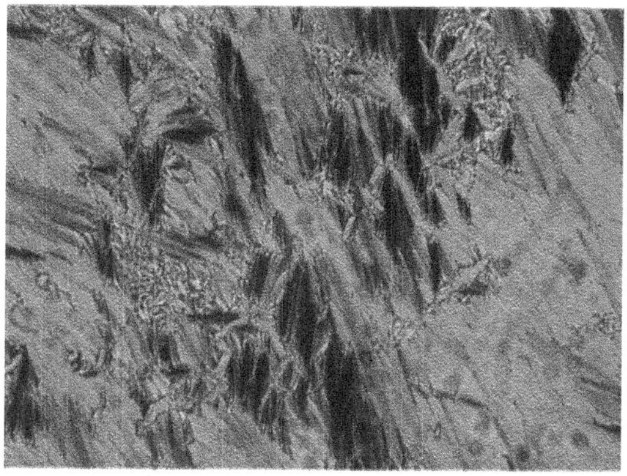

Figure 6.7 Fan-shaped texture of a smectic A liquid crystal between crossed polarizers.

Figure 6.8 Cholesteric oily strikes (*a*) and fan shape (*b*) textures of a mixture of a chiral dopant CB15 and the liquid crystal mixture E48 (Merck) at $c_w = 50\%$. The sample is seen between crossed polarizers. The layer thickness is 6 μm and the surfaces are treated with planar alignment polymer coatings and rubbed antiparallel.

Equally important to the specific texture for identification are the changes that occur in the phase transitions that take place between these textures. Texture can vary considerably within a single phase either in a single sample or in different ones. Even a typical smectic A fan texture has a number of variations. Alignment can vary in a single sample, giving a mixture of textures. Unlike the totally mixed aligned texture, this texture can be useful in identifying mesophases. Textures for chiral mesophases can be different from those seen in racemic compounds. For example, the chiral nematic (cholesteric) phase shows oily streaks, planar and focal conic fans as shown in Figure 6.8. Other types of mesogens, such as discotics and lyotropics, have their own characteristic textures. All these variations provide means for identifying mesophases.

6.6 LIQUID CRYSTAL POLYMERS

Liquid crystal polymers are a group of chemically complex and structurally distinct materials that have unique properties and are utilized in diverse applications. Discussion of the detailed chemistry is beyond the scope of this book. Liquid crystal polymers are composed of extended, rod or disc-shaped, and rigid molecules (Stegemeyer, 1994; Collings and Hird, 1997). In terms of molecular arrangement, as with liquid crystals themselves, these materials do not fall within any conventionally liquid, amorphous, crystalline, or semicrystalline classifications. In the melt condition, whereas other polymers are randomly oriented, liquid crystal polymer molecules become aligned in highly ordered configurations and show liquid crystalline phases. As solids, this molecular alignment remains. The molecules form domain structures having characteristic intermolecular spacing. Liquid crystal polymers can be described as a normal flexible polymer with mesogenic groups incorporated into their chains. In order to display liquid crystal characteristics for normally flexible polymers, rod-like or disc-like elements (called mesogens) must be incorporated into their chains. The placement of the mesogens plays a large role in determining the type of liquid crystal polymer that is formed. Main-chain liquid crystal polymers are formed when the mesogens are themselves part of the main chain of a polymer. The mesogens form stiff regions along the chain that allow the polymer to orientate in a manner similar to ordinary liquid crystals, and thus display liquid crystal characteristics. There are two distinct groups of main-chain polymer liquid crystals, differentiated by the manner in which the stiff regions are formed.

The first group of main-chain polymer liquid crystals is characterized by stiff, rod-like monomers. These monomers are typically made up of several aromatic rings which provide the necessary size. The second and more prevalent group of main-chain polymer liquid crystals is different because it incorporates a mesogen directly into that chain. The mesogen acts just like the stiff areas in the first group. Generally, the mesogenic units are composed of two or more aromatic rings that provide the necessary restriction on movement that allows the polymer to display liquid crystal properties. The stiffness necessary for liquid crystallinity results from restrictions on rotation caused by steric hindrance and resonance. Another characteristic of the mesogen is its axial ratio. The axial ratio is defined as the length of the molecule divided by the diameter. These molecules must be at least three times long as they are wide to display the characteristics of liquid crystals. If stiff elements are used they are separated or "decoupled" by a flexible bridge called a spacer. Decoupling of the mesogens provides for independent movement of the molecules, which facilitates proper alignment.

Conversely, side-chain polymer liquid crystals are formed when the mesogens are connected as side chains to the polymer by a flexible "bridge," as shown in Figure 6.9. Main-chain polymer liquid crystals often cannot show mesogenic behavior over a wide temperature range. Side-chain polymer liquid crystals, however, are able to have large range of mesogenic behavior. Side-chain polymer liquid crystals have three major structural components: the backbone, the spacer, and the mesogen. The versatility of side-chain liquid crystal polymers arises because these structures can be varied in a number of ways.

Other factors influencing the mesomorphic behavior of polymers include the presence of long flexible spacers, a low molecular weight, and regular alternation of rigid and flexible units along the main chain. By changing the nature of the mesogenic group and the spacers or by modifying their geometry, length or position, and regularity, a large variety of liquid crystal polymers with tailored properties can be obtained. Liquid crystal polymers exhibit the same phases as conventional liquid crystals (low molar mass mesogens), which are described above. However, the melting point of liquid crystal polymers is usually much higher (100–300°C) than that of low molar mass mesogens. Many liquid crystal polymers exhibit a glass transition as in conventional plastics. Glass transition temperatures for the material classes of interest are below 100°C. The important point is that liquid crystal polymers can be aligned using the same methods as used for conventional liquid crystals. The optical components composed of liquid crystal polymers can

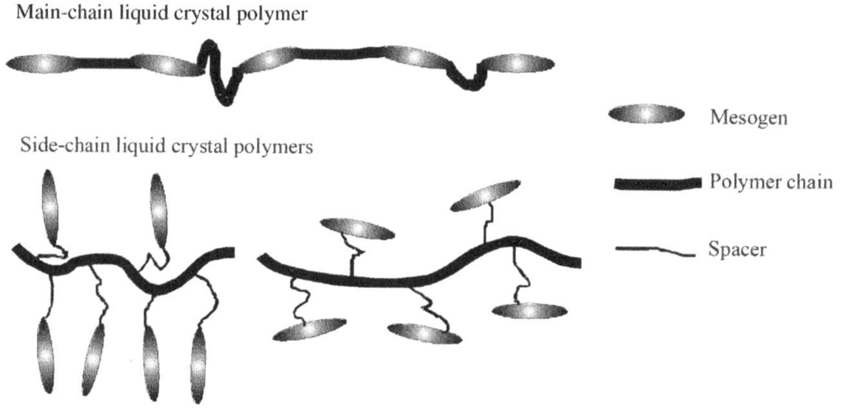

Figure 6.9 Main-chain and side-chain polymers with mesogens incorporated to form liquid crystalline phases.

be prepared by filling a prepared cell with liquid crystal monomer (liquid crystal with polymerizable groups) or polymer. The cell is then heated to the nematic phase. After reorientation of the liquid crystal, there are two methods of freezing the nematic phase: via in situ polymerization or via vitrification. In the vitrification method, liquid crystal polymers with a high molar mass that exhibit a glass transition are used. In this method the aligned liquid crystal is simply cooled below the glass temperature transition. Of course, for practical use, this glass transition must be significantly above ambient temperature. The main problem is the high viscosity of the materials. They must be heated to a very high temperature in order that they may be aligned in a uniform texture.

For the in situ polymerization method the aligned monomer is irradiated with UV light for polymerization. The curing process can be controlled by adding adequate quantities of photo-initiator or inhibitor in the monomer liquid crystal and by adjusting the UV intensity. One of the advantages of this method is that the material and cell do not have to be heated at such a high temperature (the liquid crystal monomer can have a lower melting temperature then liquid crystal polymers). Also, once the monomer is polymerized, it is very stable and its orientation can no longer be disturbed.

We now discuss the properties of acrylates (Broer, 1995; Broer and Heynderickx, 1990; Broer and Mol, 1991a,b; Broer et al., 1988, 1989a,b,c, 1993). Photopolymerization of aligned liquid crystalline monoacylate monomers and liquid crystalline diacrylate monomers is an elegant method for forming oriented and structured liquid crystal polymers. The liquid crystalline acrylates can easily be aligned using conventional alignment techniques such as by rubbing polyamides, and high thermal stability of the fixed orientation of the cured film is obtained. Figure 6.10 shows the chemical structure of typical representants (commercialized by Merck). The molecules are formed as a mesogenic unit with acrylate groups on both sides. The acrylate groups are separated from the central aromatic core by flexible methyl spacers of variable size. Some photo-initiator can be added to the liquid crystalline monomer in order to start the free-radical chain polymerization while illuminating with UV light. Alignment, photopolymerization kinetics, birefringence,

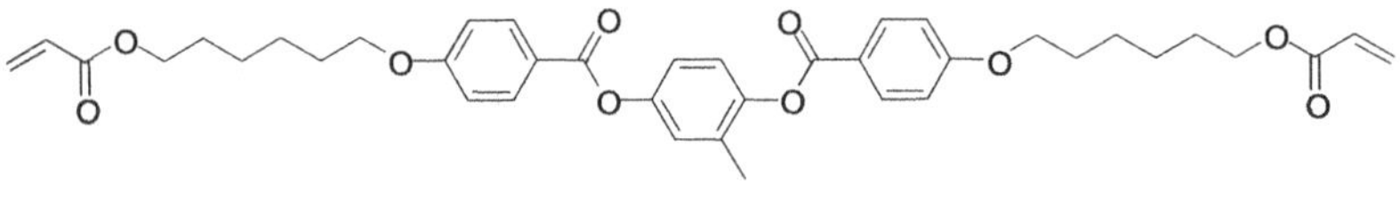

RM-82

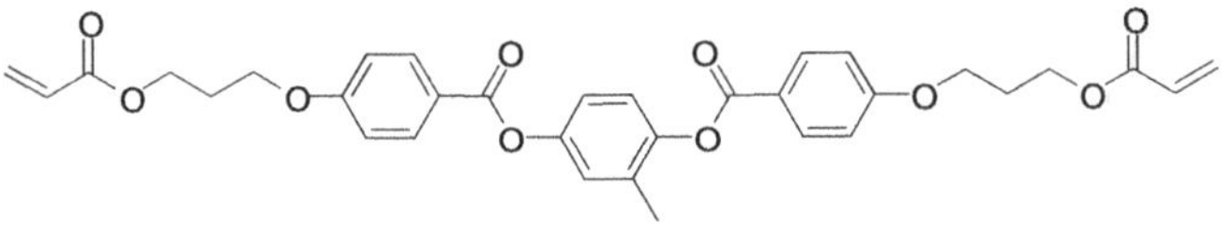

RM-257

Figure 6.10 Chemical structure of two liquid crystalline acrylates having side groups of different lengths (6-carbon spacers for RM-82 – Phillips nomenclature C6M – and 3-carbon spacers for RM-257).

temperature dependency, and phase transitions of these materials have already been reported extensively (Broer and Heynderickx, 1990; Broer and Mol, 1991a,b; Broer et al., 1989a,b,c, 1993; Schultz and Chartoff, 1998). RM-82 is nematic from 86° to 116°. For lower temperatures the material begins to crystallize rapidly. Mixtures of mesogenes generally show an extended nematic range. After polymerization in the nematic phase the molecules form an ordered crosslinked network. As the mesogenic units are uniaxially oriented along the rubbing direction in the nematic phase, the material shows the same optical properties as conventional liquid crystals. The effective temperature dependence of the birefringence for well-processed material is found to be extremely low for temperatures below 90°C.

Liquid crystal polymers form a solid and they are of course not suited for electro-optical applications as the molecules are linked together and consequently cannot be switched by applying any electric or magnetic field, as is necessary in liquid crystal displays. However, the fabrication of thermally stable birefringent optical (static) components could lead to promising applications for those materials. Based on the curing process of the liquid crystals and the thermal stability of the cured film, one can fabricate films with variations in the molecular orientation.

6.7 BIREFRINGENCE IN ISOTROPIC MATERIALS

Birefringence can occur in plastic components when polymer chains align during processing and under stress. For plastics where birefringence would disturb the optical properties, studies indicate that various process parameters may be adjusted for low birefringence. The quality is usually tested as the extinction ratio as viewed through crossed polarizers. The proportionality between birefringence and applied stress is the material's stress-optic coefficient. This is important in laser optics, compact discs, polarizer-light applications, and precision optics, which must minimize directional variations in the refractive index (Keyes, 2001). Table 6.2 qualifies different materials with respect to their sensitivity to stress birefringence. The data are taken from Ning (2000).

Molecular orientation of a polymer will cause the material to be birefringent too and the magnitude of the birefringence depends on the degree of molecular orientation and on the characteristics of the chemical bonds. A double bond has a

Table 6.2 Stress birefringence for different plastics and glass.

Polymer	Stress Birefringence
Acrylic	Fairly low
Styrene	Moderate
NAS	Low
PC	High
COC	Low
Glass	Very low

much greater polarizability than a single bond. Polarizability changes when bonds are stretched or bond angles are deformed by the application of stress, causing a birefringent effect that is clearly connected with the principal stress axes. This type of birefringence is sometimes called deformation birefringence, but this can be misleading when dealing with polymers because orientation is often produced by deformation. If a stress is maintained for an extended period, even a glassy polymer may suffer molecular conformational changes, causing a contribution to orientation birefringence. This is strictly different from stress birefringence, which should really refer only to reversible instantaneous changes that occur on loading or unloading. Note that with rubbery materials the major change in birefringence on application of stress is caused by bond orientation and that this also develops almost instantaneously. As an example we have shown in Section 1.1 a photograph (Fig. 1.1) of a stressed plastic part. For details on the subject the reader is referred to the specialized literature (Föppl and Mönch, 1972; Rohrbach, 1989; Dally and Riley, 1991).

6.8 FORM BIREFRINGENCE

Form birefringence may occur when the material contains two (or more) phases with different refractive indices, and is the result of distortion of the electromagnetic field at the interface. Thus, parallel rods or lamellae of one phase within a matrix of different refractive index provide suitable morphologies. The two phases in block copolymers often separate in this manner, are often caused to be anisotropic by processing (e.g., extrusion), and the period is often of the required order (Campbell and White, 1989; Hamley, 2000). We will develop here the general formulas following the ideas given in Yeh and Gu (1999). These lead to very useful formulas for the refractive index for different types of structures. One should be aware of the concept that is behind these formulas. The formulas can be applied to materials if the sizes of the composite parts are much smaller than the wavelengths of light. Usually there is no regular arrangement assumed because that can cause strong interferences. The composite can be composed of layers, rods, or just elliptical particles in a matrix. The reason why we treat this in detail here is to show the differences in origins of birefringence between form birefringence and anisotropic molecular polarizability.

For the calculation we consider that the form birefringence of a composite medium consists of uniformly distributed and aligned ellipsoids. This is of advantage because for an arbitrary shape of the body to be polarized the additional field produced within the material itself becomes a complicated function of position. A homogeneous field results only when the body has the shape of an ellipsoid. The ellipsoids are made of isotropic material with refractive index n_2 and embedded in an optically isotropic host medium of refractive index n_1. Both indices n_1 and n_2 might be complex values. To find the expressions for the refractive index of the composite, one considers a single dielectric ellipsoid made of material with a dielectric susceptibility χ. The dielectric ellipsoid is placed in a uniform electric field E_0 in vacuum. The electric field induces a dielectric polarization $\mathbf{P}$. The internal field is dependent on the permittivity of the ellipsoid, and also on the shape of the ellipsoid, through the depolarization factor Q_α. The three depolarization factors of an ellipsoid correspond to the three axial directions and add up to unity. The total field inside the ellipsoid is given as the external field reduced by the induced dielectric polarization corrected with the depolarization factor. To obtain a simple description, the principal coordinate system of the ellipsoid is chosen as the coordinate system. One finds for the electric field components that

$$E_x = E_{0x} - \frac{Q_x}{\varepsilon_0} P_x, \qquad E_y = E_{0y} - \frac{Q_y}{\varepsilon_0} P_y, \qquad E_z = E_{0z} - \frac{Q_z}{\varepsilon_0} P_z, \tag{6.1}$$

where Q_α are the depolarization factors in the principal axes of the ellipsoid in the x-, y-, and z-directions. Note that $\mathbf{E}$ is the electric field inside the dielectric ellipsoid, whereas $\mathbf{E}_0$ is the constant external electric field outside at infinity. For isotropic materials the polarization $\mathbf{P}$ is related to the local electric field $\mathbf{E}$ by the following definition:

$$\begin{pmatrix} P_x \\ P_y \\ P_z \end{pmatrix} = \varepsilon_0 \begin{pmatrix} \chi & 0 & 0 \\ 0 & \chi & 0 \\ 0 & 0 & \chi \end{pmatrix} \begin{pmatrix} E_x \\ E_y \\ E_z \end{pmatrix}. \tag{6.2}$$

For anisotropic materials the situation is more complicated (Cox et al., 1998). Using Equations (6.1) and (6.2) we obtain the induced polarization in the ellipsoid as a function of the external field $\mathbf{E}_0$:

$$P_x = \frac{\varepsilon_0 E_{0x}}{1/\chi + Q_x}, \qquad P_y = \frac{\varepsilon_0 E_{0y}}{1/\chi + Q_y}, \qquad P_z = \frac{\varepsilon_0 E_{0z}}{1/\chi + Q_z}. \tag{6.3}$$

The composite is usually made of different substances that occupy some volume. The definition of a composite can be described by defining volume percentages. We introduce therefore the induced dipole moment p of the whole ellipsoids of volume V. This can be written in its components as

$$p_\alpha = V \varepsilon_0 E_{0\alpha} \frac{1}{1/\chi + Q_\alpha} \qquad (\alpha = x, y, z). \tag{6.4}$$

We now consider the effective refractive index of the composite medium. The total induced polarization can be obtained by adding all the individual induced dipole moments and then dividing by the volume of the composite medium. This leads to

$$P_\alpha = \frac{\sum p_\alpha}{V} = f \varepsilon_0 E_{0\alpha} \frac{1}{1/\chi + Q_\alpha} \qquad (\alpha = x, y, z), \tag{6.5}$$

where the summation is over all the ellipsoids. The fill factor $f\,(0 \leq f < 1)$ represents the volume fraction between filler and the matrix material. With the induced polarization, we can now define the principal effective dielectric susceptibilities as

$$P_\alpha = \varepsilon_0 \chi'_{\alpha\alpha} E'_\alpha \qquad (\alpha = x, y, z), \tag{6.6}$$

where E'_α is the effective electric field inside a large ellipsoid of composite medium that contains many ellipsoids. From the discussion above, the effective electric field is known as

$$E'_\alpha = E_{0\alpha} - \frac{Q_\alpha}{\varepsilon_0} P_\alpha \qquad (\alpha = x, y, z). \tag{6.7}$$

Using Equation (6.5) for the induced polarization in the composite medium, P_α, we obtain

$$E'_\alpha = E_{0\alpha} - \frac{Q_\alpha}{\varepsilon_0} f \varepsilon_0 E_{0\alpha} \frac{1}{1/\chi + Q_\alpha} = E_{0\alpha}\left(1 - \frac{fQ_\alpha}{1/\chi + Q_\alpha}\right) \qquad (\alpha = x, y, z). \tag{6.8}$$

Now we substitute P_α from Equation (6.5) and E'_α from Equation (6.8) into Equation (6.6) and eliminate $E_{0\alpha}$ to obtain

$$f \varepsilon_0 E_{0\alpha} \frac{1}{1/\chi + Q_\alpha} = \varepsilon_0 \chi'_{\alpha\alpha} E_{0\alpha}\left(1 - \frac{fQ_\alpha}{1/\chi + Q_\alpha}\right) \qquad (\alpha = x, y, z), \tag{6.9}$$

and further on we find the susceptibility

$$\chi'_{\alpha\alpha} = \frac{f}{1/\chi + Q_\alpha(1-f)} \qquad (\alpha = x, y, z). \tag{6.10}$$

Rewritten in the components of the dielectric tensor, we find

$$\varepsilon_{\alpha\alpha} = 1 + \chi'_{\alpha\alpha} = 1 + \frac{f}{1/\chi + Q_\alpha(1-f)} = \frac{1/\chi + Q_\alpha(1-f) + f}{1/\chi + Q_\alpha(1-f)} \qquad (\alpha = x, y, z). \tag{6.11}$$

This equation can now be generalized for the case where the ellipsoids of refractive index are in a host medium of refractive index n_1. This is done by replacing χ and $\chi'_{\alpha\alpha}$ in the following way:

$$\chi \rightarrow \left[\frac{n_2^2}{n_1^2} - 1\right] \quad \text{and} \quad \chi'_{\alpha\alpha} \rightarrow \left[\frac{n_\alpha^2}{n_1^2} - 1\right]. \tag{6.12}$$

Thus one obtains that the effective principal refractive indices are given by

$$\frac{n_\alpha^2}{n_1^2} - 1 = \frac{f\,[(n_2^2/n_1^2) - 1]}{1 + (1-f)Q_\alpha[(n_2^2/n_1^2) - 1]} \qquad (\alpha = x, y, z), \tag{6.13}$$

where f is again the fill factor defined as the fraction of volume occupied by the ellipsoids. The depolarization factors Q_α account for the local screening of the electric fields due to the dielectric structure (shape) of the ellipsoids. The subscript α refers to the principal axes x, y, z of the ellipsoids. The depolarization factor Q_α depends on the principal axis. For general ellipsoids with three different semi-axes, the depolarization factors can be written using elliptical integrals and are all different. In this case the composite medium has biaxial anisotropy. For ellipsoids of revolution (spheroids) with z as the symmetry axis, the depolarization factors depend on the ellipticity m, given by the ratio of the lengths in the z- and x-directions. In this case the ellipsoids are cylindrically symmetric about the z-axis. The uniaxial symmetry implies that $Q_x = Q_y$, leading to a composite medium with uniaxial anisotropy. Accordingly, depolarization factors related to the ordinary and extraordinary waves may be defined. These depolarization factors are designated Q_o and Q_e, respectively. Simple analytic expressions for the depolarization factors can be found in Table 6.3 (Yeh and Gu, 1999).

Using the equations in Table 6.3, we can obtain the effective refractive indices of differently structured media as shown in Figure 6.11. For parallel cylinders the same equations give

$$n_e^2 = fn_2^2 + (1-f)n_1^2 \qquad \text{and} \qquad n_0^2 = \frac{f[n_2^2 - n_1^2] + n_1^2 + n_2^2}{2 + (1-f)[(n_2^2/n_1^2) - 1]}. \tag{6.14}$$

For the layered media with refractive indices n_1 and n_2 one finds and n_e and n_o as

$$n_o^2 = (1-f)n_1^2 + fn_2^2 \qquad \text{and} \qquad \frac{1}{n_e^2} = (1-f)\frac{1}{n_1^2} + f\frac{1}{n_2^2}. \tag{6.15}$$

Table 6.3 Depolarization factor *Q* for different geometries of composites.

Structure Type	Depolarization Factors
Parallel cylinders, $m \gg 1$	$Q_c = Q_e = 0,\ Q_a = Q_b = Q_o = \frac{1}{2}$
Prolate spheroids, $m > 1$ (cigar-like, ellipsoids)	$Q_c = Q_e = \frac{1}{m^2-1}\left\{\frac{m}{\sqrt{m^2-1}}\ln\left(m+\sqrt{m^2-1}\right) - 1\right\}$ $Q_a = Q_b = Q_o = \frac{1}{2}(1 - Q_c)$
Spheres, $m = 1$	$Q_a = Q_b = Q_c = \frac{1}{3}$
Oblate spheroids, $m < 1$ (pancake)	$Q_c = Q_e = \frac{1}{1-m^2}\left\{1 - \frac{m}{\sqrt{1-m^2}}\arccos(m)\right\}$ $Q_a = Q_b = Q_o = \frac{1}{2}(1 - Q_c)$
Parallel plates, $m \ll 1$	$Q_c = Q_e = 1,\quad Q_a = Q_b = Q_o = 0$

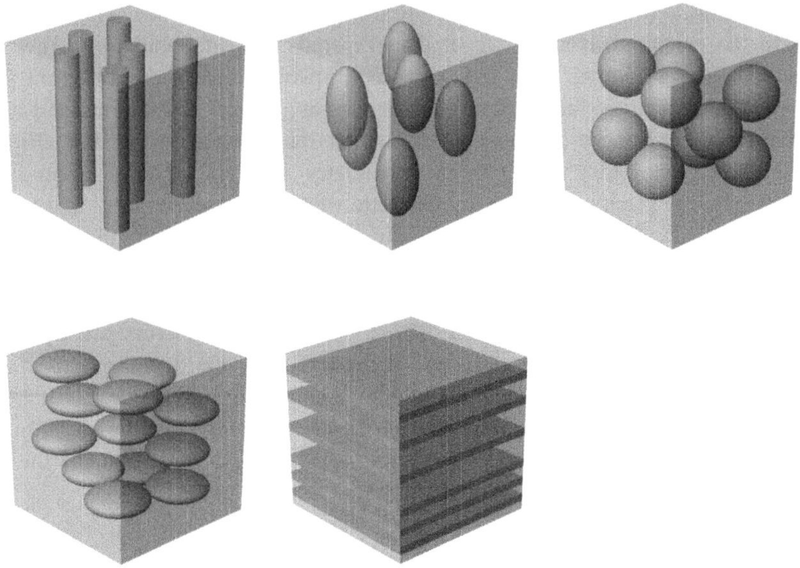

Figure 6.11 Composite materials: cylinders, ellipses, spheres, oblates, and plates. Note that the volume fraction of composition is not the same for different structures.

To give an example we assume that f is 0.5. For the values of the refractive indices of the basic materials we take $n_1 = 1.56$ and $n_2 = 1.72$. Using Equation (6.15), we obtain $n_o = 1.642$ and $n_e = 1.634$, giving a negative birefringence of $\Delta n = -0.0078$. This is a convenient way of designing birefringent compensation films for displays (Yeh and Gu, 1999).

Once again, the formulas can only be applied to materials if the size of structure elements in the composite is much smaller than the wavelengths of light. No regular arrangement is permitted because that can cause strong interference. The concept has to be strictly distinguished from the molecular polarizability discussed in the next chapter.

6.9 ORDER-INDUCED BIREFRINGENCE

The main source of birefringence in polymers and liquid crystals is the order in the system. For polymers this can be introduced by stretching the polymer network or shearing, and so is mainly due to the manufacturing process of plastic parts. For liquid crystals and liquid crystal polymers the order is present due to their molecular interaction potentials. There are three different types of order: positional order, bond orientational order, and orientational order. Positional order describes the extent to which the position of an average molecule or group of molecules shows translational symmetry. Bond orientational order describes the order of centers of nearest-neighbor molecules without requiring a regular spacing along that line. Orientational

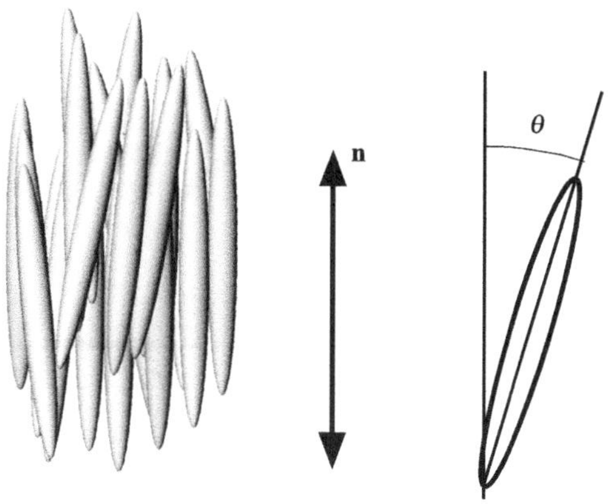

Figure 6.12 The nematic director indicates the average orientation of a small volume containing several molecules.

order measures the tendency of the molecules to align along the director on a long-range basis. To quantify how much order is present in a material, an order parameter S is defined. In the nematic phase the molecules are mainly rod-like, with their long axes aligned approximately parallel to one another. In homogeneous nematic liquid crystal, the director is a constant. In inhomogeneous nematic liquid crystal, it becomes a function of the position. A unit vector that represents the long axis of each molecule is defined. The director **n** is the statistical average of the unit vectors over a small volume element around a point. The order parameter S is defined as a statistical average of the directions of the long axes of individual molecules from the direction of the director as seen in Figure 6.12. In the nematic phase, the order parameter is defined as the statistical average of the angle between each molecule and the director:

$$S = \frac{1}{2}\langle 3\cos^2\theta - 1\rangle \tag{6.16}$$

where θ is the angle between the molecules and the director. The brackets denote an average over all of the molecules in the sample. In an isotropic liquid, the average of the $\cos^2\theta$ term is $\frac{1}{3}$, and therefore the order parameter is equal to zero. For a perfect crystal, the order parameter evaluates to one. Typical values for the order parameter of a liquid crystal range between 0.3 and 0.9. The imperfect order is a result of kinetic molecular motion. The order parameter changes with temperature. This is illustrated in Figure 6.13 for a typical nematic liquid crystal material. For some of the very well characterized materials there are approximate formulas that describe the order parameters as a function of temperature. For the nematic cyanobiphenyl 5CB, a good approximation is

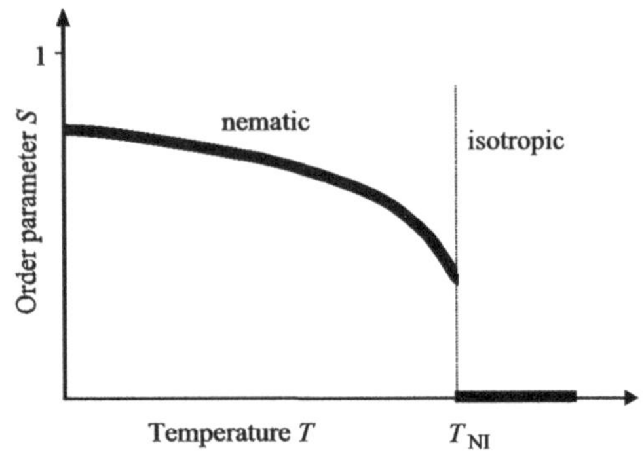

Figure 6.13 Typical temperature dependence of the order parameter for a nematic liquid crystal. T_{NI} is the phase transition temperature, nematic-isotropic.

given by the formula (Khoo and Wu, 1993)

$$S = \left(1 - \frac{T}{T_{NI}}\right)^{0.142} \quad \text{(approximate order parameter for 5CB).} \qquad (6.17)$$

The given approximation is only valid for temperatures T sufficiently smaller than the phase transition temperature T_{NI}. In a lot of practical cases this assumption is fulfilled and Eq. (6.17) gives a reasonable approximation. It does not give a exact model of the behavior at transition that shows a jump in the order parameter for first order phase transitions.

Figure 6.13 gives an illustration of the temperature behavior of the order parameter for a nematic–isotropic phase transition. To illustrate the effect of the order parameter on a molecular scale, Figure 6.14 shows the arrangement of

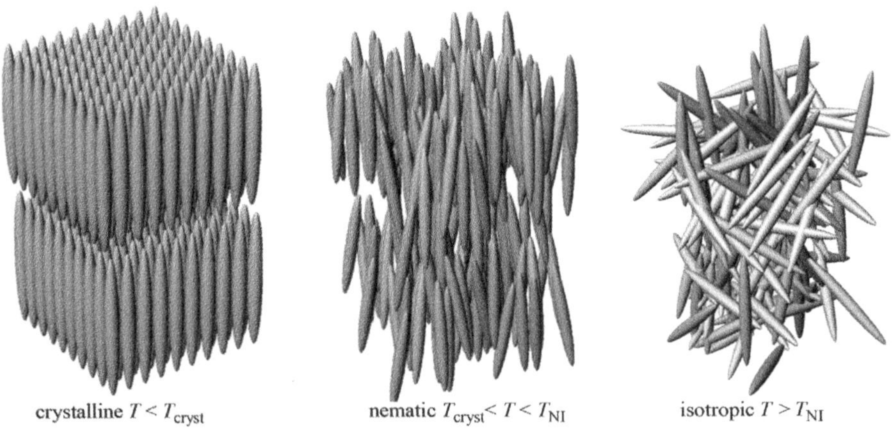

Figure 6.14 Temperature influence on the order in a nematic system. At high temperature in the isotropic phase no order is present and the order parameter is zero. Lowering the temperature leads to orientational order in the nematic phase, and if crystallization occurs, positional and orientational order is found.

molecules within a small volume. The isotropic state shows no order and becomes orientational ordered in the nematic phase. When crystalline or higher order smectic phases appear, the material has additional order, as positional or bond orientational order. We are mainly interested in optical properties and will continue here to discuss the link between the order parameter, and the optical and dielectric anisotropy. We start with a discussion of the properties of a single molecule. In an isotropic fluid or gas, the refractive index dispersion is described by the well-known Lorentz–Lorentz formula (Böttcher, 1952):

$$\frac{\varepsilon_{\text{iso}} - 1}{\varepsilon_{\text{iso}} + 2} = \frac{\rho N_{\text{A}}}{3\varepsilon_0 M}\alpha_{\text{iso}} \tag{6.18}$$

where N_{A} is number of molecules per mole (Avogadro's number), M is the molar mass, ε_0 is the dielectric constant, α_{iso} is the mean polarizability, and ε_{iso} is the dielectric permittivity of the media. Equation (6.18) is obtained in this model by considering the polarizability of the molecules at the center of a spherical cavity surrounded by continuous isotropic dielectric. For optical frequencies, Equation (6.18) can be modified by using the equality $\varepsilon = n^2$:

$$\frac{n_{\text{iso}}^2 - 1}{n_{\text{iso}}^2 + 2} = \frac{\rho N_{\text{A}}}{3\varepsilon_0 M}\alpha_{\text{iso}}. \tag{6.19}$$

For anisotropic fluids like liquid crystals, this model has to be modified. The additional problem arises that the local field is anisotropic. There are a few theoretical approximations to describe the local field in liquid crystals. For a more quantitative description based on experimental results, it seems to be sufficient to use the equations of Vuks (Vuks, 1966; Li, 2004). These equations are relatively often applied in optical studies on liquid crystals, although the local field in this approximation is assumed to be isotropic. In a simplified view we can treat the molecule as a rotational body with two different electrical polarizabilities. We assume therefore uniaxial properties of the molecule. Two different directions are distinguished, leading to two equations:

$$\frac{n_{\text{e}}^2 - 1}{\overline{n^2} + 2} = \frac{\rho N_{\text{A}}}{3\varepsilon_0 M}\overline{\alpha_{\text{p}}} \quad \text{and} \quad \frac{n_{\text{o}}^2 - 1}{\overline{n^2} + 2} = \frac{\rho N_{\text{A}}}{3\varepsilon_0 M}\overline{\alpha_{\text{s}}}. \tag{6.20}$$

The extraordinary refractive index n_{e} and the ordinary refractive index n_{o} are related here to the molecular polarizabilities parallel and perpendicular to the molecule α_{p} (p, parallel) and α_{s} (s, german "senkrecht"). An averaged refractive index represents the local field correction, with

$$\overline{\varepsilon} = \frac{1}{3}\left(\varepsilon_{\text{p}} + 2\varepsilon_{\text{s}}\right) \quad \text{and} \quad \overline{n^2} = \frac{1}{3}\left(n_{\text{e}}^2 + 2n_{\text{o}}^2\right). \tag{6.21}$$

This average is based on the assumption of a rotational symmetric body with two main axes n_{e} and n_{o}. For an elongated ellipsoid in Cartesian coordinates the index

Table 6.4 Approximate dependence of some properties of the liquid crystal material on the order parameter.

Parameter	Nomenclature	Proportionality
Elastic constant	K_{ii}	S^2
Birefringence	Δn	S
Dielectric anisotropy	$\Delta\varepsilon$	S
Magnetic anisotropy	$\Delta\chi$	S
Viscosity anisotropy	$\Delta\eta$	S

ellipsoid has two axes that are n_o long and one of length n_e. This leads to the weight factors in front of the refractive indices. Because the electric field has to be averaged, the dielectric susceptibilities are of importance and therefore the square of the refractive index appears. To obtain results for the nematic liquid crystal as a macroscopic ensemble, we need to link the molecular properties to the director of the nematic liquid crystal. The director is the statistical average of the molecular unit vectors over a small volume element. Using the order parameter S, the averaged polarizability parallel α_l and perpendicular α_t to the director can be expressed as

$$\overline{\alpha}_l = \overline{\alpha} + \frac{2}{3}\left(\alpha_p - \alpha_s\right) S \qquad \overline{\alpha}_t = \overline{\alpha} - \frac{1}{3}\left(\alpha_p - \alpha_s\right) S, \tag{6.22}$$

where $\overline{\alpha} = \frac{1}{3}\left(\alpha_p + 2\alpha_s\right)$ is the average polarizability and α_l and α_t are called the longitudinal and transversal polarizabilities of the small volume defining the director. From Equations (6.20) and (6.22) one finds

$$\frac{n_e^2 - n_o^2}{\overline{n^2} + 2} = \frac{\rho N_A}{3\varepsilon_0 M}\left(\alpha_p - \alpha_s\right) S. \tag{6.23}$$

The term on the left-hand side of Equation (6.23) is a measure of the birefringence. We can conclude that the birefringence of a liquid crystal is determined by the orientational order S, by the anisotropy of the molecular polarizability $(\alpha_p - \alpha_s)$, and by the reciprocal molar volume ρ/M. Equation (6.23) can also serve to measure the order in the system by determining the birefringence.

The order parameter influences all basic quantities of ordered systems. Table 6.4 shows the dependence of other important material characteristics on the order parameter (Crawford, 2002).

6.10 OPTICAL PROPERTIES OF LIQUID CRYSTALS AND ORIENTED POLYMERS

In this section we discuss the basic optical properties of birefringent organics from a technical point of view. Every application needs different material parameters to be optimized. For display applications the material choice is extremely important with respect to electrical, mechanical, and optical properties. For static optical components, one is mainly interested in a high birefringence and good optical quality, that is, transmission and uniformity of the liquid crystal material. If a nematic phase is used, it is also important that the liquid crystal has a nematic phase over a large temperature range if the final device is used in difficult ambient conditions. When the liquid crystal is not switched, the elastic properties, viscosity, and electric properties are less relevant. However, the viscosity of the liquid crystal should not be too high to be able to fill cells and the elastic properties should behave normally to speed up the texture relaxation and avoid sensitivity to mechanical disturbances. The main parameters for optical applications have already been defined in Section 6.4. These are the refractive index and birefringence, dispersion (Abbe value), absorption (transmittance), dichroism, and scattering (haze).

6.10.1 Scattering

Thick nematic liquid crystal samples with thicknesses above 50 μm generally look milky. This indicates that light is being scattered. The scattering is due to thermal fluctuation, which causes a random variation of the director in all points in the liquid crystal. The scattered intensity I_{scat} may be expressed proportionally as (Gennes and Prost, 1993)

$$I_{scat} \propto \frac{\Delta\varepsilon^2 d}{\lambda^4} \tag{6.24}$$

where d is the thickness of the liquid crystal slab, $\Delta\varepsilon$ the dielectric anisotropy at optical frequencies, and λ the wavelength of light. As $\Delta\varepsilon$ is strongly related to the birefringence, one expects strongly increasing scattering losses for higher birefringent materials. The scope of this book is to discuss the optical properties of structured thin films of liquid crystals. Macroscopic scattering effects that are linked to the fluctuations of the liquid crystal are not of further interest here.

6.10.2 Refractive Index

Liquid crystals for optics are mainly used at visible wavelengths and should therefore be transparent. If near UV or IR applications are envisaged, special materials often have to be used. There are only a few materials that can be used in the ultraviolet because of absorption and the risk of decomposition. High birefringence liquid crystal mixtures like biphenyls, with the prominent examples of E7 and BL006 (from Merck), are transparent above 400 nm and can be used in the infrared too. The compositions of relevant mixtures are given in Table 6.5. Better transmission in the UV is obtained with mixtures based only on cyclohexylcyclohexanes like ZLI 1695 or ZLI 1167 (from Merck) (Stegemeyer, 1994). Unfortunately they do have a limited nematic

Table 6.5 Chemical composition of selected liquid crystal mixtures.

5CB	4-n-Pentyl-4′-Cyanobiphenyl
E7	Ternary eutectic mixture of 47% 5CB (pentylcyanobiphenyl), 25% 7CB (heptyloxycyanobiphenyl), 18% 8OCB (octyloxycyanobiphenyl), and 10% T15
ZLI 1167	Ternary eutectic mixture of three p-cyano-p′-alkylcyclohexylcyclohexanes with propyl, pentyl, and heptyl as the alkyl groups CCH-3/CCH-5/CCH-7 36:34:30
ZLI 1132	Mixture of three trans-4-alkyl-(4-cyanophenyl)cyclohexanes with propyl, pentyl, and heptyl as the alkyl groups and trans-4-pentyl-(4′-cyanobiphenyl-4)cyclohexane 24/36/25:15

range and cannot be used for practical applications. A good compromise are mixtures based on phenylcyclohexanes like ZLI 1132 (Merck), which is a valuable candidate with high transmission in the near UV and with a nematic range from −6 to 70°C. We have selected six different materials from the data delivered by the supplier (Merck). The materials have either high birefringence or good transmission, and large temperature ranges for the nematic phase. The relevant parameters of these materials are summarized in Table 6.6.

All commercial liquid crystals consist of aromatic rings that have saturated bonds: as CH_3, CH_2, and C_6H_{12}. Such bonds show typical electronic transitions and have absorptions in the deep UV with resonance wavelengths of $\lambda_0 < 180$ nm. Unsaturated bonds in benzene rings C_6H_6 have transitions that show absorption in the UV at different wavelengths. Usually there are several wavelengths, with a first absorption band at $\lambda_1 \sim 180$–210 nm and a second at $\lambda_1 > 250$ nm. A model that takes into account several absorption bands is particularly suited for the description of the optical properties. The following equation can be used to model the molecular polarizabilities as a function of wavelengths (Khoo and Wu, 1993)

$$\alpha_{\mathrm{p,s}}(\lambda) \approx \sum_i (f_{\mathrm{p,s}})_i \frac{\lambda^2 \lambda_i^2}{\lambda^2 - \lambda_i^2}. \tag{6.25}$$

The wavelength λ_i gives the position of the absorption band and the factor $(f_{\mathrm{p,s}})_i$ its strength and form. Usually, three bands are assumed, with their resonance wavelengths at $\lambda_0 \sim 120$, $\lambda_1 \sim 200$, and $\lambda_2 \sim 280$ nm. In off-resonance cases the one band model is sufficient. Then, only one resonance wavelength is assumed to be present in the UV. One obtains for the molecular polarizabilities the simplified formula

$$\alpha_{\mathrm{p,s}}(\lambda) \approx f_{\mathrm{p,s}} \frac{\lambda^2 \lambda_{\mathrm{p,s}}^2}{\lambda^2 - \lambda_{\mathrm{p,s}}^2}, \tag{6.26}$$

Table 6.6 Selected properties of liquid crystals and liquid crystal mixtures.

	Units	K15 (5CB)	E7 (BL001)	ZLI 1132 ML1002	ZLI 1167	BL 006	MLC 6608	RM 82 (C6M)	CLM4039 **SLM90032**	CLM4070 **SLM90034**	Glass
		Pure substance: pentylcyanobiphenyl	Mixture of three components						Cholesteric side chain polymer	Cholesteric side chain polymer	BK7
Refractive index, n											
Ordinary, n_o		1.5442 (515 nm 25°C)	1.5231 (577 nm 20°C)	1.493 (589 nm 20°C)	1.46 (589 nm 35°C)	1.53 (589 nm 20°C)	1.4748 (589 nm 20°C)	1.53 (589 nm 20°C)	1.520 (crosslinked RT)	1.531 (crosslinked RT)	1.517 (587.6 nm)
Extraordinary, n_e		1.736 (515 nm 25°C)	1.75 (577 nm 20°C)	1.6326 (589 nm 20°C)	1.52 (589 nm 35°C)	1.816 (589 nm 20°C)	1.5578 (589 nm 20°C)	1.67 (589 nm 20°C)	1.607 (crosslinked RT)	1.645 (crosslinked RT)	
Abbe value, V_d		76 (ord), 67 (ext)		130 (ord), 86 (ext)				80 (ord), 80 (ext)			64.17
Phase transition temperature											
Nematic–isotropic	°C	35.3	60.5	71	83	113	90	116	140	135	
Nematic–crystalline	°C	24	−10	<−40	32	−15		83 (T_m)	3 (T_g)	−2 (T_g)	
Elastic constant											
$K11$	pN	6.4	12	10.1		17.9	16.7				
$K22$	pN	3	9	5.6			7.0				
$K33$	pN	10	19.5	19.7		33.5	18.1				
Dielectric constant (1 kHz 20°C)											
ε parallel		19.7	19.6	15	7.5 (35°C)	22.8					
ε perpendicular		6.7	5.1	4.7	3.6 (35°C)	5.5	186 (20°)				
Viscosity	mPas		39 (20°C) 161 (0°C)	28 (20°C) 110 (0°C)	30 (35°C)	71 (20°C)		200 (75°C)	400 (100°C)	300 (100°C)	
Advantages		Pure substance	Often used as reference material	High UV transmittance	Low birefringence	High birefringence and high T_{NI}	Negative dielectric anisotropy $\Delta\varepsilon = -4.2$	Polymerizable mesogen	Polymerizable cholesteric mixture ($\lambda_p = 390$ nm)	Polymerizable cholesteric mixture ($\lambda_p = 700$ nm)	

Sources: Broer and Mol (1991a,b), Khoo and Wu (1993), Stegemeyer (1994), Yeh (1999), Wang (2004); product specifications form Merck KG, Darmstadt, Germany.

taking into account that the resonance frequencies for parallel (p) or perpendicular (s) electric fields might be different. Their magnitude depends on the initial and final excitation states of the molecules, the transition, and the governing selection rules. In many cases f_p is larger than f_s for rod-like molecules. Often, λ_p becomes longer than λ_s and a positive birefringence results. But this is not generally true. Equation (6.20) link the refractive indices to the molecular polarizabilities as

$$\frac{n_{e,o}^2 - 1}{\overline{n^2} + 2} = \frac{\rho N_A}{3\varepsilon_0 M}\alpha_{p,s} \qquad \text{with} \qquad \overline{n^2} = \frac{n_e^2 + 2n_o^2}{3}. \tag{6.27}$$

From that one can equate n_e and n_o as

$$n_{e,o} = \sqrt{1 + \frac{\dfrac{\rho N_A}{\varepsilon_0 M}\alpha_{p,s}}{1 - \dfrac{1}{9}\dfrac{\rho N_A}{\varepsilon_0 M}(2\alpha_s + \alpha_p)}}. \tag{6.28}$$

This can be developed in a Taylor series for small $(N_A \alpha)$. If only the first-order term is considered, one gets a direct proportionality between the refractive index and the molecular polarizabilities:

$$n_{e,o} \approx 1 + \frac{\rho N_A}{2\varepsilon_0 M}\alpha_{p,s}. \tag{6.29}$$

We assume that averaging over the statistical order, which transforms the molecular polarizability into the dependence for the liquid crystalline phase, will not change this dependence. Then one can replace the polarizability by Equation (6.25) and obtain

$$n_{e,o} \approx 1 + \sum_i (g_{e,o})_i \frac{\lambda^2 \lambda_i^2}{\lambda^2 - \lambda_i^2}. \tag{6.30}$$

The temperature dependence of g determines the temperature dependence of the refractive index. It contains the order parameter because of the statistical average. Equation (6.30) involves two adjustable parameters for each refractive index: the position of the resonance λ_i and its strength g_i. By measuring the refractive index of a liquid crystal at two distinct wavelengths and at a constant temperature, these two parameters can be evaluated. As an example we show in Figure 6.15 the refractive index of 5CB. The chemical formula for 5CB is

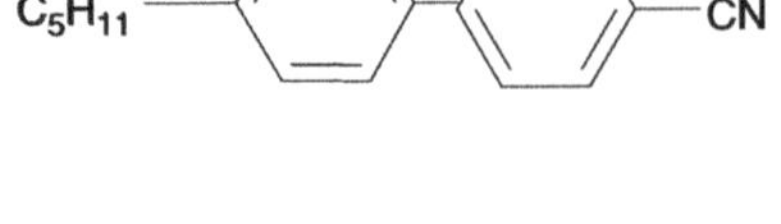

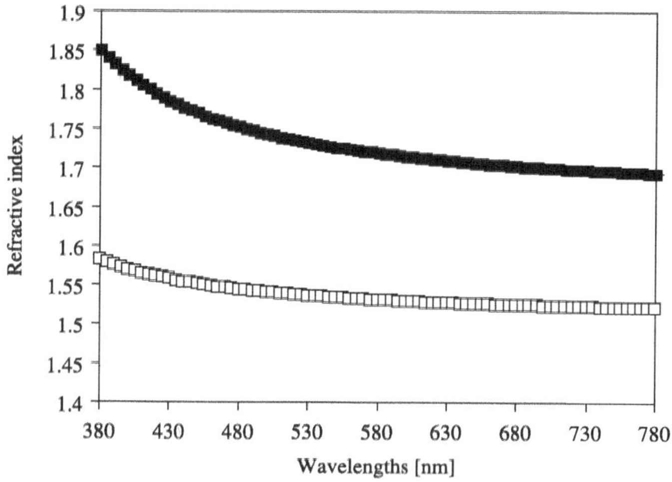

Figure 6.15 Refractive index for 5CB for 25.1°C calculated with the heuristic two-band model.

With a two band-model the refractive index can be calculated at 25.1°C using the following formulas:

$$n_e = 1 + n_{e0} + g_{e1}\frac{\lambda^2\lambda_1^2}{\lambda^2 - \lambda_1^2} + g_{e2}\frac{\lambda^2\lambda_2^2}{\lambda^2 - \lambda_2^2}$$

and

$$n_o = 1 + n_{o0} + g_{o1}\frac{\lambda^2\lambda_1^2}{\lambda^2 - \lambda_1^2} + g_{o2}\frac{\lambda^2\lambda_2^2}{\lambda^2 - \lambda_2^2}, \qquad (6.31)$$

where $\lambda_1 = 210$ nm, $\lambda_2 = 282$ nm, $n_{e0} = 0.4552$, $g_{e1} = 2.33 \times 10^{-6}\,\text{nm}^{-2}$, $g_{e2} = 1.4 \times 10^{-6}\,\text{nm}^{-2}$, $n_{o0} = 0.4136$, $g_{o1} = 1.35 \times 10^{-6}\,\text{nm}^{-2}$, $g_{o2} = 0.47 \times 10^{-6}\,\text{nm}^{-2}$ (Khoo and Wu, 1993). For the one-band model the refractive index dispersion is

$$n_{e,o} \approx 1 + g_{e,o}\frac{\lambda^2\lambda_{e,o}^2}{\lambda^2 - \lambda_{e,o}^2}. \qquad (6.32)$$

At an off-resonance region where λ is much larger than $\lambda_{e,o}$, Equation (6.32) can be expanded into a power series with the argument $\lambda_{e,o}/\lambda$ as

$$n_{e,o} = A_{e,o} + \frac{B_{e,o}}{\lambda^2} + \frac{C_{e,o}}{\lambda^4} + \cdots, \qquad (6.33)$$

where $A_{e,o}$, $B_{e,o}$, and $C_{e,o}$ are constants. For simplicity only the first two terms are usually kept for fitting the experimental data. Equation (6.33) is the equivalent of the well-known Cauchy formula for isotropic fluids. Precise measurements of the

Table 6.7 Parameters to determine the refractive indices of selected liquid crystal mixtures.

Liquid Crystal Mixture	λ^* (nm)	G (10^{-6} nm^{-2})
5CB	244	2.6
E7	250	3.06
ZLI 1132	198	3.15

refractive index are rare. For practical use a dispersion relation for the birefringence is important. Based on the models developed above, several formulas can be assumed. The difficulty is that in applications, mixtures of different liquid crystal molecules are used and the resonance wavelengths represents in the formulas several resonance frequencies. A very useful formula for the off-resonance region of the spectra is

$$\Delta n = G(T)\frac{\lambda^2\lambda^{*2}}{\lambda^2 - \lambda^{*2}}, \tag{6.34}$$

where $G(T)$ is a temperature-dependent constant and λ^* is the mean or "average" resonance wavelength. Some values are given in Table 6.7 (Khoo and Wu, 1993).

6.10.3 Anisotropic Absorption

We have seen that materials can show birefringence which results in different refractive indices for different direction of propagation and polarization of light. There are materials that exhibit in addition to the birefringence a different absorption along different directions of the polarization. A well known example were such effects are used are the sheet polarizers from Polariod. A common picture to describe such elements is the assumption that molecules or crystals that show two different absorption coefficients are mixed into a host material that can be oriented. The performance of the final component with respect to their contrast depends on the order parameter of the host system and the dichroic ratio of the dye. The dichroic ratio is the difference of the molecular absorption of the molecule along two principle axis. The order parameter describes the quality of the orientation of such principle axis in the structure. It is possible to fabricate optical components using dye mixtures with high dichroism and inducing very high order. The order is induced by stretching or compressing polymer films. The result are for instance high efficient sheet polarizers with extinction ratio of more than 100 and single component transmission of nearly 50%. Consult Chapter 7.4 for details. The direction of transmittance in such a system can be along the ordinary refractive index or along the extraordinary refractive index leading to O-type or E-type polarization components. The Polaroid sheets HN22 and HN38 are for instance O-type polarizers. The same concept can be used with liquid crystal as the orienting host material. There are liquid crystal mixture available with moderate contrast ratio. Usually the performance is less convincing due to the relatively low order parameter of typically $S = 0.8$ for commercial liquid crystal mixtures.

The description of the optical properties can be done in the following way. We assume a component that is uniaxial and has planar orientation. The optical axis is perpendicular to the propagation direction. Two directions of different absorption are found and we write down the transmitted intensity for the two different polarization directions with the help of an absorption coefficient along the extraordinary direction α_e and ordinary direction α_o. One finds

$$I_o = I_0 e^{-\alpha_o d} \tag{6.35a}$$

$$I_e = I_0 e^{-\alpha_e d} \tag{6.35b}$$

where I_o and I_e are the transmitted intensities in ordinary and extraordinary direction respectively. The transmitted intensity depends also on the thickness d of the layer. The absorption coefficients can be brought in relation to the refractive index. We know already that the linear polarization state that passed a polarization component suffers a retardation. If we assume that we are well aligned with the principled axis of the component we can write down the Jones matrix for this situation that corresponds to that of an retarder with refractive index n_1 and n_2. In Table 4.2 we read

$$\boldsymbol{J} = \begin{pmatrix} e^{-i2\pi n_1 d/\lambda} & 0 \\ 0 & e^{-i2\pi n_2 d/\lambda} \end{pmatrix} \tag{6.36}$$

Up to now we have identified the refractive indices n_1 and n_2 with a real value and therefore the Jones matrix in Equation (6.36) does not influence the transmitted intensity. If the refractive index is allowed to be complex absorption in both principle directions can be simulated. If we set for the refractive index

$$n_1 = n_e + i\,k_e \tag{6.37a}$$

$$n_o = n_o + i\,k_o \tag{6.37b}$$

the Jones matrix (6.36) becomes

$$\boldsymbol{J} = \begin{pmatrix} e^{-2\pi k_e d/\lambda}\, e^{-i2\pi n_e d/\lambda} & 0 \\ 0 & e^{-2\pi k_o d/\lambda}\, e^{-i2\pi n_o d/\lambda} \end{pmatrix} \tag{6.38}$$

To determine the relation between the complex part of the refractive index k_i and the absorption coefficient α_i we consider the case of a incident light with the Jones vector $\boldsymbol{V}_i$ of intensity I_0

$$\boldsymbol{V}_i = I_0 \begin{pmatrix} 0 \\ 1 \end{pmatrix} \tag{6.39}$$

Passing though the element described by Equation (6.38) and calculate the intensity by using Equation (4.3) we get

$$I_o = I_0 e^{-4\pi k_o d/\lambda} \tag{6.40a}$$

A similar equation follows for the case of transmission of the extraordinary component as

$$I_e = I_0 e^{-4\pi k_e d/\lambda} \tag{6.40b}$$

If we compare Equations (6.40) and (6.35) we identify the following relation between the absorption coefficients and the imaginary parts of the refractive indices:

$$\alpha_e = \frac{4\pi k_e}{\lambda} \quad \text{and} \quad \alpha_o = \frac{4\pi k_o}{\lambda} \tag{6.41}$$

Note that in this notation the wavelengths plays an important role when conversion between the absorption coefficients and the complex refractive index has to be done.

As an example we calculate typical values for a commercially available dichroic liquid crystal mixture from Merck. The mixture ZLI 4714 is based on the host material ZLI 3950 and contains 3.78% of dye. Such and liquid crystal mixture gives a contrast ration of CR = 13 for a 14 μm thick cell. The order parameter of the mixture is given as $S = 0.81$ at room temperature. The contrast ration corresponds to the ratio of the transmitted intensity between ordinary and extraordinary direction. We assume the the higher transmittance is with the ordinary direction hence a O-type behavior as usually found for liquid crystal guest host systems. The calculation gives the difference between the absorption coefficients. We start with the dichroic ratio that is given as the ratio between the transmitted ordinary and extraordinary intensity

$$\begin{aligned} \frac{I_o}{I_e} &= e^{(\alpha_e - \alpha_o)d} = 13 = e^{(\alpha_e - \alpha_o)14\mu\text{m}} \geq \alpha_e \alpha_o = \frac{\ln (I_o/I_e)}{d} \\ &= \frac{\ln 13}{14\ \mu\text{m}} = 0.183 \frac{1}{\mu\text{m}} \end{aligned} \tag{6.42}$$

To determine the values for the absorption coefficients separately, transmission measurements with polarized light are necessary. Typical values for the absorption coefficient are

$$\alpha_o \approx 0.04 \qquad \alpha_e \approx 0.22 \tag{6.43}$$

For a reference wavelengths of $\lambda = 500$ nm these values lead to complex parts of the refractive index as

$$k_e = \frac{\alpha_e \lambda}{4\pi} = -0.008 \quad \text{and} \quad k_o = -\frac{\alpha_o \lambda}{4\pi} = 0.0016 \tag{6.44}$$

Note that the absorption coefficient used here are that of the mixtures that is composed out of the liquid crystal and the dye. To determine the dichroic ratio of the dye one has to introduce the concentration c. In our case for ZLI 4714 this concentration is $c_w = 3.8\%$ in weight as given by the manufacturer (Merck KG

Darmstadt Germany). To obtain different values of extinction the concentration of the dye can be changed but the solubility of the dye is limited and high concentration are difficult to obtain.

Dichroic effects are used also in some display application. Because the liquid crystal can be reoriented with electric fields it is possible to build liquid crystal displays that uses dichroism. (Blinov and Chigrinov, 1994). The optical axis, which corresponds for a O-type system to the absorption direction, can be turned into the propagation direction and the dichroic behavior is switched off. The liquid crystal layer becomes transparent with the absorption coefficient corresponding to α_o. The electric driving allows to choose intermediate states which alters the transmission for one polarization. In such a way, a switchable polarizer can be realized as it was used in Chapter 1. Combining two components with orthogonal orientation of its absorption direction can lead to polarization independent switchable light shutters. Although there are advantages like high clear transmission, systems based on dichroism are used only in particular applications because of the limited contrast ratio (Basturk, 1989).

REFERENCES

Ampiatec (2003) http://amptiac.alionscience.com/pdf/1999MaterialEase8.pdf.

Baron, M. (2001) Definitions of basic terms relating to low molecular mass and polymer liquid crystals, *Pure Appl. Chem.* **73**, 845–895.

Bass, M. (ed.) (1994) *Handbook of Optics, Devices, Measurements & Properties*, Vol. 2, 2nd edn, McGraw-Hill, New York.

Basturk, N. and Grupp, J. (1989) Liquid Crystal Guest Host Devices and Their Use as Light Shutters in "Large Area Chromogenics: Materials and Devices for Transmittance Control", Lampert, C., et al. eds. SPIE Institute Series vol. IS4 557–576.

Bentec (2003) http://www.bentecservices.com/manufacturing.htm.

Blinov, V.G. Chrigrinov (1994) *Electrooptic Effects in Liquid Crystal Materials*, Springer, New York, ISBN 0-387-94708-6.

Borrelli, F. (ed.) (1999) *Microoptics Technology: Fabrication and Applications of Lens Arrays and Devices*, Marcel Dekker, New York.

Böttcher, C.J.F. (1952) *Theory of Electric Polarisation*. Elsevier, Amsterdam.

Broer, D.J., H. Finkelmann, and K. Kondo (1988) In-situ photopolymerization of an oriented liquid-crystalline acrylate, *Makromol. Chem.* **189**, 185–194.

Broer, D.J., R.A.M. Hikmet, and G. Challa (1989a) In-situ photopolymerisation of oriented liquid crystalline acrylates 4, *Macromol. Chem.* **190**, 3201–3215.

Broer, D.J., G.N. Mol, and G. Challa (1989b) In-situ photopolymerization of an oriented liquid-crystalline acrylate, *Macromol. Chem.* **190**, 19–30.

Broer, D.J., J. Boven, G.N. Mol, and G. Challa (1989c) In-situ photopolymerization of oriented liquid-crystalline acrylates, 3. Oriented polymer networks from a mesogenic diacrylate, *Macromol. Chem.* **190**, 2255–2268.

Broer, D.J. and I. Heynderickx (1990) Three-dimensionally ordered polymer networks with a helicoidal structure, *Macromolecules* **23**, 2474–2477.

Broer, D.J. and G.N. Mol (1991a) In-situ photopolymerisation of oriented liquid crystalline acrylates 5, *Macromol. Chem.* **192**, 59–74.

Broer, D.J. and G.N. Mol (1991b) Anisotropic thermal expansion of densely cross-linked oriented polymer networks, *Polym. Eng. Sci.* **31**, 625–631.

Broer, D.J., J. Lub, and G.N. Mol (1993) Synthesis and photopolymerization of a liquid-crystalline diepoxide, *Macromolecules* **26**, 1244–1247.

Broer, D.J. (1995) Creation of supramolecular thin film architectures with liquid-crystalline networks, *Mol. Cryst. Liq. Cryst.* **261**, 513–523.

Campbell, D. and J.R. White (1989) *Polymer Characterization, Physical Techniques*, Chapman & Hall, London.

Chandrasekhar, S. (1992) *Liquid Crystals*, Cambridge University Press, Cambridge.

Challa, G. (1993) *Polymer Chemistry: An Introduction*, Ellis Horwood, London.

Collings, P.J. and M. Hird (1997) *Introduction to Liquid Crystals*, Taylor and Francis, London.

Cox, S.J., V.Y. Reshetnyak, and T.J. Sluckin (1998) Effective medium theory of light scattering in polymer dispersed liquid crystal films, *J. Phys. D: Appl. Phys.* **31**, 1611–1625.

Crawford, G.P. (2002) *Liquid Crystal Materials*, Brown University, Division of Engineering, available at http://www.engin.brown.edu/courses/en292s16/lectures.htm.

Dally, J.W. and W.F. Riley (1991) *Experimental Stress Analysis, Part III Optical Methods for Stress Analysis*, 3rd edn, McGraw-Hill, New York.

Daoud, M. and C.E. Williams (eds) (2005) *Soft Matter Physics*, Springer, Heidelberg.

Demus, D. and L. Richter (1987) *Textures of Liquid Crystals*, Wiley VCH, Weinheim.

Demus, D., J.W. Goodby, G.W. Gray, H.W. Spiess, and V. Vil (eds) (1998) *Handbook of Liquid Crystals*, 4 vols, Wiley VCH, Weinheim.

Dierking, I. (2003) *Textures of Liquid Crystals*, Wiley VCH, Weinheim.

Föppl, L. and E. Mönch (1972) *Spannungsoptik*, 3rd edn, Springer, Berlin.

de Gennes, P.G. and J. Prost (1995) *The Physics of Liquid Crystals*, 2nd edn, Oxford University Press, New York.

GSoptics (2002) GSoptics website: www.gsoptics.com/custom_optics/charts.html.

Hamley, I.W. (2000) *Introduction to Soft Matter*, Wiley, New York.

Keyes, D.L. (2001) Polymers for photonics, *Photonics Spectra* October, 131–134.

Khoo, I.C. and S.T. Wu (1993) *Optics and Nonlinear Optics of Liquid Crystals*, World Scientific, Singapore.

Kléman, M. (1983) *Points, Lines and Walls*, Wiley, Chichester.

Li, J. and S.T. Wu (2004) Self-consistency of Vuks equation for liquid crystal refractive indices, *J. Appl. Phys.* **11**, 6253–6258.

Ning, A. (2000) *Plastics vs. Glass Optics: Factors to Consider* (part of SPIE "Precision Plastic Optics" short course note), unpublished.

Norland Optical Adhesives Datatsheet (2005) Norland Products Inc. 2540 Route 130, Suite 100 P.O. Box 637 Cranbury, NJ 08512 USA.

Pedrotti, L.S. and F.L. Pedrotti (1993) *Introduction to Optics*, 2nd edn, Prentice Hall, New York.

Pelzl, G., S. Diele, and W. Weissflog (1999) Banana-shaped compounds – a new field of liquid crystals, *Adv. Mat.* **11**, 707–724.

Rohrbach, C. (ed.) (1989) Handbuch für experimentelle Spanungsanalyse, Chapter D2, VDI Verlag, Düsseldorf.

Schultz, J.W. and R.P. Chartoff (1998) Photopolymerization of nematic liquid crystal monomers for structural applications: molecular order and orientation dynamics, *Polymer* **39**, 319–325.

Stegemeyer, H. (ed.) (1994) *Liquid Crystals* (Topics in Physical Chemistry Vol. 3), Steinkopf, Darmstadt.

Stegemeyer, H. (1999) *Lyotrope Flüssigkristalle. Grundlagen – Entwicklung – Anwendung*, Steinkopf, Darmstadt.

Vuks, M.F. (1996) Determination of the optical anisotropy of aromatic molecules from the double refraction of crystals, *Opt. Spectrosk.* **20**, 244.

Yeh, P. and C. Gu (1999) *Optics of Liquid Crystal Displays*, Wiley, New York.

Young, R.J. and P.A. Lovell (1991) *Introduction to Polymers*, 2nd edn, CRC Press, Boca Raton.

7

PRACTICAL POLARIZATION OPTICS WITH THE MICROSCOPE

7.1 INTRODUCTION

Polarized light microscopy is the standard tool for the inspection of liquid crystals and polymers and was developed for mineralogy (Viney, 1986; Kaminsky, 2004). It has all the benefits of bright field microscopy and offers a wealth of information about local orientations, which is simply not available with any other optical microscopy technique. Polarized light microscopy can distinguish between isotropic and anisotropic materials. The technique exploits optical properties of anisotropy to reveal detailed information about the structure and composition of materials. Anisotropic materials have optical properties that vary with the orientation of incident light with the orientation of the specimen or its crystallographic axes. Anisotropic materials act as beam splitters and divide light rays into two parts. The technique of polarizing microscopy exploits the interference of the split light rays, as they are reunited along the same optical path to extract information about the materials.

We will provide an introduction to polarized light microscopy by addressing the following subjects. The basic setup will be given first. After comments on the production of polarized light, the essential rule of transmission between polarizers will be reviewed. This will allow the explanation of the interference colors of birefringent objects. The two main illumination techniques will be presented: orthoscopy and conoscopy. More sophisticated aspects such as the combination of birefringent plates and retardation measurements by compensation will be considered. A method to determine the character of birefringence will be explained. New methods of local polarization mapping are discussed at the end. For more detailed information about other microscope techniques the reader is referred to the literature (Birchon, 1961; Hallimond, 1970; Francon, 1971; Beyer, 1988; Microscopy, 2006).

7.2 MICROSCOPE CHARACTERISTICS

Determining fine detail in liquid crystal textures and microscopic confinements requires an imaging system capable of providing spatial information across small distances. Most microscopes in current use are known as compound microscopes, where a magnified image of an object is produced by the objective lens, and this image is magnified by a second lens system (the ocular or eyepiece) for viewing. Thus, the final magnification of the microscope is dependent on the magnifying power of the objective lens times the magnifying power of the ocular lens system. Objective magnification powers range from $4\times$ to $150\times$. Lower magnification is impractical on a compound microscope stand because of spatial constraints in image correction and illumination. Higher magnification is impractical because of limitations in working distances and light-gathering ability. Ocular magnification ranges are typically $8\times - 12\times$, $10\times$ oculars being most common. As a result, a standard microscope will provide a final magnification range of from $\sim 40\times$ to $\sim 1500\times$. Figure 7.1 shows the components of a modern microscope. The lamp provides the

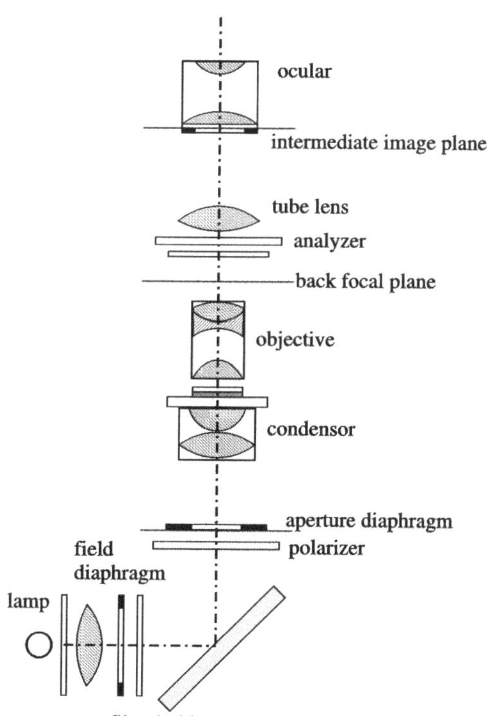

Figure 7.1 Principle components of a modern microscope with Köhler illumination setup. Several characteristic planes are shown that are important to understand image formation.

necessary light intensity. The light is distributed with the help of a Köhler illumination containing two diaphragms. The field diaphragm controls the field of illumination and the condenser aperture diaphragm is used to change the angle of the light cone that illuminates the scene. A photograph of a modern research microscope that contains all these elements is shown in Figure 7.2. We see the lamp house on the left and the rotational table. An objective revolver allows a fast change of magnification without losing focalization. Resolution is defined as the ability to distinguish two very small and closely spaced objects as separate entities. Optical resolution is determined by certain physical parameters that include the wavelength of light, and the light-gathering power of the objective and condenser lenses. The Abbe resolution criterion is based on the size and the focal lengths of the lens that will capture the light. The acceptance angle 2α of the lens is a very important parameter. The resolution is usually defined as half of the minimal focal spot diameter (diffraction limited spot) that can be created with a lens,

$$d_{\text{res}} = \frac{d_{\text{diff lim}}}{2} = 0.61\frac{\lambda}{\text{N.A.}} \qquad (\text{with N.A.} = n\sin\alpha), \tag{7.1}$$

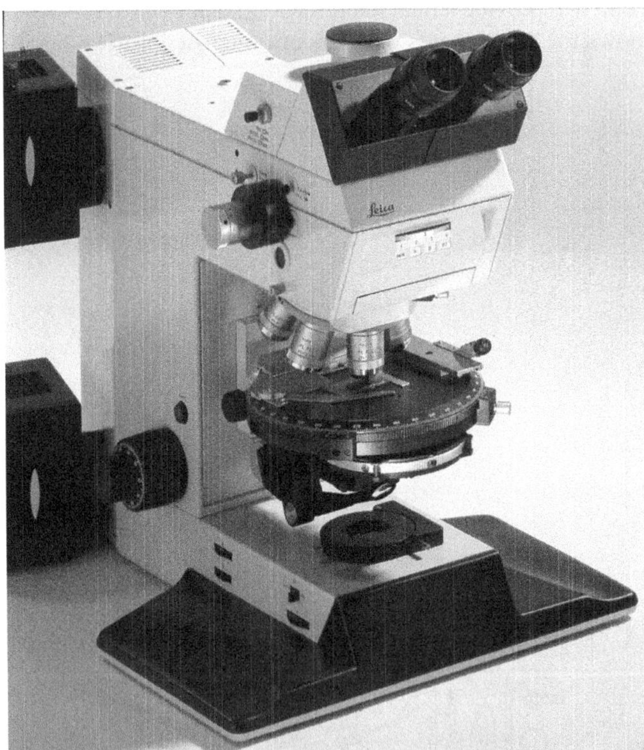

Figure 7.2 A modern research microscope with polarization equipment. The polarizer and analyzer are rotatable and a rotational stage allows the specimen to be turned under observation. (Photograph courtesy of LEICA Microsystems A.G.)

Table 7.1 Theoretical resolution limit for achromatic microscope objectives (λ = 500 nm).

Magnification of the Objective	N.A.	Theoretical Resolution [μm]
4×	0.10	3.05
10×	0.25	1.22
40×	0.65	0.47

where n is the refractive index of the surrounding medium, λ the wavelengths of light, and N.A. $= n \sin\alpha$ is the numerical aperture of the lens. The resolution will be expressed in the same units as the wavelength of the light. Alpha is one half the acceptance angle of the lens and n is the index of refraction of the medium between the specimen and the lens. The numerical aperture (N.A.) is essentially a value that describes the quality of a lens. Remember, this is a theoretical value, which is the best resolution possible. The practical resolution will always be less due to optical aberrations. Table 7.1 gives some values for different typical objectives. The two-stage optical systems can lead to empty magnification. A good rule of thumb to prevent this is that the effective magnification of an objective should be chosen to be its numerical aperture times 1000. So a 40× objective that has an N.A. of 0.65 is best used with an effective magnification of 650×. If you magnify beyond this you will only get empty magnification.

However, it is not only the objective that limits the resolution. The illuminating condenser has a resolution limit too. The combination of both gives the actual resolution limit. A simple mathematical equation defines the smallest distance ($d_{\min}$) separating two very small objects:

$$d_{\min} = 1.22\frac{\lambda}{\text{N.A.}_{\text{objective}} + \text{N.A.}_{\text{condensor}}}. \tag{7.2}$$

This is the theoretical resolving power of a light microscope. The wavelengths of light is given by λ. In practice, specimen quality usually limits $d_{\min}$ to something greater than its theoretical lower limit. The N.A. of each objective lens is inscribed in the metal tube, and ranges from 0.25 to 1.4. The higher the N.A., the better the resolution. Higher N.A. values result in shorter working distances. Very often this is not suitable when working with sandwiched liquid crystal cells. The numerical aperture can reach values above one if an immersion fluid such as immersion oil is used between the specimen and the objective. Special objectives are needed for this observation method.

From Equation (7.2) one realizes that the N.A. of the condenser is as important as the N.A. of the objective lens in determining resolution. One of the apertures in the illumination path, the condenser diaphragm, allows control of the N.A. of the illumination. Closing of the condenser diaphragm and decreasing the N.A. results in a loss of resolution. In practice, at full aperture and with good oil immersion lenses (N.A. = 1.4 for both the condenser and the objective), it is possible to be able to resolve slightly better than 0.2 μm. Working distances are about 0.1 mm. From Equation (7.1) it is clear that shorter wavelength light (bluer light) provides better resolution. The human eye is best adapted for green light. Most manufacturers of microscopes correct their simplest lenses (achromats) for green light. In order to match the numerical apertures, the condenser aperture diaphragm must be adjusted. For most studies in polarized light, the diameter of the condenser aperture should be set to about 90% of the objective numerical aperture. The interaction of light with the glass in a lens produces aberrations. Spatial aberrations (e.g., spherical aberrations) can be corrected by using lenses having different curvature on their surfaces, and chromatic (i.e., color) aberrations can be minimized by using multiple kinds of glass in combination. The objective lenses in most microscopes are achromats, and best suited for imaging with green light. Green filters narrow the bandwidth of the light, and make achromat objectives reasonably effective for most routine uses. The achromat lenses are not suitable for critical high-resolution imaging with white light, because red and blue light do not focus in the same plane as green light. Chromatic aberrations will degrade resolution in color images obtained with achromatic objectives. Color photomicrography aimed at the highest level of resolution and image clarity should be performed with totally corrected apochromatic objective lenses. Fluorite lenses offer intermediate levels of correction, better than achromats but not as good as apochromats. Fluorite lenses are well suited for fluorescence microscopy because of their high transmittance of shorter wavelength light. Higher levels of correction make objective lenses more expensive. The oculars in most microscopes are designed to work optimally with the objective lenses from the same manufacturer. Each manufacturer makes some of the color and spatial corrections in the objective and the remainder of the corrections in the ocular. Mixing brands will usually result in a degraded image. When one looks into a microscope, the magnified and corrected image seen through the oculars is a virtual image (as opposed to a real image). The ocular is not suitable for the generation of photographic or video images through the microscope. For photography or video microscopy, it is necessary to use a projection lens that generates a corrected real image. An additional characteristic of microscopic imaging is the limited depth of focus. The depth of field is the area in front of and behind the specimen that will be in acceptable focus. The optical sections caused by the limited depth of focus of the microscope are thin and usually of the order of several micrometers. This occurs in all optical devices and is dependent on a number of parameters. The most influential is again the numerical aperture N.A.

The diagram in Figure 7.3 shows a lens at its full aperture opening. The horizontal line shows the range of acceptable focus for a given spot diameter; the spot diameter is shown as a point. The criterion for acceptable focus is ultimately dependent on the circle of minimum confusion. The circle of minimum confusion is determined by all the optical aberrations. In a practical sense the acceptable focus is dependent on effective magnification. The higher the magnification of an

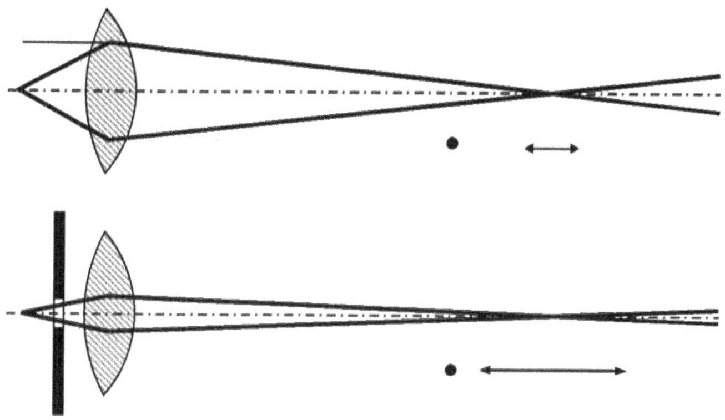

Figure 7.3 Depths of field, shown by arrows, for lenses with higher and lower numerical apertures. The depths of field are increased if the aperture diaphragm is closed. The black point indicates the dimension of the circle of minimum confusion.

object the more critical the focus. In the lower part of Figure 7.3, the numerical aperture of the lens is stopped down by an aperture. As the rays of light are now at a shallower angle, the range of focus is increased. Lenses with a short focal length will have a very tight depth of field, and a lens with a long focal length will be much deeper. The wavelength of the light is also a factor. Optical theory gives, for the depth of field, the equation

$$\delta z_{\mathrm{field}} = \frac{\lambda}{\mathrm{N.A.}^2}. \tag{7.3}$$

The depth of field deals with the focus plane of the specimen. On the other side of the lens is the focus plane of the image. The range of acceptable focus for the image is called the depth of focus. It is essentially the same as the depth of field but for one important difference, that being magnification. With higher magnification the depth of field becomes shorter, but higher magnification increases the depth of focus for the image. This is because the magnification is done with a projection lens. The depth of focus determines the thickness of the section in focus for a given objective and magnification and determines the maximum thickness of the section that will permit simultaneous observation throughout its depths. In particular instances it may be more accurate to determine the axial depth of focus from the formula (Beyer and Riesenberg, 1988)

$$\delta z_{\mathrm{focus}} = n\left(\frac{\lambda}{2 \cdot \mathrm{N.A.}^2} + \frac{150}{M \cdot \mathrm{N.A.}}\right) \quad \text{in } [\mu\mathrm{m}], \tag{7.4}$$

where n is the refractive index within the specimen, λ the wavelengths of light used expressed in micrometers, and M the overall magnification of the microscope. Table 7.2 presents data for a typical light microscope. Note that the practical

Table 7.2 Theoretical depths of focus for typical microscope magnifications.

Magnification of the Microscope (Objective Times Ocular 10×)	N.A.	Theoretical Depths of Focus [μm]
40×	0.1	175
100×	0.25	10
400×	0.65	1.2
1000×	1.30	0.4 (calculated with $n = 1.5$)

depth of field can be much better than the equation would predict. This is because the manufacturer often stops down the iris diaphragm to obtain superior results. These values do not take into account the accommodation ability of the eye. If the limits of accommodation of the eye are assumed to be from 250 mm to infinity, the added depth of focus δz_{eye} becomes

$$\delta z_{\text{eye}} = n\frac{250{,}000}{M^2} \quad \text{in } [\mu\text{m}]. \tag{7.5}$$

Note that in photomicrography the accommodation phenomena are absent. Typical values of added depths of focus due to accommodation of the eye are given in Table 7.3.

An essential factor in producing a good image with a light microscope is obtaining adequate levels of light in the specimen, or object plane. It is not only necessary to obtain bright light around the object, but for optimal imaging, the light should be uniform across the field of view. Typical modern polarized (and brightfield) microscopes have a lamphouse, which contains a 50 to 100 W high-energy tungsten-halogen lamp, attached to the base of the microscope. A transformer providing direct current (DC) voltage to the lamp is usually built directly into the microscope base and is controlled by a potentiometer positioned near the lamp switch in the bottom of the base (the lamp voltage control). Between the lamphouse and the microscope base is a filter cassette that positions removable color correction, heat, and neutral density filters in the optical pathway.

The best way to illuminate the specimen involves the use of a lens system known as a condenser. The front element of the condenser is usually a large, flattened

Table 7.3 Added depths of focus due to accommodation.

Magnification of the Microscope (Objective Times Ocular 10×)	Theoretical Depths of Focus [μm]
40×	156
100×	25
400×	1.56
1000×	0.375 (calculated with $n = 1.5$)

lens that sits directly beneath the specimen. Its placement on a movable rack provides the means to focus the light beam coming past the object and maximize the intensity and control the uniformity of illumination. Two apertures in the illumination system allow the regulation of the diameter of the illumination beam. The condenser aperture diaphragm allows on increase in contrast, but at the cost of worsening resolution. The field diaphragm does not affect resolution as dramatically and is regularly adjusted for optimal illumination. Optimal illumination of a specimen is achieved by using a variation of Köhler illumination, where the filament of the light source is in focus at the rear focal plane of the objective lens (Beyer and Riesenberg, 1988). The entire base system is designed to be vibration free and to provide the optimum light source for the illumination. Köhler illumination creates an evenly illuminated field of view while illuminating the specimen with a very wide cone of light. Figure 7.4 shows two schemas for the illumination. Parallel and divergent rays emerging from the lamp are considered separately. Two conjugate image planes are seen: One contains an image of the specimen and the other an image of the field diaphragm. The condenser aperture diaphragm controls the angle of the illumination cone that passes through the microscope optical train.

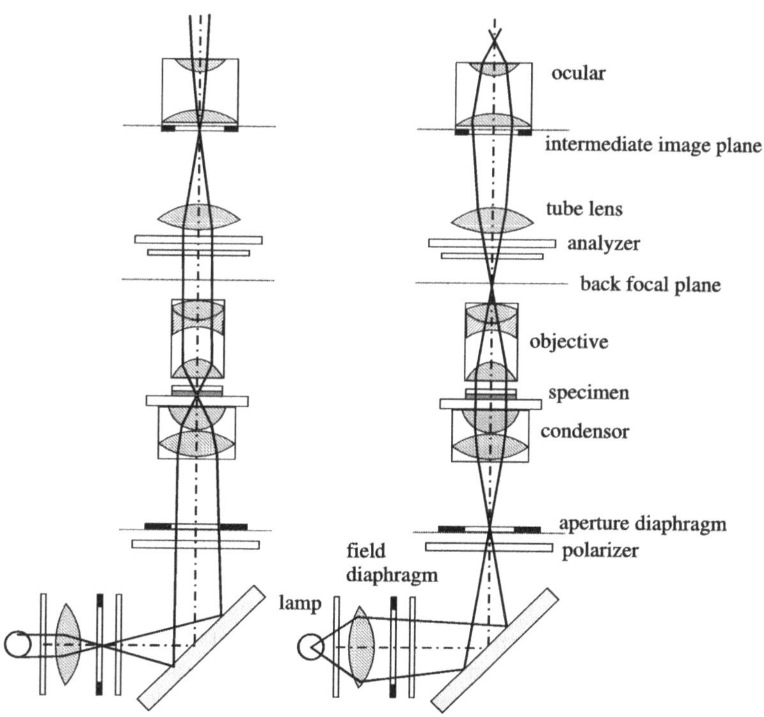

Figure 7.4 Schema for illumination in the microscope. The filament of the lamp is imaged into the aperture diaphragm. The condenser provides a bright, well-illuminated specimen.

Reducing the opening size of this iris diaphragm decreases the cone angle and increases the contrast of images observed through the eyepieces. It should be noted, however, that the condenser aperture diaphragm is not intended as a mechanism to adjust the intensity of illumination, which should be controlled by the voltage supplied to the lamp or suitable filters. Some polarized light microscopes are equipped with a fixed condenser (no swing-lens) that is designed to provide a compromise between the requirements for conoscopic and orthoscopic illumination. Variation in the degree of illumination convergence can be accomplished by adjusting the condenser aperture diaphragm or by raising or lowering the condenser (although the latter technique is not recommended for critical examinations). In general, the modern microscope illumination system is capable of providing controlled light to produce an even, intensely illuminated field of view, even though the lamp emits only an inhomogeneous spectrum of visible, infrared, and near-ultraviolet radiation.

7.3 POLARIZATION MICROSCOPE

The polarized light microscope is designed to observe and photograph specimens that are visible primarily due to their optically anisotropic character. In order to accomplish this task, the microscope must be equipped with both a polarizer, positioned in the light path somewhere before the specimen, and an analyzer (a second polarizer), placed in the optical pathway between the objective rear aperture and the observation tubes or camera port. The types of polarizers that may be used are described in Section 7.4. The microscope illustrated in Figure 7.2 is equipped with all of the standard accessories for examination of birefringent specimens under polarized light. Compared to standard brightfield microscopes, the polarized light microscope contains additional components. These include the polarizer and analyzer, strain-free objectives and condenser, a circular graduated stage capable of 360° rotation, and an opening in the microscope body or intermediate tube. A rotating specimen stage facilitates orientation studies with centration of the objectives and stage with the microscope optical axis to make the center of rotation coincide with the center of the field of view. Stress in assembly can produce optical effects under polarized light, therefore strain free objectives are necessary. An eyepiece fitted with a cross wire graticule allows the center of the field of view to be marked and makes it possible to define a center of rotation. The Bertrand lens enables easy examination of the objective rear focal plane. This allows accurate adjustment of the illuminating aperture diaphragm and observation of interference figures in the conoscopic method. A slot near the back focal plane allows the insertion of compensators/retardation plates between the polarizers, which are used to enhance optical path differences in the specimen and to analyze the type of birefringence. Figure 7.2 is a photograph of the LEICA DM RX upgradeable microscope, which can be used for transmitted polarized light orthoscopic/conoscopic applications. It can additionally be equipped with incident light contrast methods, including interferometry, microhardness testing, and microphotometry. The stand is particularly stable mechanically, and shows solid outer dimensions. An integrated 12 V, 100 W power supply provides enough light for standard observations. Furthermore, illumination is stabilized to prevent any fluctuations of current resulting from changing the brightness or color temperature.

We have made the argument that for special observational methods it is very useful to have a lamp with spectral lines and near-UV emission, as used for fluorescence observations. Then, in combination with suitable interference filters, observation in monochromatic light, at reasonable intensities, becomes possible. A good choice is the mercury vapor lamp, which shows main spectral lines at 405, 436, 546, and 580 nm. Note that the intensity of the lamp is extremely high (50 or 100 W) and that a special filter set has to be used for direct observation.

Basic adjustments have to be made before starting observations. There are three main tasks: centering of objectives, adjustment of diaphragms, and setting up of Köhler's illumination. The first step in the alignment process is to center the microscope objectives with respect to the condenser, the field of view, and the optical axis of the microscope.

7.3.1 Centering of Objectives

Early polarized light microscopes utilized fixed stages, with the polarizer and analyzer mechanically linked to rotate in synchrony around the optical axis. Although this configuration was cumbersome, it had the advantage of not requiring coincidence between the stage axis and the optical axis of the microscope. Modern polarized light microscopes are often equipped with specially designed 360° rotatable circular stages, similar to the one shown in Figure 7.2. The circular stage features a goniometer divided into 1° increments, and has two verniers placed 90° apart, with click (detent or pawl) stops positioned at 45° steps. Use of a precision ball-bearing movement ensures extremely fine control over the verniers, which allows angles of rotation to be read with an accuracy near 0.1°. A clamp is used to secure the stage so then specimens can be positioned at a fixed angle with respect to the polarizer and analyzer. The most critical aspect of the circular stage alignment on a polarizing microscope is to ensure that the stage is centered within the viewfield and the optical axis of the microscope. In general, microscopes are designed to allow adjustment of either the stage or the objectives to coincide with the optical axis, but not both. Some designs have objectives that are in a fixed position in the nosepiece with an adjustable circular stage, and others lock the stage into position and allow centration of the objectives. A pair of small setscrews in the nosepiece of most research-grade polarizing microscopes allows centering of individual objectives. Each objective should be independently centered to the optical axis, according to the manufacturer's suggestions, while observing a specimen on the circular stage. Some microscopes provide for individual objective centration, but other centration systems operate on the nosepiece as a unit. When the stage and the objectives are properly centered, a specific specimen detail placed in the center of a cross-hair reticle should not be displaced more than 0.01 mm from the microscope optical axis after a full 360° rotation of the stage.

7.3.2 Köhler Adjustment

Operationally, it is easy to obtain optimal illumination for brightfield by first placing any specimen on the stage and focusing on the object. The main point is to focus the field aperture into the field of view. To see the image of the field aperture it is recommended to completely close it. Only a small zone of the field is illuminated. Next, raise or lower the condenser until the edges of the field aperture diaphragm are clearly focused. Do not refocus the objective on the specimen while you are adjusting the condenser. If the field aperture is in focus, a centering of the condenser should be carried out. Two screws in the condenser holder allow the image of the field aperture to be moved within the field of view. When the microscope is properly illuminated, both the object and the edges of the field diaphragm should be in the same plane of focus and the field diaphragm should be centered in the field of view. The field diaphragm should be opened so that its edges obscure the periphery of the field of view.

The list of operation steps reads as:

1. Adjust the focus to see the specimen on the stage.
2. Completely close the field diaphragm.
3. Move the condenser up or down until the border or the iris hexagon is clear and neat.
4. Center with the centering device (screw).
5. Open the field diaphragm until the tip of the hexagon touches the field limit.

Modern research-grade polarized light microscopes have binocular or trinocular observation tube systems. The eye tubes are usually adjustable for a range of interocular distances to accommodate the interpupillary separation of the microscopist (usually between 55 and 75 mm). Many polarized light microscopes are equipped with an eyepiece diopter adjustment, which should be made to each of the eyepieces individually. Some microscopes have a graded scale on each eyepiece that indicates the position of the eye lens with respect to the main body of the eyepiece. Other models hold the body of the eyepiece in a fixed position securely in the eye tube with a pin and slot. If a graded marking is present in the ocular, the first step in diopter adjustment is focus on this. The specimen should be focussed through the ocular with graded markings to see the markings and the specimen well focussed. At this point, refocus the second ocular (do not use the microscope coarse or fine focus mechanisms) until the specimen is in sharp focus for both eyes separately. These settings will vary from user to user, so record the position of the eye lenses if the eyepiece has a graded scale for quick return to the proper adjustment.

7.4 POLARIZERS

Essential polarization elements in polarization microscopes are the polarizers and wave-plates. The most widely used material is Polaroid film, invented by Land in 1932. Polaroid is the trade name for the most commonly used dichroic material. It selectively absorbs light from one plane, typically transmitting less than 1% through a sheet of polaroid. It may transmit more than 80% of light in the perpendicular plane. The word "Polaroid" usually refers to polaroid H-sheet, which is a sheet of iodine-impregnated polyvinyl alcohol (Shurcliff, 1962). A sheet of polyvinyl alcohol is heated and stretched in one direction while softened, which has the effect of aligning the long polymeric molecules in the direction of stretch. When dipped in iodine, the iodine atoms attach themselves to the aligned chains. Light waves with electric fields parallel to these chains are strongly absorbed because of the dissipative effects of the electron motion in the chains. The direction perpendicular to the polyvinyl alcohol chains is the "pass" direction. Polarizers produced on this basis are available as thin films from different suppliers. Characteristic values of some models are listed in Table 7.4. For characterization, the transmission of a pair of parallel and crossed polarizers is usually given. Key figures of quality of a polarizer are the transmission and the extinction ratio. To judge quality, the spectral characteristic (dispersion) is of interest. Figure 7.5 shows spectral characterizations for samples from different manufacturers. Transmission for a single polarizer and the extinction ratio are shown. Note that the scale for the extinction ratio in Figure 7.5*b* is logarithmic and the extinction is in the order of 1000. Very highly efficient polarizers are available for display applications. They have excellent characteristics in the visible range. Sheet polarizers for the near-UV and IR regions are rare. Very good polarizers nearly reach the physical limits and have single

Table 7.4 Typical characteristics of selected polarizers usually used for display and polarized light applications.

	Thickness	Transmission			Efficiency		
	[μm]	Single	Parallel	Crossed	[%]	Hue a	Hue b
Nitto Denko							
G1220DU	205	41.5	34	0.018	99.95	−1	3
EG1224DU	205	43	37.1	0.01	99.99	−1.3	3.2
EG1425DU	210	43.5	38.5	0.02	99.95	−1.5	3.5
QE 10-39	215	39	30.5	0.05	99.9	0.5	−3
Sanritz							
H-C2-1218	215	42.7	32.65	4.22	87.805	−0.55	−7.5
H-C2-1112S	215	38	28.39	2.2	92.5	−2.15	−3.6
LL-C2-8212	154	44.27	38.07	1.409	96.565	−0.55	−1.4
LL-98	215	46.52	39.63	3.9266	90.53	−1.29	1.21
LL-8312	154	48	40.5	10.25	77.4	−1.15	−3.45

Data compiled from the manufacturers' datasheets, Polaroid, 1990; Nitto Denko, 1996; Sanritz, 2000.
a, b are the color coordinates in NBS System (Wyszecki 2000)

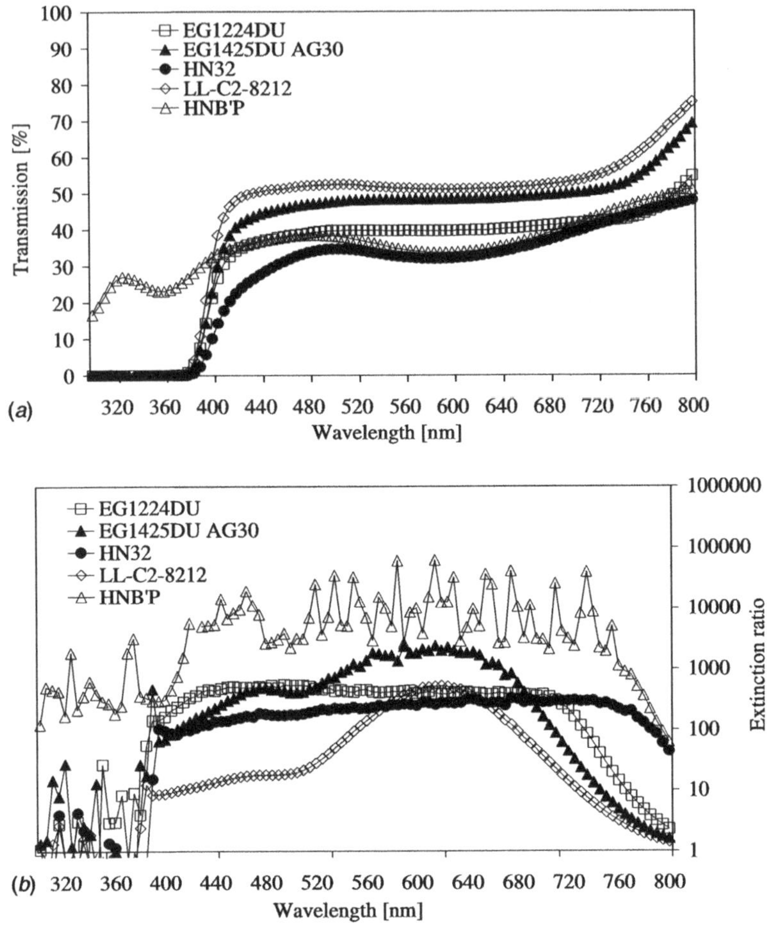

Figure 7.5 (*a*) Transmission and (*b*) extinction ratio of selected thin-film polarizers from different manufacturers. The measurements represent the transmission through a single polarizer and through two crossed polarizers. The EG types are from Nitto Denko Corp., HN types are from Polaroid-3M, and the LL type is from Sanritz Corporation.

component transmission of 50% at 99.99% extinction ratio. For UV light with wavelengths between 270 and 400 nm, only a few polarizers are available. Only the HNP'B (Polaroid 3M) linear polarizer is useful in the near-UV. The spectral range of this component is from 275 to 750 nm, with an extinction ratio higher than 100. Modern components use wire-grid principles and have an extended spectral range (Moxtek, 2004). These polarizers can be used in the visible and infrared at the same time. They are fabricated by high-resolution lithographic patterning technology and a combination of multilayer structures. High purity (99.99%) aluminum is deposited on a glass substrate and lithographically etched to form a wire-grid (aluminum rib) structure on a glass substrate. If the pitch is less than $\frac{1}{2}$ of the shortest wavelength in air of the light to be polarized, then light with polarization perpendicular to the wire is transmitted. The other polarization of light sees a mirror-like surface and is almost totally reflected. About 5–10% of the incident light of both polarizations will be absorbed in the aluminum wires, the same absorption that would occur for an unpatterned aluminum mirror surface.

Thus, the typical grating period is less than 0.2 μm for use with $\lambda = 400$ nm violet light. The wire-grid thickness is typically between 0.1 and 0.2 μm, but has to be on a substrate with a typical thickness of 500 μm. A more detailed description of the wire-grid polarizers' structure describing the wavelength and incidence angle dependence of both the transmitted and reflected light intensities can be found in the work of Perkins and colleagues (2000). A very exciting polarizer based on thin polymer films is the so-called DBEF (Weber et al., 2000). The basic principle is multilayer interference, which occurs when a multitude of layers of different refractive index and well-designed thickness are put together. If, in addition, one of the layers can be made birefringent, the result is a polarization device that reflects one polarization and transmits the other. This will be discussed in more detail in Chapter 11. The effect is based on Bragg reflection phenomena in multilayer structures. For very high extinction ratios, crystal optical components might be used. Crystal components are thick (several centimeters), and the extinction ratio is extremely sensitive to the alignment of the component. Small angular deviations of the light send through such a component might produce transmission. Glan–Thompson polarizers polarize over only a 4° field of view. Therefore they are only used for special tasks and are not commonly used in polarization microscopy. To qualify the quality of observations in polarization microscopy, one can refer to common defects of polarization elements. The following list gives the basic errors that might occur when imperfect elements are used (Axometrics, 2006).

- Polarizers have nonideal diattenuation, because $T_{max} < 1$ and $T_{min} > 0$.
- Polarizers usually have some retardation; there is a difference in optical path length between the transmitted (principal) eigenpolarization and the small amount of the extinguished (secondary) eigenpolarization. For example, sheet polarizers and wire-grid polarizers show substantial retardation when the secondary state is not completely extinguished.
- The polarization properties vary with angle of incidence. Birefringent retarders commonly show a quadratic variation of retardation with angle of incidence, which increases along one axis and decreases along the orthogonal axis.
- The polarization properties vary with wavelength.
- For polarizers, the accepted state and the transmitted state can be different. Consider a polarizing device formed from two linear polarizers, one behind the other, oriented rotated by a small angle. Incident light linearly polarized within the direction of the first polarizer has the highest transmittance for all possible polarization states and is the accepted state. The corresponding exiting beam is linearly polarized at the angle given by the position of the second polarizer.
- The eigenpolarizations of the polarization element may not be orthogonal. Such a polarization element is referred to as inhomogeneous. A polarizer may

transmit linearly polarized light at 0° without change of polarization, while extinguishing linearly polarized light oriented at 85°.

- A polarization element may depolarize. A polarizer or retarder with a small amount of depolarization, when illuminated by a completely polarized beam, will have a small amount of unpolarized light in the transmitted beam. A partially polarized beam can no longer be extinguished by an ideal polarizer. Depolarization results from surface roughness, bulk scattering, random strains and dislocations, and thin-film microstructure.

A serious source of error is the birefringence in other optical components within the microscope. Unwanted birefringence in microscope objectives can arise primarily by two mechanisms. The first is "natural" birefringence, which is an artifact of the inherent anisotropic character of glasses, crystals, and other materials used to make the lenses. To circumvent this problem, manufacturers choose strain-free optical glass or isotropic crystals to construct lens elements. The second type is "strain" birefringence, which occurs when multiple lenses are cemented together and mounted in close proximity with tightly fitting frames. Strain birefringence can also occur as a result of damage to the objective due to dropping or rough handling. Stress-free objectives are marked P or POL. When both the objectives and the condenser are stress- and strain-free, the microscope viewfield background appears as a deep solid black when observed without a specimen between crossed polarizers.

7.5 POLARIZATION COLORS

To seriously discuss color phenomena, one has to take into account three major aspects: perception of color by the eye, light transmission of a birefringent specimen between two polarizers, and the spectrum of illumination. The spectrum of the lamp is important, but does not originate particular information about the specimen under observation. It is sufficient to know that quantitative analysis requires a detailed knowledge of the illumination source. When color effects are discussed it is necessary to know how the color reception of the eye can be described. We leave obtaining details of this to the interested reader (Völz, 2002; Kuehni, 2003) and give only a basic introduction into the tristimulus color perception model.

7.5.1 Color Perception

The CIE (Commission Internationale de L'Éclairage) colorimetric system, developed in the late 1920s, is based on experiments establishing which lights are matching. It does not have an explicit connection to the appearance of stimuli. The relative success of this assumption has strengthened the belief in its applicability. Today, models of color vision are usually expressed in terms of cone sensitivity functions. These are believed to be linearly related to color-matching functions, and vice versa. Models of color vision have become more sophisticated by considering surrounds and other experimental conditions, but the fundamental assumption of the direct relationship between color matching and color appearance remains. Brightness and lightness and their differences are expressed in terms of luminance or luminous reflectance, even though this does not consider the spectral brightness component. These are empirical approximations with a degree of validation from the correlation with visual data. In 1860, Maxwell provided the first set of measured functions derived using a visual colorimeter. Between 1920 and 1930 J. Guild (1930) and W. D. Wright (1928–29) built improved equipment to measure these functions, and the color-matching functions of various observers were determined. These data were considered by the CIE for standardization. In this form the functions were standardized as the CIE 1931 2° standard observer color-matching functions applicable to a field of view of 2°. In 1964 the CIE 1964 10° standard observer color-matching functions were added, applying to a field of view of 10° and based on measurements (Wyszecki and Stiles, 2000). With these definitions of standard observers, as well as definitions of illuminants and of the reflectance factor defining the reflecting properties of object colors, a system was available for quantitative colorimetry.

The retina of the human eye has two categories of light receptor: rods, which are active in dim light and have no color sensitivity, and cones, which are active in bright light and provide us with our ability to discriminate color. One can identify three types of cones, which are sensitive to red, green, and blue (R, G, and B). The eye/brain discriminates color by processing the relative stimuli in the three sensors. In order to quantify human color vision, the CIE has established a set of imaginary "red," "blue," and "green" primary colors that, when combined, cover the full gamut of human color vision; that is, a combination of the three can match any monochromatic light source. The combinations of these three "light" sources required to match monochromatic (spectral) light are determined experimentally as mentioned above. The curves for these combinations are shown in Figure 7.6 and are called the Color Matching Functions for the Standard

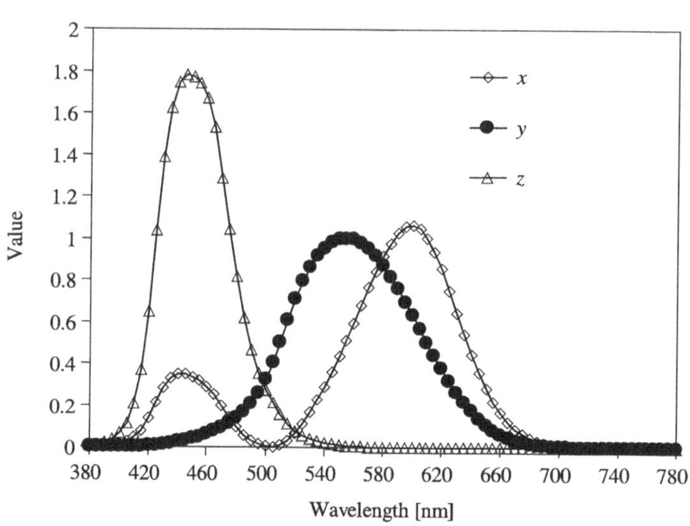

Figure 7.6 CIE 1931 2° standard observer color-matching functions.

Table 7.5 Color parameters for selected wavelengths.

Wavelengths [nm]	$\bar{x}$	$\bar{y}$	$\bar{z}$
405	0.0232	0.0006	0.1102
436	0.3285	0.0168	1.623
485	0.058	0.1693	0.6162
546	0.3597	0.9803	0.0134
580	0.9163	0.87	0.0017
633	0.5419	0.217	0

Colorimetric Observer. They are designated $\bar{x}, \bar{y}$, and $\bar{z}$. With these curves, color coordinates can be directly read. As an example, Table 7.5 shows the color coordinates for monochromatic spectral lines of a mercury lamp and some other often-used wavelengths.

The color matching functions are used to derive the *XYZ* tristimulus values that uniquely define an object's colorimetry; two objects with the same tristimulus values have identical color appearance when viewed under the same illumination conditions. For transparent objects, the *X*, *Y*, and *Z* tristimulus values are calculated by integrating (summing) the product of the spectral transmission $T(\lambda)$, the illuminant (the light source) $S(\lambda)$, and the corresponding color-matching function over the whole visible spectral range from 380 nm to 780 nm.

$$X = \int_{380}^{780} S(\lambda)T(\lambda)\bar{x}(\lambda)\,d\lambda, \qquad Y = \int_{380}^{780} S(\lambda)T(\lambda)\bar{y}(\lambda)\,d\lambda,$$
$$Z = \int_{380}^{780} S(\lambda)T(\lambda)\bar{z}(\lambda)\,d\lambda. \tag{7.6}$$

Although the tristimulus values uniquely define an object's color, they do not define the eye's response to the color, which depends on the environment and the eye's adaptation. If two objects with different spectral reflectivity have the same color appearance (tristimulus values) under one light source, they are said to be metamerically matched. If they exhibit a marked difference under another light source they suffer from viewing illuminant sensitivity, frequently called metamerism.

The *XYZ* tristimulus values are important because they form the basis of the widely used CIE 1931 chromaticity diagram, shown in Figure 7.7. This diagram is based on normalized tristimulus values, *x*, *y*, and *z*, where

$$x = \frac{X}{X+Y+Z} \qquad y = \frac{Y}{X+Y+Z} \qquad z = \frac{Z}{X+Y+Z} = 1 - x - y. \tag{7.7}$$

This normalization (division by $X+Y+Z$) removes the brightness from the diagram so that only two coordinates, *x* and *y*, are needed to define chromaticity.

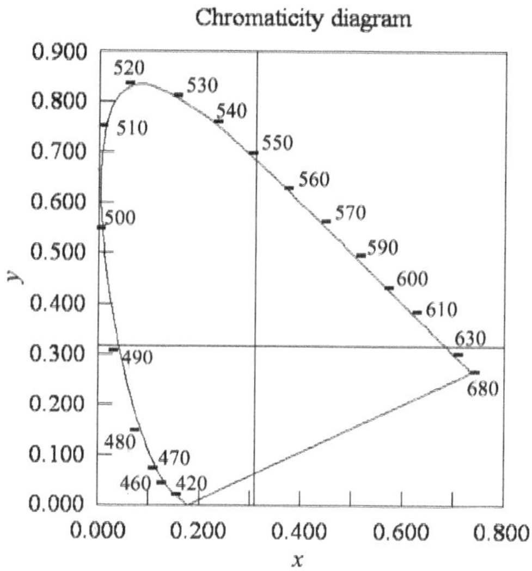

Figure 7.7 CIE 1931 diagram. The pure colors that are given by wavelengths are at the outer perimeter of the curve. All perceptible colors of the eye are found inside the curve.

The $z = 1 - x - y$ would be redundant. As Y is related to luminance, colors are sometimes expressed as xyY tristimulus values. All perceptible colors are lying in a horseshoe line limited area, as seen in Figure 7.7. The horseshoe line starting at 420 nm on the lower left and wrapping around the top to 680 nm on the right is called the spectrum locus. It represents the pure spectral colors. The straight line connecting the endpoints of the horseshoe is called the purple boundary. The full gamut of human vision lies within this figure. The vertical axis gives an approximate indication of the proportion of green; the horizontal axis moves from blue on the left to red on the right. The location of white depends on the illuminant color temperature. Some typical values are given in Table 7.6.

In spite of all the useful characteristics of the CIE xy chromaticity diagram, it lacks one very important characteristic. If the distance between any arbitrary two points is the same as the distance between another point pair, the perceived distance will not be the same. In the worst case, if the perceived distances are the same, actual distances can differ as much as 20 times. In order to correct this, researchers are trying to find a perceptually uniform color space. CIE proposed two alternatives as improvements over the CIE xyY space. These are CIE LUV and CIE LAB.

Table 7.6 Color coordinates of selected light sources.

Incandescent lamp	2856 K	$x = 0.448$, $y = 0.407$
Direct sunlight	5335 K	$x = 0.336$, $y = 0.350$
Overcast sky, D65	6500 K	$x = 0.313$, $y = 0.329$
Red laser light at 633 nm		$x = 0.700$, $y = 0.300$

7.5.2 Transmitted Intensity Between Polarizers

There are two polarizing filters in a polarizing microscope: the polarizer and the analyzer. The polarizer is situated below the specimen stage, usually with its permitted vibration direction rotatable through 360°. The analyzer is sited above the objectives and can be moved in and out of the light path as required. Under standard observation conditions, when both the analyzer and polarizer are in the optical path, they are usually positioned with their permitted vibration directions at right angles to each other. The polarizer and analyzer are said to be crossed. No light passes through the system and a dark field of view is present in the eyepieces. If an anisotropic specimen illuminated by white light is viewed with the crossed polarizers off the extinction position, colors, known as interference colors, will be seen. These colors result from unequal transmission by the analyzer of the various components of white light. What wavelengths are seen depends on the retardation or path difference produced in the crystal for each of the wavelengths. To understand color generation it is sufficient to discuss the case of uniaxial materials. The basic equation for light transmittance of uniaxial anisotropic materials between crossed polarizers is given by Equation (4.14), which reads

$$I = \frac{I_0}{2}\sin^2(2\varphi)\sin^2\left(\frac{\pi\Delta n(\lambda)d}{\lambda}\right). \tag{4.14}$$

Here Δn is an effective birefringence of the material under observation, λ is the wavelength of light, d is the thickness of the sample, and φ is the angle that the optical axis of the uniaxial material makes with the polarizer. The product Δnd is called retardation and its value is dependent on the wavelength because of the dispersion of the birefringence $\Delta n(\lambda)$. The relative transmission (relative transmitted intensity) $T(\lambda)$ becomes maximum for an orientation of the optical axis at 45°. At $\varphi = 45°$ the equation for the relative transmission becomes

$$T(\lambda) = \sin^2\left(\frac{\pi\Delta n(\lambda)d}{\lambda}\right). \tag{7.8}$$

Imagine a crystal that has a thickness and birefringence that produces a retardation of 550 nm, viewed 45° off extinction. With the equation above, we can determine the transmission by the analyzer of the components to construct a curve of transmitted intensity by the analyzer versus wavelength of light. Figure 7.8 shows plots for three different retardation values at 200, 550, and 800 nm. The sequence of colors that results from increasing retardation is one of the basic facts of optical microscopy with polarized light. If there is zero retardation, the light recombines in the same orientation as it had originally and is blocked by the upper polarizer. Equation (7.8) leads to zero. Isotropic materials have the same refractive index in all directions, produce zero retardation, and appear black. For low retardations, Equation (7.8) does not have a minimum in the visible and no visible wavelengths are blocked. One sees part of the whole spectrum and a neutral gray or white.

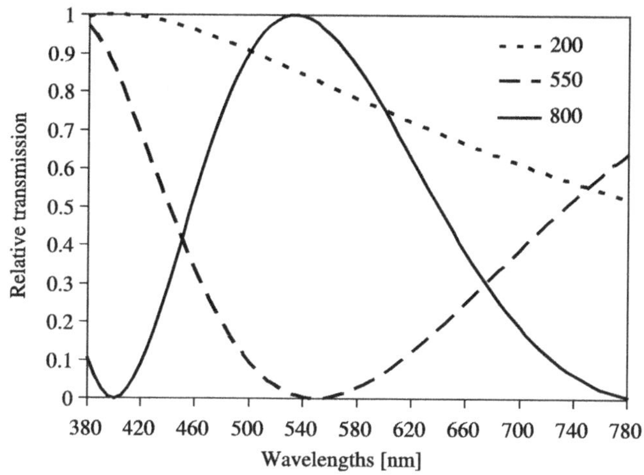

Figure 7.8 Transmission of a uniaxial anisotropic specimen for different retardations (200, 550, and 800 nm) between crossed polarizers and oriented with its optical axis at 45° with respect to the polarizer.

As retardation reaches 400 nm, the first transmission minimum reaches the blue end of the spectrum and the transmitted color shifts to yellow, then red. At 550 nm the minimum is in the green, and we see magenta, the result of mixing red and violet. Each multiple of 550 nm will result in a magenta hue. The sequence of colors from one magenta to the next is termed an order. The colors that appear within the first order are listed with their retardation in Table 7.7.

An anisotropic material with an interference color of first white produces a retardation of about 200 nm. This results in a transmission curve shown by the dotted line in Figure 7.8. Note that this low-order white results from all wavelengths being transmitted, but less of the orange and red colors and more of the violet, blue, and green colors, and thus appears more bluish to the eye. Figure 7.8 shows additionally the transmission curve of deep red for 550 nm and green at 800 nm retardation. For very large retardations, several wavelengths are blocked, while others are transmitted. As more and more windows in the spectrum appear, the colors become progressively more pale. Finally they approach white, but a warmer off-white rather than the cold neutral white of low retardation. Table 7.8 gives some more information on color appearance for higher orders.

If the thickness of the crystal is known, the birefringence for the privileged directions can be determined from the interference color. This becomes important when one looks at thin slabs of anisotropic materials, which might have a thickness of only a few microns. Thus, for example, a planar liquid crystal layer in a thin film of 5 μm thickness that shows 1° red interference colors has a birefringence of about 0.1. Note that this may not be the absolute birefringence, but only the birefringence for the fast and slow vibration directions exposed by the orientation of the crystal (tilt angle).

Table 7.7 Detailed color nomination for the first-order colors.

Retardation [nm]	Color Description
0	Black
40	Iron-gray
97	Lavender-gray
158	Grayish-blue
218	Clear gray
234	Greenish-white
259	Almost pure white
267	Yellowish-white
275	Pale straw-yellow
281	Straw-yellow
306	Light yellow
332	Bright yellow
430	Brownish-yellow
505	Reddish-orange
536	Red
551	Deep red
565	Purple
575	Violet
589	Indigo

Some crystals, and especially minerals, show colors that are not part of the standard interference color sequence. Such colors are termed anomalous. There are three main effects leading to such anomalous colors. First, they can result when the mineral is strongly colored and its natural color combines with the interference color. Colors of such minerals are produced by absorption. Secondly, the refractive index of the material can change remarkably with wavelength (dispersion), so that different wavelengths satisfy different interference criteria. In some cases the mineral may be isotropic for some wavelengths but not for others. Thirdly, the

Table 7.8 Higher order color description with general properties.

Retardation [nm]	Order	Colors	Notes
0	Zero	Black	
0–550	First	Gray, white, yellow, red	Neutral colors are cold, yellows dull
550–1100	Second	Violet through spectrum to red	Purest colors, though not totally pure
1100–1650	Third	Violet through spectrum to red	Have a "fluorescent" appearance
1650 and up	Fourth and higher	Mostly greens and pinks	Colors become more washed out with increasing retardation

privileged directions of the material vary with wavelength, so the mineral never goes completely to extinction at any one time. Anomalous interference colors tend to be dark. Blue, violet, green and brown are all possible (Zollinger, 1999).

7.5.3 Color Space for Birefringent Colors

Liquid crystal cells can be used to generate electrical birefringent color effects (Ohkubo et al., 1994). We would like to demonstrate the color change of a liquid crystal slab for different thicknesses. For the calculation we have to take into account the dispersion of the materials. We chose E7 (Merck) with the dispersion parameters described in Chapter 6. The thickness of the layer is varied between 0 and 5 μm, which leads to a maximum retardation of 1.125 μm (at $\lambda = 550$ nm). With that range we cover a wide span of colors as described in Table 7.8. In Figure 7.9, the interesting point to note is the rate of change when the thickness is increased. The curve starts at about the neutral or white point in A and moves on to the yellow red at point B for 1 μm thickness. In this first-order branch, all violet interference colors were successively addressed and the curve ends up at C where the thickness is 1.3 μm. The second-order interference colors appear if point D at 3 μm slab thickness was passed. That corresponds to a retardation Δnd of more than 550 nm. The curve continues to the green at point E. Larger retardation does not lead to a better color performance and we do see that the area that can be addressed for color production is limited. Therefore, birefringent color displays are rare and are only used for applications where high transmission is needed.

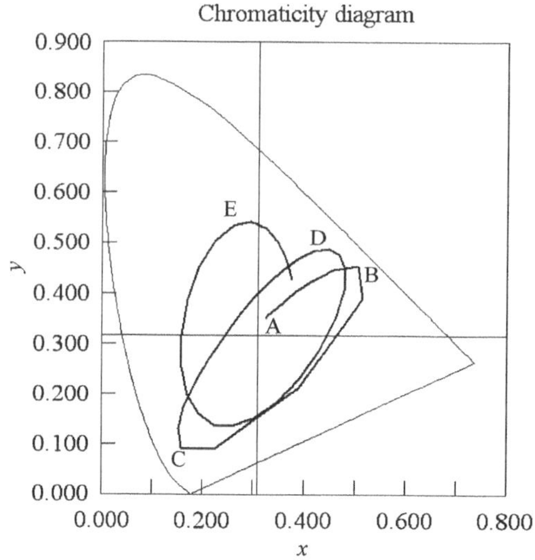

Figure 7.9 Colors created by liquid crystal layers of varying thickness. The retardation increases linearly with increasing layer thickness from A to D. The cell thickness is changed from 0 at point A to 1 μm at B, 1.3 μm at C, 3 μm at D, and 4.5 μm at E. The dispersion material parameters for E7 were used.

7.6 COMPENSATION AND RETARDATION MEASUREMENT

Practically all research-level polarized light microscopes are equipped with a slot in the body tube above the nosepiece and the objective rear focal plane, but beneath the analyzer. The purpose of this slot is to house a compensator or retardation plate in a specific orientation with respect to the polarizer and analyzer transmission axis vibration directions. The slot dimension has now been standardized (DIN specification) to 20 × 6 mm. Optical anisotropy is studied in the polarized light microscope with accessory plates that are divided into two primary categories: retardation plates, which have a fixed optical path difference, and compensators, which have variable optical path lengths. Retardation plates or compensators can be employed to determine the relative retardation or optical path differences that are introduced into the optical system by specimen birefringence. The terms "relative retardation," used extensively in polarized light microscopy, and "optical path difference (OPD)," are both formally defined as the relative phase shift between the orthogonal polarizations, expressed in nanometers. The terms retardation and OPD are collective and referred to as the specimen birefringence. It is obvious that specimens having differing thickness and refractive index gradients can display identical optical path differences or relative retardations. Furthermore, if either the birefringence or the thickness of a specimen is known, the other parameter can be easily determined.

Retardation plates are composed of optically anisotropic quartz, mica, and gypsum minerals. They are ground to a precise thickness and mounted between two optical windows having flat (plane) faces, which are designed to introduce a fixed amount of retardation between the orthogonal polarizations. More recently, several manufacturers have shifted to the application of a highly aligned and stretched linear organic polymer to produce anisotropic retardation plates. In most cases the optical axis of the retardation material is confined to the surface plane of the anisotropic plate. Incident rays of linearly polarized light enter the plate perpendicular to the optical axis and are separated into orthogonal components that follow the same trajectory through the plate. However, due to refractive index differences introduced by the anisotropic retardation material, one of the polarizations is shifted in phase (retarded) relative to the other. The wavefronts emerging from a retardation plate or compensator are polarized either in a linear fashion or with varying degrees of ellipticity, depending upon the degree of relative phase retardation. The three most common retardation plates produce optical path length differences of an entire wavelength (ranging between 530 and 570 nm), one-half wavelength (260–280 nm), or a quarter wavelength (137–150 nm). In addition, a variable optical path length can also be obtained by utilizing a tapering wedge-shaped design that covers a wide spectrum of wavelengths (up to six orders or about 3000 nm). A wavefront passing through a full-wave (one wavelength) retardation plate remains linearly polarized upon emerging and retains the same vibration plane. In contrast, the one-half wavelength plate rotates the plane of linearly polarized light by 90°, and the quarter-wavelength plate converts linearly polarized light into circularly polarized light (and vice versa). Retardation plates that introduce less than a quarter wavelength of phase shift produce elliptically polarized light. A full-wave plate is often referred to as a first-order red plate, because it produces the interference color having a first-order red (magenta) color. If the plate originated in Germany, it will probably be labeled "Rot I." The first-order retardation plate is a standard accessory that is frequently utilized to determine the optical sign (positive or negative) of a birefringent specimen in polarized light microscopy. Quarter-wavelength retardation plates are usually fashioned from quartz or muscovite crystals. Depending upon the manufacturer, quarter-wavelength plates may be marked Mica, Glimmer, $1/4\ \lambda$, or $\delta = 147$ nm. A primary consideration when using compensators and retardation plates is to establish the direction of the slow axis vibration vector. By convention, this direction will be Northeast–Southwest and will be marked "slow," z', or γ. Compensators and retardation plates are also commonly used for qualitative applications, such as control of background illumination or to improve the contrast and visibility in weakly birefringent specimens. In a properly adjusted polarized microscope (when the polarizer and analyzer transmission azimuths are exactly 90° apart) the background appears very dark. Details from specimens that are very thin, or have low levels of birefringence, are often obscured. In many cases, the visibility and contrast in these specimens can be improved if a small amount of compensation (5–10 nm) is introduced in order to increase the intensity of the background. Compensators can also be used to fine-tune the level of phase displacement between the orthogonal wavefronts (ordinary and extraordinary waves) with a target of increasing visibility and contrast of specimen detail.

7.6.1 Fast and Slow Direction

The lambda plate can be used to determine the fast and slow direction in a crystal. To do this, the crystal must be aligned on the microscope stage with one of its privileged directions oriented at 45° to the polarizing direction of the microscope. For compensation one introduces a lambda plate and compares two orthogonal position of the specimen. We will discuss a real case to illustrate the principle. Imagine that a birefringent layer produces 100 nm of retardation. Figure 7.10 shows that layer as the first element on the left. Light travels from the left to the right. In the microscope this would correspond to first-order gray interference color between crossed polarizers. Inserting the compensator with its slow direction, γ, parallel to the slow direction in the crystal will produce an additional 550 nm of retardation for the wave that had vibrated in the slow direction in the birefringent layer. This case is outlined in Figure 7.10*a*. The slow direction corresponds to the direction of higher refractive index. Simply adding the retardation gives 550 nm + 100 nm = 650 nm. The corresponding interference color for 650 nm is second-order blue. In this case we say that addition has occurred. When the orientation of the optical axis of the specimen is changed, the slow direction is perpendicular to the slow direction of the lambda plate. This time, as seen in Figure 7.10*b*, the wave that had vibrated in the fast direction in the birefringent film is slowed by the compensator, and the wave that had vibrated in the slow

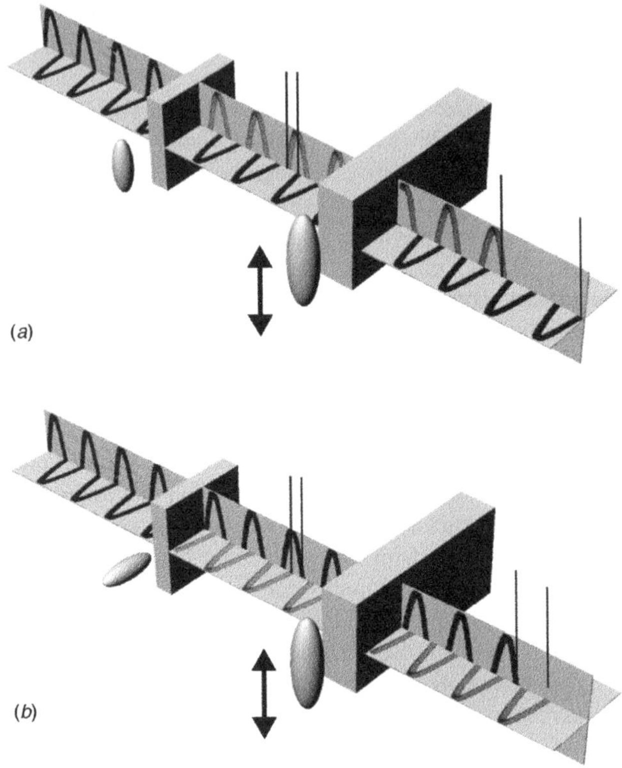

Figure 7.10 Compensation of birefringence with a retardation plate of defined retardation $\Delta = \lambda$ (550 nm). (*a*) Addition of phase shift leads to larger retardation. (*b*) Subtraction gives a decrease in retardation.

direction in the crystal becomes the fast wave in the compensator. Thus the previously fast wave will be slowed down and the previously slow wave will have overtaken it by 450 nm. In this case, we say that subtraction has occurred, as 550 nm − 100 nm = 450 nm. This 450 nm corresponds to first-order yellow. Subtraction occurs when the fast direction in the crystal is aligned parallel to the slow direction of the compensator, hence parallel to the γ-direction indicated on the compensator. The two different positions of the specimen lead to distinct, different interference colors. The addition position leads to blue, the subtraction position to yellow. The interference colors allow, together with the knowledge of the compensator's axis orientation, the determination of the direction of the fast and slow axis. Compensation is sometimes difficult because no useful retardation might be accessible to perform the measurements, especially if homogeneous systems with high birefringence are studied. The technique is therefore limited.

7.6.2 Determination of Retardation in Uniaxial Materials

In a practical sense compensation can be used to determine the retardation in uniaxial materials when the optical axis is oriented perpendicular to the observation direction. In the case of homeotropic alignment, there is no retardation to measure because the specimen behaves optically isotropic for normal incidence. For a tilted optical axis an effective birefringence is determined that depends on the tilt angle. If the sample has a complicated internal structure only summary results can be determined, which need to be interpreted carefully. If we study the case of a birefringent slab, and the optical axis is oriented perpendicular to the observation direction, we do have the situation of a birefringent plate between crossed polarizers where Equation (4.14) for the intensity modulation applies. The methodology is as follows. First one looks for an extinction position of the specimen. That position defines, relative to the polarizer orientation, the direction of the optical axis. The optical axis is now either parallel to the analyzer or to the polarizer. Now the sample is rotated by 45°. Now Equation (7.8) gives the relative transmission. The compensator represents a certain phase shift Δ that leads to a modification of Equation (7.8). If d is the thickness of the sample, n_1 and n_2 the refractive indices, and Δ the retardation of the compensator, Equation (7.8) becomes

$$T = \sin^2\left(\frac{\pi((n_1 - n_2)d \pm \Delta)}{\lambda}\right). \tag{7.9}$$

The plus or minus in front of the compensator's retardation Δ applies if the compensator is in the addition or subtraction position. Compensation of retardation is only possible for the case when the compensator's optical axis is in the subtraction position. By changing the retardation one is either able to find the zero order, which is black in white light illumination, or not. If not, there might be two reasons. First, the birefringence of the compensator may be too small to compensate the retardation given by the sample. A different compensator or a modified technique then has to be applied. A second possibility is that the combination sample–compensator is in the addition position. In this case the sample should be rotated by 90° to bring it into the subtraction position and then compensation retried. The retardation has to be read at the compensator. There are different models that allow compensation up to 30λ. Very convenient for studies with polymers and liquid crystals are models that have a retardation range of about 5λ. They still have rather high precision and a reasonable range of compensation. Note that the retardation $(n_1 - n_2)d$ can have a positive or negative sign. The subtraction and addition argumentation hold further, but one has to consider that the sign of the birefringence is not evaluated here. There is a way to determine the sign of the birefringence of uniaxial materials by conoscopy. We will discuss this after introducing the conoscopic method.

7.7 CONOSCOPY

7.7.1 Principle of Operation

Diffraction phenomena can be made visible with a polarizing microscope in the conoscopic setup (Conoscopy 2006, Derochette 2005). The so-called conoscope visualizes the directional distribution of diffracted light according to the rules that apply for transformation of directions into locations in the conoscopic figure. For the resulting conoscopic pattern (i.e., directions image), it does not matter how the directional distribution of light has been generated, if it is caused by multidirectional illumination followed by modulation in a (liquid) crystal or by diffraction. To show the power of the method, we show in Figure 7.11 two conoscopic images for particular liquid crystal structures. In Figure 7.11*a*, a homeotropic texture (E7 Merck, 50 μm thick, 40×, N.A. = 0.6) is examined in transmission in white light between crossed polarizers. In Figure 7.11*b*, a picture taken in reflection mode from a blue phase II is presented (E48 50%CB15 (Merck), oil immersion, N.A. = 1.3, crossed polarizers, 550 nm).

The examination of a birefringent slab in parallel light between crossed polarizers reveals its optical character in one direction only. Very important additional information may be obtained by passing a strongly convergent beam of light through the birefringent slab to examine its optical character in many directions at one and the same time. This can be done by viewing between crossed polarizers not the image of the slab or the object image, but another optical image formed in the principal focus of the objective. This image is called, variously, the directions image (as opposed to the object image), the image in convergent light, or the interference figure. When the microscope is used in this way it is said to behave as a conoscope, and the observations are described as conoscopic as they are made in convergent light. The divergent rays pass through the objective and converge again in the upper focal plane of the objective to form a small real image.

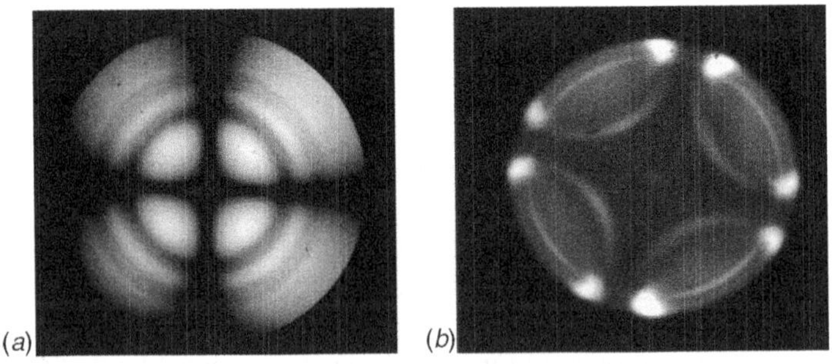

Figure 7.11 (*a*) Liquid crystal in homeotropic texture (E7) under conoscopic observation. The photograph is taken in white light and the contrast of the interference fringes is decreased towards larger observation angles. (*b*) An interference image observed in reflection from a liquid crystalline blue phase is shown at 550 nm (CB15, E48 50%). Particular reflection directions are visible as so-called Kossel lines.

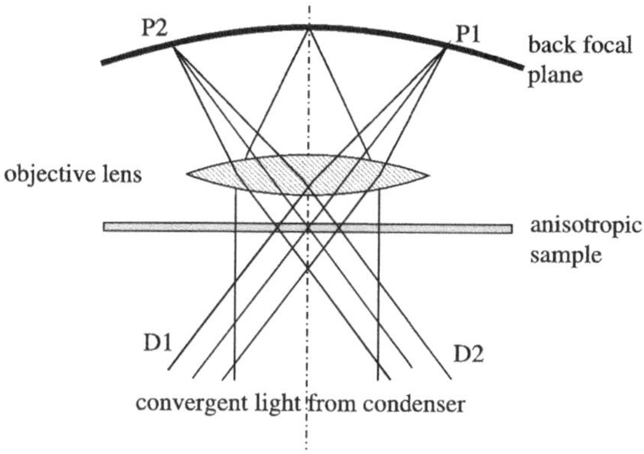

Figure 7.12 The incoming elementary parallel beams are focused to form spots in the back focal plane of the lens. The location of these spots is a function of the direction of propagation of the incoming beam (i.e., angle of inclination and azimuth).

The schematic ray tracing of Figure 7.12 illustrates the basic concept of conoscopy: transformation of a directional distribution in the front focal plane into a lateral distribution (named directions image) appearing in the back focal plane (which is curved in most cases). The incoming elementary parallel beams are converging in the back focal plane of the lens with the distance of their focal point from the optical axis being a (monotonous) function of the angle of beam inclination. A parallel beam of direction D1 is focused in the objective focal plane at a point P1. The same is true for another direction D2 focused at point P2 due to the properties of the focal plane of a lens. Each point of the focal plane thus receives the light of one direction only. If we can examine this focal plane, we could infer the properties of each point of the figure from the properties of the associated light direction. Thus, conoscopy is not looking at the image of the crystal produced by the objective lens in the usual way with a normal eyepiece (orthoscopic image), but at what happens in the focal plane of the objective lens. Liquid crystals are often observed in sandwiched cells. If conoscopy is used for multilayer structures one has to respect the fact that the external angle is not equal to the internal angle. At the outer boundary of the sample, refraction occurs when no special measures are taken. Figure 7.13 presents the details. The light enters, usually from air, and is refracted towards the axis of the microscope, which decreases the effective angle inside the sample. If the refractive index of the sample is known, one is able to calculate the maximum internal angle for a given numerical aperture of the objective. If the refractive index of the surrounded material is given by n_i, the internal conoscopic angle can be calculated as

$$\Theta_\mathrm{i} = 2 \cdot \sin^{-1}\left(\frac{\mathrm{N.A.}}{n_\mathrm{i}}\right) \tag{7.10}$$

where N.A. is the numerical aperture of the objective.

Table 7.9 gives some values if a sample is observed in air and the cover glass has a refractive index of $n_\mathrm{i} = 1.5$. When index-matching oil is used, this effect does

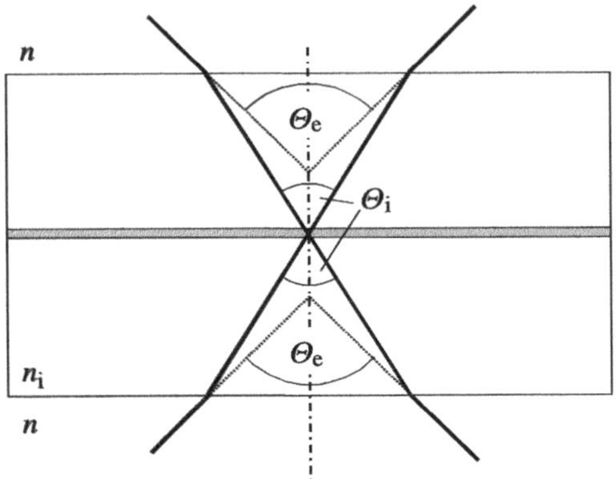

Figure 7.13 The conoscopic internal angle is different from the external because of refraction. The thin layer (gray) is observed under an angle Θ_i, and the numerical aperture of the objective gives Θ_e. In the case of immersion with index-matching fluids, this effect disappears.

not appear, because refraction at the surface is inhibited. In the microscope with illumination by a condenser the objective lens alone is sufficient for generation of the directions image, and no other means is required. The elementary parallel beams coming from the crystal are focused by the lens to form the directions image in the back focal plane of the lens. The simplest method to observe the directions image is to remove the ocular and directly view the image in the focal plane through the analyzer. The most usual method employed is to insert in the body tube an auxiliary lens – the Bertrand lens – most often situated above the analyzer. The Bertrand lens images the interference figure into the focal plane of the ocular. Then the interference figure can be seen magnified. Figure 7.14 shows a schematic ray trace in the microscope. The Bertrand lens is found after the field lens. The condenser aperture diaphragm should be open wide to access a maximum of angles. Very high numerical aperture condensers with N.A. >0.9 are usually used. For special techniques like rotational table methods that is not possible because of the long working distance required. The loss of accessible angle is often compensated by the additional information available by rotating the sample. It often happens that in a mounted sample the domains are so small that more than one appears in the field. In this case it is necessary to isolate the interference figure of a selected domain. A diaphragm should be used to illuminate only the desired domain.

Table 7.9 Some characteristics of objectives used for conoscopy, including their internal conoscopic angle.

Numerical aperture, N.A.	0.4	0.6	0.9	1.3 Oil
Magnification	20×	50×	100×	100×
External conoscopic angle Θ_e	46°	73°	128°	120
Internal conoscopic angle with cover glass of $n_i = 1.5$, Θ_i	30°	46°	72°	120
Working distance [mm]	12	6	0.3	0.2

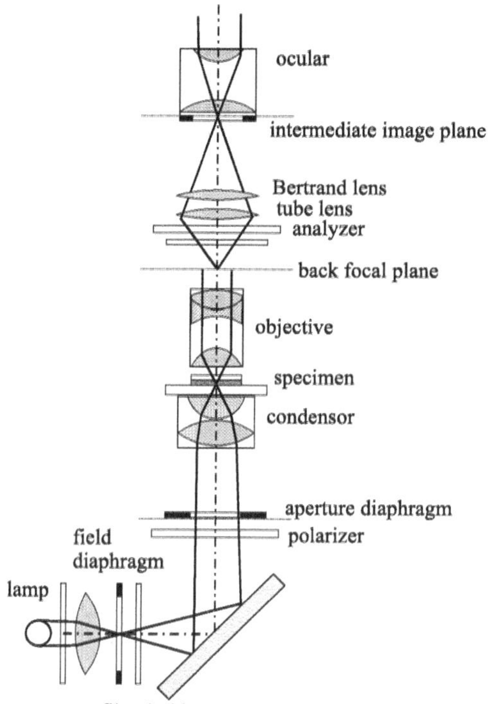

Figure 7.14 A microscope light trace for conoscopic observation. Compared to orthoscopic observation, an additional lens, the Bertrand lens, is placed above the field lens and makes an image of the back focal plane in the intermediate image plane. The illumination is not changed with respect to Figure 7.4.

Research polarized light microscopes may have a provision for centration and focusing of the Bertrand lens.

To put the polarization microscope with crossed polarizers in the conoscopic mode, the following steps should be taken:

1. Switch to the highest power objective lens. The numerical aperture should be as high as possible (observation in air >0.9, observation in an immersion fluid >1.3).
2. Switch in the condenser lens. Adjust for Köhler illumination. Take care that the N.A. of the condenser lens is at least that of the objective or higher. Open the aperture diaphragm completely to assure illumination through all angles.
3. Switch in the Bertrand lens. Focus the Betrand lens.

7.7.2 Uniaxial Material Conoscopy

Different classes of anisotropic materials react differently and we will only discuss the appearance of conoscopic figures for uniaxial crystals. Imagine that we have a liquid crystal in homeotropic orientation or a crystal of some uniaxial mineral oriented in a position such that the optical axis is perpendicular to the stage. In the conoscopic mode we will find a centered uniaxial interference figure as shown in Figure 7.11. It consists of dark bands that cross the field of view to form a black cross called isogyres. Depending on the thickness and birefringence of the birefringent layer, one may also see bands of color that form circles that are concentric with the center of the cross. These color bands are called isochromes. At the intersection of the bars of the cross is the melatope, where light that has traveled along the optical axis exits. For a perfectly centered uniaxial interference figure, with the optical axis exactly perpendicular to the microscope stage, the cross will not change position or orientation during a full 360° rotation of the stage. As the thickness and birefringence of the crystal increase, the number of isochromes seen in the interference figure increases.

We know from Chapter 2 that the variation of refractive index with direction of incident light is represented by the index ellipsoid. To determine the refraction indices and vibration directions of a light beam moving through a crystal, one considers sections in the uniaxial index ellipsoid perpendicular to the light direction. Let as examine now the complete uniaxial figure perpendicular to the optic axis. In Figure 7.15 a uniaxial index ellipsoid is shown. Three different directions of light propagation are represented with their corresponding section of index ellipsoid. For each direction, the distance covered by light in the crystal is different, being at a minimum for the direction perpendicular to the crystal section A. Moreover, the birefringence is also direction-dependent according to the section in the index ellipsoid. For each direction in B and C, a different effective refractive index is

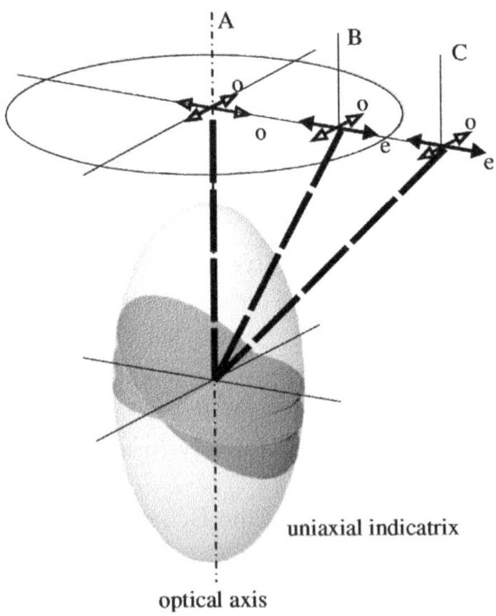

Figure 7.15 Illustration of the appearance of conoscopic figures in uniaxial material.

found. The calculation of the effective index was carried out in Section 2.7. The retardation depends on the light propagation direction due to the effective refractive index and the different path lengths through the birefringent layer because of the oblique propagation. The section for the case when the propagation is along the optical axis is a circle in the index ellipsoid. The associated light beam parallel to the optical axis is focused at the center of the conoscopic figure (point A). This point is always dark, because birefringence for this direction is zero (optical axis direction). The section with a more oblique propagation leading to B is more inclined, its indices are parallel to the polarizer and analyzer main directions. The light focuses on point B, and the indices are now n_o and n'_{eff}. There is also extinction for this beam due to the condition that the indices are parallel to the polarizer and analyzer. This is easily seen when in Equation (4.14) the value for $\varphi = 0$ is used. The section with even more inclination of the propagation direction leading to C is similar to the preceding one with a greater index n''_{eff}. Point C on the interference figure should be dark for the same reason as in B. We can draw on Figure 7.15 the orientation of refraction indices, as shown in Figure 7.16.

As the index ellipsoid for uniaxial material has circular symmetry around the optical axis, the interference figure should also have a circular symmetry around the pole of the optical axis (center of figure). As can be seen in Figure 7.16, the ordinary index n_o is always tangential to the concentric circles, and the extraordinary indices n_{eff}' are always radial with a magnitude increasing from center to periphery. The value of this n_{eff}' index is constant on a circle. All the points with indices parallel to the directions of the polarizer and analyzer are dark. They form the isogyre. The isogyre is invariant by rotation because, for symmetry reasons, the projection of the indices of refraction is also invariant by rotation. All points on a circle have the same indices of refraction with different orientations. For symmetry reasons, the optical

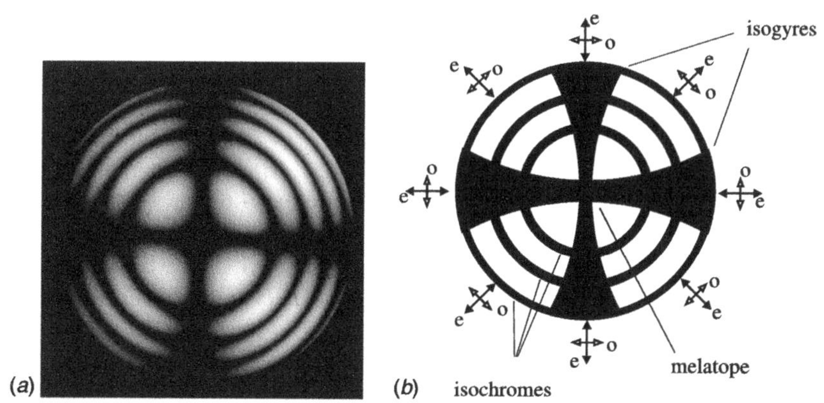

Figure 7.16 (*a*) A centered uniaxial interference figure is shown (optical axis figure). A homeotropic aligned liquid crystal layer from E7 is observed at 550 nm wavelength and with an objective 40× N.A. of 0.60. The polarizer and analyzer are oriented at 0 and 90°, respectively. (*b*) Illustration of the local axis direction.

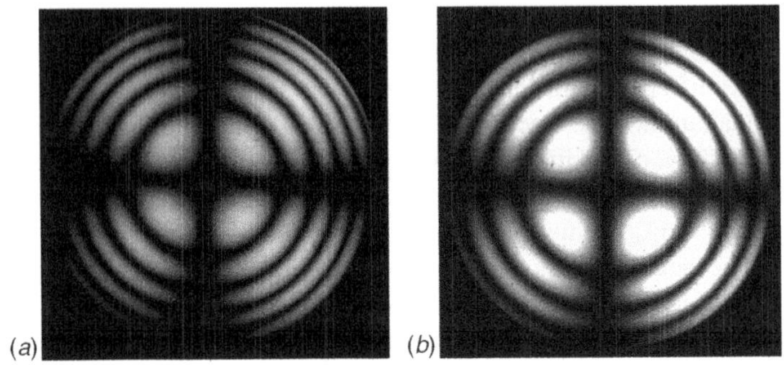

Figure 7.17 Optical axis figures from homeotropic E7 (40× N.A. 0.6) for two different wavelengths (interference filter bandwidth 5 nm): (*a*) 485 nm and (*b*) 633 nm.

path in the anisotropic layer section is also constant, because the angle of the incident light is the same for all points on a circle, so the retardation is constant. It can be seen that extinction could occur for all the points of a circle for the same wavelength of the light. This is the isochrome. Because isochromes are defined by equal retardation, their position changes when the retardation is changing or equally the retardation is held fixed and the wavelength of light is varied. Increasing retardation leads to more tightly spaced isochromes. And again, for the same sample, smaller wavelengths will produce a larger number of isochromes. Figure 7.17 shows this effect for the homeotropic sample of E7. The isochromes for a shorter wavelength are closer.

7.7.3 Determination of Optic Sign

A uniaxial interference figure can be used to determine the optic sign of a birefringent material. In the uniaxial interference figure, there will usually be an area close to the melatope that shows an interference color of first-order gray. Sometimes this is difficult to identify with liquid crystals because of their large birefringence. One determines the sign of the birefringence by observing the effects of inserting the first-order red (or a 550 nm compensator) on this gray area close to the melatope. This effect is illustrated in Figure 7.18. If the birefringent material is optically positive, the NE and SW quadrants of the interference figure turn second-order blue, because addition will have occurred in these quadrants. The NW and SE quadrants will turn first-order yellow, because subtraction will have occurred in these quadrants. If one remembers where the yellow color occurs, drawing a line connecting the yellow quadrants would cross the slow direction in the compensator, like the vertical stroke on a plus sign.

If the birefringent material is optically negative, the NE and SW quadrants of the interference figure turn first-order yellow, because subtraction will have occurred in these quadrants. The NW and SE quadrants will turn second-order blue, because addition will have occurred in these quadrants. If one remembers where the yellow color occurs, drawing a line connecting the yellow quadrants would be

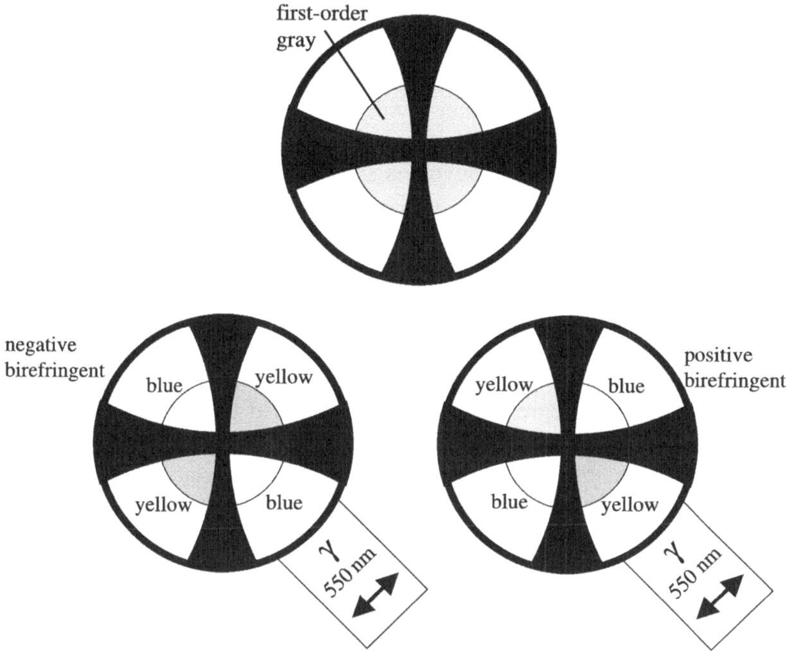

Figure 7.18 To determine the sign of the birefringence one inserts a first-order retardation plate (550 nm). Close to the melatope, where the retardation is small, one observes a characteristic color shift. The direction of the fast axis of the retarder plate is important.

parallel to the slow direction in the compensator, like the single stroke on a minus sign. Because the first-order yellow contains no blue light, it is possible to visualize this color effect when the sample is viewed under blue light. In Figure 7.19 light with a wavelength of 485 nm is used to record the uniaxial interference figures.

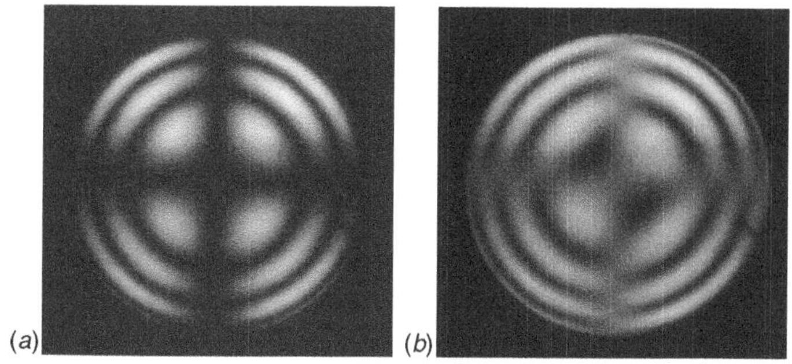

Figure 7.19 Conoscopic figure of a homeotropic sample of the nematic liquid crystal mixture E7. The sample is 20 μm thick and observed with blue light at 485 nm wavelength (100× N.A. 0.9). (*a*) The conoscopic interference image. (*b*) The conoscopic interference image when a retardation plate of 500 nm is applied. Dark zones are visible in the center at NW and SE.

A homeotropic sample of E7 and 20 μm thickness is used. In Figure 7.19*a* the uniaxial interference figure without retarder shows isogyres. If the first-order red (550 nm retardation) is put in, the NE and SW stay bright; hence they are blue. In the NW and SE, dark zones become evident; these areas are yellow and when observed in blue light become dark. It is interesting to see that due to subtraction and addition the position of the isogyres is shifted.

7.7.4 Off-Centered Uniaxial Interference Figure

Perfectly centered optical axis figures are rarely observed and it becomes necessary to work with more or less off-center figures. In particular, if liquid crystal alignment has to be studied, the off-axis angle is one of the most important quantities. For homogeneous textures it represents the pretilt angle of the liquid crystal. Off-axis interference figures result when the optical axis is not parallel to the axis of the microscope. Two possibilities are considered to illustrate cause and effect in such interference figures. Figure 7.20 shows a positive uniaxial case in which the index ellipsoid is inclined with respect to the reference surface at different angles. The drawing shows the birefringent slab in a position in which the plane of vibration of the analyzer is a plane of symmetry of the interference figure. The metalope is still visible in the field of view. As the sample is rotated, the optical axis describes a cone, and the metalope, a circle. Figure 7.21 shows the effect of rotating the sample in a plane. The arms of the isogyres move across the field, essentially as straight bars, paralleling the trace of the planes of vibration of the upper and lower polarizer. This depends, of course, very much on the quality of the sample. If the metalope is completely outside the field of view, the cross is no longer visible. Figure 7.22 illustrates that effect and shows an example. The crystal plate in this case is uniaxial negative, as indicated by the outline of the indicatrix, with a higher birefringence. The same effect is observed. For higher birefringence the point of emergence of the optical

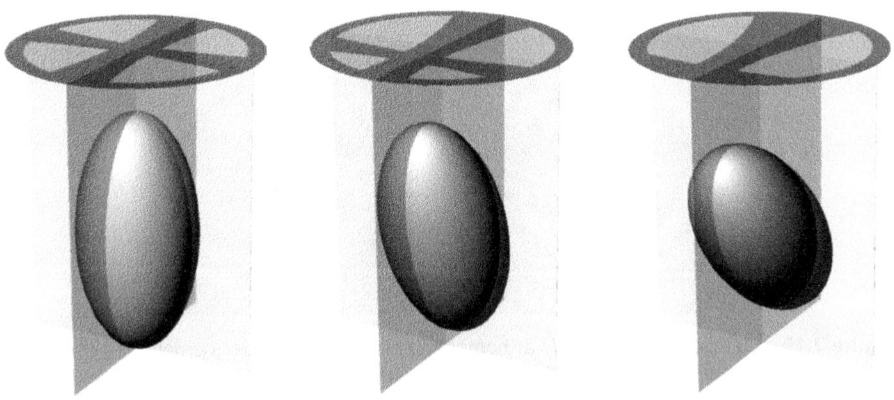

Figure 7.20 Appearance of off-axis interference figures for different inclinations of the indicatrix. The melatope is in the field of view of the microscope.

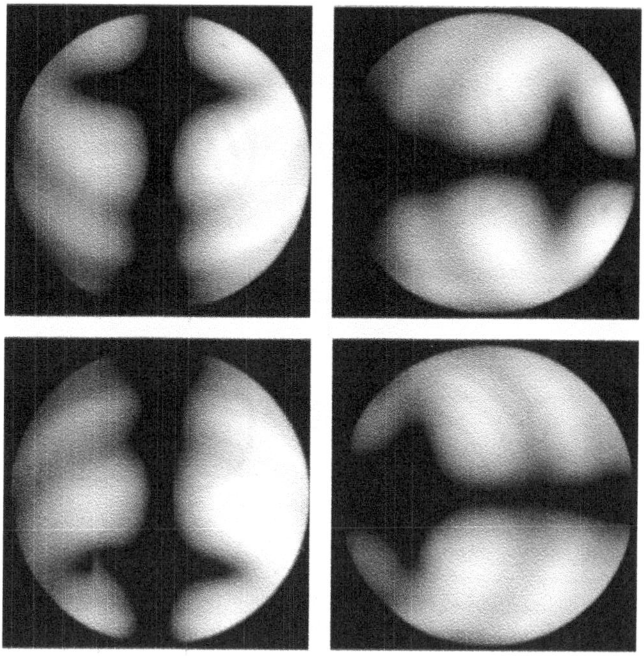

Figure 7.21 Clockwise rotation of an off-centered optical axis figure. The sample is a hybrid aligned high-pretilt sandwich of E7. The thickness is 6 μm. The quality of the sample does not allow the sharp isogyres to be observed.

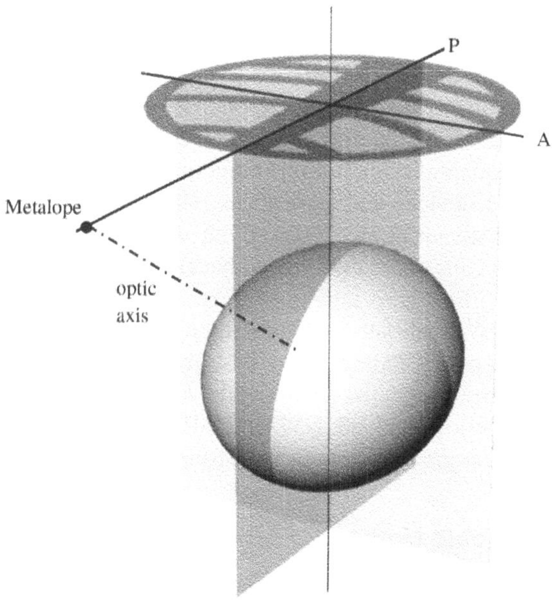

Figure 7.22 Off-centered optical axis figure for a negative birefringent uniaxial material. The melatope is outside of the field of view of the microscope.

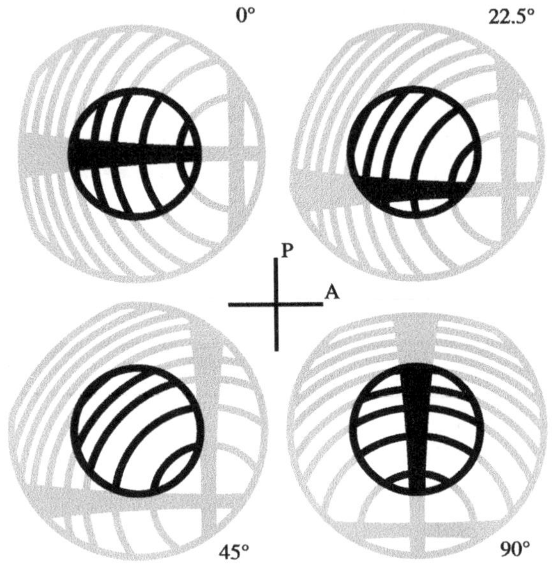

Figure 7.23 Clockwise rotation of an off-centered optical axis figure. The melatope lies outside the field of the microscope. In black, the interference figure within the field of view, and in gray the melatope outside the field of view.

axis describes a circle that is outside the field of view. As the anisotropic plate is rotated on the microscope stage, the point of emergence of the optical axis describes a circle outside the field of view. A study of the motion of the bars permits the location of the approximate position of emergence of the optical axis. This depends very much on the quality of the sample. For example, the isogyres and isochromes in Figure 7.23 show that the point of emergence of the optical axis lies in the lower right-hand quadrant. Such a technique might be used to study the pretilt angle of liquid crystal devices and at surfaces (Crossland et al., 1976; Van Horn and Winter, 2001; Nastishin et al., 2001). Even with an off-centered interference figure, the optical sign can still be determined because one can keep track of the isogyres and still divide the figure into quadrants. Again, for an optically positive crystal, all NE and SW quadrants will turn blue and the NW and SE quadrants will turn yellow, both colors replacing the first gray color present before insertion of the compensator. Similarly, for an optically negative crystal, all NE and SW quadrants will turn yellow and all NW and SE quadrants will turn blue, both colors replacing the first gray color present before insertion of the compensator.

7.7.5 Rotational Table Methods

In a routine examination of liquid crystal and polymer structures, orthoscopic and conoscopic techniques are used to determine the birefringence and the orientation of the optical axis. Polymers and liquid crystals are less ordered than minerals. This leads to a degeneration of the interference image and often does not allow the methods known from crystal analyzers to be used. The sample size in addition is usually much larger than that known from the inspection of minerals. It is therefore interesting to use equipment that will permit controlled rotation of the specimen into a desired position. Any device that can be attached to the microscope stage to permit rotation about one or more axis than the axis of the microscope is called a universal stage or stage goniometer. In former times it was possible to obtain such devices with up to five axes of rotation. Today they are difficult to find.

A simple version of such a stage with two additional rotational axes is shown in Figure 7.24. This universal stage can be used with standard microscopes when long working distance objectives (>15 mm) and condensers (>15 mm) are used. The long working distances limits the numerical aperture and usually the maximum is N.A. = 0.6. This represents an internal angle of 46°.

To show where such a device is of use we present here an example of an analysis of a hybrid liquid crystal texture with pretilt. The liquid crystal is confined between glass plates. The angles given in the interference figure are not the true angles inside the structure. If one is able to rotate the optical axis cross in the center, one can read the angle directly. In addition, the determination of the sign of birefringence becomes more convenient in the optical axis position. Figure 7.25 presents the example. The hybrid texture with homeotropic and pretilted planar alignment leads to an interference figure where the melatope is outside the field of view. When the sample is rotated, the melatope can be brought into the optical axis position and the angle can be read directly at the stage. More information can be found in more specialized literature (Birchon, 1961; Wahlstrom, 1969; Schuhmann and Kornder, 1973).

Figure 7.24 Rotational table that allows rotation of samples about two axes.

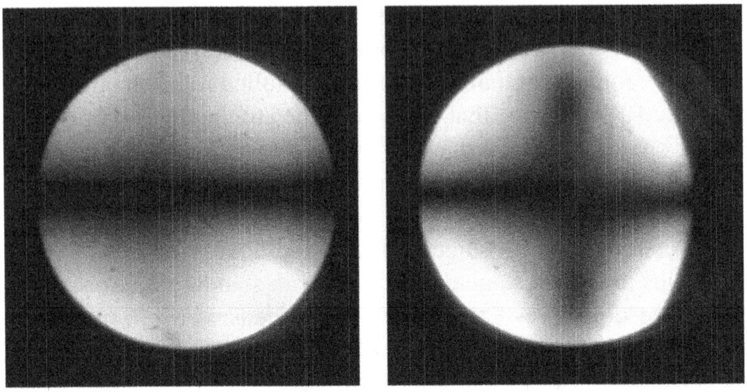

Figure 7.25 With the rotational table the melatope of the structure can be brought into the optical axis. The example shown here allows the direct determination of an angle that is linked to the internal structure of a hybrid aligned liquid crystal film (white light, E7, 9 μm, 20× N.A. 0.4, inclination angle 19°).

7.8 LOCAL POLARIZATION MAPPING

In liquid crystal texture the birefringence or dichroism can vary rapidly in time and space, revealing subtle changes in molecular arrangements occurring at a submicroscopic scale. In order to enhance the contrast in polarized light microscopy, it is important to reduce to a minimum the contribution to the image that arises from scattered light. It is usual to characterize this by an extinction ratio, which is defined as the ratio of light intensity transmitted between parallel polarizers to that transmitted when the polarizers are crossed. Ideally, one would like the extinction coefficient to be infinite. However, it is a fundamental consequence of the image formation properties of a conventional microscope that the extinction coefficient is finite, even if perfect polarizers are used (Kubota and Inoué, 1959). Typically, a value of 10^3 is obtained.

Different techniques have been proposed to increase the spatial resolution (Massoumian et al., 2003) using confocal microscopy or the contrast by polarization mapping (Oldenbourg and Mei, 1995; Kaminsky et al., 2004). The techniques for fast and sensitive measurements of two-dimensional birefringence patterns have significantly improved during the past two decades. These methods came originally from biological applications with relatively low retardations and contrast (Rowe et al., 1995). In the mid-1990s, the first imaging measurement systems with rotated optical polarization elements were reported (Otani et al., 1994; Bajor, 1995; Pezzanitti and Chipman, 1995; Zhu et al., 1999). Devices with a variable retarder and/or beam splitting (Oldenbourg and Mei, 1995; Laude-Boulesteix et al., 2004) were also proposed. The systems measure distributions of birefringence by use of a liquid-crystal universal compensator. The first commercially available systems now allow the determination of spatially varying polarization properties for contrast enhancement (CRI, 2005). The measurement range is limited to the

250 nm retardation well suited for biological applications but often too small for liquid crystal texture characterization. The aim of the measurement systems is to determine locally absorption, dichroism, the direction of the optical axis, and the retardation.

A complete analysis is only possible when the Jones or Mueller matrix is completely determined at each position in the field of view. The Jones matrix is a 2×2 matrix with complex entries and contains eight independent values. A basic measurement scheme is given in Gerrard and Burch (1994). These methods involve passing beams of light through the device and through various polarizers and phase plates, and then measuring the intensity of each beam that emerges. Four different polarization states at the input are analyzed in linear or circular light at the exit. Overall, 10 measurements have to be made to obtain all parameters with proper sign. In this case Jones matrices are determined, which means that depolarization effects are not considered. To obtain information about depolarization effects, the Mueller matrix analysis has to be performed (Dahl, 2001). To obtain high-quality measurements it is important to use high-extinction-ratio polarizing elements. On the other hand, the electronic imaging equipment has to be sensitive to also detect low-intensity values. Otherwise the calculation of the matrix parameters becomes undefined and fails if divisions have to be performed during calculation.

With assumptions on the matrices it is possible to work with simplified measurement schemes. One of the commercially available systems (PolScope from CRI instruments; CRI, 2005), for instance, contains in series a monochromatic light source, a variable elliptical polarizer, a condenser, a specimen, an objective lens, and a right- or left-circular analyzer and imaging detector. The variable retarder at the entrance generates arbitrary polarization states that will be analyzed by the circular polarizer. The variable elliptical polarizer and analyzer can be made from a pair of liquid crystal retarder cells with 45° between their principal axes and a linear polarizer. Instead of the liquid crystal retarders, a pair of variable compensators, such as Berek or Babinet–Soleil compensators, or a pair of electro-optical or piezo-optical modulators can be used for this purpose.

A different method to increase resolution is by using an interference microscope with polarized light (Sickinger et al., 1999; Dändliker et al., 2004). The advantage here arises from the phase measurement of the light field, which is in principle not limited in lateral resolution as is the intensity measurement (Rockstuhl et al., 2004). The experimental setup of the interferometer can be a classical Mach Zehnder configuration mounted on a microscope stand (Sickinger et al., 1999). The light source is usually a laser coupled onto a fiber. A fiber coupler splits the wavefield into a reference and an object arm with an adjustable energy ratio. This permits the optimization of the contrast of the interference fringes for ease in determination of the phase. In the object arm the light that exits the fiber illuminates a sample. The light from the exit of the fiber is collimated onto the object such that the illuminating wave sufficiently resembles a plane wave (its Gaussian waist is much larger than the transversal extension of the object). The transmitted light enters a magnification stage that consists of a telescopic system, with the first lens being a variable objective and the second lens having a fixed focal length. After recombination, the resulting intensity distribution is imaged on camera. The phase distribution of the wavefield in the object arm can be obtained with the help of a five-frame interference algorithm, where five frames of the two-dimensional intensity are recorded, each frame being consecutively shifted in phase by adding an additional phase delay of a quarter of a wavelength into the object arm. For phase field measurements, the resolution in the x–y-plane is classically given by an object-plane equivalent of a pixel of the CCD camera. Wavefields with a resolution down to 20 nm have been measured with a second magnification step with such instruments (Rockstuhl et al., 2004). We show examples of measurements with such a device in Chapter 8.

REFERENCES

Axometrics (2006) Axometrics, Inc., 515 Sparkman Drive, Huntsville, AL 35816, USA. Available at http://www.axometrics.com/reference.htm.

Bajor, A.L. (1995) Automated polarimeter-macroscope for optical mapping of birefringence, azimuths, and transmission in large area wafer. Part I. Theory of the measurement, *Rev. Sci. Instrum.* **66**, 2977–2990.

Beyer, H. and Riesenberg, H. (1988) *Handbuch der Mikroskopie*, 3rd edn, Verlag Technik, Berlin.

Birchon, D. (1961) *Optical Microscope Technique*, Georges Newnes, London.

CRI (2005) Technical documentation, CRI Polscope, LOT Oriel GmbH and Co. KG, Darmstadt, Germany.

Crossland, W.A., Morrissy, J.H., and Needham, B. (1976) Tilt angle measurements of nematic phases of cyanobiphenyls aligned by obliquely evaporated films, *J. Phys. D* **9**, 2001–2014.

Dahl, I. (2001) How to measure the Mueller matrix of liquid-crystal cells, *Meas. Sci. Technol.* **12**, 1938–1948.

Dändliker, R., Märki, I., Salt, M., and Nesc, A. (2004) Measuring optical phase singularities at subwavelength resolution, *J. Opt. A: Pure Appl. Opt.* **6**, S189–S196.

Francon, M. and Mallick, S. (1971) *Polarization Interferometers*, Wiley, London.

Gerrard, A. and Burch, J.M. (1994) *Matrix Methods in Optics*, Dover, New York.

Hallimond, A.F. (1970) *The Polarizing Microscope*, 3rd edn, Vickers Instruments, York.

Kaminsky, W., Claborn, K., and Kahr, B. (2004) Polarimetric imaging of crystals, *Chem. Soc. Rev.* **33**, 514–525.

Kubota, H. and Inoué, S. (1959) Diffraction images in the polarizing microscope, *J. Opt. Soc. Am.* **49**, 191–198.

Kuehni, R.G. (2003) *Color Space and its Division*, Wiley, New York.

Laude-Boulesteix, B., De Martino, A., Drévillon, B., and Schwartz, L. (2004) Mueller polarimetric imaging system with liquid crystals, *Appl. Opt.* **43**, 2824–2832.

Massoumian, F., Juskatis, R., Neil, M.A.A., and Wilson, T. (2003) Quantitative polarized light microscopy, *J. Microscopy* **209**, 13–22.

Microscopy (2006) Avabilable at http://www.microscopy.fsu.edu/.

Moxtek (2004) Moxtek Inc. 452 West 1260 North, Orem, Utha 84052, USA.

Nastishin, Y.A., Dovgyi, O.B., and Vlokh, O.G. (2001) Optical conoscopy of distorted uniaxial liquid crystals: computer simulations and experiment, *Ukr. J. Phys. Opt.* **2**, 98–106.

Nitto Denko (1996) Nitto Denko Corporation Tokyo Japan, Catalog EN22.

Ohkubo, K., Ohtsubo, J., and Izumi, N. (1994) Color modulation properties of a liquid-crystal device, *Appl. Opt.* **33**, 5895–5901.

Oldenbourg, R. and Mei, G. (1995) New polarised light microscope with precision universal compensator. *J. Microscopy* **180**, 140–147.

Otani, Y., Shimada, T., Yoshizawa, T., and Umeda, N. (1994) Two dimensional birefringence measurement using the phase shifting technique, *Opt. Eng.* **33**, 1604–1609.

Perkins, R.T., Hansen, D.P., Gardner, E.W., Thorne, J.M., and Robbins, A.A. (2000) Broadband wire grid polarizer for the visible spectrum, assigned to Moxtek. U.S. Patent 6,122,103, 2000. Application, No. 337970 filled on 1999-06-22.

Pezzanitti, J.L. and Chipman, R.A. (1995) Mueller matrix imaging polarimetry, *Opt. Eng.* **34**, 1558–1568.

Polaroid Corporation (1990) Technical data sheets. Available at http://www.apioptics.com/.

Schuhmann, H. and Kornder, F. (1973) In Rinne-Berek, *Anleitung zur allgemeinen und Polarisations-Mikroskopie der Festkörper im Durchlicht*, 3rd edn, E.Schweizerbart'sche Verlagsbuchhandlung, Stuttgart.

Rockstuhl, C., Salt, M., and Herzig, H.P. (2004) Theoretical and experimental investigation of phase singularities generated by optical micro- and nano-structures, *J. Opt. A: Pure Appl. Opt.* **6**, S271–S276.

Rowe, M., Pugh Jr, E., Tyo, J., and Engheta, N. (1995) Polarization-difference imaging: a biologically inspired technique for observation through scattering media, *Opt. Lett.* **20**, 608–610.

Sanritz Corporation (2000) Technical datasheet 2000–04.D1.

Shurcliff, W.A. (1962) *Polarized Light*, Harvard University Press, Cambridge.

Sickinger, H., Schwider, J., and Manzke, B. (1999) Fiber based Mach–Zehnder interferometer for measuring wave aberrations of microlenses, *Optik* **110**, 239–243.

Van Horn, B.L. and Winter, H.H. (2001) Analysis of the conoscopic measurement for uniaxial liquid-crystal tilt angles, *Appl. Opt.* **40**, 2089–2094.

Viney, C. (1986) Using the optical microscope to characterize molecular ordering in polymers, *Polym. Eng. Sci.* **26**, 1021–1032.

Völz, H.G. (2002) *Industrial Color Testing*, 2nd edn, Wiley-VCH, Weinheim.

Wahlstrom, E. (1969) *Optical Crystallography*, Wiley, New York.

Weber, M.F., Stover, C.A., Gilbert, L.R., Nevitt, T.J., and Ouderkirk, A.J. (2001) Giant polarization optics, *Science* **287**, 2451–2456.

Wyszecki, G. and Stiles, W.S. (2000) *Color Science*, 2nd edn, Wiley, New York.

Zhu, Y., Koyama, T., Takada, T., and Murooka, Y. (1999) Two dimensional measurement technique for birefringence vector distributions: measurement principle, *Appl. Opt.* **38**, 2225–2231.

Zollinger, H. (1999) *Color. A Multidisciplinary Approach*, Helvetica Chimica Acta, Zürich.

8

OPTICS OF LIQUID CRYSTAL TEXTURES

8.1 INTRODUCTION

The texture of a liquid crystal between two substrates is a complicated, inhomogeneous distribution of the director, which can include defects. Such defect textures are described in detail in several excellent publications (Kleman, 1983; Lavrentovich et al., 2001; Lavrentovich, 2003). Usually the texture is observed under the microscope, which does not give direct access to director data. Observed between crossed polarizers, colors and zones of different intensity are seen, which are result of light propagation through the entire thickness of the liquid crystal layer. Modern techniques allow the local director distributions to be probed in three dimensions (Smalyukh et al., 2001). In conventional polarizing microscope observations, the main problem is interpreting the images under the microscope in the "language" of liquid crystal physics to determine the texture, hence the local orientation of the director.

In this chapter we summarize the principles of texture determination by optical means. We start by introducing the basic concepts of elastic deformations for the main class of liquid crystals in use, the nematics. Three useful techniques to determine the texture of liquid crystal will be presented: the guess, conformal mapping, and the rigorous method. In the following chapter we present the basic concepts of optics for uniform textures between flat substrates, and we then discuss spatially inhomogeneous systems. Some of the problems can be reduced in the dimensionality. That is evident for gratings, but also possible for rotationally symmetric structures. We will also discuss, briefly, confined structures and the influence of

Polarized Light in Liquid Crystals and Polymers. By Toralf Scharf

inhomogeneous director deformation on the resolution of liquid crystal optical devices.

8.2 CALCULATION OF LIQUID CRYSTAL DIRECTOR DISTRIBUTIONS

8.2.1 Basic Equations

In a uniform state, a liquid crystal has a structure characterized by position-independent order parameters. When constraints are imposed on the sample by limiting its surfaces (e.g., with the walls of a container) or by external fields (magnetic, electric, and so on) acting on the molecules, the liquid crystal structure becomes deformed. As deformations cost energy, they are accompanied by restoring forces that oppose the deformations. The constant of proportionality between a deformation and the corresponding restoring force is known as the elastic constant. In a crystalline solid, as is well known, the material undergoes homogeneous strain under deformation and a restoring force arises to oppose the change in distance between neighboring points. This restoring force cannot exist in a positionally disordered material (e.g., liquids). However, if the material is orientationally ordered, restoring torques may arise to oppose the orientational deformations (curvatures). In those phases of liquid crystals that have both the (partial) positional and orientational orders, the restoring force and torque may arise simultaneously. The phenomena that liquid crystals exhibit under the influence of external fields are usually distinguished by two characteristics (Singh 2001):

1. The energy involved per molecule in producing these effects is small compared to the strength of intermolecular interaction, and
2. The characteristic distances involved in these phenomena are large compared to molecular dimensions.

In describing these long-length phenomena, it is convenient to regard the liquid crystal as a continuum with a set of elastic constants rather than on a molecular basis. The understanding of the elastic constants of liquid crystals is important for a number of reasons. In the first place they appear in the description of virtually all phenomena where the variation of the director (a unit vector along the direction of alignment of molecules) is manipulated by external fields and confinements. Secondly, they provide unusually sensitive probes of the microscopic structure of the ordered state. Knowledge of the elasticity of liquid crystals is also necessary for the study of order parameter fluctuations and defect stability in them.

As the free energy of a system with broken continuous symmetry is invariant with respect to spatially uniform displacements that take the system from one point in a ground state manifold to another, the free energy density is expanded in terms of the spatial derivatives of the order parameter fields. The elastic continuum theory deals only with small spatial derivatives. Consequently, only the lowest order terms in the expansion are taken into account. The contribution to $f(r)$ due to deformation is called the elastic free energy density $f_d(r)$, which by definition is function of the spatial derivatives of the order parameter fields. In addition to $f_d(r)$, we may have contributions to $f(r)$ due to external fields (if any) on the sample. Thus, in general

$$f(r) = f_0 + f_d(r) + f_e(r) \tag{8.1}$$

where f_0 and $f_e(r)$ represent the free energy density of a spatially uniform state and contribution due to external fields, respectively. As the liquid crystal molecules are generally diamagnetic, electrically polarizable, and anisotropic in their magnetic and electric properties, the application of a field usually helps to align molecules. Therefore, $f_e(r)$ is opposite in sign to $f_d(r)$. Often, due to competition between the alignments favored by the field, deformed structures are created. Experimentally, it is possible to fix the orientation of the molecules at a boundary by surface treatment of the container walls and therefore to generate desired deformations with outside external fields.

Uniaxial nematic liquid crystal molecules can be visualized as a system consisting of rotationally symmetric ellipsoids, the orientation of which is described by a unit vector **n** called the director. For a uniaxial system with basis vector **n** along the axis of continuous rotation symmetry, the constraints of this symmetry should be satisfied by all system parameters. These requirements lead to the well-known Frank expression for the nonchiral uniaxial nematics (Gennes, 1993):

$$f_d = \tfrac{1}{2}K_{11}(\nabla \cdot \mathbf{n})^2 + \tfrac{1}{2}K_{22}(\mathbf{n}(\nabla \times \mathbf{n}))^2 + \tfrac{1}{2}K_{33}(\mathbf{n} \times (\nabla \times \mathbf{n}))^2. \tag{8.2}$$

This is a standard expression for the elastic free energy density associated with distortion of the director field $\mathbf{n}(\mathbf{r})$ for a nematic liquid crystal in the absence of chiral or polar effects. The term $K_{11}(\nabla\mathbf{n})^2$ is called splay, $K_{22}(\mathbf{n}\nabla \times \mathbf{n})^2$ is twist, and $K_{33}(\mathbf{n} \times \nabla \times \mathbf{n})^2$ is bend. Equation (8.2) has three bulk elastic constants, K_{11}, K_{22}, and K_{33} (unit [N]), associated with the deformations. The complete equation would contain surface elasticity terms, but as surface terms do not, in general, contribute to bulk properties, they can be dropped. They do, however, play an important role in the surface properties of liquid crystals. Information about anchoring energy can be derived from their measurements. The cholesteric liquid crystals exhibit helical ordering on a macroscopic scale. This leads to an additional term in Equation (8.2) that is linear. The equation for the free energy density becomes

$$f_d = k_2(\mathbf{n}(\nabla \times \mathbf{n})) + \tfrac{1}{2}K_{11}(\nabla \cdot \mathbf{n})^2 + \tfrac{1}{2}K_{22}(\mathbf{n}(\nabla \times \mathbf{n}))^2 + \tfrac{1}{2}K_{33}(\mathbf{n} \times (\nabla \times \mathbf{n}))^2, \tag{8.3}$$

where k_2 is a new elastic constant measured in units [N/m]. Taking $q_0 = k_2/K_{22}$, the (bulk) free energy density for a cholesteric phase can be written as

$$f_\mathrm{d} = \tfrac{1}{2}K_{11}(\nabla\cdot\mathbf{n})^2 + \tfrac{1}{2}K_{22}(\mathbf{n}(\nabla\times\mathbf{n}) + q_0)^2 + \tfrac{1}{2}K_{33}(\mathbf{n}\times(\nabla\times\mathbf{n}))^2 - \tfrac{1}{2}K_{22}q_0^2. \quad (8.4)$$

The twist of a cholesteric liquid crystal has full pitch of $2\pi/q_0$, but as $\mathbf{n}(\mathbf{r})$ and $-\mathbf{n}(\mathbf{r})$ are physically indistinguishable, the physical period of repetition is π/q_0. The twist term in Equation (8.4) can be considered to be a contribution due to deformation from this uniform helical structure.

As f_d must be positive in order to give stability for the uniformly aligned state, all the K values must be positive. Their values can be obtained from a theory that links the continuum theory with microscopic theory including the molecular parameters (Wu, 2005). From experimental data on the Frank elastic constants, one finds that, for rod-shaped molecules, K_{22} is always the smallest of the three. For most of the materials one finds

$$0.5 < \frac{K_{33}}{K_{11}} < 3.0 \qquad \text{and} \qquad 0.5 < \frac{K_{22}}{K_{11}} < 0.8. \quad (8.5)$$

However, for the discotic nematic, K_{22} is the largest and K_{33} the smallest; that is, $K_{22} > K_{11} > K_{33}$. The value of these constants is found to depend on the molecular length-to-width ratio, but not on dipole moment, if any, of the constituent molecules. For nematics that show a smectic phase at lower temperatures, a presmectic stiffening (increase of K_{22} and K_{33}) is observed (Singh, 2001). With the free energy density the total distortion free energy ΔF can be calculated by integrating over the volume:

$$\Delta F = \int f_\mathrm{d}\,\mathrm{d}r. \quad (8.6)$$

One can consider different deformations. We will use the example of a planar cell as shown in Figure 8.1 for illustration.

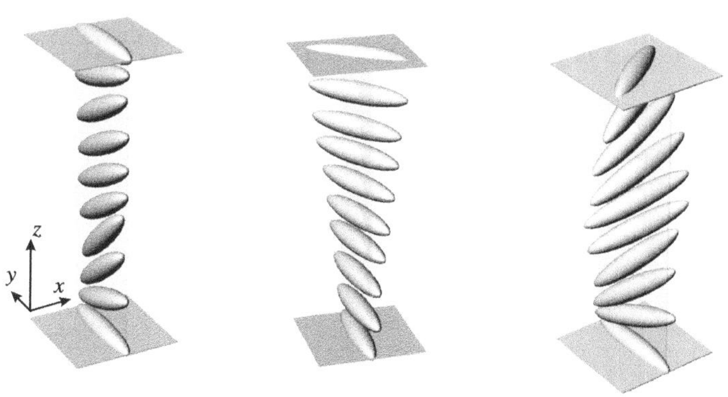

Figure 8.1 Tilt, twist, and mixed deformation in a planar-aligned liquid crystal texture.

In the tilt-only deformation the nematic director has no x-components and is only tilted towards the z-axis with a tilt angle θ. It can be expressed as

$$\mathbf{n} = \begin{pmatrix} 0 \\ \cos\theta \\ \sin\theta \end{pmatrix}. \tag{8.7}$$

This describes the deformation profile where an initial director distribution $\mathbf{n}(z)$ is parallel to the y-axis. If θ depends only on z, the free energy density can be simplified as follows by using

$$\nabla\cdot\mathbf{n} = \cos\theta(z)\frac{\mathrm{d}\theta(z)}{\mathrm{d}z} \quad \text{and} \quad \nabla\times\mathbf{n} = \begin{pmatrix} \sin\theta(z)\dfrac{\mathrm{d}\theta(z)}{\mathrm{d}z} \\ 0 \\ 0 \end{pmatrix} \tag{8.8}$$

to find

$$f_\mathrm{d} = \frac{1}{2}(K_{11}\cos^2\theta(z) + K_{33}\sin^2\theta(z))\left(\frac{\mathrm{d}\theta(z)}{\mathrm{d}z}\right)^2. \tag{8.9}$$

A second prominent example is the twist mode. If one considers only deformations along the z-axis, which is of course not the general case, the equation can be simplified. Only deformations that leave the director parallel to the x-axis are considered. The director can be represented as

$$\mathbf{n} = \begin{pmatrix} \cos\varphi \\ \sin\varphi \\ 0 \end{pmatrix}, \tag{8.10}$$

which leads, in the case of z-only dependence of the twist angle φ, to

$$\nabla\cdot\mathbf{n} = 0 \quad \text{and} \quad \nabla\times\mathbf{n} = \begin{pmatrix} -\cos\varphi(z) \\ -\sin\varphi(z) \\ 0 \end{pmatrix}\frac{\mathrm{d}\varphi(z)}{\mathrm{d}z}. \tag{8.11}$$

The deformation free energy can then be written as

$$f_\mathrm{d} = \frac{1}{2}K_{22}\left(\frac{\mathrm{d}\varphi(z)}{\mathrm{d}z}\right)^2. \tag{8.12}$$

If both modes are combined, the director can be expressed with

$$\mathbf{n} = \begin{pmatrix} \cos\varphi\cos\theta \\ \sin\varphi\cos\theta \\ \sin\theta \end{pmatrix}. \tag{8.13}$$

The angles φ and θ are the twist and tilt angle, respectively. As both parameters depend only on z, the equation for the free energy become

$$f_\mathrm{d}=\frac{1}{2}K_{11}\cos^2\theta(z)\left(\frac{\mathrm{d}\theta(z)}{dz}\right)^2+\frac{1}{2}K_{22}\cos^4\theta(z)\left(\frac{\mathrm{d}\varphi(z)}{dz}\right)^2$$
$$+\frac{1}{2}K_{33}\sin^2\theta(z)\left(\frac{\mathrm{d}\theta(z)}{dz}\right)^2+\frac{1}{2}K_{33}\cos^2\theta(z)\sin^2\theta(z)\left(\frac{\mathrm{d}\varphi(z)}{dz}\right)^2, \qquad (8.14)$$

where we have used

$$\nabla\cdot\mathbf{n}=\cos\theta(z)\frac{\mathrm{d}\theta(z)}{\mathrm{d}z}$$

and

$$\nabla\times\mathbf{n}=\begin{pmatrix}\sin\varphi(z)\sin\theta(z)\dfrac{\mathrm{d}\theta(z)}{\mathrm{d}z}-\cos\varphi(z)\cos\theta(z)\dfrac{\mathrm{d}\varphi(z)}{\mathrm{d}z}\\ -\cos\varphi(z)\sin\theta(z)\dfrac{\mathrm{d}\theta(z)}{\mathrm{d}z}-\sin\varphi(z)\cos\theta(z)\dfrac{\mathrm{d}\varphi(z)}{\mathrm{d}z}\\ 0\end{pmatrix}. \qquad (8.15)$$

A different notation can be found that separates the twist and tilt angle derivatives:

$$f_\mathrm{d}=\frac{1}{2}(K_{11}\cos^2\theta(z)+K_{33}\sin^2\theta(z))\left(\frac{\mathrm{d}\theta(z)}{dz}\right)^2$$
$$+\frac{1}{2}\cos^2\theta(z)(K_{22}\cos^2\theta(z)+K_{33}\sin^2\theta(z))\left(\frac{\mathrm{d}\varphi(z)}{dz}\right)^2. \qquad (8.16)$$

The equations above allow uniform textures to be calculated if boundary conditions are taken into consideration. To do so, one has to find the energy minimum of the structures including boundary conditions. Exhaustive examples including weak anchoring problems are given in Barbero and Evangelista (2001).

Expressions for the minimization of the energy become very complicated for problems when more than one spatial coordinate is involved. However, by inspection it already becomes evident that for problems where the elastic constant can be assumed to be unique, great simplification can be made. In the so-called one-constant approximation where $K_{11}=K_{22}=K_{33}=K$, Equation (8.3) becomes

$$f_\mathrm{d}=\tfrac{1}{2}K((\nabla\cdot\mathbf{n})^2+(\mathbf{n}(\nabla\times\mathbf{n}))^2+(\mathbf{n}\times(\nabla\times\mathbf{n}))^2). \qquad (8.17)$$

With this equation it becomes possible to discuss two-dimensional director configurations that are linked to defect structures. We will now discuss some basic examples, limiting our attention to two-dimensional problems in the one-constant approximation. Let us consider only the x–y-plane. We assume a director distribution that can be described as a function of the angle φ, which depends on the spatial coordinates x and y:

$$\mathbf{n} = \begin{pmatrix} \cos\varphi(x, y) \\ \sin\varphi(x, y) \\ 0 \end{pmatrix}. \tag{8.18}$$

The divergence and curl of such a director field are given by

$$\nabla \cdot \mathbf{n} = -\sin\varphi(x, y)\frac{\partial\varphi(x, y)}{\partial x} + \cos\varphi(x, y)\frac{\partial\varphi(x, y)}{\partial y}$$

and

$$\nabla \times \mathbf{n} = \begin{pmatrix} 0 \\ 0 \\ \cos\varphi(x, y)\dfrac{\partial\varphi(x, y)}{\partial x} + \sin\varphi(x, y)\dfrac{\partial\varphi(x, y)}{\partial y} \end{pmatrix}, \tag{8.19}$$

which leads in Equation (8.17), after some algebra, to

$$f_{\mathrm{d}} = \frac{K}{2}\left(\left(\frac{\partial\varphi(x, y)}{\partial x}\right)^2 + \left(\frac{\partial\varphi(x, y)}{\partial y}\right)^2\right). \tag{8.20}$$

To find this final result it is helpful to note that the term $(\nabla\mathbf{n})^2$, splay, is a number that is squared, the twist term $(\mathbf{n}\nabla \times \mathbf{n})^2$ is zero in this case, and that the bend term $(\mathbf{n} \times \nabla \times \mathbf{n})$ leads to a scalar product. Equation (8.20) serves as the basis for analysis of common bulk defect structures. Finding the function $\varphi(x, y)$ that minimizes Equation (8.6) can be done by the calculus of variations (Courant and Hilbert, 1931). We will not develop the whole theoretical background here, but will just give the derivation of the key equation for our case. The derivation is based on the assumption that the angle $\varphi(x, y)$ is composed of two terms: a function that minimizes Equation (8.6) given as $\varphi_0(x, y)$, and a second term that undergoes small variations $\varphi(x, y)$ multiplied with a constant β. One writes

$$\varphi(x, y) = \varphi_0(x, y) + \beta\varphi(x, y). \tag{8.21}$$

The constant β is assumed to be small. With

$$\frac{\partial\varphi(x, y)}{\partial x} = \frac{\partial\varphi_0(x, y)}{\partial x} + \beta\frac{\partial\varphi(x, y)}{\partial x} \quad \text{and} \quad \frac{\partial\varphi(x, y)}{\partial y} = \frac{\partial\varphi_0(x, y)}{\partial y} + \beta\frac{\partial\varphi(x, y)}{\partial y}, \tag{8.22}$$

Equation (8.20) becomes

$$f_d = \frac{K}{2}\left(\left(\frac{\partial \varphi_0(x,y)}{\partial x} + \beta\frac{\partial \varphi(x,y)}{\partial x}\right)^2 + \left(\frac{\partial \varphi_0(x,y)}{\partial y} + \beta\frac{\partial \varphi(x,y)}{\partial y}\right)^2\right). \tag{8.23}$$

After inserting f_d from Equation (8.23) into Equation (8.6) for the total energy, it has to be differentiated with respect to β. That yields the following expression

$$\frac{\partial \Delta F}{\partial \beta} = \iint_G \frac{K}{2}\left(2\frac{\partial \varphi(x,y)}{\partial x}\left(\frac{\partial \varphi_0(x,y)}{\partial x} + \beta\frac{\partial \varphi(x,y)}{\partial x}\right)\right.$$
$$\left. + 2\frac{\partial \varphi(x,y)}{\partial y}\left(\frac{\partial \varphi_0(x,y)}{\partial y} + \beta\frac{\partial \varphi(x,y)}{\partial y}\right)\right) \mathrm{d}x\,\mathrm{d}y. \tag{8.24}$$

G is the area of integration. In the limit of β going to zero one obtains

$$\lim_{\beta\to 0}\frac{\partial \Delta F}{\partial \beta} = \iint_G K\left(\frac{\partial \varphi(x,y)}{\partial x}\frac{\partial \varphi_0(x,y)}{\partial x} + \frac{\partial \varphi(x,y)}{\partial y}\frac{\partial \varphi_0(x,y)}{\partial y}\right)\mathrm{d}x\,\mathrm{d}y. \tag{8.25}$$

The result has to be set equal to zero:

$$\iint_G K\left(\frac{\partial \varphi(x,y)}{\partial x}\frac{\partial \varphi_0(x,y)}{\partial x} + \frac{\partial \varphi(x,y)}{\partial y}\frac{\partial \varphi_0(x,y)}{\partial y}\right)\mathrm{d}x\,\mathrm{d}y = 0. \tag{8.26}$$

The left-hand side can now be integrated by parts. The integration leads to an integral over the domain boundary S and to a double integral over the area G (Courant and Hilbert, 1993). The result is

$$\int_S \varphi(x,y)\frac{\partial \varphi_0(x,y)}{\partial x}\,\mathrm{d}y + \int_S \varphi(x,y)\frac{\partial \varphi_0(x,y)}{\partial y}\,\mathrm{d}x$$
$$-\iint_G\left(\frac{\partial^2 \varphi_0(x,y)}{\partial x^2} + \frac{\partial^2 \varphi_0(x,y)}{\partial y^2}\right)\varphi(x,y)\,\mathrm{d}x\,\mathrm{d}y = 0. \tag{8.27}$$

As there must always be a boundary where $\varphi_0(x, y)$ is fixed, on the boundary $\varphi(x, y)$ must be equal to zero. The first two terms are evaluated on the boundary and because on the boundary $\varphi(x, y)$ is zero, the first term in Equation (8.27) vanishes. $\varphi(x, y)$ is otherwise arbitrarily chosen, so the only way to make the integral over the area zero is by setting the term in brackets equal to zero. Thus the function $\varphi_0(x, y)$ that minimizes ΔF must satisfy the equation

$$\frac{\partial^2 \varphi_0(x,y)}{\partial x^2}+\frac{\partial^2 \varphi_0(x,y)}{\partial y^2}=0. \tag{8.28}$$

This is Laplace's equation in two dimensions. At this point we want to repeat the conditions needed to obtain this equation. We have assumed a two-dimensional problem in the one-constant approximation where the elastic anisotropy is neglected. All methods for solving Laplace's equation are suited to determining the equilibrium director profile. In what follows, we consider the equilibrium configuration and we write, instead of φ_0, only φ.

For rotationally symmetric problems it is recommended to express Equation (8.28) in polar coordinates. With $x = \rho\cos\alpha$ and $y = \rho\sin\alpha$ Laplace's equation reads as

$$\frac{\partial^2 \varphi}{\partial \rho^2}+\frac{1}{\rho}\frac{\partial \varphi}{\partial \rho}+\frac{1}{\rho^2}\frac{\partial^2 \varphi}{\partial \alpha^2}=0, \tag{8.29}$$

where $\rho = \sqrt{x^2+y^2}$ and $\alpha = \arctan(y/x)$. There are several solutions to this equation. The one without dependence on ρ is of special interest and can be written as

$$\varphi(\alpha) = m\alpha + \varphi_c \qquad \text{with } m = \pm\tfrac{1}{2},\ \pm 1,\ \pm\tfrac{3}{2}\cdots \tag{8.30}$$

and where m and φ_c are constants. One restriction on the function $\varphi(\alpha)$ is that it has to be single valued. As the director in a liquid crystal can point in either one of two directions, this means that $\varphi(\alpha)$ must change by some multiple of π when α is increased by 2π. This is the reason why the constant m must change by a multiple of $\frac{1}{2}$ in Equation (8.30). It is not difficult to visualize the director configurations for some of these so-called axial or wedge disclinations. If m, which is called the strength of the disclination, is positive, then the director rotates counterclockwise in traversing a counterclockwise path around the disclination. If m is negative, the director rotates clockwise in traversing this same counterclockwise path. For different values of m, the angle of the rotation for the director is different. With $m = \pm\frac{1}{2}$ one gets for one full loop around the disclination an angle of π. If $m = \pm 1$ then the director rotates by 2π. Figure 8.2 illustrates different director configurations that are solutions of the posed problem. To discuss particular problems, for instance for the discussion of interaction of defects, one should not forget that the superposition principle holds for Laplace's equation. That means that director profiles can be constructed by adding different solutions of the type in Equation (8.30). The director profile given by the angle φ would then read (Singh, 2002):

$$\varphi = \sum_i m_i \arctan\left(\frac{y-y_i}{x-x_i}\right) + \text{const.} \tag{8.31}$$

The strengths of each defect are given by m_i, and the position is defined by the coordinates (x_i, y_i).

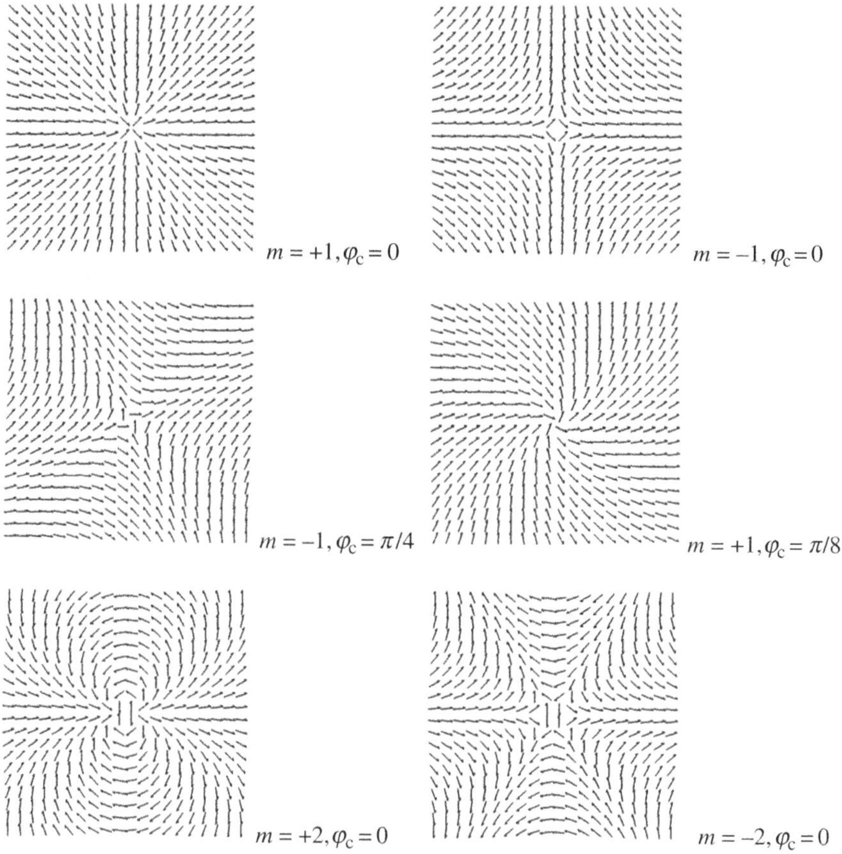

Figure 8.2 Director visualization for different defect configurations that fulfill Equation (8.30).

There are other disclinations besides axial (wedges) ones that can form in nematic liquid crystals. In axial disclinations, the rotational axis of the director in traversing a loop around the disclinations is parallel to the disclination. In twist disclinations, the rotation axis is perpendicular to the disclination. Due to the fact that the director twists, an entirely new class of disclinations forms in chiral nematic liquid crystals. Likewise, the spatial periodicity of both chiral nematic and smectic liquid crystals allows for defects in the periodic structure in addition to defects in the director configuration. These additional defects are quite different and resemble dislocations in solids (Voloschenko and Lavrentovich, 2000).

8.2.2 Conformal Mapping

When adjacent regions on a substrate surface of a nematic liquid crystal cell are subjected to different anchoring conditions, the boundary between the regions constitutes a disclination line attached to the surface substrate, that is, a surface disclination.

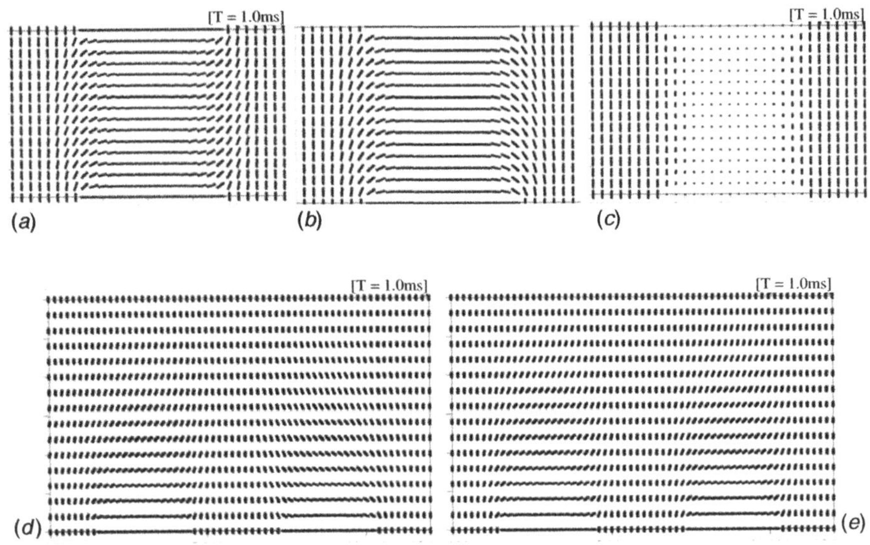

Figure 8.3 Possible liquid crystal director conformations for cells and grids with symmetric a, b, c and asymmetric d, e boundary conditions.

Disclinations sticking at a solid–nematic interface also occur when the substrate is not flat or has sharp corners or edges (Uche et al., 2005). The director configurations around the surface disclinations associated with modified flat substrates for nematics can be analyzed using a function theory method (Okano et al., 1994). In the following we assume that the director field is two-dimensional; The state of alignment in the liquid crystal cell depends only on the x- and y-coordinates, and the director $\mathbf{n}$ is in the x–y-plane. We assume that the surface anchoring of the liquid crystal is always strong. Figure 8.3 shows different liquid crystal cell structures, first as a single cell conformation and then as a periodic cell that fulfils these conditions. The single cell structure in Figure 8.3*a*, *b*, *c* are symmetric and the anchoring conditions change at the bottom of the cell between homeotropic and planar. The anchoring conditions of the structures in Figure 8.3*d* and *e* are asymmetric and so are the corresponding grid conformations. All the shown structures are theoretically possible. Whether conformations appear in reality depends on the corresponding free energy of the distortion. In the following, we will present a director field calculation for a single asymmetric cell, made with the method of conformal mapping.

Consider a cell in which the liquid crystal is confined between two parallel-plane substrates at distance d. Let the x-axis lie in the bottom substrate and the y-axis perpendicular to that plane, as shown in Figure 8.4. The director has the form given in Equation (8.18), $\mathbf{n} = (\cos\varphi, \sin\varphi, 0)$, and is confined in the (x, y) plane. It forms an angle $\varphi(x, y)$ with the x-axis. We impose the boundary conditions for the director so that we have on one substrate surface uniform alignment, and on the other a modified substrate as shown in Figure 8.4. The alignment conditions change from homeotropic to planar and from planar to homeotropic at positions $x = -a$ and

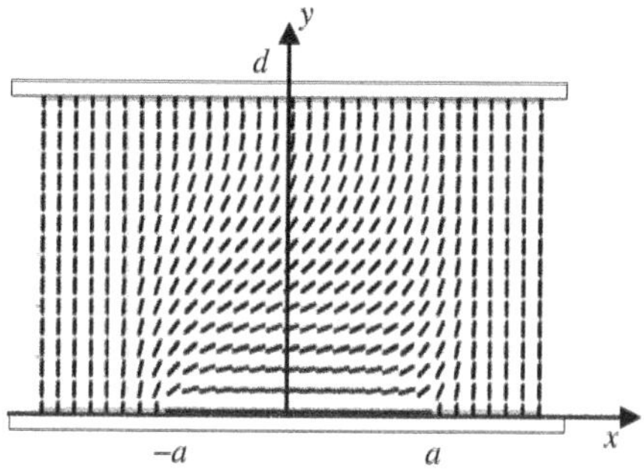

Figure 8.4 Nematic director field for a liquid crystal cell of modified substrates with one alignment-patterned surface. The planar alignment is imposed on the lower substrate from $x = -a$ to $x = a$.

$x = a$, respectively. In the one constant approximation must satisfy the equilibrium configuration of the director, in which ΔF is locally minimal, the two-dimensional Laplace's equation,

$$\frac{\partial^2 \varphi(x, y)}{\partial x^2} + \frac{\partial^2 \varphi(x, y)}{\partial y^2} = 0 \tag{8.28}$$

subjected to boundary conditions. One now introduces a complex variable $z = x + iy$ and considers a regular function Φ in the complex z-plane. Then the real and imaginary parts of Φ are harmonic functions satisfying Laplace's equation (8.28). Here we assume that the imaginary part of Φ satisfies the given boundary conditions. However, in general, it is not an easy task to determine the required regular function directly. Therefore one uses the method of conformal mapping (see for instance the work of Churchill, 1960). By means of a suitable regular function $w = f(z) = u + iv$ one maps the domain $0 \leq y \leq d$ of the complex z-plane conformal onto the upper half of the complex w-plane. If we can find a regular function $\Psi(w)$ in the w-plane that satisfies the boundary condition in the w-plane, then transforming back into the original z-plane we have the required solution $\varphi(x, y)$ of the boundary value problem as the imaginary part of $\Phi(z) = \Psi(f(z)) = \psi(x, y) + i\varphi(x, y)$, which maps the x-axis ($y = 0$) and the line $y = d$ onto the u-axis, as shown in Figure 8.5. For the special problem we discuss here in the w-plane, a regular function, whose imaginary part is compatible with the boundary conditions, is found to be

$$\Psi(w) = \tfrac{1}{2} \cdot [\log(w - e^{-\pi a/d}) - \log(e^{\pi a/d} - w)]. \tag{8.32}$$

This is because the logarithmic function for complex variables $\log(z) = \ln|z| + i\arg(z)$ changes the imaginary part from 0 to π when changing the variable z from positive to negative values; that is, $\mathrm{Im}(\log(-a)) = \pi$ and $\mathrm{Im}(\log(a)) = 0$ for a real and $a > 0$. The distinction between the director positions θ and $\theta \pm \pi$

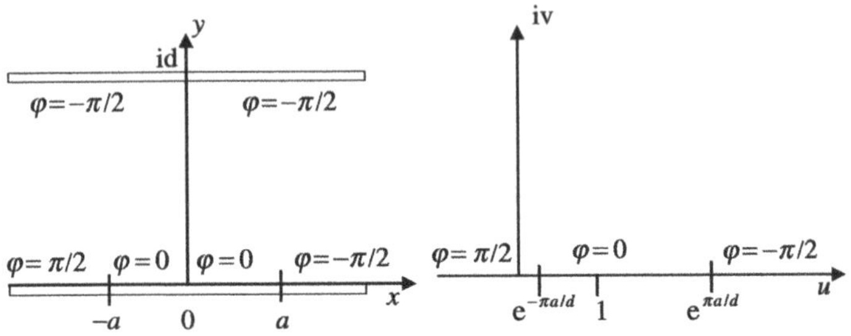

Figure 8.5 The complex *z*- and *w*-plane with boundary conditions before and after conformal mapping.

is irrelevant for the angle of the nematic liquid crystal director. Therefore there are different possibilities for creating the boundary conditions and functions $\Psi(w)$ that fulfil these boundary conditions. Different director configurations are possible, where one is energetically favorable. The configuration shown in Figure 8.4 is the one with the lowest free energy (Geurst et al., 1975; Okano et al., 1994). Backward transformation of the function $\Psi(w)$ gives

$$\Phi(z) = \Psi(\mathrm{e}^{\pi z/d}) = \tfrac{1}{2}\cdot[\log(\mathrm{e}^{\pi z/d} - \mathrm{e}^{-\pi a/d}) - \log(\mathrm{e}^{\pi a/d} - \mathrm{e}^{\pi z/d})]. \tag{8.33}$$

After simplification, one finds finally the tilt angle profile $\varphi(x, y)$ as the imaginary part of $\Phi(z)$ as

$$\varphi(z) = \tfrac{1}{2}\arctan\left(\frac{\sin(\pi y/d)\sinh(\pi a/d)}{\cos(\pi y/d)\cosh(\pi a/d) - \cosh(\pi x/d)}\right). \tag{8.34}$$

The model allows extension to gratings with more than one period. Using Equation (8.34), one is able to calculate director configurations for arbitrary cell thicknesses and sizes of the alignment patterned structure. This technique also allows analytical solutions to be found for more complicated problems such as surface modulations or other confinements as tubes and wedges. The angle φ can be either a twist or a tilt angle.

As a last example we would like to give the approximate solution for a modulated substrate surface. We are looking for a solution of the problem when the surface has a sinusoidal modulation and the director lies either flat on the surface or is perpendicular to it. We assume that φ describes a tilt angle with the x axis. By inspection one finds that the equation

$$\Phi(z) = \pi + \arctan\left[\frac{1}{h}(h - \mathrm{Im}\ w)\cos(\mathrm{Re}\ w)\right], \tag{8.35}$$

with transformation

$$w = \frac{1}{h}\{(h - \text{Im } z)[z - \text{i}\sin(\text{Re } z)] + z\text{Im } z\}, \tag{8.36}$$

leads to the following form of the director field:

$$\varphi(x, y) = \pi + \arctan\left[\frac{(h - y)(h + \sin x)\cos x}{h^2}\right]. \tag{8.37}$$

The proposed model function does not fulfil Laplace's equation and is strictly speaking no solution of the problem. However, it gives a reasonable approximation of the director profile. The parameter h describes the thickness of the layer and the surface structure has a period of 2π. The amplitude of the modulation is one. Figure 8.6 shows a visualization of the director. Without limitations, constants can be added to the director field that can rotate the director from a homeotropic to a planar configuration. Both cases are shown. The director configuration in the one constant approximation can be useful for judging whether a texture is stable or not. It allows the calculation of the free energy by using Equation (8.20). Integration over the entire domain gives the elastic energy stored in the system as predicted by Equation (8.6). In this way, qualitative information can be obtained to compare different texture models.

8.2.3 Defect States

Nematic defects refer to regions where the nematic director **n** is not uniquely defined. One distinguishes between point or line defects, where the singular region is limited to a point or line, respectively. The classification of defects is based on group-theoretical theorems. The essential group-theoretical quantity assigned to a defect is the integer topological charge. This is defined in terms of the molecular orientational ordering around the defect. Isolated defects with nonzero topological charges are topologically stable, in the sense that one cannot get rid of them by local continuous transformations without affecting the director field far from the defect site. According to their topological charge, defects fall

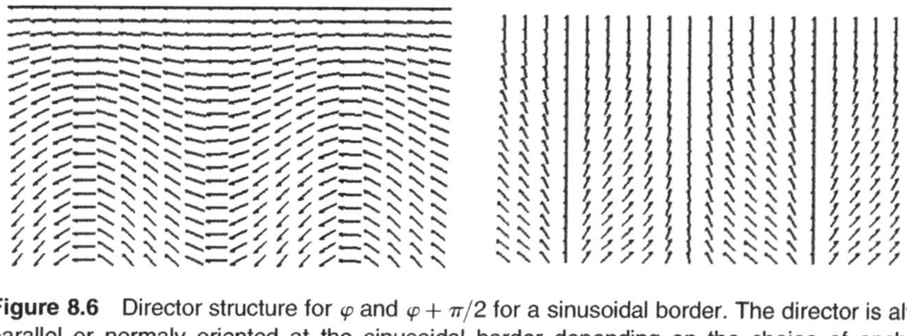

Figure 8.6 Director structure for φ and $\varphi + \pi/2$ for a sinusoidal border. The director is always parallel or normaly oriented at the sinusoidal border depending on the choice of anchoring conditions being planar or homeotropic, respectively.

into different topological classes. The law of the conservation of topological charges is analogous to the conservation laws of electric and other physical charges. However, one cannot attribute a sign to a topological charge because of the "head to tail" invariance of the director **n**; therefore, it is usually taken to be positive. So, a system consisting of two defects of charge 1 can transform into new states with total charges of either 2 or 0. The latter case corresponds to the annihilation of defects, in which a defectless (also called vacuum) state is obtained. In general, this phenomenon is advantageous in the strongly established nematic phase, because the presence of defects is extremely costly energetically. To predict the potential cancellation of interacting defects, one introduces the strength m, also called the winding number or Frank index, as already introduced above. This reveals the rotation of the director **n** on encircling the defect counterclockwise. Point defects are characterized by integer m. For line defects, half-integer values are also allowed because of the "head to tail" invariance. Defects of opposite strengths may annihilate each other. One conventionally refers to the singularities with plus/minus strength signs as defect/antidefect. Note that line defects of integer m are topologically unstable because they can deform continuously into defectless states through the so-called "escape along the third space direction" (Kléman, 1983). Defects may appear because of topological, energetic, or dynamic reasons. For fixed boundary conditions, the total topological charge of all defects is a conserved quantity. Their value is determined by the topological properties at the enclosing boundary. The number of defects in the equilibrium state depends on the energy balance between the elastic, surface, and eventual external field forces. An example of dynamically generated defects is represented by a sudden temperature-driven change from the isotropic to the nematic phase in a bulk (nonconstrained) liquid crystal. The bulk nematic phase is characterized by a preferred (symmetry breaking) direction in space, but in the isotropic phase no direction is preferred. If the quench is fast enough, different symmetry-breaking directions emerge in different regions of the nematic liquid. Consequently, a multidomain structure is formed, where the director orientations of different domains have been triggered by random thermal fluctuations. Numerous defects are expected to form at the walls of the domains. With time, the concentration of defects gradually decreases in order to reduce the free energy of the system. More precisely, the system gradually evolves into a defectless state, because, for the unconstrained liquid crystal, the topological charge is zero. The region in which degree of order is apparently perturbed by the defect is referred to as the "core" of the defect. The simplest hypothesis is that the core or defect or disclination is an isotropic liquid, therefore the core energy is proportional to the temperature difference from the phase transition temperature multiplied by the Boltzman constant. The core radius can be estimated to be in the order of 30–50 nm (Oswald and Pieranski, 2000).

8.2.4 Rigorous Director Simulation

Simulation of the director fields for all kinds of geometries is only possible with rigorous methods that take anisotropy, boundary conditions, and external forces into account. The simulations of the reorientation of the nematic liquid crystal in spatially varying electric fields require knowledge of cell structure. In particular for the case of a switchable cell, this represents a stack of at least three different layers, that is, the electrodes, the dielectric layer, and the liquid crystal itself. The dielectric layer is a polymer that is treated to give well-defined anchoring conditions at the substrate surfaces. Often, strong anchoring is assumed where the liquid crystal molecules at the upper and lower boundaries are held fixed. For periodic arrangements only one single period is calculated, and periodicity at the lateral boundaries can be imposed. The electrodes on the top of the cell and on the bottom can be structured. The director profile has to be calculated with respect to the electric field and the elastic deformation inside the liquid crystal bulk. The flexo-electric effect, flow effects, and higher-order elastic terms are not taken into account in standard simulation procedures. The concept is the same as mentioned above in that one looks for the minimum of the free energy density in a certain volume. External electric fields are represented by the term

$$f_e = \tfrac{1}{2}\mathbf{E}\cdot\mathbf{D} \qquad \text{with } \mathbf{D} = \boldsymbol{\varepsilon}\varepsilon_0\mathbf{E}. \tag{8.38}$$

The magnetic field contribution can be expressed in a similar way (Blinov and Chigrinov, 1996). The electric and elastic energies tend to balance in order to minimize the global energy density, and, with Equation (8.1), the equation becomes in the case of a nonchiral liquid crystal

$$f_d = \tfrac{1}{2}K_{11}(\nabla\cdot\mathbf{n})^2 + \tfrac{1}{2}K_{22}(\mathbf{n}(\nabla\times\mathbf{n}))^2 + \tfrac{1}{2}K_{33}(\mathbf{n}\times(\nabla\times\mathbf{n}))^2 + \tfrac{1}{2}\varepsilon_0\mathbf{E}\boldsymbol{\varepsilon}\mathbf{E}. \tag{8.39}$$

Equation (8.39) gives the energy density for the bulk of the liquid crystal in the standard Frank–Oseen model, where n is the director field, K_{ii} are the elastic constants, $\boldsymbol{\varepsilon}$ the dielectric tensor of the liquid crystal, and $\mathbf{E}$ the electric field. The dielectric tensor changes when reorientation of the liquid crystal takes place. The equilibrium is obtained for the minimum of the energy in the volume. To minimize the free energy, a Lagrange multiplier method is used. A viscosity γ describes the viscous relaxation of the director field in time and is set as an isotropic variable. Then one can write (Kitamura, 1995)

$$\gamma\frac{\partial n_i}{\partial t} = \frac{\mathrm{d}}{\mathrm{d}x_j}\left(\frac{\partial f}{\partial n_{i,j}}\right) - \frac{\partial f}{\partial n_i} + \alpha n_i, \qquad i, j = 1, 2, 3. \tag{8.40}$$

The constant α is the Lagrange multiplier and $n_{i,j}$ is the derivation of the ith director component with respect to the jth coordinate $n_{i,j} = \mathrm{d}n_i/\mathrm{d}x_j$. Several methods might be used to solve this equation (Ge et al., 2005). The program used throughout this book is LCD Master (Shintech, 2005), which is based on using a finite-element algorithm to calculate the director profile. The director profile is calculated for a given error criterion. The calculation is time-dependent, because of the viscous behavior of the liquid crystal. After an initial director configuration is set, the electric field is switched on. The electric field interacts with the dielectric anisotropy of the liquid crystal and creates a deformation of the director field. This deformation increases the elastic energy. In the equilibrium state, the electric and the elastic energies tend to balance in order to minimize the global energy density.

8.3 OPTICAL PROPERTIES OF UNIFORM TEXTURES

For uniform textures, plane wave methods of light propagation lead to the correct results. These are methods for stratified media as presented in Chapter 4. Simplest to treat are cases when the light is incident normal to the surface. For oblique incidence, the 4×4 matrix methods usually have to be used. We will discuss here only normal incidence problems, which are sufficient for the understanding of most of the optical effects appearing in texture analysis. The propagation direction is fixed for all examples, as it would be in the microscope, from bottom to top. We consider only normal incidence. We will use the Jones matrix formalism discussed in Section 4.2 for calculations. Our main interest is in finding formulas for the transmission between crossed and parallel polarizers. The textures we will discuss are shown in Figure 8.7. A sandwich cell with two aligning surface substrates is assumed. The homeotropic texture has its optical axis perpendicular throughout the liquid crystal cell. For tilted planar alignment the optical axis is oblique but uniform. Hybrid aligned cells consist of two differently aligned surface substrates, causing a change of director direction angles in the propagation direction. A twisted structure shows rotation of the optical axis along the propagation direction.

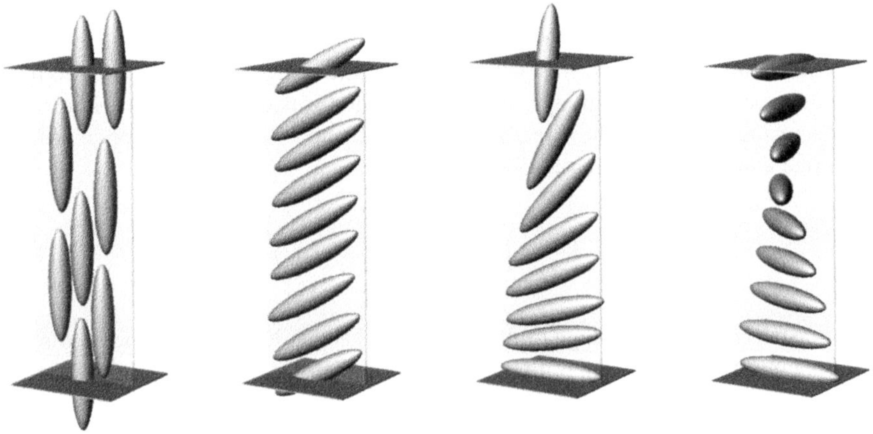

(a) (b) (c) (d)

Figure 8.7 Uniform nematic textures for which the plane wave modeling of light propagation can be applied: (*a*) homeotropic, (*b*) tilted planar, (*c*) hybrid, and (*d*) twisted.

8.3.1 Homeotropic Nematic Texture

The homeotropic texture can be described with the Jones matrix (Table 4.2, Section 4.2):

$$\mathbf{J}^{\text{homeotrop}} = \mathrm{e}^{-\frac{2\pi i}{\lambda} n_o d} \begin{pmatrix} 1 & 0 \\ 0 & 1 \end{pmatrix}. \tag{8.41}$$

Between crossed polarizers and under an angle of orientation φ of the sample, this leads to an output Jones vector

$$\mathbf{V}_{\text{out}} = \boldsymbol{P}_{\text{vertical}} \boldsymbol{R}(-\varphi) \boldsymbol{J}_{\text{homeotrop}} \boldsymbol{R}(\varphi) \boldsymbol{P}_{\text{horizontal}} \mathbf{V}_{\text{in}} \tag{8.42}$$

and further on

$$\begin{pmatrix} V_x^{\text{out}} \\ V_y^{\text{out}} \end{pmatrix} = \mathrm{e}^{-\frac{2\pi i}{\lambda} n_o d} \begin{pmatrix} 0 & 0 \\ 0 & 1 \end{pmatrix} \begin{pmatrix} \cos(\varphi) & -\sin(\varphi) \\ \sin(\varphi) & \cos(\varphi) \end{pmatrix} \begin{pmatrix} 1 & 0 \\ 0 & 1 \end{pmatrix} \times \begin{pmatrix} \cos(\varphi) & \sin(\varphi) \\ -\sin(\varphi) & \cos(\varphi) \end{pmatrix} \begin{pmatrix} 1 & 0 \\ 0 & 0 \end{pmatrix} \begin{pmatrix} V_x^{\text{in}} \\ V_y^{\text{in}} \end{pmatrix}. \tag{8.43}$$

After multiplication of the matrices from right to left, one obtains

$$\begin{pmatrix} V_x^{\text{out}} \\ V_y^{\text{out}} \end{pmatrix} = \mathrm{e}^{\frac{-2\pi i}{\lambda} n_o d} \begin{pmatrix} 0 & 0 \\ 0 & 0 \end{pmatrix} \begin{pmatrix} V_x^{\text{in}} \\ V_y^{\text{in}} \end{pmatrix}. \tag{8.44}$$

Because the matrix has only zero components, the resulting Jones vector has only zeros too. The intensity is calculated for all incident Jones vectors and gives

$$\mathbf{I} = \mathbf{V}_{\text{out}} \mathbf{V}_{\text{out}}^* = |V_x^{\text{out}}|^2 + |V_y^{\text{out}}|^2 = 0. \tag{8.45}$$

This means that under crossed polarizers for normal incidence, the homeotropic texture is always black, which is a characteristic of this texture. If the angle of incidence is altered there will be transmission due to the birefringence at off-normal angles. The texture is extremely sensitive to texture variations and if a homeotropic aligned cell is touched, the flow of the liquid crystal creates an off-axis angle of the director and brightness. Observed with the conoscopic observation technique, where light transmission under all observation angles is done simultaneously, the texture gives uniaxial conoscopic crosses as shown in Section 7.4. Between parallel polarizers the texture transmits all the light that is transmitted by the first polarizer.

8.3.2 Planar Nematic Texture

The planar texture is actually a birefringent slab. Transmission for a birefringent slab has already been calculated in Section 4.2. The result was Equation (4.14):

$$I = \frac{I_0}{2}\sin^2(2\varphi)\sin^2\left(\frac{\pi(n_1 - n_2)d}{\lambda}\right). \tag{8.46}$$

The transmitted intensity I is proportional to the incoming intensity I_0 and depends on the position of the sample φ with respect to the coordinate system of polarizers. The transmitted intensity is a sinusoidal function of the angle and of the retardation. A peculiarity of the liquid crystal planar texture is that it often has a pretilt angle with respect to the surface. If a plane wave travels through an anisotropic medium where the optical axis is inclined, it will see an effective refractive index. The value of the effective refractive index was calculated in Section 2.7 and is given by Equation (2.61). Usually, the pretilt angle θ is measured from the substrate surfaces; this needs a redefinition of Equation (2.61). The equation for the effective refractive index with pretilt θ becomes

$$n_{\text{eff}} = \frac{n_e n_0}{\sqrt{n_e^2\sin^2\theta + n_o^2\cos^2\theta}}. \tag{8.47}$$

For a tilt angle of zero, the effective refractive index is equal to n_e. If the tilt angle is 90°, as for the homeotropic texture, the effective refractive index becomes n_o. In between, values from n_e to n_o result. The intensity between crossed polarizers for a uniform planar texture with pretilt θ can be calculated by setting $n_2 = n_o$ and $n_1 = n_{\text{eff}}$ in Equation (8.46). This gives

$$I = \frac{I_0}{2}\sin^2(2\varphi)\sin^2\left[\frac{\pi d}{\lambda}\left(\frac{n_e n_0}{\sqrt{n_e^2\sin^2\theta + n_o^2\cos^2\theta}} - n_o\right)\right]. \tag{8.48}$$

As one can see, the structure of the equation has not changed when compared with Equation (8.46). The transmitted intensity is zero if φ is a multiply of $\pi/2$. Compared to a planar sample without pretilt, the texture would have a different color because the retardation is smaller. The color change of anisotropic samples is discussed in Section 7.5.

8.3.3 Hybrid Nematic Texture

Up to now we have dealt with uniform textures that have no internal structure. Uniform textures also include textures with variation of the director in the direction of propagation and uniform appearance in the two others. An example of this is the hybrid texture. Cells with a hybrid texture are made by combining homeotropic and planar surface substrates in a cell. The planar substrate might have a pretilt. When light propagates through such a system, it sees locally a birefringent plate that has the same orientational position but shows various tilt angles. The locally varying tilt angle causes a local change in retardation, and the direction of the optical axis is that of the alignment direction of the planar surface. The retardation has to be integrated over the whole cell thickness. In the Jones matrix formalism, this reads as follows. The hybrid texture is sandwiched between two polarizers and the position is fixed by φ. We divide the hybrid cell into slices of thickness δ. All have the same orientation and one finds the Jones matrix of the whole ensemble as a multiplication of all "local" Jones matrices. The local Jones matrix is that of a retarder (Table 4.3), with $n_1 = n_o$ and $n_2 = n_{eff}$:

$$\mathbf{J}^{\text{hybrid}} = \begin{pmatrix} e^{-\frac{2\pi i}{\lambda} n_o \delta} & 0 \\ 0 & e^{-\frac{2\pi i}{\lambda} \frac{n_e n_o}{\sqrt{n_e^2 \sin^2 \theta_m + n_o^2 \cos^2 \theta_m}} \delta} \end{pmatrix} \cdots \times \begin{pmatrix} e^{-\frac{2\pi i}{\lambda} n_o \delta} & 0 \\ 0 & e^{-\frac{2\pi i}{\lambda} \frac{n_e n_o}{\sqrt{n_e^2 \sin^2 \theta_1 + n_o^2 \cos^2 \theta_1}} \delta} \end{pmatrix}, \tag{8.49}$$

where we have divided the thickness into m slices to give $\delta = d/m$, and the angle θ_i is fixed but varies from slice to slice. This multiplication of matrices is easily made because all the matrices are diagonal. In the limit of δ going to zero the matrix can be written in an integral form:

$$\mathbf{J}^{\text{hybrid}} = \begin{pmatrix} e^{-\frac{2\pi i}{\lambda} n_o d} & 0 \\ 0 & e^{-\frac{2\pi i}{\lambda} \int_0^d \frac{n_e n_o}{\sqrt{n_e^2 \sin^2 \theta(z) + n_o^2 \cos^2 \theta(z)}} dz} \end{pmatrix}. \tag{8.50}$$

Note that the local tilt angle $\theta(z)$ is a function of z. Applying the standard formalism to calculate the intensity between crossed polarizers, we see that we only have to exchange the effective refractive index by the integral. Therefore, Equation (8.48) transforms to

$$I = \frac{I_0}{2} \sin^2(2\varphi) \sin^2\left[\frac{\pi}{\lambda}\left(\int_0^d \frac{n_e n_o}{\sqrt{n_e^2 \sin^2 \theta(z) + n_o^2 \cos^2 \theta(z)}} dz - d n_0\right)\right]. \tag{8.51}$$

The last thing to do is to evaluate the integral. To do so we need the exact deformation profile inside the liquid crystal layer. The planar director distribution in the hybrid cell leads to the free energy equation (8.9), which contains two elastic constants for splay and bend. The determination of the tilt angle distribution is therefore not straightforward, but is based on the minimization of the energy. In the one constant approximation, the situation is greatly simplified and we will proceed using this approximation. It leads to an equation for the tilt angle that is equal to Laplace's equation in one dimension:

$$\frac{\mathrm{d}^2\theta(z)}{\mathrm{d}z^2} = 0. \tag{8.52}$$

One solution of this equation, for a situation where we have homeotropic anchoring on one substrate with a director angle $\pi/2$ and a pretilt θ_0 on the other, is given by

$$\theta(z) = \frac{(\pi/2 - \theta_0)}{d} z + \theta_0. \tag{8.53}$$

Used in Equation (8.51) to model the tilt angle as a function of the thickness, this leads to an elliptical integral that cannot be solved analytically. Numerical methods have to be used. For the case when θ_0 is zero, the angle θ varies linearly from 0 to $\pi/2$, and the intensities can be written as

$$I = \frac{I_0}{2}\sin^2(2\varphi)\sin^2\left[\frac{\pi d}{\lambda}\left(\frac{2}{\pi}\int_0^{\pi/2}\frac{n_\mathrm{e}n_\mathrm{o}}{\sqrt{n_\mathrm{e}^2\sin^2\theta + n_\mathrm{o}^2\cos^2\theta}}\mathrm{d}\theta - n_\mathrm{o}\right)\right]. \tag{8.54}$$

For a typical example with $n_\mathrm{e} = 1.7$ and $n_\mathrm{o} = 1.5$, the integral is found to be

$$\frac{2}{\pi}\int_0^{\pi/2}\frac{n_\mathrm{e}n_\mathrm{o}}{\sqrt{n_\mathrm{e}^2\sin^2\theta + n_\mathrm{o}^2\cos^2\theta}}\mathrm{d}\theta - n_\mathrm{o} = \frac{2}{\pi}\int_0^{\pi/2}\frac{1.7\cdot 1.5}{\sqrt{2.89\sin^2\theta + 2.25\cos^2\theta}}\mathrm{d}\theta - 1.5$$
$$= 0.0953. \tag{8.55}$$

The effective retardation is reduced from $n_\mathrm{e} - n_\mathrm{o} = 0.2$ to $\Delta n = 0.0953$, that is, by about one-half.

8.3.4 Twisted Nematic Textures

The twisted nematic liquid crystal structure is well known for its polarization guiding properties. The Jones matrix has already been presented in Table 4.3. The concept for obtaining it is the same as that shown for the hybrid configuration. One slices the structure to obtain thin uniform retarders for which the Jones matrix is known and multiplies these matrices to get a closed analytical expression. Figure 8.8 illustrates the subdivision of the twisted structure. The process differs from the hybrid case in that a more complex power law for matrices applies. The calculation is performed explicitly. Let **G** be the Jones matrix of a thin

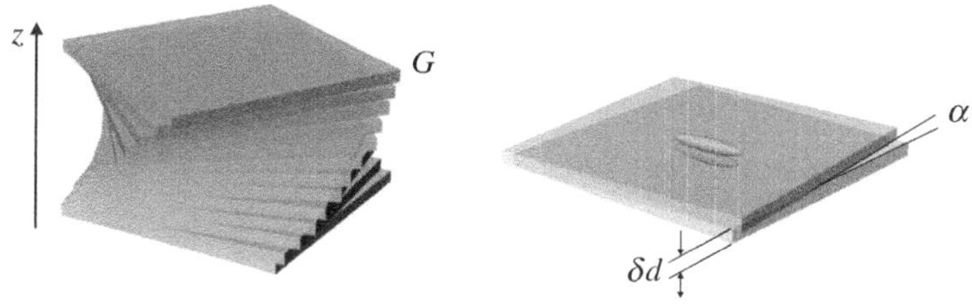

Figure 8.8 Illustration of the slicing of a twisted liquid crystal slab to calculate its optical properties.

uniaxial retarder with retardation ρ. From Table 4.3 we can write, by using $\Delta n = n_e - n_o$

$$\mathbf{G} = \begin{bmatrix} e^{-i\rho} & 0 \\ 0 & e^{i\rho} \end{bmatrix} \qquad \text{with } \rho = \frac{\pi \Delta n}{\lambda} \delta d. \tag{8.56}$$

S(α) is a rotation matrix for the angle α:

$$\mathbf{S}(\alpha) = \begin{bmatrix} \cos(\alpha) & \sin(\alpha) \\ -\sin(\alpha) & \cos(\alpha) \end{bmatrix}. \tag{8.57}$$

The Jones matrices for the first two layers are then

$$\mathbf{J}_1 = \mathbf{SGS}^{-1} \qquad \text{and} \qquad \mathbf{J}_2 = \mathbf{S}(2\alpha)\mathbf{GS}^{-1}(2\alpha) = \mathbf{S}^2\mathbf{GS}^{-2}. \tag{8.58}$$

When propagating through, they have to by multiplied:

$$\mathbf{J}_{2\cdot 1} = \mathbf{S}^2\mathbf{GS}^{-2}\mathbf{SGS}^{-1} = \mathbf{S}^2\mathbf{GS}^{-1}(\mathbf{S}^{-1}\mathrm{S})\mathbf{GS}^{-1} = \mathbf{S}^2(\mathbf{GS}^{-1})^2. \tag{8.59}$$

After the Nth layer, the matrix of the complete system can be written as

$$\mathbf{J}_\mathbf{N} = \mathbf{S}^\mathbf{N}(\mathbf{GS}^{-1})^N, \tag{8.60}$$

which leads to

$$\mathbf{J}_N = \begin{bmatrix} a & b \\ c & d \end{bmatrix},$$

where

$$a = \cos N\alpha \cos N\nu + \frac{\tan\alpha}{\tan\nu} \sin N\alpha \sin N\nu - i\frac{\sin\rho}{\sin\nu} \sin N\nu \cos(N+1)\alpha,$$

$$b = \frac{\tan\alpha}{\tan\nu} \cos N\alpha \sin N\nu - \sin N\alpha \cos N\nu - i\frac{\sin\rho}{\sin\nu} \sin N\alpha \sin(N+1)\alpha,$$

and

$$c = -b^*$$

and

$$d = a^*. \tag{8.61}$$

Here the Chebyshev's identity

$$(\mathbf{GS}^{-1})^N = \frac{\sin(N\nu)}{\sin\nu}\mathbf{GS}^{-1} - \frac{\sin(N-1)\nu}{\sin\nu} \qquad \text{with } \cos\nu = \cos\alpha\cos\rho \tag{8.62}$$

is used to rewrite the matrix components in a convenient form. Now we make some approximations for the rotational angle α and the thickness of the elementary uniaxial plates δd. N is a large number and can be set equal to $N+1$. For infinitesimal small rotations α and layer thicknesses δd we can develop the cosine function to find

$$\cos x = 1 - \frac{x^2}{2} + \cdots \rightarrow 1 - \frac{\nu^2}{2} = \left(1 - \frac{\alpha^2}{2}\right)\left(1 - \frac{\rho^2}{2}\right) = 1 - \frac{\alpha^2}{2} - \frac{\rho^2}{2} + \cdots$$

so

$$\nu^2 = \alpha^2 + \rho^2, \tag{8.63}$$

and with ρ from Equation (8.56) one reformulates

$$\nu = \alpha\sqrt{1 + \frac{\rho^2}{\alpha^2}} = \alpha\sqrt{1 + \left(\frac{\pi\Delta n\delta d}{\lambda\alpha}\right)^2}. \tag{8.64}$$

If we introduce the macroscopic twist angle $\Phi = N\alpha$ and the overall retardation

$$\beta = N\rho = \frac{\pi\Delta n}{\lambda}N\delta d = \frac{\pi\Delta n}{\lambda}d$$

one can write by multiplying Eq. 8.64 by N

$$\begin{aligned} N\nu &= N\alpha\sqrt{1 + \left(\frac{N\pi\Delta n\delta d}{N\lambda\alpha}\right)^2} \\ &= \Phi\sqrt{1 + \left(\frac{\pi\Delta n d}{\lambda\Phi}\right)^2} \\ &= \sqrt{\Phi^2 + \beta^2}. \end{aligned} \tag{8.65}$$

If we substitute $N\nu$ from Equation (8.65) and Φ, β in Equation (8.61) we obtain

$$\mathbf{J}^{\text{twisted}} = e^{\frac{i\pi d(n_e+n_o)}{\lambda}} \begin{pmatrix} a & b \\ -b^* & a^* \end{pmatrix}$$

with

$$a = \cos\Phi\cos\gamma + \frac{\Phi}{\gamma}\sin\Phi\sin\gamma - i\frac{\beta}{\gamma}\cos\Phi\sin\gamma,$$

$$b = \frac{\Phi}{\gamma}\cos\Phi\sin\gamma - \sin\Phi\cos\gamma - i\frac{\beta}{\gamma}\sin\Phi\sin\gamma,$$

and

$$\gamma = \sqrt{\beta^2 + \Phi^2} \quad \text{and} \quad \beta = \frac{\pi d(n_e - n_o)}{\lambda}. \tag{8.66}$$

This is the Jones matrix already stated in Section 4.2. The Jones matrix of Equation (8.66) can be rewritten by separating a rotation matrix at the left side in the following form:

$$\mathbf{J}^{\text{twisted}} = \begin{pmatrix} \cos\Phi & -\sin\Phi \\ \sin\Phi & \cos\Phi \end{pmatrix} \begin{pmatrix} \cos\gamma - i\frac{\beta}{\gamma}\sin\gamma & \frac{\Phi}{\gamma}\sin\gamma \\ -\frac{\Phi}{\gamma}\sin\gamma & \cos\gamma + i\frac{\beta}{\gamma}\sin\gamma \end{pmatrix}. \tag{8.67}$$

If γ becomes a multiple of π, the Jones matrix simplifies to a rotation-only matrix. The twisted liquid crystal layer acts as a perfect polarization rotator with a rotation angle Φ. If normal modes are chosen, for instance along the direction of orientation of the director at the entrance, the optical properties can be tuned by γ to obtain nearly achromatic operation conditions. A detailed discussion of this equation can be found in the work of Yeh (1999). A remarkable feature is the guiding of polarization states that can be obtained when the last matrix in Equation (8.67) breaks down to a matrix with diagonal elements only. This is the case for large β and describes the operating principle of achromatic twisted nematic cells. Between crossed polarizers the rotation matrix at the left side would in turn simulate a rotation of the analyzer. If the total twist angle is $\Phi = 90°$, this means that we can discuss the optical properties of the twisted cell as would be the case for the right matrix in Equation (8.67) between parallel polarizers. The transmission formula between crossed polarizers can be calculated as

$$\mathbf{J}^{\mathrm{twisted}} = \begin{pmatrix} 0 & 0 \\ 0 & 1 \end{pmatrix} \begin{pmatrix} \cos(\varphi) & -\sin(\varphi) \\ \sin(\varphi) & \cos(\varphi) \end{pmatrix} \begin{pmatrix} \cos\Phi & -\sin\Phi \\ \sin\Phi & \cos\Phi \end{pmatrix}$$
$$\times \begin{pmatrix} \cos\gamma - \mathrm{i}\dfrac{\beta}{\gamma}\sin\gamma & \dfrac{\Phi}{\gamma}\sin\gamma \\ -\dfrac{\Phi}{\gamma}\sin\gamma & \cos\gamma + \mathrm{i}\dfrac{\beta}{\gamma}\sin\gamma \end{pmatrix} \begin{pmatrix} \cos(\varphi) & \sin(\varphi) \\ -\sin(\varphi) & \cos(\varphi) \end{pmatrix} \begin{pmatrix} 1 & 0 \\ 0 & 0 \end{pmatrix}, \tag{8.68}$$

which leads to rather complicated expression. We leave it to the reader to calculate the complete expression and only look at the special case when the twist is $\Phi = 90^\circ$ and the sample is placed at $\varphi = 0$ (the entrance polarizer parallel with the director). Then the equation becomes

$$\mathbf{J}^{\mathrm{twisted}} = \begin{pmatrix} 0 & 0 \\ 0 & 1 \end{pmatrix} \begin{pmatrix} 1 & 0 \\ 0 & 1 \end{pmatrix} \begin{pmatrix} 0 & -1 \\ 1 & 0 \end{pmatrix} \begin{pmatrix} \cos\gamma - \mathrm{i}\dfrac{\beta}{\gamma}\sin\gamma & \dfrac{\Phi}{\gamma}\sin\gamma \\ -\dfrac{\Phi}{\gamma}\sin\gamma & \cos\gamma + \mathrm{i}\dfrac{\beta}{\gamma}\sin\gamma \end{pmatrix}$$
$$\times \begin{pmatrix} 1 & 0 \\ 0 & 1 \end{pmatrix} \begin{pmatrix} 1 & 0 \\ 0 & 0 \end{pmatrix}. \tag{8.69}$$

By multiplying the matrices one obtains

$$\mathbf{J}^{\mathrm{twisted}} = \begin{pmatrix} 0 & 0 \\ \cos\gamma - \mathrm{i}\dfrac{\beta}{\gamma}\sin\gamma & 0 \end{pmatrix}. \tag{8.70}$$

To find the intensity for arbitrary polarized light at the input one needs to calculate the Jones vector at the output:

$$\begin{pmatrix} V_x^{\mathrm{out}} \\ V_y^{\mathrm{out}} \end{pmatrix} = \begin{pmatrix} 0 & 0 \\ \cos\gamma - \mathrm{i}\dfrac{\beta}{\gamma}\sin\gamma & 0 \end{pmatrix} \begin{pmatrix} V_x^{\mathrm{in}} \\ V_y^{\mathrm{in}} \end{pmatrix}. \tag{8.71}$$

The intensity then becomes

$$I = \mathbf{V}_{\mathrm{out}}\mathbf{V}_{\mathrm{out}}^* = \left|\cos\gamma - \mathrm{i}\frac{\beta}{\gamma}\sin\gamma\right|^2 |V_x^{\mathrm{in}}|^2 = \left(\cos^2\gamma + \frac{\beta^2}{\gamma^2}\sin^2\gamma\right)|V_x^{\mathrm{in}}|^2 \tag{8.72}$$

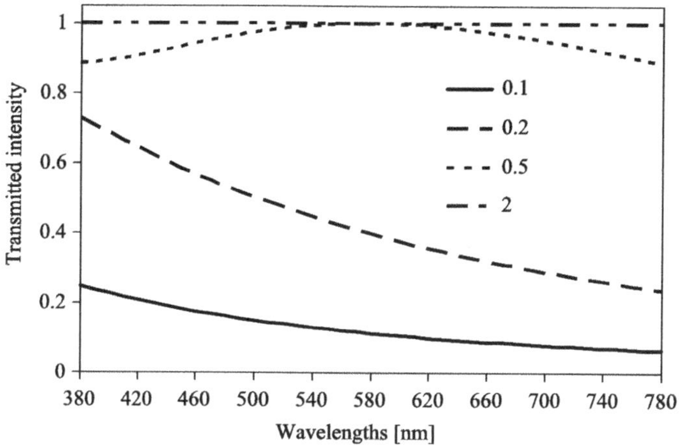

Figure 8.9 Spectral transmitted intensity for a twisted nematic cell with 90° twist angle between crossed polarizers and oriented with the director parallel to the entrance polarizers. The intensity is plotted for different values of Δnd.

and with $\gamma = \sqrt{\beta^2 + \Phi^2}$, $\beta = \pi\Delta nd/\lambda$, and $\Phi = 90°$, one obtains, finally,

$$I = \left(1 - \frac{\lambda^2}{4d^2\Delta n^2 + \lambda^2}\sin^2\sqrt{\pi^2\frac{4\Delta n^2 d^2 + \lambda^2}{4\lambda^2}}\right)|V_x^{\text{in}}|^2. \tag{8.73}$$

To illustrate the optical appearance in this case we plot in Figure 8.9 the spectral intensity for different values of Δnd. One may distinguish different behaviors for small and large values of Δnd. For values $\Delta nd = 0.1$ and 0.2 the transmission leads to a yellowish color because the blue part of the spectrum is less transmitted. However, for $\Delta nd = 0.5$, the transmission is already nearly achromatic. With a birefringence of $\Delta n = 0.2$ this corresponds to a thickness of the birefringent slab of $d = 2.5$ μm. The cell already works in the so-called guiding regime and rotates the polarization for nearly all wavelengths by 90°. This effect is used in modern display devices and allows easy color management.

8.3.5 Smectic Textures

In smectic phases texture determination through calculation is a very difficult task. Often, geometrical models are used to construct layers and determine the structure inside restricted geometries. Several phases attract our interest, in particular the smectic A and C phases. The helically structured smectic C* phase, also called the chiral ferroelectric smectic, is treated below. A special case of this last, the stabilized antiferroelectric phase, is discussed in Section 8.9.

Textures are usually drawn by illustrating the layer arrangement. Figure 8.10 shows examples of homeotropic, bookshelf, tilted bookshelf, and chevron textures.

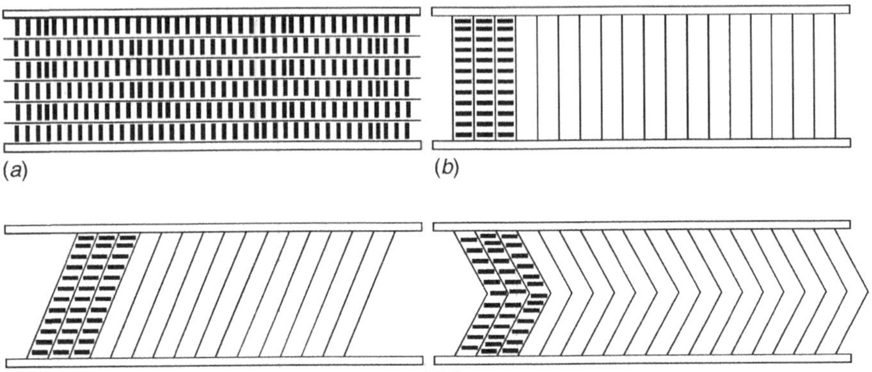

(a) (b) (c) (d)

Figure 8.10 Uniform textures in smectic phases: (*a*) homeotropic, (*b*) bookshelf, (*c*) tilted bookshelf, and (*d*) chevron textures.

When the layers are parallel to the surface substrates the texture is called homeotropic. Perfect perpendicular states are called bookshelf geometry. Inclined layers are found in the tilted bookshelf geometry and chevrons have a kink within the liquid crystal sandwich. Each texture is typical for different phases. Often, liquid crystal cells are manufactured with planar surface anchoring conditions. In smectic A phases where no tilt of the director inside the layer is present, the perfect bookshelf texture can be obtained. Optically, such a liquid crystal layer behaves exactly like a planar oriented nematic layer without pretilt. In this case, Equation (8.46) applies for calculation of the transmitted intensity. The homeotropic texture of smectic A phases is optically similar to the homeotropic nematic texture. If the molecule is tilted with respect to the smectic layers and the surface has planar anchoring conditions, smectic layers can by oriented also tilted at approximately the tilt angle. This leads to tilted bookshelf and chevron textures. Because the optical properties are given by the molecules, the optical properties are still similar to the planar nematic texture. The layer structures become important if defects are involved.

8.3.6 Ferroelectric Smectic Liquid Crystals

Nematic and smectic A liquid crystal phases are too symmetric to allow any vector order, such as ferroelectricity. The tilted smectics, however, do allow ferroelectricity if they are composed of chiral molecules. In the simplest case, the chiral smectic (smectic C*), the average long molecular axis is tilted from the layer normal by a fixed angle, but the molecules are free to rotate on the so-defined tilt cone. The molecular director in bulk ferroelectric liquid crystals adopts a helical structure. Due to the helical structure there are some particular textures that are not found in nonchiral phases. This relates mainly to the direction of the helical axis with respect to the substrates. Figure 8.11 shows two examples. A homeotropic texture would, in its unperturbed state, show a helix perpendicular to the substrates. A planar texture would

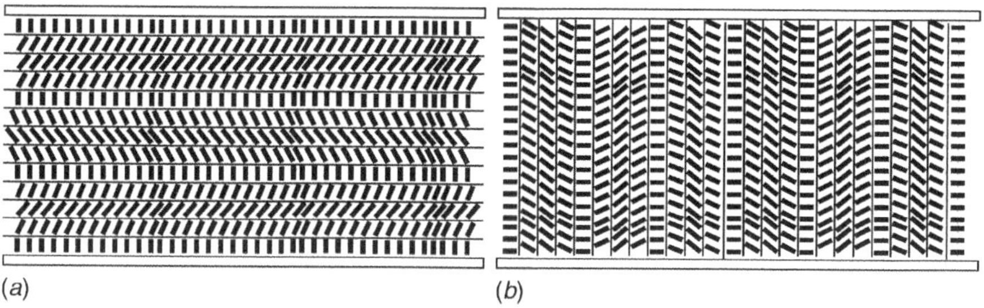

Figure 8.11 Uniform textures of ferroelectric smectic phases: (*a*) homeotropic and (*b*) bookshelf.

have a helix parallel to the substrates. When confined between closely spaced glass plates (spaced closer than the ferroelectric helix pitch), the natural helix could be suppressed as illustrated in Figure 8.12. The smectic layers were oriented approximately perpendicular to the substrate surfaces. Layer configurations that have been found to be stable are the same as for nonchiral smectic phases. Configurations optically similar to the planar nematic texture are obtained. If the helix is not suppressed, the helix leads to a rotation of the molecules from layer to layer. For short-pitch

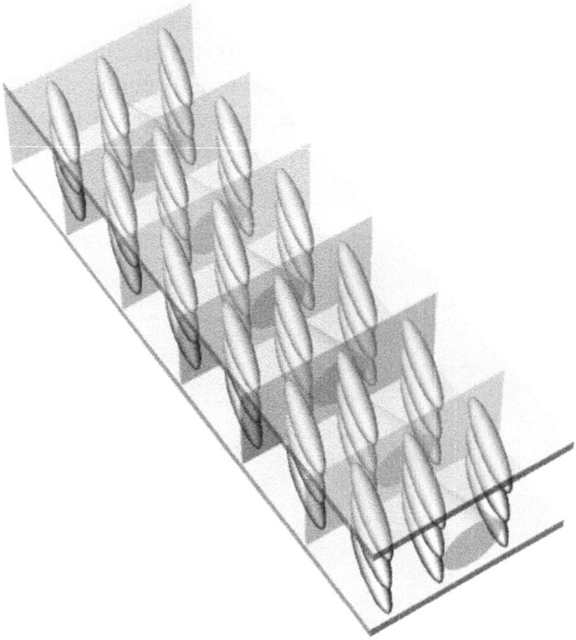

Figure 8.12 Surface stabilized texture in a smectic C* in bookshelf geometry. Note that the helix is suppressed due to the large helix-to-thickness ratio. The structure becomes similar to a bookshelf smectic C with planar alignment surfaces.

ferroelectrics with a pitch much smaller than the wavelengths of light, an averaging of the dielectric susceptibility tensor is possible (see Section 8.9 for details). This can lead to an effective refractive index tensor. The case of long pitch materials with helix axis perpendicular to the substrates is similar to a twisted nematic texture. An effective refractive index in each layer has to be assumed because of the tilt angle inside the layers. If the helix has a long pitch and its axis is oriented parallel to the substrate surfaces as in the bookshelf texture, the locally spatially varying index distribution needs special simulation tools to apply and cannot be treated within the plane wave methods. When the pitch is in the order of the wavelengths, the system shows interference effects. Such interference effects are called Bragg interferences and are treated in Chapter 11.

8.4 OPTICAL PROPERTIES OF LIQUID CRYSTAL DEFECTS

8.4.1 Point Defects in Nematic Liquid Crystal

In practice, defect structures in liquid crystals can be well monitored using a polarizing microscope. In order for this to be the case, it is necessary that defects give rise to long-range distortions in the orientational ordering, because the polarizing microscopy experiment probes only changes that appear on a scale larger than the light wavelength $\lambda > 0.5\ \mu m$. The nematic director field results in a phase difference between the ordinary and extraordinary components of the ray passing through a nematic liquid crystal Kaucic et al., 2004. This difference is then observed as an interference pattern. The liquid crystal is usually confined in the plane-parallel cell and placed between polarizer and analyzer. Figure 8.13 shows an example microphotograph of the defects we want to analyze. We assume a spatial distribution

Figure 8.13 Defects in a nematic liquid crystal with a strength of $m = 2$ and $m = -1$. A corresponding director structure can be calculated by superposition as lined out in Section 8.2 (40× Leica DMRP, white light, E7 on PMMA/RN873 10 μm).

of birefringence $\Delta n(x, y)$ and orientational angle $\varphi(x, y)$. The intensity between crossed polarizers depends on both parameters and on the wavelengths λ as

$$I = I_0 \sin^2[2\varphi(x, y)] \sin^2\left[\frac{\pi}{\lambda} d\Delta n(x, y)\right]. \tag{8.74}$$

In texture analyses very often the thickness d of the investigated sample is not very well known. For small areas of the sample a uniform thickness can be assumed. Birefringence is represented as color between crossed polarizers. Often the defect zone shows uniform color. This has to be proved by rotating the sample between crossed polarizers. Under these conditions we can set the product of birefringence $\Delta n(x, y)$ and thickness d constant in the observed field. Equation (8.74) may be simplified and now contains only the local orientational distribution of the director $\varphi(x, y)$:

$$I = I_0 \sin^2[2\varphi(x, y)]. \tag{8.75}$$

Note that the total intensity I_0 in Equation (8.75) is not the same as in Equation (8.74). In fact, this is the basic formula used for defect analysis. It links the local orientation of the director and transmitted intensity. It neglects local variations of birefringence or tilt angles, which is not always valid. It is well understood that this formula cannot be used for quantitative analysis because of the simplification on which it is based. To analyze the director we put the origin of the coordinate system at the left lower corner in Figure 8.14. Two defects are seen with different winding strengths. To calculate the director distribution we use Equation (8.31) and set defects at two posistions $P_1(8,1)$ and $P_2(5,4)$. The winding number of the defects is set to $m(P1) = -1$ and $m(P2) = 2$. It results in a director field that can

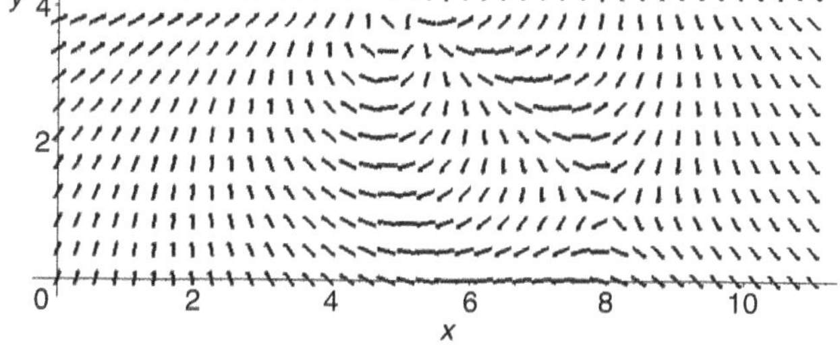

Figure 8.14 Assumed defect structure, representing an intensity distribution similar to that in Figure 8.13.

be calculated with the formula

$$\varphi = 2\arctan\left(\frac{y-4}{x-5}\right) - \arctan\left(\frac{y-8}{x-1}\right). \tag{8.76}$$

The transmitted intensity is calculated with Equation (8.75) and leads to Figure 8.15. When Figures 8.13 and 8.15 are compared, a very good qualitative agreement is found. Note that due to the intensity formula of Equation (8.75), the number of black brushes is four times the winding number of the defect. For the defect at P_2 we have set $m = 2$ and obtain 8 dark brushes leaving the core of the defect.

An additional important point in defect simulation is the appearance of the core. The core is a very-small-sized object, typically of 30–50 nm diameter, surrounded by a high refractive index gradient. Such a configuration cannot be simulated with ray tracing or matrix methods. A three-dimensional simulation has to be performed to approach such problems. To obtain a consistent view we compare three-dimensional FDTD results and a matrix approach after Berreman for a defect of strength $m = 1$ and $\varphi_0 = \pi/4$. The results are shown in Figure 8.16. The director configuration is seen under crossed polarizers with incident TM polarization (E_x component of the electric field). A field of 2.5 μm × 2.5 μm is shown. At a wavelength of 550 nm, and for $n_e = 1.8$ and $n_o = 1.5$, the intensity in TE (E_y component of the electrical field) is plotted after traveling through a thickness of 550 nm. The Berreman matrix method gives infinitely good resolution. The FDTD method shows the core to be 270 nm insize. When observed with a microscope, one is limited by the lateral resolution, which is about equal to the wavelength of the light used for observation and is larger than 550 nm. It becomes clear that the investigations on the core structure are not possible using standard polarizing microscopy observations.

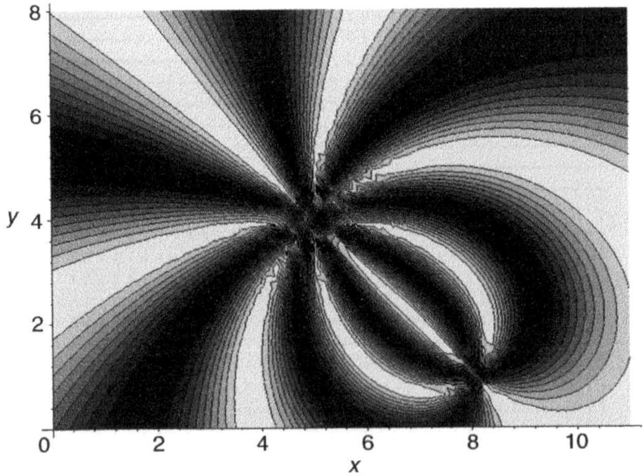

Figure 8.15 The interference texture corresponding to defect structures shown in Figure 8.14. The transmitted intensity of light is plotted.

(*a*) (*b*) (*c*)

Figure 8.16 (*a*) Director field and transmitted intensity simulations for a defect winding number $m = 1$. (*b*) The result for the Berreman matrix method, which has infinite resolution. (*c*) The FDTD method includes strong index gradients and gives a finite diameter of the area that would be identified as the defect's core. It is approximately 270 nm large (size of simulation box 2.5 μm × 2.5 μm × 0.5 μm).

8.4.2 Bulk Twist Disclination Lines in Nematics

Surface defects are one class that is obtained when surface structured substrates are used or special functions have to be obtained (Cheng, 1981). More commonly defects would arise in the bulk of the liquid crystal layer. Singular twist director fields are discussed in the textbook by Gennes (1993) and can be obtained from the classic equations of nematostatics that use the one elastic constant approximation. Accurate measurement of the propagation of light through nematic liquid crystal films containing twist disclination lines is predicted by the FDTD method. As we have seen already, the Berreman-matrix-type method fails to predict lateral scattering effects caused by a high rate of lateral gradients in the optic axis in the vicinity of the disclination lines. The FDTD method can be applied to the understanding of the fundamental relationships between optical response and nematic liquid crystal textures in which various types of defect coexist. Studies made, for example, by Hwang and Rey (2005, 2006) are three-dimensional light propagation analyses in a very limited volume, but have shown that the FDTD is highly sensitive to the strength of the disclination and that it can easily distinguish the nature of commonly observed twist lines. Based on the present and earlier results, it can be concluded that the FDTD method will be an indispensable tool for simulating devices or effects in which the optical signals contain textural information that is due to twist and wedge disclination lines of variable order. Improved observation methods have to be used to detect director structures (Voloschenko and Lavrentorich, 2000; Smalyukh et al., 2001) and light fields in the vicinity of disclinations (Rockstuhl et al., 2004).

8.5 SURFACE LINE DEFECTS IN NEMATICS

Surface disclination lines (Porte, 1977) are one of the most interesting effects in liquid crystal research and they lead regularly to new display configurations

(Brown et al., 2000). Investigations are usually carried out with the polarizing microscope and detailed structure analysis is difficult. At alignment patterned surfaces, multistable alignment can be obtained in the plane of the liquid crystal layer as azimuthal multistability (Lee and Clark, 2001; Kim et al., 2002; Behdani et al., 2003) or out of the plane (Scharf et al., 2005) as zenithal multistability. Note that here one has to distinguish between texture multistability and surface mediated multistability. Although texture multistability is obtained with one and the same surface and identical anchoring conditions, surface multistability is caused by a change of alignment at the substrate. This can lead, of course, to different textures too. Figure 8.17 gives an example of texture multistability in a liquid crystal cell. One substrate has varying alignment conditions. Planar aligned zones are surrounded by homeotropic alignment. The direction of the planar alignment is not specified as it was made by isotropic silicon oxide evaporation. Homeotropic areas are made by silanization (Park et al., 2005). At the border of each planar area, a line defect arises. These defects might have different signs and support different configurations of the director field inside the structured area. Multistable configurations are found that are visible between crossed polarizers as pixels with varying intensity distribution over the pixel area (Varghese 2005a,b). The multistable configurations are a result of surface pinned defects (Cheng, 1981).

8.5.1 Hybrid-Homeotropic Alignment Borders

It is interesting to have a closer look at the border between homeotropic and planar aligned domains. The deformation is purely elastic. The interesting quantity is the widths of the deformations as observed optically. To do so, we consider a sample

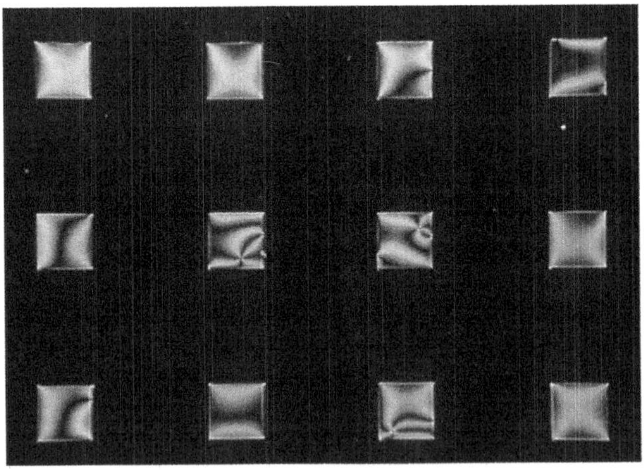

Figure 8.17 Multistable texture configurations in surface alignment textured liquid crystal cell. The sample has a homeotropic and planar alignment patterned substrate. In the cell, the alignment patterning gives hybrid and homeotropic aligned domains (E7 6 μm; size of the squares 100 μm × 100 μm). Different stable configurations can be seen where the optical axis of the whole pixel is oriented in different directions.

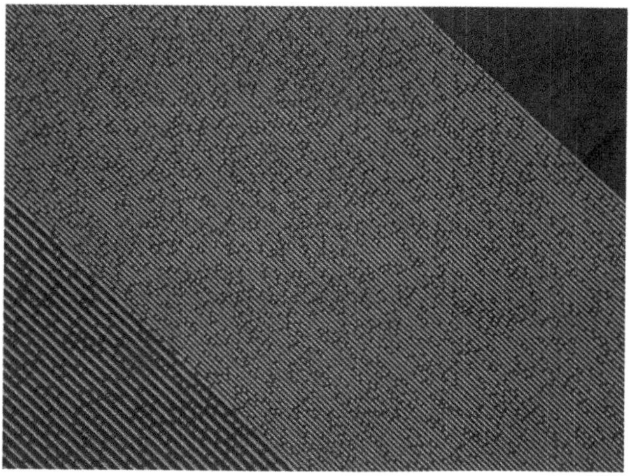

Figure 8.18 Textures in defect mediated systems. The black stripes are homeotropic textures and the bright areas have hybrid texture. The width of the stripes on the bottom left is 8 μm, in the center 4 μm, and on the upper right 2 μm. The cell thickness is 6 μm, filled with E7 (crossed polarizer, sample oriented at 45° with respect to the polarizers).

with different spacing of parallel oriented stripes as shown in Figure 8.18. The sandwich cell contains an alignment patterned surface that creates homeotropic and planar aligned domains. The superstrate has strong homeotropic anchoring. A possible internal structure is visualized in Figure 8.3*d*,*e*. The texture picture makes evident that for a certain spatial frequency the defect structure leads to a uniform texture. This is the case for a spacing of domains well below the sample thickness. In our example, a 2 μm grating can no longer be resolved as a patterned structure when implemented in a 6 μm thick cell. The elastic forces average out the orientation over a distance comparable to the structure period. We shall look at a single boundary between the homeotropic and planar aligned border. Figure 8.19 shows intensity photographs between crossed polarizers with different orientations of the sample. The size of the area is 240 μm × 140 μm. The defect lines mark the transition between hybrid and homeotropic texture. If the defect line is oriented parallel to one of the polarizers, the defect shows only weak contrast. Orientation at 45° leads to high contrast at the border. This indicates that the transition does not show twist deformations. A further proof of this was found by studying the resulting texture on patterned gratings with periods below 1 μm (Scharf et al., 2005). Uniform alignment results in an orientation of the director along the border. The width of the disclination line can be estimated by inspection and is 4 μm. To quantify the width it is useful to measure the phase profile. With a Mach–Zehnder interference microscope (Sickinger et al., 1999), one can measure the phase profile for one direction of polarization. The results of these measurements at two different positions along an edge are shown in Figure 8.20. The measurements are carried out with light of wavelength $\lambda = 633$ nm. A phase change of 1.3λ can be found.

(a)

(b)

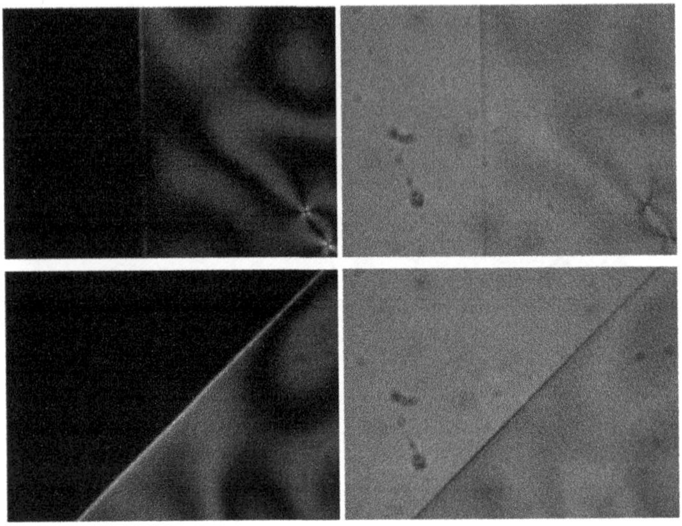

Figure 8.19 Intensities at the homeotropic planar alignment transition border observed between crossed and parallel polarizers. On the left the view between crossed polarizers and on the right between parallel polarizes. (*a*) The sample is aligned along one of the polarizers. (*b*) The sample is rotated by 45° with respect to the polarizer direction. (E7, 6 μm thick, picture size 240 μm × 140 μm.)

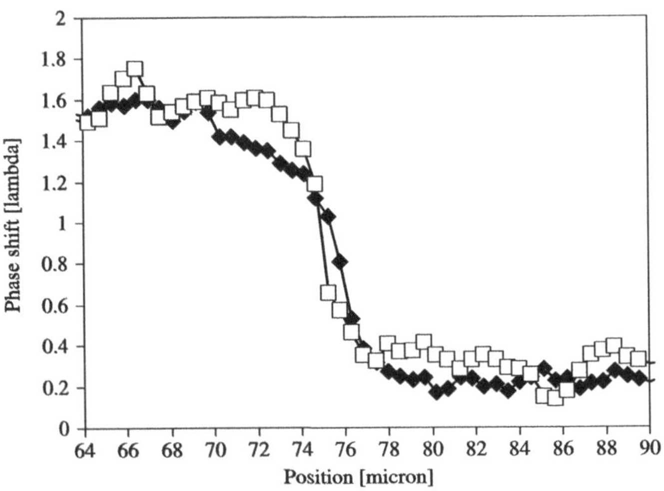

Figure 8.20 Phase profile measured at the transition zone between homeotropic and hybrid alignment at two different positions of the sample. The type of defect cannot be specified. The thickness of the liquid crystal sandwich is 6 μm. The cell is filled with E7. The transition zone reaches from position 72 to 77 and is 4 μm, less than the thickness of the cell.

A hybrid aligned cell filled with E7 of 6 μm thickness would produce a maximum retardation of (see Section 8.3) $\Delta n_{\mathrm{eff}} d = 0.8$ μm, which is about 1.26 times the wavelength of observation. The reorientation from hybrid to homeotropic alignment can therefore be seen as complete. The slope of the curve shows us that the transition zone has a width of about 4 μm. That corresponds well with the intensity line widths in the photographs in Figure 8.19. The full phase shift is expected only for structures larger than 4 μm. That can be proved by having a look at Figure 8.18 again. A fine grating with 4 μm period cerates then an almost uniform texture. If the structure's period is smaller, the relaxation of the texture will be done within the thickness of the liquid crystal slab and the effective phase shift is decreased. Then the fine structure becomes invisible.

8.5.2 Defect Lines on Photo-Aligned Surfaces with Hybrid Texture

A very convenient method to produce defect lines at surfaces is the photo-alignment technique (Schadt et al., 1992; Li et al., 1997; O'Neill and Kelly, 2000; Yamaguchi et al., 2002). The surface anchoring can be considered as strong and it allows the implementation of pretilt angles. We discuss here the case of a hybrid aligned sample with a defect line formed by domains that have different alignment directions (in-plane) of the director. The most interesting case is when directors in each domain are perpendicular to each other. Figure 8.21 shows an example of such a defect line in a 9-μm-thick cell filled with the liquid crystal mixture ZLI 3376-100 from Merck ($\Delta n = 0.0846$). The photo-alignment was made with a commercial polymer LPP JP 265 (from Rolic-Ciba). The UV alignment was carried out with polarized light at normal incidence, which results in a texture without pretilt. Two exposures with different polarization directions where made through a photomask. The countersubstrate was treated with a polymer RN 783 from Nissan that gives homeotropic alignment. The width of the line in the microscope is estimated to be 10 μm and 5 μm for perpendicular

(a) (b)

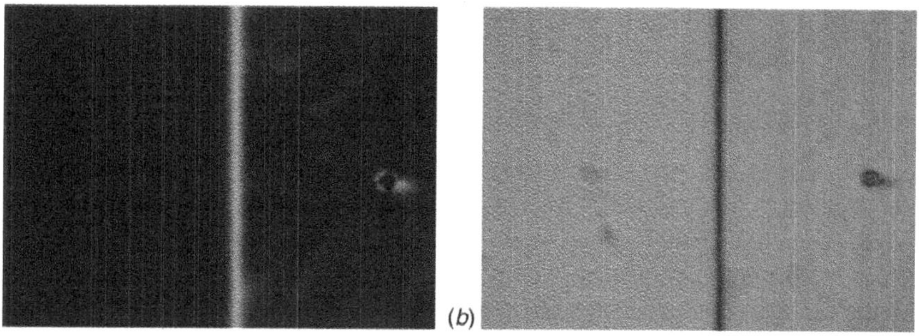

Figure 8.21 Photograph of a defect line of two different planar aligned domains at one surface in a hybrid liquid crystal of 9 μm thickness (ZLI 3376-100, RN 783, LPP JP 265, picture size 240 μm × 140 μm). The sample is oriented along one of the polarizers. (*a*) The picture is taken between crossed polarizers; (*b*) the sample is seen between parallel polarizers.

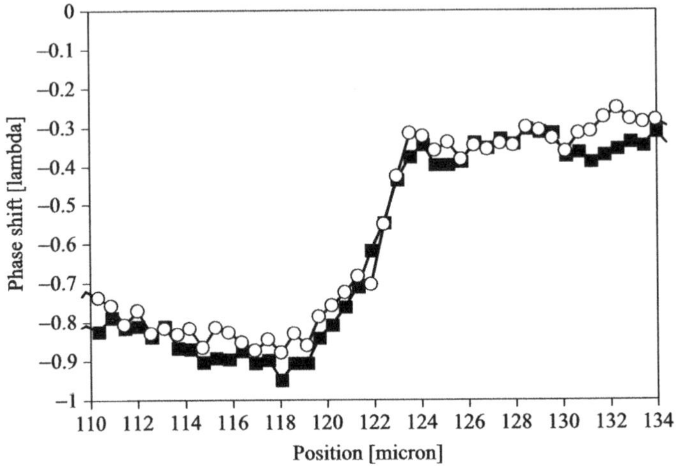

Figure 8.22 Phase step at a hybrid–hybrid defect line between perpendicular oriented domains. The cell is 9 μm thick (ZLI 3376-100, $\Delta n = 0.0846$). The width of the transition zone is estimated to be 6 μm.

and parallel polarizers, respectively. As clearly seen in Figure 8.21, the disclination line appears bright, indicting possible twist deformation. The zones alongside are only distinguishable when a compensator is used to determine the direction of the optical axis. One side will be in the addition position and the other in the subtraction position. See Chapter 7 for details on using compensation techniques. Otherwise they show the same behavior between polarizers.

To compare different defect configurations, the phase step is measured. The results are presented in Figure 8.22. The phase step is approximately 0.5λ. The transition zone shows a width of 6 μm. The measurements are carried out in linear polarized light and the situation can be compared to observation between parallel polarizers. The width of the disclination line compares well with the measurement. The twist type of disclination leads to transmission at the disclination line and to scattering. For line width measurements, observation between crossed polarizes is not recommended because it leads to blurred and overexposed images. It is more convenient to measure the line between parallel polarizers when the defect line is dark.

8.5.3 Defect Lines on Nanostructured Surfaces

Next we will present the effect of elastic deformation on a nanostructured surface. The situation is fundamentally different because the alignment on the surface is frustrated. A sandwiched sample of E7 (Merck) is prepared with a planar aligned surface by rubbing and a nanostructured multistable surface as a countersubstrate. The multistable surface is made by a replication technique in epoxy photopolymer (Scharf et al., 2004). The structure of the surface is given in Figure 8.23. Squares of 2 μm × 2 μm are filled with a grating of 400 nm period and arranged to obtain frustrated alignment conditions. Such a structure has two easy axes of alignment at 45° with respect to the arrangement of squares. If the countersubstrate is planar

Figure 8.23 Checkerboard structure creates multistable alignment. The structure is composed from squares with 2 μm side lengths and filled with a grating of 400 nm period.

aligned with the alignment direction oriented along the arrangement of squares, domains of different twist direction will form. Between crossed polarizers there will be no visible difference. Figure 8.24 shows the micrographs of such a defect line. The texture on the right and the left are uniform twisted textures with 45°

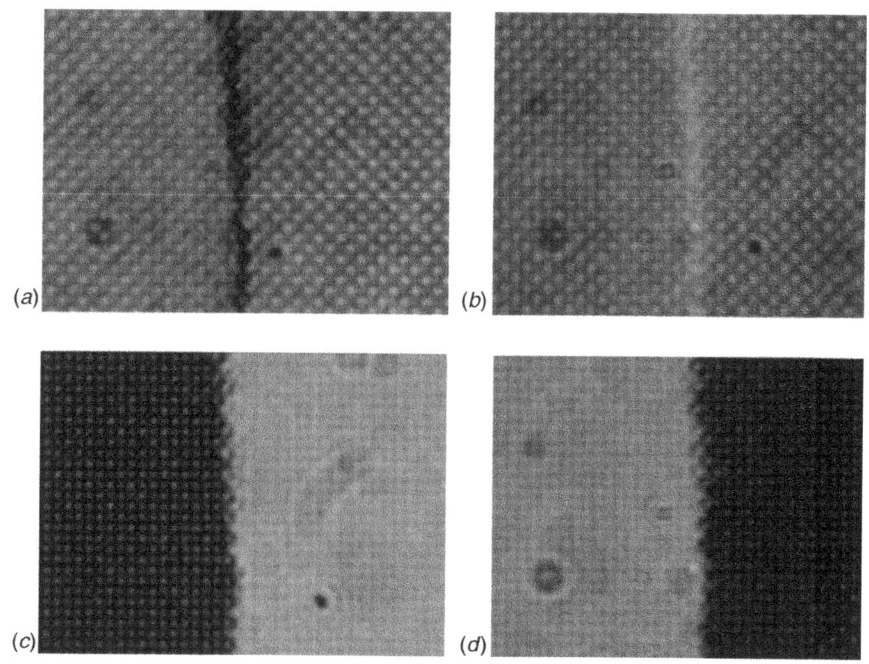

Figure 8.24 A defect line of opposite twist on a nanostructured surface. The regular structure corresponds to 2 μm × 2 μm squares (picture size 55 μm × 75 μm). The same area is seen between crossed polarizers in (*a*), parallel polarizers in (*b*), polarizers oriented at 45° and −45° in (*c*) and (*d*), respectively.

twist angle and different senses of twist rotation. The cell is 6 μm thick and filled with E7. Between crossed or parallel polarizers the texture behaves the same. The contrast is best if the sample is observed between polarizers oriented at 45° and set parallel to the rubbing direction of the polyimide surface. In this case one domain will guide the light at 45° and the other at −45°. The directions of polarizations of different domains are perpendicular if they reach the analyzer. Therefore one is blocked completely and the other is transmitted. Examining Figure 8.24 it becomes evident that in the case of frustrated surface alignment on the particular nanostructure, the width of the disclination line has the thickness of one structure element, which is 2 μm. The "resolution" is higher than in the preceding case. These is mainly caused by the effect of frustration and can be explained with a finite anchoring strength of the surface.

8.5.4 Width of a Twist Disclination Line

For surfaces with varying anchoring conditions the widths of a disclination line can be used for measuring the anchoring strengths (Li et al., 1997; Yamaguchi and Sato, 2000; Hasegawa, 2002; Skarabot et al., 2003). However, that is dangerous as the optical measurement of the widths is only an indirect method to determine the lateral extension of the director deformation. To clarify what the widths of a line defect are and how they can be used for surface anchoring measurements is the aim of this chapter.

In general it is very difficult to calculate analytically the director field of nematic liquid crystals. For two-dimensional director fields the analytical method introduced in Section 8.2 can be used. This method is restricted to particular deformations. It permits director fields to be simulated for pure twist deformation or pure splay-bend deformations. For twist deformations, pretilt angles cannot be included because they will lead to splay-bend deformations if different orientation angles are used on the same substrate surface. For the defect geometries under consideration this is sufficient.

Consider a cell in which the liquid crystal is confined between two parallel substrates, the distance between the two substrates being d. Let the x-axis lie in the lower substrate, the y-axis being perpendicular to that plane. The director $\mathbf{n}$ is in the (x, y)-plane and forms a twist angle $\varphi(x, y)$ with the x-axis:

$$\mathbf{n} = (\cos\varphi, \sin\varphi, 0). \tag{8.77}$$

We impose the boundary conditions for the director so that we have on one substrate surface uniform alignment and on the other a modified substrate. The alignment conditions are planar and the twist angle at the surface changes from $\varphi = 3/2\pi$ to $\varphi = \pi$ at position $x = 0$. This is a defect with strength $s = \pm 1$ pinned at the surface. The equilibrium director configuration must satisfy the two-dimensional Laplace's equation subject to the boundary conditions. Using conformal mapping we find for the twist angle

$$\varphi(x, y) = 2\left\{-\frac{\pi y}{4d} + \frac{1}{2}\mathrm{im}\left[\ln\left(1 - e^{\frac{\pi}{d}(x+iy)}\right)\right]\right\} + \frac{\pi}{2}. \tag{8.78}$$

Simulations of optical properties are carried out by the finite-difference time-domain method (FDTD; see Chapter 5 for details). The FDTD method delivers results that include diffraction and scattering. As an example, Figure 8.25 shows Electric field amplitude distribution for TE (E_y) incident polarization at a wavelength of 550 nm coming from below and traverse the liquid crystal layer. The corresponding director profile is shown in Figure 8.26. This corresponds to the E_y component of the electric

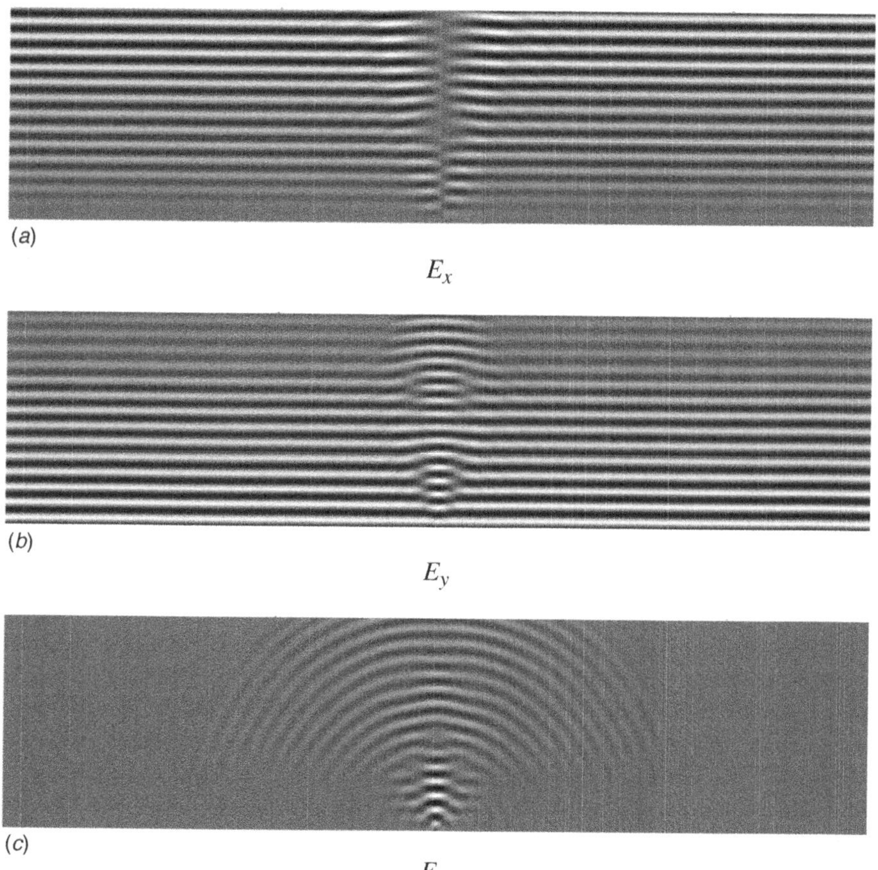

Figure 8.25 Electric field distributions for the whole layer with a thickness of 5 μm and a width of 20 μm: (*a*) E_x, (*b*) E_y, and (*c*) E_z. It is clearly seen that due to the twist the amplitude of the incoming E_y component, which is shown out of the plane in Figure 8.26, is transferred in the E_x component. E_z shows the loss of the system by diffraction effects of the defect. The E_z amplitude values are multiplied by 10 for better visualization.

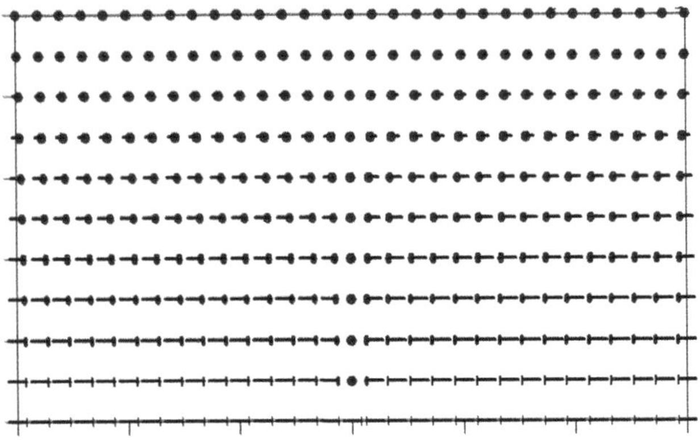

Figure 8.26 Example for strength one defect structures where the twist angle is changing. The defect is pinned at the lower surface. The field is 5 μm thick and 20 μm wide.

field in Figure 8.25. If TM polarization is analyzed we first discuss the E_x component. This is the normally white mode of operation of twisted nematic displays. It will show all imperfections. The simulation box is 5 μm high and 20 μm wide, with a sampling of 25 points per wavelength (20 nm). The spatial distribution of the birefringence causes a change of the intensity and the phase is modulated. Figure 8.25 allows a qualitative insight into light propagation around defects. We first see that the energy is transferred from E_y to E_x while propagating through the twisted structure. That is the guiding effect of polarization that is typical for such kinds of textures. In the vicinity of the defect, the E_z components, which are directed in the direction of propagation, are present. Energy that is coupled in that electric field component is lost and contributes to scattering.

To be more quantitative about the line widths one sees in observations, one should rather look at the field components after the light has traveled through the structure. As an example, the appearance between crossed and parallel polarizers is calculated in Figure 8.27. TE (E_y) polarization enters the layer and TM and TE polarization are analyzed. The liquid crystal layer thickness is 5 μm and the optical parameters for E7 are used. The defect line widths calculated with matrix methods are shown for comparison in Figure 8.28. The intensity calculated with rigorous methods (FDTD) shows particular interference effects for the parallel polarizer situation. When observed with a polarized microscope one has to take into account the resolution of the microscope objective. If, for instance, an objective 20× with a numerical aperture of 0.5 is used, the theoretical lateral resolution at 550 nm is $\delta = 0.61\lambda/\mathrm{N.A.} = 0.67$ μm. The observed intensity has to be calculated by a convolution of the microscope objective's resolution function (usually an Airy function) with the simulated intensity pattern. The period of the intensity variations is about the same and is not therefore resolved. The defect line width can be defined via the contrast. It is set to be at the difference of positions where the intensity drops to 50%. In Figure 8.27 we find for rigorous simulation, and neglecting the sidelobs, a linewidth of $\delta x_{\mathrm{rigorous}} = 1.3$ μm. For the matrix method the

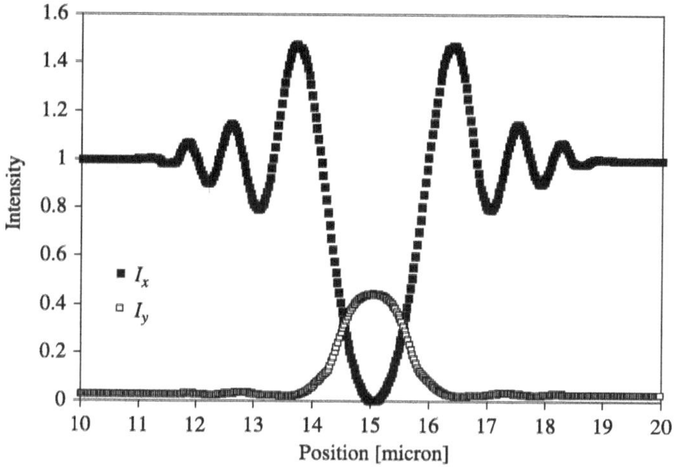

Figure 8.27 Calculated intensity of the electric field of TE and TM polarization at the output. TE (E_y electric field component) polarization is incident. The intensity profile is shown with zero intensity in the middle at defect line position $x = 15$.

width is $\delta x_{\text{matrix}} = 0.57\ \mu\text{m}$. Again one should not forget that this is not the value seen with the microscope due to the limited resolution of observation. A remarkable difference between matrix methods and rigorous simulation is noticed which results from the strong gradient of birefringence at the defect position. As a result of this, one has to state that quantitative investigations of defects, to obtain information about anchoring properties, have to be based on rigorous optical modeling.

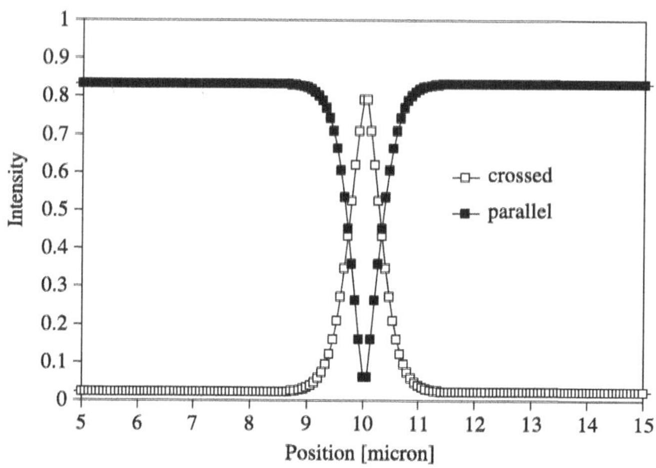

Figure 8.28 Matrix method intensity profiles are shown for the same defect texture as in Figure 8.27. The defect line position is at position $x = 10$. The cases for crossed and parallel polarizers are shown.

8.5.5 Three-Dimensional Simulations of Light Propagation in Liquid Crystal Microstructures

Bistable nematic devices fall into a particular class of display devices of significant technological interest for low-power applications, primarily because of their bistable operation. Surface gratings with a grating pitch of around or below a micrometer are the key element introducing surface bistability. Devices based on monogratings are usually operating with liquid crystal distortions in the zenithal plane and exhibit a defect state and a defect-free state (Brown et al., 1997). Azimuthal bistable nematic devices can also be made, if supporting surfaces with suitable double-periodicity surface-relief gratings are introduced (Thurston et al., 1980; Bryan-Brown et al., 1994; Kitson and Geisow, 2002). This latter class can potentially demonstrate improved optical viewing properties. A rigorous three-dimensional (3-D) optical wave propagation study for devices based on such fine features was made by Kriezis and colleagues (2002a). This was the first rigorous modeling of liquid crystal optical studies in three dimensions. The proposed numerical scheme uses the FDTD method for the region of the device that involves 3-D variation in geometry and material properties, but it uses plane-wave expansions for the lower and upper supporting layers, which are uniform along both transverse directions. This hybrid approach will greatly reduce the computational burden and will allow for realistic device parameters to be considered in the model.

8.6 DEFECTS IN SMECTIC PHASES

Smectic phases are characterized by their layer structure. For nematics there are several basic textures: the homeotropic, the planar or bookshelf texture, and the focal conic texture, which leads to fan shapes. All of these can be modified by tilting, kinking, or boundaries between differently oriented domains. The overall appearance is therefore complex and we will limit our discussion to a few cases.

We start with the smectic A. Due to their layered structure, smectic phases form focal–conic domains (Kléman, 1989). Those are domains formed by deformed smectic layers that fold around line defects, preserving equidistance of structural layers everywhere except in the vicinity of the defect lines. A typical appearance is shown in Figure 8.29. The line defect, shown as a thick line in the figure, may be an ellipse, a hyperbola, or two parabolas. The smectic layers within a focal–conic domain adopt their arrangement and appear as concentric circles. They also have the distinctive property of preserving an equal distance between them. A focal–conic domain built around an ellipse and a hyperbola is the most common type of defect in thermotropic smectic A phases.

There is also a more complicated polygonal texture. This is a texture composed of several focal–conic domains located at the boundary surfaces. When smectic phases are observed in the microscope, fan-shaped textures are most common (Demus and Richter, 1987; Dierking, 2003). These are textures formed partly by focal–conic domains with their hyperbolae lying in the plane of observation. The layers are

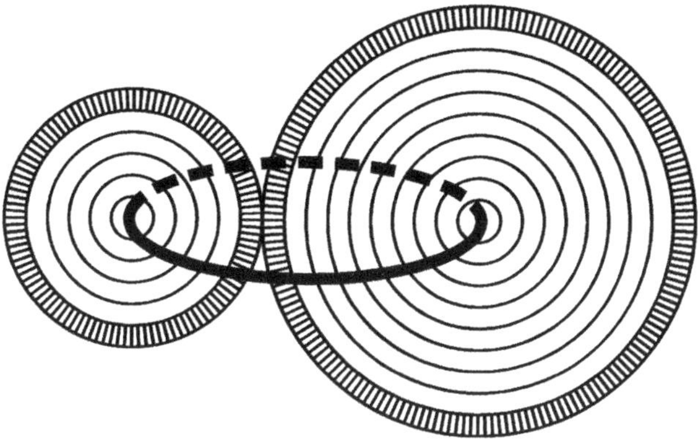

Figure 8.29 Focal–conic domain construction. A vertical section showing layers of the structure with a representation of the arrangement of the molecules within one of them.

aligned almost normal to the sample surfaces. The regular sets of hyperbolae are called "boundaries of Grandjean"; they serve as limiting surfaces between domains with different director orientations. A typical fan shape texture can be visualized by drawing layers with the arrangement of the molecules. Figure 8.30 gives an example. Optically the fan shape texture can be treated as a birefringent film with locally varying optical axis direction. We have found already that in this case Equation (8.75) gives the correct distribution of the intensity between crossed polarizers. Because the size of the texture elements, hence the domains, is large and the phase does not scatter so much as in the nematic phase, the textures appear as fans. That is because orientations of the director along either the polarizers or analyzer appear black, and in between light is transmitted. For a curved domain like that of Figure 8.30, that would mean that brushes appear that are fixed with respect to the analyzer/polarizer position and change when the sample is rotated. For a smectic A phase the molecular director is perpendicular to the smectic layers and it is easy to identify the zones of extinction. In the

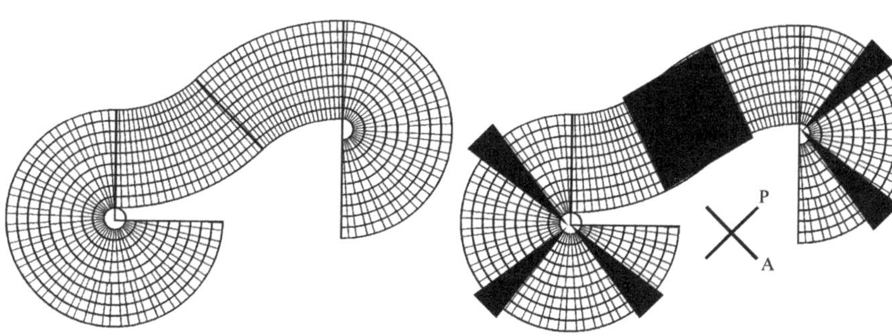

Figure 8.30 Fan shape texture with different curvature. The layer thickness is always preserved and the director of the smectic A phase is perpendicular to the layer normal.

polarizing microscope, between crossed polarizers, the zones become black when the molecular director is parallel to either the polarizer or the analyzer. If we put them at 45°, as shown on the right of Figure 8.30, we can mark the zones where the layer normal is parallel to one of the polarizers and color these areas in black. We find brushes for zones of high curvatures and dark and bright zones. The optics is similarly described as the point defects for nematics and only takes the lateral distribution of the optical axis into account.

8.6.1 Ferroelectric Liquid Crystal Domain Walls

Defect structures usually involve large changes of the molecular orientation over relatively short distances. Similar large changes in director orientation over short distances can also take place in display-type liquid crystal devices, where complex director profiles can be formed due to the finite pixel and electrode size. An example of such a director profile is the domain wall, which exists in ferroelectric liquid crystal devices (Goodby and Blinc, 1991). One type of ferroelectric liquid crystal is based on chiral smectic C* materials and can be used in surface stabilized ferroelectric liquid crystal displays (SSFLCD). In these devices the liquid crystal is confined between two glass plates coated with an alignment layer in a Very thin layer of around 1.5–2 μm thickness. The cell gap is less than the helical pitch of the chiral smectic C* helix and, therefore, the helix is suppressed by the interaction between the liquid crystal and the aligning surfaces. A further consequence is the appearance of nonzero bulk spontaneous polarization, which allows switching between two states by applying positive or negative pulses, leading to a bistable device. In such devices the chevron layer structure usually exists, where smectic layers are tilted with respect to the substrate normal and a cusp forms in the center of the cell. In essence, the structure supports two switched states and at interfaces between them domain walls form.

To analyze the propagation of light through domain walls in a rigorous and consistent way, the finite-difference time-domain method has to be chosen (Kriezis et al., 2000b). As noted above, a domain wall in a ferroelectric liquid crystal (FLC) device is typically characterized by a width comparable in size with the device thickness (around 2 μm), and so it is difficult to probe its structure in detail. The precise director profile present is therefore unknown. Modeling the structure is also difficult because the continuum mechanics of FLCs is still a developing subject. A simpler continuum model can be used, treating the chevron interface in the FLC cell as an internal surface. The structure is twist dominated, with only a few degrees of tilt. For optical simulations, the agreement between the FDTD method and the Berreman method is only acceptable in the case of normal incidence for predicting the overall shape of curves. In the case of oblique incidence in particular, the Berreman method fails to predict the correct influence of the domain wall on the normalized transmitted optical intensity. This is expected, as the integral two-dimensional director profile has to be decomposed into a number of stratified profiles, each one treated completely independently of all others and without taking lateral scattering into account.

8.7 CONFINED NEMATIC LIQUID CRYSTALS

Ordering in confined nematic liquid crystals has been subject to intensive research for many years now (Crawford and Zumer, 1996). Two main configurations have been the focus of most effort: planar cells and polymer-dispersed liquid crystals (PDLCs). In the former, the liquid crystal is sandwiched between two plane surfaces. Electrical and optical functionality is achieved through torque on the molecules, which in turn resists through elastic forces and molecular anchoring at the interfaces. In the lattercase, birefringent spherical droplets of liquid crystal are included in a polymer matrix and contain defects that can be switched between several configurations (Drzaic, 1995). The type of defect generally depends on the size and shape of the droplets. The switching characteristics of PDLCs can be modified mainly through changes in the surface anchoring energy and the intermolecular interactions. The complex variety of behaviors that surface anchoring or boundary conditions impose on the liquid crystals is not only evident in individual PDLC spheres or hexagonal droplets, but also in large, cylindrical, polymer cavities, and infiltrated opals. The effect of confined geometries and surfaces on the behavior of liquid crystals can also be observed in planar cells with corrugated surfaces.

Complex surface interactions have been studied in a variety of confined geometries: ridges (Pfohl et al., 2001; Rosenblatt, 2004), two-dimensional irregular patterns (Ibn-Elhaj and Schadt, 2001), or regular square lattices (Mertens et al., 2002; Penterman et al., 2002; Schuller et al., 2003; Uche et al., 2005). An alternative and widely configurable route to control liquid crystalline order is followed by Prakash (2005). A three-dimensional microscale architecture, the planar-spherical cell, is harnessed to influence the formation and dynamics of point defects. The resulting systems can be continuously tuned all the way from planar cells (typical of nematic display devices) to spherical voids (typical of polymer-dispersed liquid crystal devices). The behavior was confined experimentally using special microstructured liquid crystal cells. The structure sizes are comparable with those used for diffractive optical devices, discussed in detail in Chapter 10. We will therefore limit the discussion here to confined systems where the liquid crystal is surrounded by polymer materials.

8.7.1 Droplets and Capillaries

Confined liquid crystals are commonly encountered in two-phase systems such as polymer dispersed liquid crystals (PDLC) or in capillaries. Such PDLC materials are based on mixtures of a liquid crystal material and a polymer precursor. A phase separation procedure results in a dispersion of liquid crystal inclusions (droplets) inside a solid isotropic polymer matrix with control over the inclusion physical dimensions, density, and distribution (Drzaic, 1995). Potential electro-optic applications in displays or light shutters are based on light scattering from the liquid crystal inclusions, which can be controlled by an external voltage. An external voltage reorients the liquid crystal material along the direction of the applied electric field and substantially reduces the light scattering, if the refractive index of the polymer matches the ordinary refractive index. As the intended electro-optic applications rely on light scattering, the first step is to address the problem of light scattering from a single liquid crystal inclusion in an infinite surrounding polymer. Spherical inclusions are by far the most common. Subsequently, knowledge of single droplet scattering characteristics has to be used in order to predict the scattering behavior of droplet collections or distributions. Commonly that is done using effective medium theories (Cox et al., 1998) or multiple scattering techniques (Kelly and Wu, 1993). Depending on the relative droplet size-to-wavelength ratio, different approximations are engaged for the light scattering studies. For electrically small and soft liquid crystal droplets, the Rayleigh–Gans (RG) approximation (Zumer and Doane, 1986) is the default choice. Light scattering from electrically large droplets can be analyzed by employing the anomalous diffraction approximation (Zumer, 1988). For applications it is more likely to involve submicron-sized droplets. The RG approximation is frequently employed, but its applicability might not be fully justified. Several studies aim to investigate the accuracy of the RG approximation for spherical nematic liquid crystal droplets. Liquid crystal droplets show several configurations that can be identified by microscope analysis (Ondis-Crawford et al., 1991; Higgins et al., 2005). An interesting point here is to compare the optical properties with respect to the prediction of the finite-difference time-domain method. In the work of Kriezis (2002), the finite-difference time-domain method for anisotropic films was adapted to the light scattering studies of PDLC droplets. Ray tracing seems only useful for large droplets (Reyes, 1998).

Consider the case of an isolated PDLC droplet characterized by its optical dielectric tensor $\boldsymbol{\varepsilon}$ and fully surrounded by an isotropic polymer with a refractive index of n_m. A relative optical tensor with respect to the polymer material can be defined for the droplet according to $\varepsilon_r = \varepsilon/(\varepsilon n_m^2)$. The scattered electric field in the far-field zone is generally written as the product of a spherical wave and a scattering amplitude **f**, the latter being a vector function of the illumination direction (**a**) and the scattering direction (**s**):

$$E_s = f(\mathbf{a}, \mathbf{s}) \frac{e^{-i\mathbf{kr}}}{\mathbf{r}}. \tag{8.79}$$

In Equation (8.79) **k** contains the wavenumber for the surrounding polymer, so that $k = \omega\sqrt{\mu_0\varepsilon_0}n_m$. An exact expression applies to the scattering amplitude, based on an integral over the volume of the scattering object. Therefore, in a rigorous approach, it is necessary to calculate the electric field in the scattering entity. We will present a few examples first and then continue the discussion of the scattering field calculation. The FDTD method will be applied to the confinement problem. We will study light propagation through nematic liquid crystal capillaries to give a slightly different view of the problem, as has already been done by Kriezis (2002) and Wang et al. (2003). Here we will

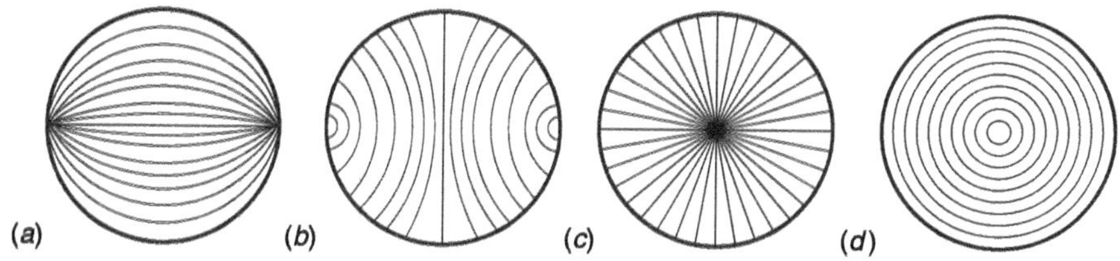

Figure 8.31 Liquid crystals can have different orientations in capillaries. Among them are the planar structures: (*a*) bipolar, (*b*) axial, (*c*) radial, and (*d*) concentric configurations.

focus on the interaction of a plane wave with the cross-section of a nematic liquid crystal capillary.

Depending on the anchoring and the type of nematic liquid crystal material, different configurations are possible inside the capillary. Figure 8.31 shows two capillaries with radial and concentric configurations, respectively. We will investigate the near-field of the light of anisotropic capillaries of different structural types (radial, concentric, and bipolar). By reasons of symmetry these capillaries have identical cross-sections along the capillary axis. We do not consider escape modes. We can therefore reduce the problem to a two-dimensional one: to the propagation of a plane wave in a cross-sectional plane. The directors of the liquid crystal molecules lie for all considered texture types in the cross-sectional plane. So, the situation is like that described in Chapter 2, and the TE and TM polarizations can be considered separately. The geometry of separation is shown in Figure 8.32. The TE polarization corresponds to a simple case, because the cross-section domain is "seen" as isotropic by the electric field due to the uniaxial anisotropy of the liquid crystal material. So we will concentrate our studies on TM polarized plane waves.

First we will give the reasons for the necessity of taking into account the anisotropy of the material. For a TM polarized plane wave passing through a

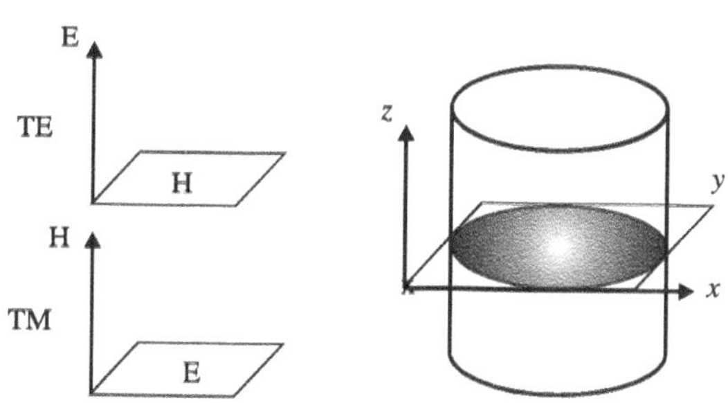

Figure 8.32 A plane wave enters the capillary cross-section; shown are the geometry and polarization definitions for the two-dimensional problem. The capillary is visualized as the cylinder on the right.

capillary of radial or concentric configuration we did calculate the diffracted light using the FDTD method in assuming the material inside the capillary to be isotropic. Therefore, the dielectric tensor ε is scalar (but locally variant) and equal to n_{eff}^2, where n_{eff} is the refractive index of the molecule director "seen" from the propagation direction of the entering plane wave. The propagation direction is fixed along the x-axis. Using this simplification one neglects the anisotropy of the material. How the results are falsified by this simplification is shown by direct comparison with rigorous FDTD simulations, which take into consideration the anisotropy of the material. Results are shown in Figure 8.33 for the capillaries of radial and concentric configurations. The plane wave enters from the left side and its wavelength is 400 nm. The figure shows the calculated intensities of the TM polarized wave propagating through the liquid crystal filled capillary. The capillary has a radius of 1 μm. The refractive indices of the anisotropic material are $n_o = 1.5$ and $n_e = 1.8$. The refractive index outside the capillary is $n = 1.0$. The significantly different distributions of intensity for the anisotropic and isotropic capillaries, in particular for the radial configuration, emphasize the necessity of taking into account the anisotropy of the material.

We compare now different director configurations. All results shown in the following are calculated with rigorous FDTD simulations for TM polarized light. Figure 8.34 shows calculated intensity and phase distributions in an area around the capillary for different capillary configurations. The radius of the capillary is 2λ. The wavelength is equal to $\lambda = 550$ nm. The ordinary and extraordinary refractive

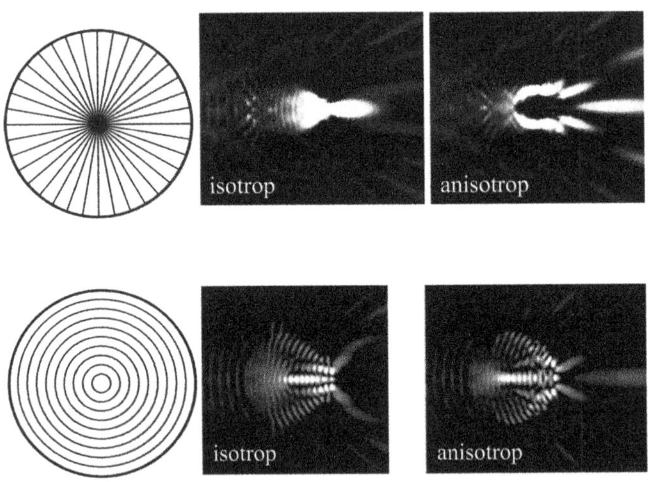

Figure 8.33 Calculated intensities for TM polarized light propagating through a liquid-crystal-filled capillary of radial and concentric configuration with a radius of 1 μm, $\lambda = 400$ nm. Left: assumed as isotropic but inhomogeneous (n_{eff} "seen" from the propagation direction). Right: anisotropic with $n_o = 1.5$ and $n_e = 1.8$. (The surrounding n is 1.0, and the size of the simulation domain 4.0 μm × 6.2 μm.)

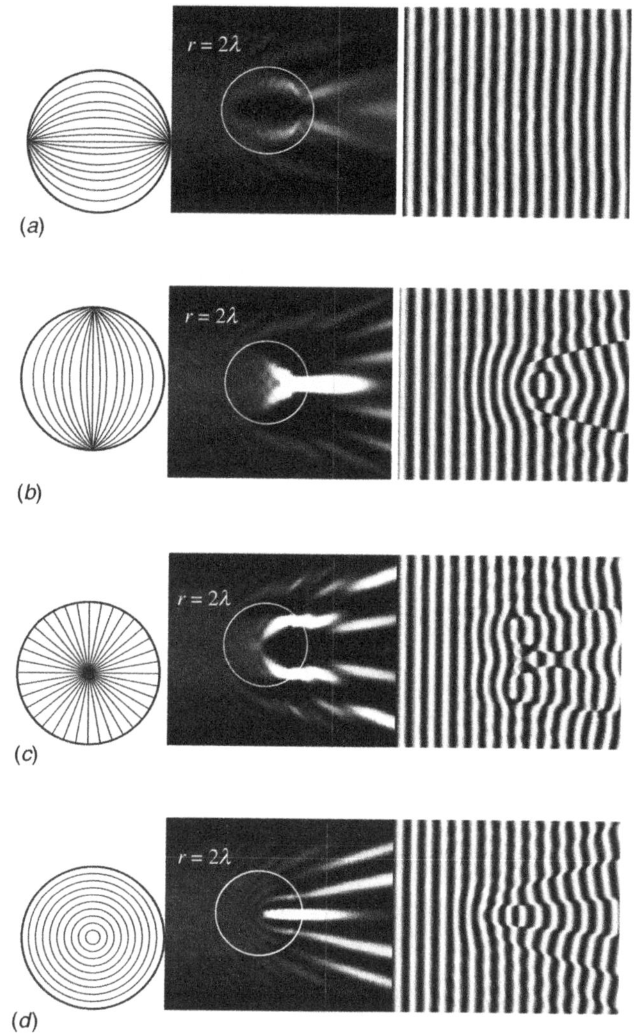

Figure 8.34 Calculated intensity and phase distributions for differnt capillary configurations: (*a*) bipolar(1), (*b*) bipolar(2), (*c*) radial, and (*d*) concentric. Left to right for each of (*a*)–(*d*): Director configuration in the liquid crystal filled capillary, intensity (area 4.5 × 5.6 μm), phase profile for $\lambda = 550$ nm. The TM-polarized plane wave comes from left, $n_o = 1.5$, $n_e = 1.8$, environment $n = 1.5$.

indices of the liquid crystal material are assumed to be $n_o = 1.5$ and $n_e = 1.8$. The refractive index outside the capillary is $n = n_o = 1.5$. Because of the higher refractive index of the ambient medium, which was adapted to the liquid crystal material, we do not have strong reflection at the capillary as is the case for $n = 1.0$, shown in Figure 8.33. The capillary of bipolar configuration with the average direction of the director parallel to the incident light wave (Fig. 8.34*a*) has almost no influence on the phase field of the entering TM polarized light. However, the bipolar configuration perpendicular to the incident light wave (Fig. 8.34*b*) shows an optical behavior that is similar to that of a cylindrical isotropic lens. The effects of the radially and concentrically configured anisotropic capillaries (Fig. 8.34*c* and *d*) are more surprising. The phase fields are more perturbed and the field behind the radial capillary is characterized by some significant phase singularities.

Focusing now on the radial configuration, we will investigate the light diffraction by capillaries of different diameters. Again we use the FDTD method in two dimensions for TM polarization. Figure 8.35 shows the simulation results. If the radius of the capillary is decreased, the perturbation to the phase field is reduced. If one analyzes the far-field (Bohley, 2004), the influence on the intensity distribution is not as

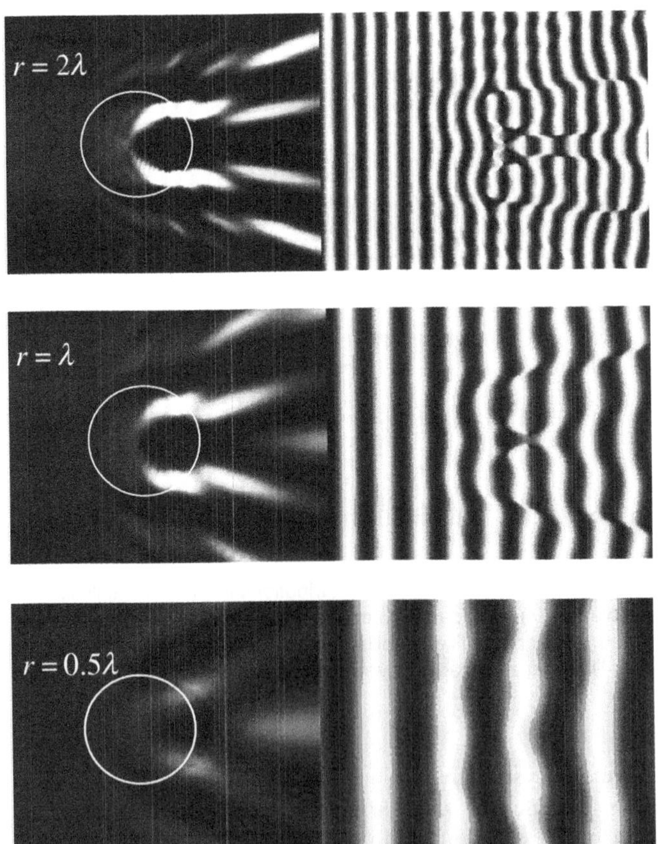

Figure 8.35 Optical properties of capillaries filled with nematic liquid crystal with radial director configuration. From left to right: intensity, and phase of magnetic field ($\lambda = 550$ nm, TM polarization, $n_o = 1.5$, $n_e = 1.8$, environment $n = 1.5$).

high, as it is assumed from the intensity distributions in the near-field (Kriezis et al., 2002b). In particular, for the capillary of radius 2λ, the near-field intensity is strongly diffused by the capillary but the far-field remains strongest in the propagation direction of the incident wave. To evaluate the light scattering of an ensemble of confined nematic droplets one has to put the results of the rigorous field simulations into a scattering model. An exact expression applies to the scattering amplitude of Equation (8.79) based on an integral over the volume of the scattering object (Kriezis, 2002), given by

$$f(\mathbf{a},\mathbf{s}) = -\frac{k}{4\pi}\oint_V \mathbf{s}\times\mathbf{s}\times((\varepsilon_r(\mathbf{r})-\mathbf{1})\mathbf{E}(\mathbf{r}))\mathrm{e}^{ik\mathbf{r}\cdot\mathbf{s}}\,\mathrm{d}V, \tag{8.80}$$

where $\varepsilon_r = (\varepsilon n_m^2)$ and $k = \omega\sqrt{\mu_0\varepsilon_0}n_m$. In Equation (8.80), the total electric field $\mathbf{E}(\mathbf{r})$ inside the object is involved, which is normally unknown. For PDLC droplets that are characterized by low refractive index contrast and, furthermore, their size being sufficiently small, so that the maximum phase shift introduced by a diameter crossing is small, one can state

$$\left|\frac{n_{e,o}}{n_m}-1\right| \ll 1 \qquad \text{and} \qquad 2kR\left|\frac{n_{e,o}}{n_m}-1\right| \ll 1. \tag{8.81}$$

In Equations (8.81) n_o and n_e are the ordinary and extraordinary liquid crystal refractive indices, respectively. Assuming that these two conditions are fulfilled for both refractive indices, one can apply the RG approximation and substitute the unperturbed incident plane wave expression for the unknown internal field (Bohren and Hoffman, 1983). This is also known as the Born approximation. It treats the scattering PDLC object as a collection of radiating dipoles, which are excited solely by the incoming plane wave. Interaction with each other is omitted. The incident plane wave polarized along an arbitrary direction $\mathbf{p}$ is written as

$$\mathbf{E}_{\mathrm{inc}}(\mathbf{r}) = \mathbf{p}\,\mathrm{e}^{-ik\mathbf{r}\cdot\mathbf{a}}. \tag{8.82}$$

Substituting Equation (8.82) into Equation (8.80) results in

$$f(\mathbf{a},\mathbf{s}) = -\frac{k^2V}{4\pi}\mathbf{s}\times\mathbf{s}\times(S(\mathbf{k}_s)\mathbf{p}), \tag{8.83}$$

where

$$S(\mathbf{k}_s) = \frac{1}{V}\oint_V(\varepsilon_r(\mathbf{r})-1)\mathrm{e}^{i\mathbf{k}_s\cdot\mathbf{r}}\,\mathrm{d}V \qquad \text{with } \mathbf{k}_s = k(\mathbf{s}-\mathbf{a}). \tag{8.84}$$

Equation (8.84) is identified as the three-dimensional Fourier transform of the tensor quantity $(\varepsilon(\mathbf{r})-\mathbf{1})$. Using the vector identity for the triple product in Equation (8.83) results in the more useful expression for the scattering amplitude,

$$f(\mathbf{a}, \mathbf{s}) = -\frac{k^2 V}{4\pi}(S(\mathbf{k}_s)\mathbf{p} - \mathbf{s} \cdot (\mathbf{s} \cdot S(\mathbf{k}_s)\mathbf{p})). \tag{8.85}$$

Based on Equation (8.85) and the volume integration of Equation (8.84), one can calculate the scattering amplitude for any given combination of illumination direction (**a**), incident polarization (**p**), and observation (scattering) direction (**s**). Often it is important to assess the accuracy with respect to a related problem with known analytical solution. Scattering from a dielectric sphere can be described by the Mie theory (Bohren and Hoffman, 1983). Such a comparison can provide an estimate of the error introduced by the approximations as well as for the unavoidable numerical discretization error present in the FDTD method. This was done in the work of Kriezis (2002) and leads to the conclusion that for PDLC droplets in the range $0.5 < kR < 5$ the RG method provides a correct general estimate to the variation of the differential scattering cross-section. The FDTD method was combined with a very fine grid was used to provide an accurate numerical calculation for the differential cross section. Typical local and average errors for the RG method in the above range are usually around 10–20%. However, higher errors have also been recorded, in particular for radial droplets where average errors of around 40% have been found.

8.8 INSTABILITIES IN LIQUID CRYSTALS

Investigation of the interaction of electromagnetic waves with periodic structures can be of predictional (design) and analytical character (structure determination). The progress of rigorous optical methods allows a better understanding of the optics in structures with periodicities of optical wavelength size. For such structures, conventional computation methods (plane wave, ray-trace) fail, because they are not able to consider effects like scattering and lateral interaction between optical fields. Such discussion is interesting for instabilities in liquid crystals. Analyzing the optics of a dissipative pattern formation in nematic liquid crystals by electrohydrodynamic convection in thin planar cells has been performed in the work of Bohley et al. (2005). Such electrohydrodynamic convection patterns are an appropriate model system for periodic anisotropic structures, because the wavenumbers are, to a certain extent, controllable by the excitation frequencies, but the maximal director amplitude can be adjusted with the amplitude of the voltage (de Gennes, 1993). Additionally, the system is not disturbed by a structured electrode.

The structure and dynamics of electrohydrodynamic convection have been extensively studied in the last decades. Principally, an electric field excites a periodic convection roll pattern, which is related to a regular director deflection pattern. The pattern of the simplest structures is spatially periodic in one preferred direction in the cell plane. This preference is determined by the alignment axis of the nematic at the parallel cell plates. In general, it is assumed in a model that the director of the nematic liquid crystal molecules remains in a plane (no twist deformation). This is correct in most situations. Electrohydrodynamic convection can occur in different modes, depending on the electric excitation frequency. Generally, the velocity field in the convection rolls deflects the homeotropic or planar ground state of the liquid crystal director in the cell. Near the onset of convection, one usually assumes a periodic sine modulation of the deflection in the cell plane and a simple sinusoidal profile of the deflection angle in the direction normal to the cell plates.

Historically, the investigation of electrohydrodynamic convection patterns by optical means started with orthoscopic microscope methods such as shadowgraph method, where the deflection of light rays in the spatially modulated electrohydrodynamic convection pattern is observed (Rasenat et al., 1989). With the shadowgraph method, the wavelength of the pattern, defects, and more complex patterns can be examined. Laser diffraction of liquid crystal gratings has been studied theoretically and experimentally (Carroll, 1972). This method allows the quantitative determination of the period and the orientation of the pattern as well as its dynamics, and in principle also the quantitative determination of the director deflection amplitudes. Zenginoglou and Kosmopoulos have analyzed diffraction theoretically with an advanced geometric optics method (Zenginoglou and Kosmopoulos, 1988) and a literalized wave propagation approach (Zenginoglou and Kosmopoulos, 1997). John and colleagues investigated the laser diffraction caused by electrohydrodynamic convection with a ray-trace method based on crystal optics (John et al., 2003). They compared the results with those obtained using an analytical approach, which is valid for thin cells, small maximal director angles, and for larger spatial periods, as they are typical for the conduction regime. Up to now, the optics of conduction patterns have mainly been investigated. Although the periodicity of the structure is not taken into account in the ray-trace and in the analytical approaches, it is included in the FDTD calculation. This periodicity is more important if the period of the structures is small, because in this case the interaction between neighboring units of the grating is stronger. The ray-trace approach provides a good approximation for intensity and phase of the electric field, but fails for smaller periods of the structure. Generally, the results obtained with these approaches match well with those obtained with the FDTD for thin cells and small deflection angles if the structure period is not too small (Bohley et al., 2005). The results obtained with the rigorous FDTD and the analytical respective ray-trace methods differ strongly for higher deflection angles and smaller periods. Under these circumstances, a light phase dislocation effect can be detected in the near of the electrohydrodynamic convection cell exit.

We would like to present an example of a typical simulation for roll instabilities in nematics. The electric and magnetic fields for light propagation are computed for a region of 50×25 μm, with a periodic FDTD method with Mur-boundary conditions for a wavelength of $\lambda = 600$ nm and TM linear polarized light. The structure of the director field is calculated by the formula

$$\varphi = \varphi_0 \sin\left(\frac{x}{a}\right) \sin\left(\frac{z}{a}\right), \tag{8.86}$$

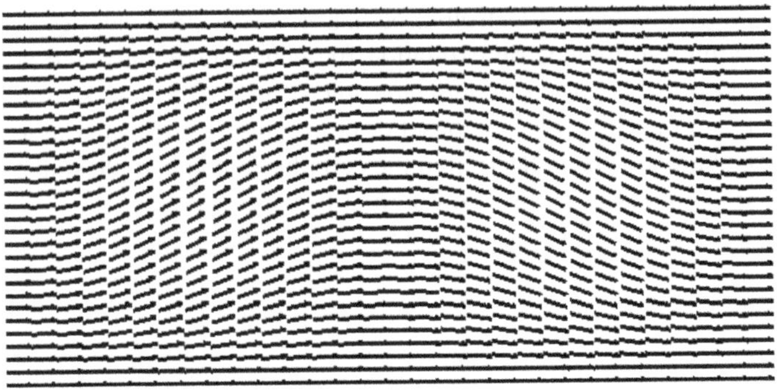

Figure 8.36 Typical director structure for roll instabilities in nematic liquid crystals. The director field is calculated using Equation (8.86).

where $\varphi_0 = 0.5$ and the cell thickness $a = 25$ μm. The cell thickness a is also the period of the rolls. Figure 8.36 shows the typical director structure found with such analytical modeling.

The optics of such structures is conveniently simulated with the FTDT method, although simulation times are long because of the large domain that has to be considerd. The refraction indices are $n_o = 1.5$ and $n_e = 1.65278$, similar to those found in the literature for such materials (Amm et al., 1998). A reduced accuracy was used for such a large field size of simulation. We used a sampling of wavelength/10. The main contribution to transmission is in the incident polarization (E_x of the electric field) and is plotted in Figure 8.37. The scattered field in the other component E_z is less than 7%, compared to the main contribution in E_x, E_y is zero.

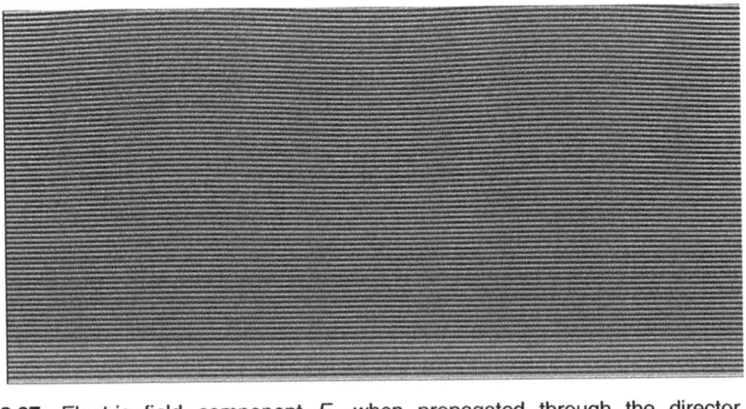

Figure 8.37 Electric field component E_x when propagated through the director field in Figure 8.36. It is clearly seen that the wavefront forms regular patterns at the output and that the light propagation behaves gently and does not show singularities.

8.9 DEFORMATION OF LIQUID CRYSTAL DIRECTORS BY FRINGING FIELDS

The effect of fringing fields on the optical properties of a liquid crystal display at the electrode edges clearly impacts on the operation of an adaptive optics element over some parts of the wavefront. Quantifying this effect is important in device characterization. Only a few experimental studies have been carried out to date (Storrow et al., 1998; Bahat-Treidel et al., 2005; Fan-Chiang et al., 2005). The test elements in which fringing can be studied quite easily are electrically controlled birefringence (ECB) devices. The commercially produced device used here for demonstration has antiparallel rubbed polyimide alignment layers, transparent indium tin oxide electrodes etched to produce an overlap area of 1 cm^2 in the center of the substrates, and was filled with the nematic liquid crystal substance 5CB. The material parameters are given in Chapter 6.

The liquid crystal device includes spacers that define its thickness, and the thickness across the area is kept constant. The thickness of the liquid crystal test element was determined to be 6 μm. For an ECB device below threshold, the ordinary and extraordinary refractive indices throughout the device are n_e and n_o (parallel and perpendicular to the liquid crystal orientation direction, respectively). When a voltage above threshold is applied, the optical properties are more complex. The maximum phase shift that a liquid crystal device can impose on an incident wavefront

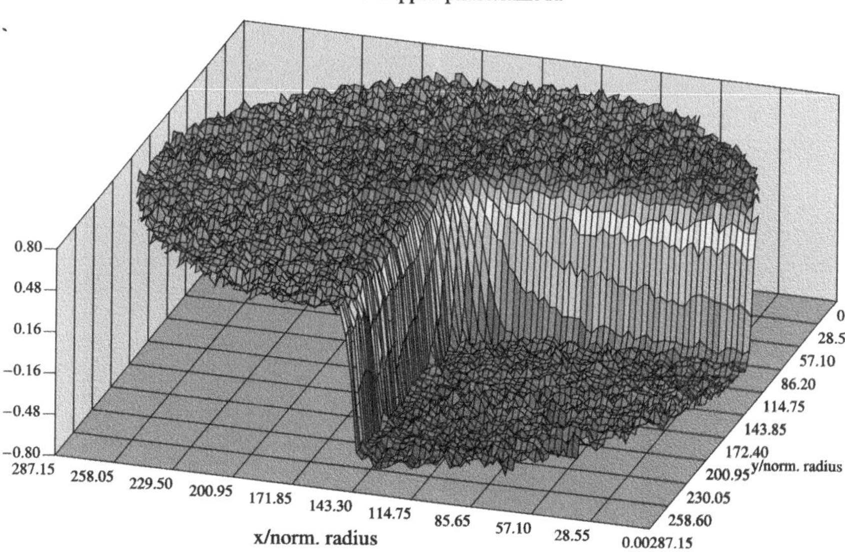

Figure 8.38 Phase profile at a corner of a liquid crystal pixel device. The cell is 6 μm thick and filled with 5CB. A voltage of $V = 0.9$ Vrms is applied. The two directions where the fringing field is along and perpendicular to the alignment show slightly different phase profiles. The illustration covers an area of 290 μm × 290 μm.

occurs for light polarized along the optical axis of the element. Although the optical properties of nematic liquid crystal devices are well understood when a voltage is applied in a direction perpendicular to the substrates and well within the electrode overlap area, it is considerably more difficult to deduce the optical properties at an electrode edge. The field profile and orientation of the director are interdependent and are in turn influenced by the dielectric and elastic properties of the liquid crystal. Even predicting the field profile when a linear, isotropic, homogeneous dielectric fills a device is difficult. When the dielectric is anisotropic and the field influences the orientation of the dielectric in a nonlinear manner, as is the case here, the problem must be solved numerically. It is therefore instructive to approach the issue of quantifying the optical influence of fringing fields from an experimental point of view.

For the measurements shown, we used a phase shifting interferometer working with polarized light (Sickinger et al., 1999). This interference microscope measures

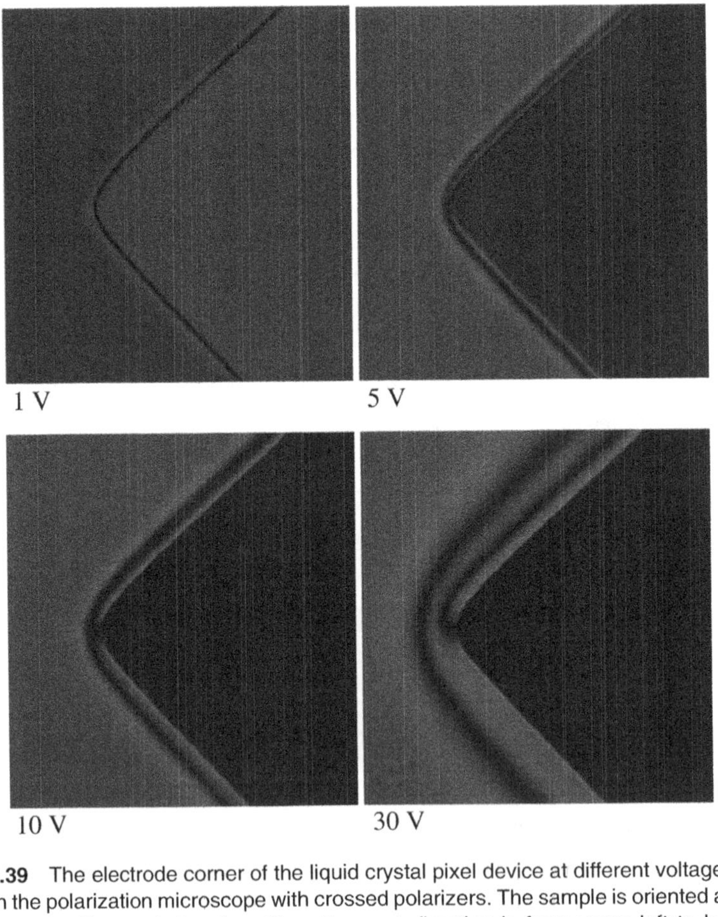

Figure 8.39 The electrode corner of the liquid crystal pixel device at different voltages seen at 633 nm in the polarization microscope with crossed polarizers. The sample is oriented at 45° with respect to their alignment direction. The alignment direction is from upper left to lower right. Voltages are given in rms for a square wave at 1 kHz. The area shown is 300 μm × 300 μm.

the phase, which is not subject to resolution limitations. Let us first look at the corner of the electrode area. Here both situations, alignment parallel and perpendicular with respect to the electrode edge, are found. Figure 8.38 shows a measurement for the cell described. The voltage can be varied and is in this case given as $V = 0.9\ V_{rms}$ at 1 kHz square wave. We see flat areas with wavefront distortion less than 0.05 rms wavelength and the transition zone between them. The switched electrode area with less retardation is at the lower level, and the higher level represents the nonswitched zone. It is interesting to compare this figure with the corresponding microscope pictures shown in Figure 8.39. The microphotographs were taken between crossed polarizers. The optical axis of the cell was oriented at 45°. We see that in the microscope picture for 1 V in Figure 8.39, a black line appears that indicates the transition zone between the reoriented and the nonswitched part. The line has a small extension. Its width is about 5 μm. The liquid crystal layer is not yet reoriented completely, which is indicated by the transmission in the observed area. At higher voltages the retardation of the liquid crystal cell is decreased further. The images for 1 V, 5 V, 10 V, and 30 V show that the reorientation is nearly complete, which results in extinction between crossed polarizers. The switched area becomes black. For high reorientation states at 30 V, the fringes in the picture indicate that the distortion is extended far into the nonswitched zone. Directions along and perpendicular to the rubbing direction show distinctly different behavior. If the electrode etch is along the alignment, one fringe is observed that propagates into the nonswitched area at higher voltages. If the director of the alignment is perpendicular to the electrode edge, a set of two fringes develops.

To see how the fringing fields extend in the nonswitched area in both directions, one can compare phase profiles at different voltages. The phase profile

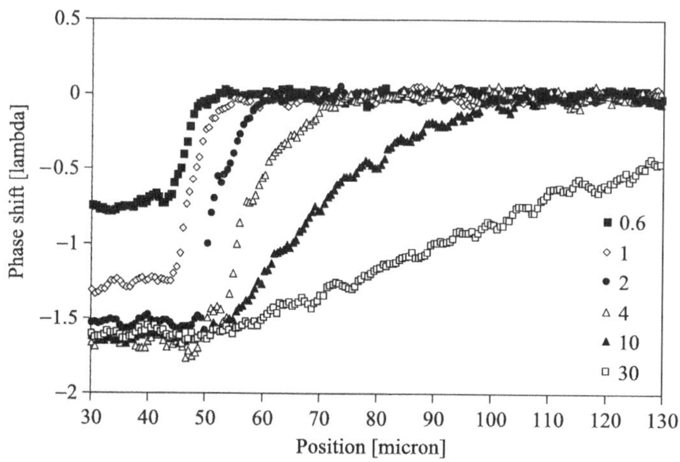

Figure 8.40 Measured phase shift if the director is aligned parallel with respect to the electrodes. Different voltages from 0.6 to 30 V are applied. The electrode edge is at position 43 ($\lambda = 633$ nm).

is interferometrically measured with a Mach–Zehnder interferometer and a phase shifting technique at 633 nm. Figures 8.40 and 8.41 show the results of such measurements. The advantage of this is that no reconstruction of the phase profiles from intensity measurements is necessary, which often leads to difficulties in reconstruction of the phase. All the information about the deformation is contained in this measurement, as the local phase profiles have to be considered as a sum of optical effects that happens when light propagates through the anisotropic layer. The phase profiles are taken in polarized light and that corresponds to a situation between parallel polarizers. Figures 8.40 and 8.41 show that the fringing field can extend the zone of reorientation enormously. For voltages of 0.6 V and 1 V, the phase shift does not exceed the maximum value of 1.5 wavelengths. The width of the zone where the major phase shift (90%) happens varies. For 0.6 V and 0.7 wavelengths, phase shift, the width is 6 μm for parallel and 8 μm for perpendicular orientations, respectively. At 2 V, the maximum phase shift of 1.5 wavelengths is almost reached, and the width of the zone is 20 μm in the parallel orientation and 17 μm for the perpendicular. Not that due to the measurement technique, errors might appear if the polarization is rotated out of the plane of the measurements. This is particularly important for the case where the alignment is initially parallel to the alignment direction. The fringing field is then perpendicular to the director orientation and will turn the director out of plane. This falsifies the measured phase profile. We see that the extension in the lateral direction is always larger than the thickness of the liquid crystal layer. If the voltage is further increased, the extension can reach values of several tenths of a micron. Such effects will definitely limit the performance of passive switchable devices. More details from an application point of view are described in Chapter 10.

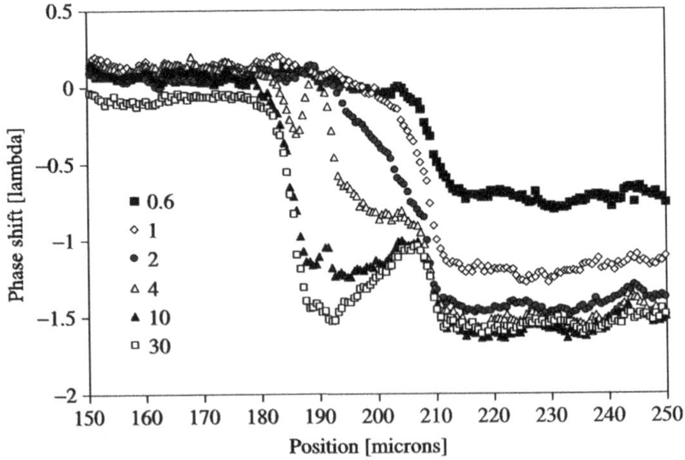

Figure 8.41 Measured phase shift for a perpendicular alignment to the electrodes. The electrode edge is at position 210 ($\lambda = 633$ nm). Different voltages from 0.6 to 30 V are applied.

8.10 RESOLUTION LIMIT OF SWITCHABLE LIQUID CRYSTAL DEVICES

For determining the resolution limit, different criteria might be used. Useful error criteria can include the diffraction efficiency of gratings. However, the implemented phase distribution is the desired quantity of the device for realizing an optical function like contrast or retardation. For visualization the transmittance between polarizers is the best tool. In this section the transmittance and phase profile, calculated using different methods, are discussed. If one assumes slowly varying distributions of the refractive index profile, one can work in the classical limit. The unusual spatial distribution of the liquid crystal director field makes a proof of applicability necessary. The Eikonal approximation requires that the transverse change of the refractive index has to be small compared with the product of the refractive index n_o and the wavenumber k_0. In the case of a medium with a birefringence Δn_{eff}, this condition reads

$$|\mathrm{grad}(\Delta n_{\mathrm{eff}})|_{\mathrm{perpendicular}} \ll n_o \cdot k_0, \tag{8.87}$$

where the left term is the maximum value of the gradient perpendicular to the direction of propagation. For the cases studied in Section 8.9, the gradient can be calculated in the following manner. The transition zone δx for the major phase jump was evaluated to be approximately 8 μm for 0.6 V for a cell thickness of 6 μm. The birefringence of 5CB may be found in Table 6.6 as $\Delta n_o = n_e - n_o = 0.19$. The lateral gradient is therefore $\Delta n/\delta x = 0.19/8/\mu\mathrm{m} = 0.02/\mu\mathrm{m}$. On the other side of Equation (8.87) we calculate with $n_o = 1.5442$ and $\lambda = 633$ nm

$$n_o k_o = n_o \frac{2\pi}{\lambda_0} = 1.5442 \frac{2\pi}{0.633\ \mu\mathrm{m}} = 15/\mu\mathrm{m} \gg 0.02/\mu\mathrm{m}, \tag{8.88}$$

which is fulfilled. If the reorientation were smooth no correction would have to be taken into account. The fringing fields are strong at the electrode's edges and that might lead to much higher gradients. In the case of defect structures when the lateral dimension of a director change are about 50 nm. The gradients are strong and reach values of $\Delta n/\delta x = 0.19/0.05/\mu\mathrm{m} = 3.8/\mu\mathrm{m}$. Looking at the values above, the question arises why one should use rigorous simulation methods if the gradients are weak? When the optics of devices is studied, sampling is usually done to probe locally the optical properties. If this sampling is too rough, the simulation fails. In the thin optics approach for diffractive elements, rays are sent through devices and the phases and amplitude modifications are considered. It is assumed that these rays stay within their sampling volume and do not cross each other. Plane wave methods stay valid when such crossing is not observed. The liquid crystal layer shows a gradient index behavior that provokes such crossing at very high sampling rates. For instance, with a 6 μm thickness of the liquid crystal layer and a 6-μm-wide zone of phase jump, the elements behave like a prism with a deviation angle of about 10°. Crossing of rays would therefore happen at the exit surface of a device if its sampling is below 1 μm. The situation turns critical if devices are studied that have periods or spatial elements that are only several microns wide. Then plane wave methods can no longer be used for simulations, and rigorous modeling has to be considered.

Let us come back to our problem of the resolution of a liquid crystal device. The elastic deformation of the liquid crystal leads to gradients of the refractive index. For electrically switchable systems with electrode edges, the effect of elastic deformation is superposed with the fringing electric field at the edges. In principle the resolution is only defined by the elastic deformation. It is therefore not the optics that limits the resolution of liquid crystal devices for photonics, but the deformation of the liquid crystal director and the fringing field at the electrode edges.

To obtain a consistent view on this subject it is worth comparing cases with textures defined by surface defects and electrically switchable systems. We define two systems to be compared as follows. A liquid crystal cell of a certain thickness d has at position 0 two surface defects that lead to a deformation profile as shown in Figure 8.42. A planar domain and a homeotropic domain are brought together. The second system is a planar aligned cell with two electrodes superposed at position 0. The thickness is also d. Although the first has no electrical field switched on, the second shows field dependence. It is interesting to see the difference between the two systems and to compare the intensity profiles at the output for different voltages. The simulation of light propagation in Figure 8.43 shows several expected results. For textures without twist the polarization stays within the plane of incidence. When twist is present, transfer of energy in perpendicular electric field components occurs. This is the case in Figure 8.44. For both cases the deformation zone of the field mediated case is much larger than the defect mediated case.

To be more quantitative, the amplitude and retardation are plotted for two cases at different voltages in Figures 8.45 and 8.46. Figure 8.45 shows the amplitude and retardation for the case when only tilt deformation is present. The polarization of light always stays in the same plane and all effects can be described by discussing

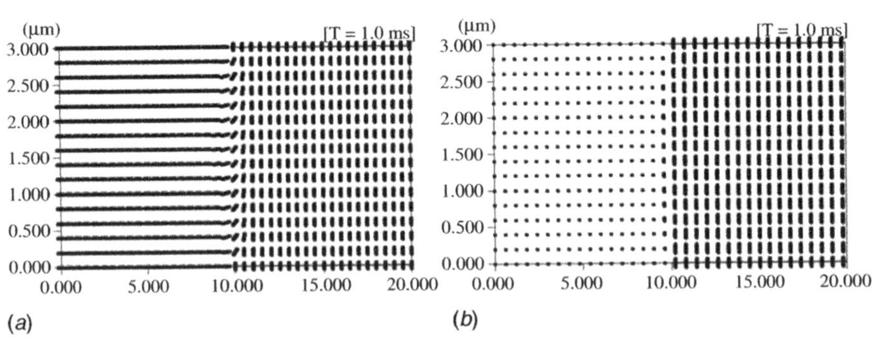

Figure 8.42 Liquid crystal texture creating a high refractive index gradient. The texture is mediated by surface defects: (*a*) A tilt deformation is dominant; (*b*) twist is induced due to the surface defects.

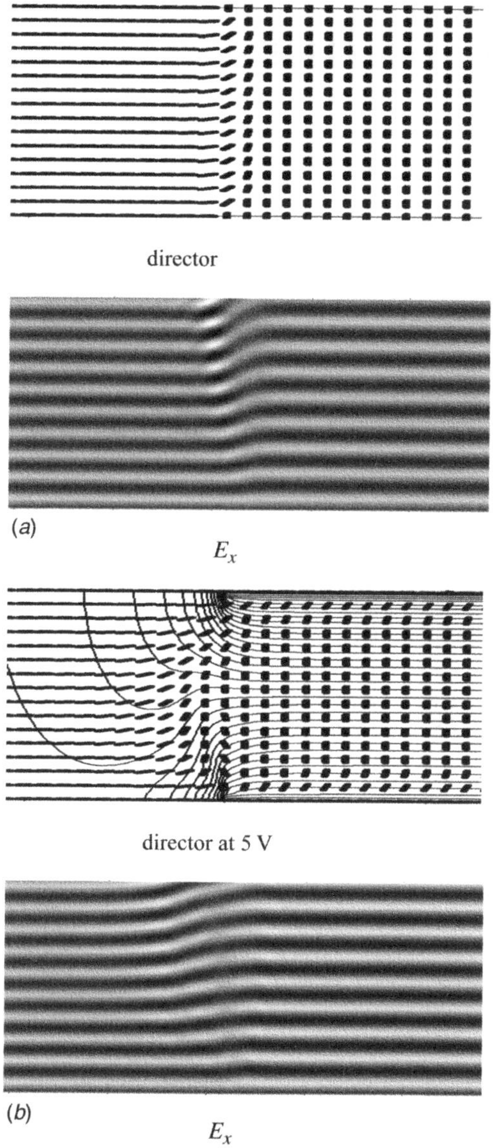

Figure 8.43 Electromagnetic waves propagating through selected textures: (*a*) Tilt-only configuration for the defect mediated case and (*b*) the field forced reorientation with 5 V applied between the electrodes (TM polarization E_x; the field is 3 μm × 14 μm).

the incident polarization TE. The applied electric field deforms the liquid crystal director and the result of this deformation is easily visibly in the phase profile. The defect situation shows the smallest extension of the transition zones. In Figure 8.45 one measures 1 μm for the widths for a 3-μm-thick layer. For the

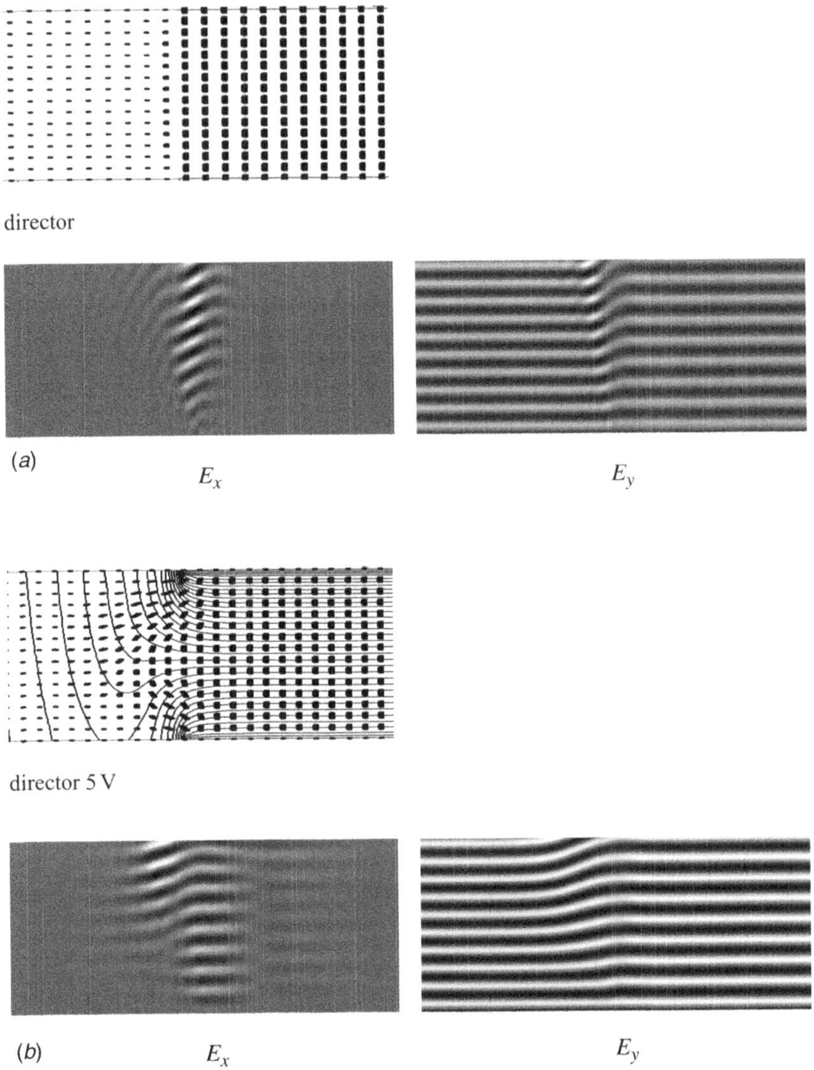

Figure 8.44 Electromagnetic waves propagating through selected textures: (*a*) Twist-only configuration for the defect mediated case and (*b*) the field forced reorientation, with 5 V applied between the electrodes (TE polarization E_y; the field is 3 μm × 14 μm).

field reorientation case the values are approximately 2 μm, 3 μm, and 4 μm for 2, 5, and 10 V, respectively. The maximum phase shift is already reached below 5 V so that the transition zone corresponds to the cell thickness. If higher voltages are applied, the zone gets larger.

More complicated is the case when twist deformation occur. Now energy is transferred to the perpendicular polarization. Figure 8.46 shows the simulated amplitude

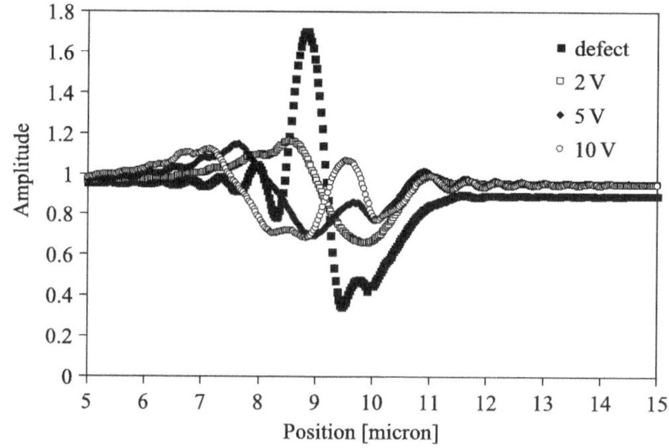

Figure 8.45 Amplitude and phase shift for tilt configuration at different voltages, simulated with the rigorous method for one polarization component E_x (as seen between parallel polarizers); E_y is zero and E_z is two orders of magnitude smaller.

and phase values. The phase is given for TM polarization as it is the incident polarization and therefore the polarization that is disturbed. The zone of the deformation width can be estimated to be 1 μm for the defect configuration, 2 μm at 2 V, 3 μm at 5 V, and 11 μm at 10 V. We find more or less the same values as for the tilted case. The difference now is that twist occurs and discussion of the optical properties is less obvious. This becomes clear if we look at the amplitude for the polarization TE. This amplitude represents the intensity when observed between crossed polarizers and oriented with the alignment along one polarizer. We would find a transmitted intensity modulation already as large as 4 μm at 2 V. This represents the double of the phase profile deformation and is caused by the in-plane reorientation of the liquid crystal director due to transversal electric field components in the fringing fields.

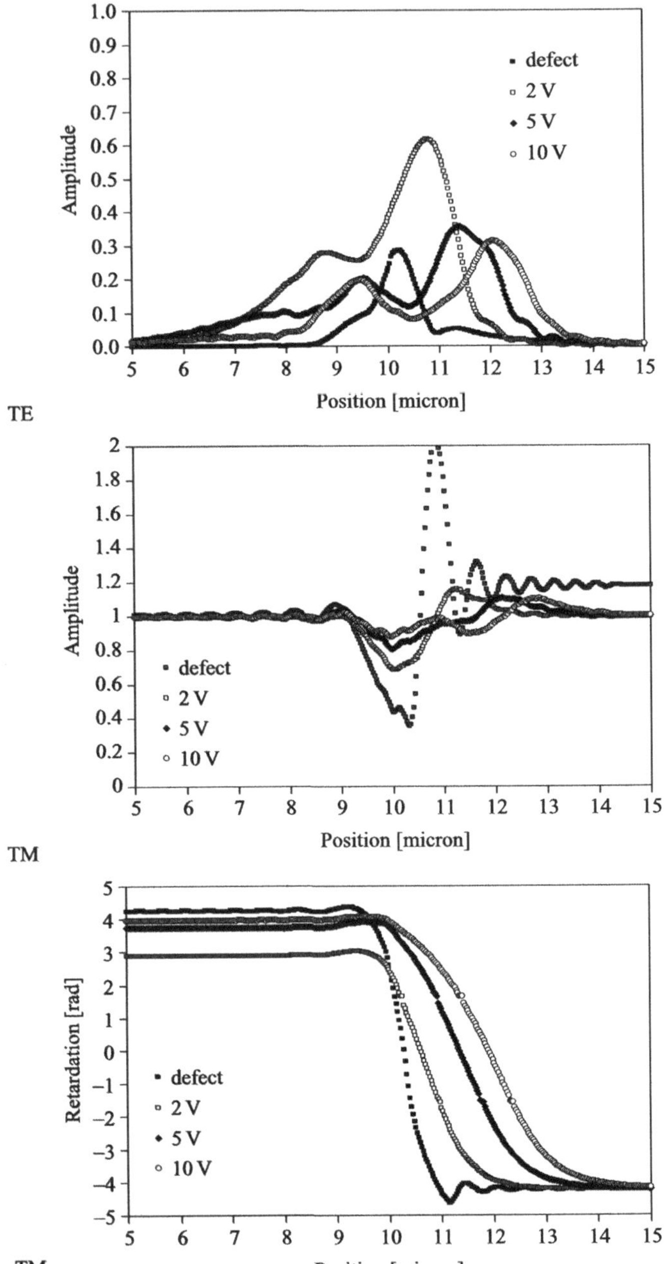

Figure 8.46 The twist configuration is analyzed, with amplitude and phase shift corresponding to the texture in Figure 8.42*b* on the right. The amplitude and phase for different voltages is simulated with the rigorous method. The input polarization is TM (E_y). TE and TM are analyzed, but the retardation is only given for TM. Due to the rotation of the polarization in the transition zone, energy is transferred into TE polarization (polarization rotation in twisted structures).

How critical this is for a concrete device should be judged for each individuals optical system. However, it will definitely reduce resolution.

In summary, we can say that the resolution limit of low-voltage-driven liquid crystal photonic devices is the cell thickness; this can only be reduced when surface defects are considered. The design of switchable optical components can be done by modeling the transition zone either as a linear phase ramp (Yu et al., 2004) or by using a kernel function that can change for different voltages (Apter et al., 2004).

8.11 SWITCHING IN LAYERED PHASES

8.11.1 Smectic C Switching

In smectic C phases the director is tilted with respect to the layer normal. The texture of smectic C phases resembles very much that of smectic A phases because of the layer structure they have in common. When the molecule is chiral, successive smectic C layers show a gradual change in the direction of tilt always lying on the surface of a hypothetical cone of angle 2θ. The angle around the circle of precession is known as the azimuthal angle. A helical structure is formed in the chiral smectic C (SmC*) mesophase, with the pitch being the distance along the z-axis needed to reach the same molecular orientation, as illustrated in Figure 6.4. This helical structure results in a spontaneous polarization. The polarization vector is perpendicular to the director and contained in the layer plane. Therefore, all possible directions for the vector are at a tangent to the circle of intersection of the cone with the plane. A bulk SmC* sample, free to develop its helical structure, will not show ferroelectric behavior, as the spontaneous polarization will average to zero over one pitch (because polarization vectors go around an entire circle and cancel each other out). Typically, the pitch includes many layers in such a material. The value of the pitch can be as low as several hundred nanometers and causes strong interference effects like those in cholesteric liquid crystal phases. Molecular ordering and electro-optic switching in chiral smectic liquid crystal phases is discussed in detail in several textbooks (Goodby and Blinc, 1991; Lagerwall, 1999). In the case of ferroelectric switching in chiral smectic C* phases, the molecular director **n** moves about a cone of angles that is centered on the horizontal axis. Each molecule has a ferroelectric dipole **P**, which is perpendicular to its length. This is shown in Figure 8.47.

When an electric field **E** is applied to the cell, there is an interaction between **E** and **P**, which forces the director to move around the cone to a point of equilibrium. If the ferroelectric liquid crystal is constrained in a thin layer between suitable aligning surfaces, the cone is suppressed and the precessing of the **n** director is suppressed to two states. In this case, applying an electric field will switch the director (and therefore the optical axis) from one side of the cone to the other. The angle between the two switched states is the switching angle 2θ, which is twice the liquid crystal tilt angle. In such switching the liquid crystal layer acts as an optically uniaxial medium with its optical axis in the plane of the layer. Hence, it behaves like a

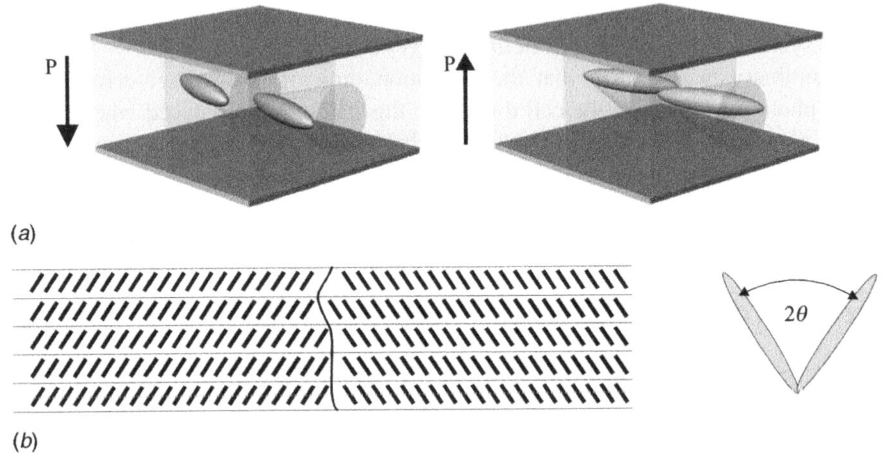

Figure 8.47 Principle of bistable ferroelectric switching. (*a*) The director is moving on a cone and the polarization vector P is oriented parallel to the external electric field. (*b*) If seen from above, the director takes two different positions. The angle between the two different positions is twice the tilt angle 2θ.

switchable waveplate whose fast and slow axes can be in two possible states separated by the switching angle 2θ, and whose retardation Δnd depends on the thickness d and birefringence Δn of the material. This is not only true for ferroelectric switching in the chiral smectic C* phase, but also for electroclinic switching in chiral smectic A phases. We can use Jones matrices to model the optical behavior of such a ferroelectric liquid crystal layer. However, it is easier to apply the formula for the intensity of a birefringent plate between crossed polarizers. This was given by Equation (8.74). We have to respect the new definition of our tilt angle θ with respect to the in-plane orientational angle φ in the original equation. In the cell we have two states and if we set one parallel to the polarizer at $\varphi = 0$, the other will move the optical axis to $\varphi = 2\theta$. As the first state with $\varphi = 0$ shows complete extinction, the intensity transmitted when switched becomes

$$I = I_0 \sin^2(4\theta) \sin^2\left(\frac{\pi}{\lambda} d\Delta n\right). \tag{8.89}$$

Hence, maximum transmission for intensity modulation occurs when $\Delta nd = \lambda/2$ and $\theta = \pi/4$. The device can also be used for phase modulation. Here the aim is to achieve maximum change of retardation. This is the case if the two axes are interchanged; hence the director is rotated by $\theta = \pi/2$.

8.11.2 Antiferroelectric Switching

In the late 1980s, a different arrangement of the molecules in their layer plane was discovered. The phase is known as the antiferroelectric liquid crystal (AFLC) phase.

This phase occurs in some materials at a temperature below the ferroelectric liquid crystal phase. These materials are chiral and possess a spontaneous polarization. The difference is that in the antiferroelectric liquid crystal phase, the director is tilted in the opposite direction in alternate layers. The director is again tilted, thereby always lying on a theoretical cone. In each subsequent layer the director is tilted in the opposite direction and the spontaneous polarization points in the opposite direction. However, the director still precesses around the z-axis. For antiferroelectric liquid crystals, the pitch is the distance for the director to precess 180° instead of 360°, as for ferroelectric liquid crystals. This is because, due to the opposite tilt in adjacent layers, the director has also gone around half of the cone. The great application potential of antiferroelectric liquid crystals has been demonstrated in sophisticated flat-panel display prototypes. By proper molecular design, one can develop a new generic class of antiferroelectric materials that presents an elegant solution to contrast problems imposed by defects (Lagerwall et al., 2001). Their optical properties make them unique, not only among liquid crystals, but also among electro-optical materials in general. This generic class is called orthoconic. Normal surface-stabilized antiferroelectrics are optically positive biaxial crystals, with an effective optical axis along the smectic layer normal. When the tilt directions in adjacent smectic layers are made perpendicular to each other, the material becomes negatively uniaxial, with the optical axis lying perpendicular to the smectic layer normal. The electro-optic effect in such a material is based on the fact that the optical axis can be switched between three mutually orthogonal directions, corresponding to zero, negative, or positive values of the applied electric field. To understand this, it is instructive to deduce the optical properties of the general surface-stabilized antiferroelectric state. We know from Chapter 2 that in the case when we are in the principal coordinate system the dielectric tensor can be written as Equation (2.23):

$$\boldsymbol{\varepsilon} = \begin{pmatrix} \varepsilon_1 & 0 & 0 \\ 0 & \varepsilon_2 & 0 \\ 0 & 0 & \varepsilon_3 \end{pmatrix}. \tag{8.90}$$

The resultant index of refraction is not a tensor, but is linked to this value by the usual dependence $\varepsilon_x = n_1^2$, $\varepsilon_y = n_2^2$, $\varepsilon_z = n_3^2$. In the nematic phase two values degenerate. For rod-like molecules at optical frequencies, $\varepsilon_{\text{parallel}}$ are always larger than $\varepsilon_{\text{perpendicular}}$; hence the optical axis is along the director and is the slow axis. In the corresponding smectic phases without order inside the smectic layers (smectic A, smectic C), the optical properties are similar but the electro-optical properties are different. For the smectic A phase the director is normal to the layer, and in the smectic C phase the director is tilted at a certain angle relative to the layer normal. Although the dielectric tensor has the same characteristic in smectic A and nematic, a smectic C phase is biaxial in symmetry, although the biaxiality is small. One must additionally distinguish between the eigenvalues ε_1, ε_2, and ε_3 in the reference system where the angle θ is rotated with respect to the layer normal. Because one ε value is much larger then the other two, one can, in a good approximation, consider a smectic C phase as optically uniaxial. The principal coordinate system is along the director, which forms the optical axis. This optical axis can in principle move on a cone with constant tilt angle θ. The tilt direction is essentially the same from layer to layer in a tilted smectic phase. This state of order is called synclinic.

We know that the dielectric tensor for the synclinic state is represented by Equation (8.90) in the local reference frame, where the 2-axis is along the director, the 3-axis is perpendicular to the tilt plane, and the 1-axis is perpendicular to both, as given in Figure 8.90. Let us transform it to the laboratory frame in which the 3-axis is along the smectic layer normal and the 1-axis is along the layer. We make in addition the assumption that we can treat the optical properties in the local system as uniaxial in that $\varepsilon_1 = \varepsilon_3 < \varepsilon_2$. We thus turn both axes anticlockwise around the 3-axis. The dielectric tensor representing the synclinic state in the laboratory system is obtained by a similarity transformation

$$\boldsymbol{\varepsilon}_\theta = \boldsymbol{R}\boldsymbol{\varepsilon}\boldsymbol{R}^{-1} \tag{8.91}$$

where $\boldsymbol{R}$ is the rotation matrix:

$$\boldsymbol{R}(\theta) = \begin{pmatrix} \cos\theta & \sin\theta & 0 \\ -\sin\theta & \cos\theta & 0 \\ 0 & 0 & 1 \end{pmatrix}. \tag{8.92}$$

The result is

$$\boldsymbol{\varepsilon}_\theta = \begin{pmatrix} \varepsilon_1\cos^2\theta + \varepsilon_2\sin^2\theta & (\varepsilon_2 - \varepsilon_1)\sin\theta\cos\theta & 0 \\ (\varepsilon_2 - \varepsilon_1)\sin\theta\cos\theta & \varepsilon_1\sin^2\theta + \varepsilon_2\cos^2\theta & 0 \\ 0 & 0 & \varepsilon_1 \end{pmatrix}. \tag{8.93}$$

An antiferroelectric phase is qualified by successive layers of difference inclination. This state is called anticlinic, as shown in Figure 8.48. If one regards the tensor for the oppositely inclined state being transformed to the laboratory frame, we use the same transform Equation (8.91), only θ has to be replaced by $-\theta$. This only affects the two components ε_{xy} and ε_{yx}, which both change sign. Because the layer structure is much smaller, then the light sees the zero field anticlinic structure as a homogeneous medium with dielectric properties that are the average of those representing the two sets of layers with opposite tilt. As the two tensors have now been transformed to the same reference system, we can just add them, which means that we add their respective components. This yields

$$\boldsymbol{\varepsilon}(\theta) = \frac{\boldsymbol{\varepsilon}_\theta + \boldsymbol{\varepsilon}_{-\theta}}{2} = \begin{pmatrix} \varepsilon_1\cos^2\theta + \varepsilon_2\sin^2\theta & 0 & 0 \\ 0 & \varepsilon_1\sin^2\theta + \varepsilon_2\cos^2\theta & 0 \\ 0 & 0 & \varepsilon_1 \end{pmatrix}. \tag{8.94}$$

This is the dielectric tensor describing the properties of the antiferroelectric state as a function of tilt angle θ, where we have assumed that the dielectric tensor is uniaxial in the local coordinate system. One realizes directly that $\boldsymbol{\varepsilon}(\theta)$ is generally biaxial and always diagonal in the laboratory system x, y, z, setting out with the optical axes in the tilt plane, but soon changing to have them perpendicular to the

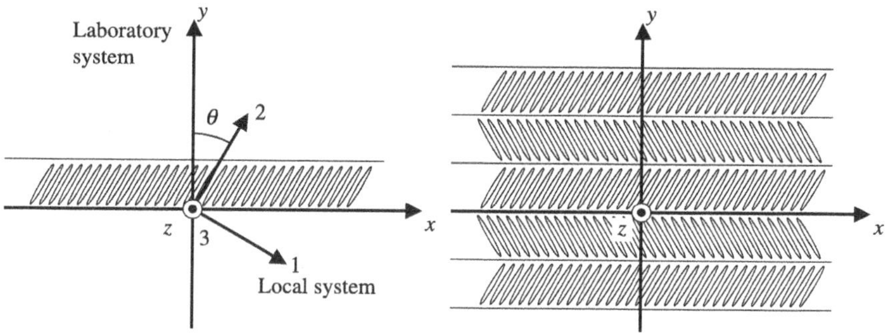

Figure 8.48 Relation between the laboratory and the local reference system for a tilted smectic layer system. On the right, an antiferroelectric state is shown.

tilt plane. The most important insight is, however, that for $\theta = 45^\circ$ the tensor becomes uniaxial. In this case one finds

$$\boldsymbol{\varepsilon}(45^\circ) = \begin{pmatrix} \dfrac{\varepsilon_1 + \varepsilon_2}{2} & 0 & 0 \\ 0 & \dfrac{\varepsilon_1 + \varepsilon_2}{2} & 0 \\ 0 & 0 & \varepsilon_1 \end{pmatrix}. \tag{8.95}$$

For optical frequencies $\varepsilon_1 \ll \varepsilon_2$, hence $\varepsilon = (\varepsilon_1 + \varepsilon_2)/2$ is always larger than ε_1. Thus the birefringence is negative and the dielectric tensor is oblate, with the optical axis along the z-direction, that is, perpendicular to the layer normal and perpendicular to the cell surface in the surface-stabilized state. This has the consequence that normal incident light is always along the optical axis in the field-free state and this state is therefore absolutely black for every crossed-polarizer setting and independent of the inhomogeneity of the smectic layer alignment. In fact, the extinction of the dark state is only limited by the quality of the polarizers. The described optical properties with negative birefringence can only be achieved in the surface-stabilized state.

REFERENCES

Amm, H., Grigutsch, M., and Stannarius, R. (1998) Spatio-temporal analysis of electroconvection in nematics, *Z. Naturforsch.* **53**, 117–126.

Apter, B., Efron, U., and Bahat-Treidel, E. (2004) On the fringing-field effect in liquid crystal beam-steering devices, *Appl. Optics* **43**, 11–19.

Bahat-Treidel, E., Apter, B., and Efron, U. (2005) Experimental study of phase-step broadening by fringing fields in a three-electrode liquid crystal cell, *Appl. Opt.* **44**, 2989–2995.

Barbero, G. and Evangelista, L.R. *Elementary Course on the Continuum Theory for Nematic Liquid Crystals*, World Scientific Publishing Company, Singapore.

Behdani, M., Soria, S., Bader, M.A., Ihlemann, J., Marowsky, G., and Rasing, T. (2003) Alignment of liquid crystals with periodic submicron structures ablated in polymeric and indium tin oxide surfaces, *Appl. Phys. Lett.* **82**, 2553–2555.

Blinov, L.M. and Chigrinov, V.G. (1996) *Electrooptic Effects in Liquid Crystal Materials*, Springer, New York.

Bohley, C. (2004) *Polarization Optics of Periodic Media*, PhD thesis, University of Neuchâtel.

Bohley, C., Heuer, J., and Stannarius, R. (2005) Optical properties of electrohydrodynamic convection patterns: rigorous and approximate methods, *J. Opt. Soc. Am. A* **22**, 2818–2826.

Bohren, C.F. and Hoffman, D.R. (1983) *Absorption and Scattering of Light by Small Particles*, Wiley, New York.

Brown, C.V., Bryan-Brown, G.P., and Hui, V.C. (1997) Modelling nematic liquid crystal alignment on asymmetric surface grating structures, *Mol. Cryst. Liq. Cryst.* **301**, 163–168.

Brown, C.V., Towler, M.J., Hui, V.C., and Bryan-Brown, G.P. (2000) Numerical analysis of nematic liquid crystal alignment on asymetric surface grating structures, *Liquid Crystals* **27**, 233–242.

Carroll, T.O. (1972) Liquid crystal diffraction grating, *J. Appl. Phys.* **43**, 767–770.

Chandrasekar, S. (1992) *Liquid Crystals*, 2nd edn, Cambridge University Press.

Cheng, J. (1981) Surface pinning of disclinations and the stability of bistable nematic storage devices, *J. Appl. Phys.* **52**, 724–727.

Churchill, R.V. (1960) *Complex Variables and Applications*, McGraw-Hill, New York.

Courant, R. and Hilbert, D. (1993) *Methoden der Mathematischen Physik*, 4th edn, Springer Verlag, Berlin.

Cox, S.J., Reshetnyak, V.Y., and Sluckin, T.J. (1998) Effective medium theory of light scattering in polymer dispersed liquid crystal films, *J. Phys. D: Appl. Phys.* **31**, 1611–1625.

Crawford, G.P. and Zumer, S. (Eds) (1996) *Liquid Crystals in Complex Geometries*, Taylor and Francis, London.

Demus, D. and Richter, L. (1987) *Textures of Liquid Crystals*, Wiley-VCH, Berlin.

Dierking, I. (2003) *Textures of Liquid Crystals*, Wiley-VCH, Weinheim.

Drzaic, P.S. (1995) *Liquid Crystal Dispersions*, World Scientific, Singapore.

Fan-Chiang, K.H., Wu, S.T., and Chen, S.H. (2005) Fringing-field effects on high-resolution liquid crystal microdisplays, *J. Disp. Technol.* **1**, 304–313.

Ge, Z., Wu, T.X., Lu, R., Zhu, X., Hong, Q., and Wu, S.T. (2005) Comprehensive three-dimensional dynamic modeling of liquid crystal devices using finite element method, *J. Disp. Technology* **1**, 194–206.

de Gennes, P.G. (1993) *The Physics of Liquid Crystals*, Clarendon Press, Oxford.

Geurst, J.A., Spruit, A.M.J., and Gerritsma, C.J. (1975) Dynamics of $s = 1/2$ disclinations in twisted nematics, *J de Physique* **36**, 653–664.

Goodby, J.W. and Blinc, R. (1991) *Ferroelectric Liquid Crystals: Principles, Properties and Applications (Ferroelectricity and Related Phenomena)*, Gordon & Breach Science Publishers Ltd, Philadelphia.

Hasegawa, M. (2002) Azimuthal anchoring energies of photoaligned nematic liquid crystals produced by small dosages of linearly polarized UV light. *Jap. J. Appl. Phys.* **41**(10B), L1167–L1168.

Higgins, D.A., Hall, J.E., and Xie, A. (2005) Optical microscopy studies of dynamics within individual polymer-dispersed liquid crystal droplets, *Acc. Chem. Res.* **38**, 137–145.

Hwang, D. and Rey, A. (2005) Computational modeling of the propagation of light through liquid crystals containing twist disclinations based on the finite-difference time-domain method, *Appl. Opt.* **44**, 4513–4522.

Hwang, D. and Rey, A. (2006) Computational studies of optical textures of twist disclination loops in liquid crystal films by using the finite-difference time-domain method, *JOSA A* **23**, 483–496.

Ibn-Elhaj, M. and Schadt, M. (2001) Optical polymer thin films with isotropic and anisotropic nano-corrugated surface topologies, *Nature* **410**, 796–799.

John, T., Behn, U., and Stannarius, R. (2003) Laser diffraction by periodic dynamic patterns in anisotropic fluids, *Europhys. J. B* **35**, 267–278.

Kelly, J.R. and Wu, W. (1993) Multiple scattering effects in polymer dispersed liquid crystals, *Liq. Cryst.* **14**, 1683–1694.

Kim, J.H., Yoneya, M., and Yokoyama, H. (2002) Tristable nematic liquid-crystal device using micropatterned surface alignment, *Nature* **420**, 159–162.

Kitamura, M. (1995) Computer simulation of director profile in three dimensional electric field, *SID Digest*, 540–543.

Kitson, S. and Geisow, A. (2002) Controllable alignment of nematic liquid crystals around microscopic posts: Stabilization of multiple states, *Appl. Phys. Lett.* **80**, 3635–3637.

Kléman, M. (1983) *Points, Lines and Walls*, Wiley, Chichester.

Kléman, M. (1989) Defects in liquid crystals, *Rep. Prog. Phys.* **52**, 555–654.

Kriezis, E.E., Filippov, S.K., and Elston, S.J. (2000b) Light propagation in domain walls in ferroelectric liquid crystal devices by the finite-difference time-domain method, *J. Opt. A: Pure Appl. Opt.* **2**, 27–33.

Kriezis, E.E., Newton, C.J.P., Spiller, T.P., and Elston, S.J. (2002a) Three-dimensional simulations of light propagation in periodic liquid-crystal microstructures, *Appl. Opt.* **41**, 5346–5356.

Kriezis, E.E. (2002) Comparative study of light scattering from liquid crystal droplets, *Micro. Opt. Tech.* **35**, 437–441.

Lagerwall, S.T. (1999) *Ferroelectric and Antiferroelectric Liquid Crystals*, Wiley-VCH, Weinheim.

Lagerwall, S., Dahlgren, A., Jägemalm, P., Rudquist, P., D'havyé, K., Pauwels, H., Dabrowski, R., and Drzewinski, W. (2001) Unique electro-optical properties of liquid crystals designed for molecular optics, *Adv. Funct. Mater.* **11**, 87–94.

Lavrentovich, O.D. and Kleman, M. (2001) "Defects and Topology of Cholesteric Liquid Crystals," In *Chirality in Liquid Crystals*, 5, Springer Verlag, New York.

Lavrentovich, O.D. (2003) "Defects in liquid crystals: surface and interfacial anchoring effects" In Arodz, H., Dziarmaga, J., and Zurek, W.H. (Eds), *Patterns of Symmetry Breaking*, NATO Science Series, II. Mathematics, Physics, and Chemistry, Vol. 127, Kluwer Academic Publishers, Dordrecht, 161–195.

Lee, B.W. and Clark, N.A. (2001) Alignment of liquid crystals with patterned isotropic surfaces, *Science* **291**, 2576–2580.

Li, X.T., Pei, D.H., Kobayashi, S., and Iimura, Y. (1997) Measurement of azimuthal anchoring energy at a liquid crystal photopolymer interface, *Jpn J. Appl. Phys.* **36**, L432–L434.

Mertens, G., Röder, T., Schweins, R., Huber, K., and Kitzerow, H. (2002) Shift of the photonic band gap in two photonic crystal liquid crystal composites, *Appl. Phys. Lett.* **80**, 1885–1887.

Okano, K., Kitahara, K., and Ushizawa, E. (1994) Surface disclinations on modified substrates, *Jpn J. Appl. Phys.* **33**, 6262–6267.

Ondis-Crawford, R., Boyko, E.P., Wagner, B.G., Erdmann, J.H., Zumer, S., and Doane, J.W. (1991) Microscopic textures of nematic droplets in polymer dispersed liquid crystals, *Appl. Phys. Lett.* **69**, 6380–6386.

O'Neill, M. and Kelly, S.M. (2000) Photoinduced surface alignment for liquid crystal displays, *J. Phys. D: Appl. Phys.* **33**, R67–R84.

Oswald, P. and Pieranski, J.P. (2000) *Les Cristaux Liquides*, Gordon and Breach, Paris.

Park, S., Padeste, C., Schift, H., Gobrecht, J., and Scharf, T. (2005) Chemical nanopatterns via nanoimprint lithography for simultaneous control over azimuthal and polar alignment of liquid crystals, *Adv. Mater.* **17**, 1398–1401.

Penterman, R., Klink, S.I., de Koning, H., Nisato, G., and Broer, D.J. (2002) Single-substrate liquid-crystal displays by photo-enforced stratification, *Nature* **417**, 55–58.

Pfohl, T., Kim, J.H., Yasa, M., Miller, H.P., Wong, G.C.L., Bringezu, F., Wen, Z., Wilson, L., Kim, M.W., Li, Y., and Safinya, C.R. (2001) Controlled modification of microstructured silicon surfaces for confinement of biological macromolecules and liquid crystals, *Langmuir* **17**, 5343–5351.

Porte, G. (1977) Surface disclination lines observed in nematic liquid crystal when the surface induces homogenously tilted alignment, *J. de Physique* **38**, 509–517.

Vijaya Prakash, G., Kaczmarek, M., Dyadyusha, A., Baumberg, J.J., and D'Alessandro, G. (2005) Control of topological defects in microstructured liquid crystal cells, *Opt. Exp.* **13**, 2201–2209.

Rasenat, S., Hartung, G., Winkler, B.L., and Rehberg, I. (1989) The shadowgraph method in convection experiments, *Exp. Fluids* **7**, 412–420.

Reyes, A.J. (1998) Ray propagation in nematic droplets, *Phys. Rev E.* **57**, 6700–6705.

Rockstuhl, C., Salt, M., and Herzig, H.P. (2004) Theoretical and experimental investigation of phase singularities generated by optical micro- and nano-structures, *J. Opt. A: Pure Appl. Opt.* **6**, S271–S276.

Rosenblatt, C. (2004) Nanostructured surfaces: Scientific and optical device applications, *Mol. Crys. Liq. Crys.* **412**, 1727–1744.

Schadt, M., Schmitt, K., Kozinkov, V., and Chigrinov, V. (1992) Surface induced parallel alignment by linearly polymerized photopolymers, *Jpn J. Appl. Phys.* **31**, 2155–2164.

Scharf, T., Gernez, C., Shlayen, A., Basturk, N., and Grupp, J. (2004) Liquid crystal alignment on replicated nanostructured surfaces, *Mol. Cryst. Liq. Cryst.* **412**, 135–145.

Scharf, T., Park, S., Padeste, C., Schift, H., Basturk, N., and Grupp, J. (2005) Liquid crystal alignment on chemical nanaopatterns, *Mol. Cryst. Liq. Cryst.* **438**, 55–65.

Schuller, C., Klopf, F., Reithmaier, J.P., Kamp, M., and Forchel, A. (2003) Tunable photonic crystals fabricated in III–V semiconductor slab waveguides using infiltrated liquid crystals, *Appl. Phys. Lett.* **82**, 2767–2769.

Shintech Corp. (2005) Available at http://www.shintech.co.jp/.

Sickinger, H., Schwider, J., and Manzke, B. (1999) Fiber based Mach–Zehnder interferometer for measuring wave abberations of microlenses, *Optik* **110**, 239–243.

Singh, Y. (2001) Elasticity of thermotropic liquid crystals, *Curr. Sci.* **80**, 1026–1035.

Singh, S. (2002) *Liquid Crystals*, World Scientific, New Jersey.

Skarabot, M., Kralj, S., Rastegar, A., and Rasing, T. (2003) Observation of twist nematic liquid-crystal lines, *J. Appl. Phys.* **94**, 6508–6512.

Smalyukh, I., Shiyanovskii, S.V., and Lavrentovich, O.D. (2001) Three-dimensional imaging of orientational order by fluorescence confocal polarizing microscopy, *Chem. Phys. Lett.* **336**(1–2), 88–96.

Storrow, G.M., Gleesony, H.F., and Murray, A.J. (1998) An analysis of the optical properties of a single-element liquid crystal device, *Pure Appl. Opt.* **7**, 1411–1423.

Thurston, R.N., Cheng, J., and Boyd, G.D. (1980) Mechanically bistable liquid crystal display structures, *IEEE Trans Electron. Devices* **27**, 2069–2080.

Uche, C., Elston, S.J., and Parry-Jones, L.A. (2005) Microscopic observation of zenithal bistable switching in nematic devices with different surface relief structures, *J. Phys. D: Appl. Phys.* **38**, 2283–2291.

Varghese, S., Crawford, G.P., Bastiaansen, C.W.M., de Boer, D.K.G., and Broer, D.J. (2005a) Four-domain twisted vertically aligned liquid crystal pixels using microrubbing, *Appl. Phys. Lett.* **86**, 181914.

Varghese, S., Crawford, G.P., Bastiaansen, C.W.M., de Boer, D.K.G., and Broer, D.J. (2005b) High pretilt four-domain twisted nematic liquid crystal display by microrubbing: Process, characterization, and optical simulation, *J. Appl. Phys.* **97**, 053101.

Voloschenko, D. and Lavrentovich, O.D. (2000) Optical vortices generated by dislocations in a cholesteric liquid crystal, *Optics Lett.* **25**, 317–319.

Wang, B., Bos, P.J., and Hoke, C.D. (2003) Light propagation in variable-refractive-index materials with liquid-crystal-infiltrated microcavities, *JOSA A* **20**, 2123–2130.

Wu, S.T. (2005) Physical properties of liquid crystals, Lecture Notes, OSE 6938C, College of Optics & Photonics, University of Central Florida.

Yamaguchi, R. and Sato, S. (2000) Increase of azimuthal anchoring energy and thermal stability at homogeneously oriented liquid crystal/polymer interfaces, *Jpn J. Appl. Phys.* **39**, L681–L683.

Yamaguchi, R., Goto, Y., and Sato, S. (2002) A novel patterning method of liquid crystal alignment by azimuthal anchoring control, *Jpn J. Appl. Phys.* **41**, L889–L891.

Yu, C.-J., Park, J.-H., Kim, J., Jung, M.-S., and Lee, S.-D. (2004) Design of binary diffraction gratings of liquid crystals in linear graded phase model, *Appl. Opt.* **43**, 1783–1788.

Zenginoglou, H.M. and Kosmopoulos, J.A. (1988) Geometrical optics approach to the obliquely illuminated nematic liquid crystal diffraction grating, *Appl. Opt.* **27**, 3898–3901.

Zenginoglou, H.M. and Kosmopoulos, J.A. (1997) Linearized wave-optical approach to the grating effect of a periodically distorted nematic liquid crystal layer, *J. Opt. Soc. Am. A* **14**, 669–675.

Zumer, S. and Doane, J.W. (1986) Light scattering from a small nematic droplet, *Phys. Rev. A* **34**, 3373–3385.

Zumer, S. (1988) Light scattering from nematic droplets: Anomalous diffraction approach, *Phys. Rev. A* **37**, 4006–4015.

9

REFRACTIVE BIREFRINGENT OPTICS

9.1 BIREFRINGENT OPTICAL ELEMENTS

As refractive optics we name here all phenomena that can be described by tracing rays through them, hence the classical optical regime without considering diffraction effects due to structures of the size of the wavelengths. The wavefront is reshaped continuously without using the periodicity of the light as in diffractive optical elements. The design tool used for such elements is polarization ray tracing. See Chapter 3 for more fundamental details on the different concepts therein. Such optical elements include lenses, prisms, and in a strict sense retarders and depolarizers that leave the wavefront steady but might change the state of polarization or retard the light when passing through. The quality of such elements is measured as aberrations for lenses, and in general as the deviation from a predefined wavefront shape. For prisms that should be an inclined plane and for a uniform retarder there should be no change of the wavefront at all.

If one analyzes the kinds of birefringent optical elements that are used in the present day and that are based on the refraction of light, one finds mainly one category: polarized beam splitters (or polarizers). These are prisms or systems composed of several prisms that separate polarizations. Optical elements for imaging are rare, and micro-optical components are under development. We do not want to review here the principles of crystal components as their design can rarely be copied with liquid crystals and polymers. The interested reader is instead referred to the literature (Shurcliff, 1962; Saleh and Teich, 1991; Hecht, 1998; Iizuka,

Polarized Light in Liquid Crystals and Polymers. By Toralf Scharf

2002). All these elements have one thing in common: They are relatively thick and measure between a few millimeters and several centimeters. One reason for this is that the relatively low birefringence in such classical optical materials as quartz causes thick elements to achieve a useful retardation. On the other hand, the element functions are based on ray-optics and often on deviation, Fresnel reflection for polarization separation, and total internal reflection, which requires certain angles and propagation distances to be realized. Classical machining and polishing are used for fabrication of such elements.

For polymer components the situation is different. Inhomogeneity in the fabrication process of plastic components does not allow birefringent elements to be formed directly. A different possibility is to use thin components of thicknesses below 1 mm. This limits very much the design capabilities and only a few applications can be envisaged where refractive optical elements can be used. The exception of course is in the display industry, which mass-produces refractive optical elements, in otherwords, the LCD. The birefringent layer is thin and, like all optical components, is for a special purpose. However, with the development of new material classes like liquid crystal polymers, new optical elements are being demonstrated. In particular, for the display industry, illumination modules are designed that allow polarization outcoupling. These are very recent developments that are based on new fabrication techniques.

We will discuss here three basic systems that include all aspects of element design. To begin, we will discuss the properties of light propagation in simple geometries like prisms and discuss a particular application for spectrometry. This shows that for system design all aspects have to be considered, including spatial and temporal coherence. Microlenses, a rather new subject, are then reviewed. Different concepts are presented that serve as beamsplitters or adaptive lenses.

9.2 FABRICATION OF REFRACTIVE COMPONENTS

One important aspect of birefringent components with organic materials is the fabrication of such elements. As we have stated above, birefringent refractive elements from polymers are not usually fabricated by the standard methods of the optical industry. We will describe here methods for the fabrication of standard liquid crystal planar cells and thick birefringent elements of liquid crystals and polymers. "Thick" means that the active layer will have a thickness that does not exceed 1 mm. With a mean birefringence of $\Delta n = 0.1$, such a thickness allows maximum retardations of 100 μm.

One has to distinguish two different concepts. First one can use liquid crystals and encapsulate them to form birefringent components that might even be actuated. That can be done with planar technology. Secondly, replication in liquid crystal polymers is possible so that stable solid-state elements are formed. Both concepts have been realized in the past and we sketch out here the basic fabrication concepts that take up the process steps from standard display manufacturing and add some new concepts.

9.2.1 Standard Planar Liquid Crystal Cell Manufacturing

We are considering here passive liquid crystal devices, which means that there are no electronics implemented on the substrates. Nevertheless, the substrate can contain electrodes for driving purposes. A liquid crystal device is composed of several layers. The main parts are two glass plates, connected by seals. An indium tin oxide (ITO) layer is used as an electrode. In order to form the electrodes, common lithographic equipment like resist coaters, steppers, and dry or wet etching equipment are used. Dry etching can provide much better line-width control, but wet etching is the faster and cheaper method because it is a batch process. A passivation layer, sometimes called a hard coat layer, based on SiO_x, is coated over the ITO to electrically insulate the surface. Polyimide is printed or spin-coated over the passivation layer as an alignment layer. The polyimide has to be polymerized and is then rubbed to give a preferred alignment direction. For homeotropic alignment polymers, rubbing is not necessary. The pretilt angle varies with the rubbing strengths and the alignment polymer.

The thickness of this layer is determined by spacers, which keep the two glass plates at a fixed distance. These consist of small glass or plastic balls or sometimes fibers. Ball or fiber spacers are sprayed on one substrate. Three main processes can be used: (1) Dry spray, which is used for high-throughput and large display manufacturing; (2) semi-dry spacer spray, which is the best method for medium and small displays that are not so high throughput; and (3) wet spacer spray, which is not used very often now, but gives a very nice spacer uniformity and low numbers of spacer clusters. In laboratory and for very thick cells, foil spacers are used that are cut and applied outside the active area. Foils with high chemical resistivity have to be used. For optical applications, thick glass substrates of severeal millimeters' thickness are of preference for obtaining uniform spacer-free areas.

The cell then has to be sealed. For large factories, screen printing is the best method of seal deposition. High throughput and high performance can be combined with this method. For smaller production volumes and higher design flexibility, seal dispensing is the best way. The seal material has to be precured in an oven before the substrate glass plates are forwarded to the assembly. After cell assembly, the final cure of the seal happens in a hot press oven. Alternatively, devices can be pressed and cured one by one. In the cell assembly process, both glass plates are aligned and combined. The position of the glass plates against each other is fixed by UV-hardened polymer spots.

The liquid crystal display fill method is usually a vacuum application. In the laboratory filling by capillary forces is often used. For vacuum filling, the liquid crystal devices are placed in a vacuum chamber mounted above the liquid crystal fluid. The chamber is then pumped down and the empty device is evacuated. The fill ports are lowered into the trough and the chamber is brought back to atmospheric pressure. The atmospheric pressure forces the liquid crystal fluid into the device. After filling the panels, the hole in the sealing is closed in a separate process step. After proper surface cleaning, polarizer foils are attached to the surfaces to the front side and the back side of the standard liquid crystal display panel. This is the last step in the main LCD fabrication.

9.2.2 Fabrication of Thick Refractive Components

The use of planar technology is perfect for retarders, thin systems, or diffractive optical components. However, retardation is limited, as is the form of the phase distribution that can be obtained. Therefore, efforts have been made to incorporate non-planar surface technologies for fabrication of optical elements. The first optical components date back to 1979 (Sato, 1979), when a Japanese group built a birefringent lens by filling a liquid crystal cell composed from a flat and a curved substrate. The curved substrate was a concave lens and formed, together with the flat counter-substrate, a cavity that was filled with liquid crystal. The alignment was done by conventional polyimide coating and rubbing. Figure 9.1 shows the arrangement of components and a photograph of a realized cell. The curved surface with planar alignment has a radius of curvature of $R = 0.1$ m and is made from BK7, with a refractive index of $n = 1.5151$. The liquid crystal lens is filled with the liquid crystal mixture E7 ($n_e = 1.709$, $n_o = 1.51$). This leads to a change of the focal length when the direction of the entrance polarization is flipped. For polarization direction parallel to the alignment, the effective birefringence of the liquid crystal is n_e and the focal length was measured at $f = 0.430$ m. The theoretical focal length for polarization along n_e is given by

$$f = \frac{R}{n_e - n_{\text{lens}}} = \frac{0.1}{1.709 - 1.5151} = 0.516 \text{ m} \tag{9.1}$$

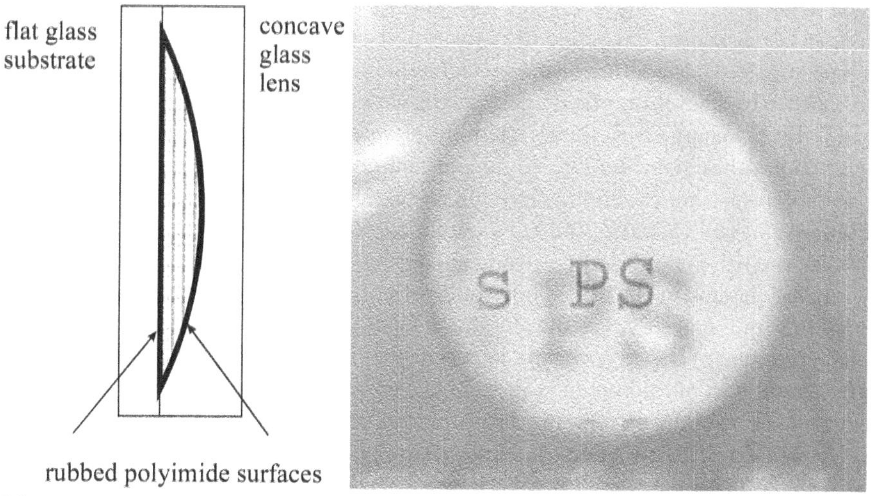

Figure 9.1 Construction and realization of a birefringent lens made by liquid crystal filling. (*a*) A section through a filled liquid crystal cell. (*b*) A photograph showing two images because of the double refraction. The diameter is 1 cm and the thickness of the liquid crystal layer is 200 μm.

If the polarization is perpendicular to the alignment, the effective birefringence of the liquid crystal is the ordinary refractive index n_o. The focal length becomes

$$f = \frac{R}{n_o - n_{\text{lens}}} = \frac{0.1}{1.51 - 1.5151} = -19.6 \text{ m} \tag{9.2}$$

The quality of the component is very good, although for very thick samples of more than 500 μm, scattering becomes important and elements look milky. Obtaining a uniform alignment of the liquid crystal is the main issue. There are difficulties when rubbing is used for alignment. Rubbing is usually done with a velvet cloth fixed on a rotated cylinder.

For controlling the rubbing parameters a rubbing length can be defined that has as its main parameters the pressure of the cylinder on the substrate, the rotational speed of the cylinder, its diameter, and the speed of movement of the substrate. See the work of Bahadur (1990) and Ishihara (2005) for details and actual trends about liquid crystal alignment. If the substrate is no longer flat, the definition of some of these parameters fails and the rubbing becomes nonuniform. That might not be a problem for "reasonable" flat substrates such as the concave lens shown in Figure 9.1, but it becomes very difficult for complicated shapes and deep structures. Examples are given in the literature (Commander et al., 2000). There are two solutions. In the first, an evaporation technique for creating the alignment layer can be used. Such techniques as SiO_x evaporation have been widely studied and show good results (Barberi et al., 1994; Oswald and Pieranski, 2000). In SiO_x evaporation a certain angle between the substrate and the evaporation source is needed. This is a critical parameter for the alignment strengths imposed. For curved surfaces this angle changes, and might change the surface anchoring conditions. A second method is the use of nanostructures for liquid crystal alignment, as suggested by Berreman (1973). Modern fabrication techniques allow the production of such structures either by photolithography (Bryan-Brown et al., 1998) or by replication (Kim et al., 2002; Scharf et al., 2004). The problem then is to find a suitable material combination that gives durable alignment and prevents degradation.

Let us discuss an example where microlenses are implemented in a sandwiched liquid crystal cell (Scharf, 2005). To obtain a good alignment on a microstructured surface, one has to implement a strong anchoring alignment layer. This can be made by treating the lens surface with polyimide or polyvinyl-alcohol and rubbing or with a nanostructure that serves as the alignment layer. When a rubbing technique is used, good alignment at the edges of the lens is difficult to obtain, although the lenses are only several tens of micrometers in height. This is easy to understand if one remembers that the rubbing is done by brushing the surface with a tissue of fibers of a certain length. The situation of nanostructures as alignment layers is different. Because the nanostructure is present everywhere on the substrate regardless of the nature of the microstructure or any height difference, the alignment is very good and the strength can be adjusted by optimizing the nanostructure geometry. As an example, Figure 9.2 shows a replicated microlens array immersed in the liquid crystal mixture E7. The observation wavelengths is 550 nm that lead to

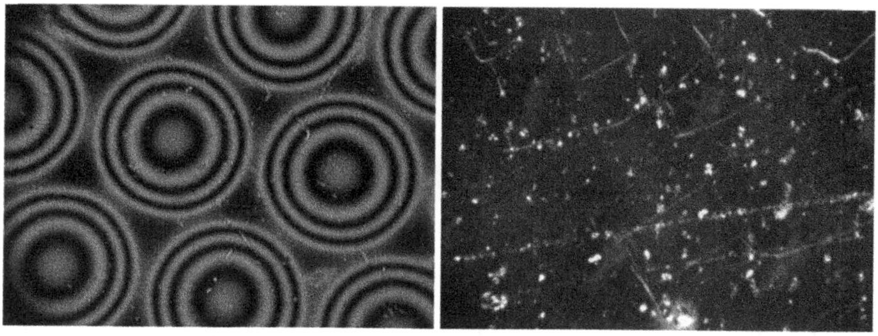

Figure 9.2 Quality of alignment of microstructured surfaces with a nanostructured alignment layer incorporated into a sandwiched cell. A planar aligned array is shown between crossed polarizers oriented under 45° and 0° with respect to the alignment direction. The alignment quality is best checked if the sample is oriented parallel to one of the polarizers (right). The exposure conditions are adapted to see the scratches and imperfections.

pronounced interference fringes. The alignment directions of the substrates are parallel. For one polarization we obtain nearly index matching between the refractive index of the lens materials NOA 81 (Norland adhesive epoxy, UV curable, refractive index $n = 1.56$) and the liquid crystal mixture E7 ($n_e = 1.75$ and $n_o = 1.532$). The lens function is obtained for the other polarization direction. The photographs are taken between crossed polarizes when the optical axis is along one polarizer and at 45°. Shown is a hexagonal array of microlenses with a diameter of 145 μm and a pitch of 150 μm. A particularity of the fabrication of polymerized components is the possibility of using rubbed surfaces for alignment together with replication (Stapert et al., 2003). For polymerizable liquid crystal mesogens it becomes possible to fill cavities with surfaces that were treated with polyimide and rubbed. The liquid crystal layer thickness reaches up to 88 μm. During polymerization the liquid crystal mesogens form a solid network and allow separation from the aligning surface. This is additionally promoted by the use of a surfactant. A solid-state birefringent element is formed and can be incorporated into other devices for further use.

For birefringent prisms the situation concerning the alignment is less critical as they are usually not microstructured at the surface. Assembling can be carried out by using spacers of different thickness and for alignment standard planar technology can be used. In this case the surfaces are rubbed and the liquid crystal cells are mounted in the conventional way afterwards. As an example, birefringent wedges of 200 μm thickness are shown in Figure 9.3. The alignment is homogenous and of good quality. It is also possible to form twist elements with optical quality.

Alternative methods for device fabrication include polymerization-induced phase separation and polymer stabilization (Masuda 1998a,b). Active elements are usually made by structuring electrodes, as in the case of switchable microlenses (Nose and Sato, 1989; Scharf et al., 2000), incorporation of microstructures and switch the liquid crystal in the whole cell (Ren et al., 2005a; Prakash et al., 2005) or by using field meshing as the most advanced technology (Kornreich et al., 1984;

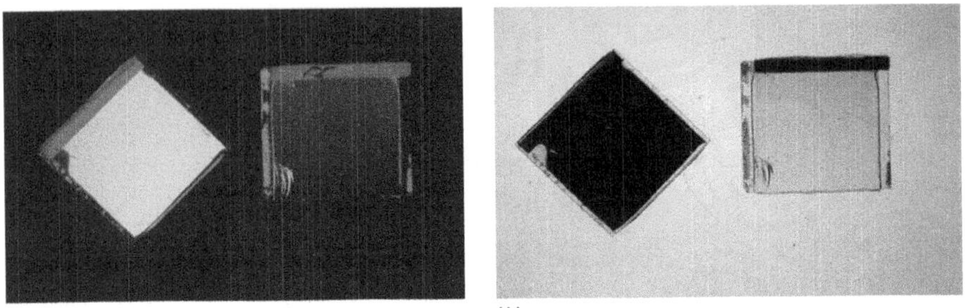

(a) (b)

Figure 9.3 Twisted and planar liquid crystal polymer systems with wedge geometry. The cells are 25 mm × 25 mm and the thickness varies from almost zero to 200 μm. (*a*) The twisted structure becomes transparent between crossed polarizers, but the parallel cell does not change the polarization state (*b*) parallel polarizers lead to a black appearance of the twist structure. The parallel cell stays transparent.

Brinkley 1988; Loktev et al., 1998; Naumov and Vdovin, 1998b). All of these are based on planar technology with rubbed surfaces for alignment. The difference is in the use of special techniques for electrode structuring, making multilayer structures, or even evaporating spatially different transparent conductive layers to influence the electric field distribution. We shall discuss this in detail with an application in Section 9.5.

A different method of producing lens-like structures is the use of phase separation due to polymerization. When a mixture of liquid crystal and monomers is irradiated with UV light, phase separation can appear that leads to a liquid-crystal-rich phase and a polymer-rich phase (Drzaic, 1995). The polymer-rich phase is further polymerized and becomes solid. Spatial inhomogeneous illumination can produce optical elements (Masuda et al., 1998a; Ren et al., 2005a). The process is difficult to control. The concept might also be used to change the basic parameters of the liquid crystal material itself. To change the mechanical restoring forces and the elastic properties, a polymer dispersed system might be advantageous (Masuda et al., 1998b). The active media might even be replaced by a polymer dispersed liquid crystal (Ren et al., 2005b).

9.3 OPTICAL PROPERTIES OF MODIFIED BIREFRINGENT COMPONENTS

Conventional birefringent components for optics are made from birefringent materials and have well-defined optical axes. They might be composed of several elements to obtain special functions. Modified birefringent components take advantage of the fact that with liquid crystal and polymer technology, new elements can be envisaged with nonuniform distribution of the optical axis. The simplest example is the twisted nematic cell. It has a nonuniform optical axis distribution. The axis rotates by a certain angle. The cell as a whole, therefore, cannot easily be described and special formalisms are used to treat the optics correctly. Using twist and other "nonconventional" textures might allow the creation of elements with advanced features. Examples are tunable Fabry–Perot etalons with particular internal liquid crystal textures (Patel, 1991; Lee et al., 1999).

9.3.1 Polarization Converters

One of the most interesting elements to change the state of polarization is the so-called polarization converter. The polarization converter is an element that creates particular spatial distributions of polarizations. Elements can be made with nanostructured surfaces and metal stripes or spatial varying polarizers (Bomzon et al., 2001; Niv et al., 2003; Levy et al., 2004). We will discuss elements fabricated with liquid crystal technology.

Several configurations have been proposed and realized in order to tune the polarization at the outgoing beam in particular (Asada and Nishiwaki, 1994; Stalder, 1996; Stalder and Schadt, 1996a,b; Honma et al., 2000). The most important element is the θ-cell polarization converter. The entrance and exit plates of the cell are linearly and circularly rubbed, respectively. The direction of the linear rubbing on the entrance plate determines the cell axis. Each liquid crystal molecule chain is characterized by a twist angle (i.e., the angle between the orientation of the molecules at the entrance and at the exit plates), which is a function of the angular position with respect to the cell axis. When the polarization-guiding conditions are met, a linearly polarized beam incident on the entrance plate, propagating parallel to the θ-cell normal and with electric field vector parallel or perpendicular to the cell axis, experiences a rotation of its polarization direction by the twist angle. This phenomenon occurs for a broad range of wavelengths (twisted nematic cell optics in the waveguide limit). Having a closer look at the cell textures in their stable configuration, as in Figure 9.4, one realizes that there are two parts with different senses of rotation. On the top half the rotation is clockwise, but on the lower part it is characterized by counterclockwise rotation. In this case, a defect line running

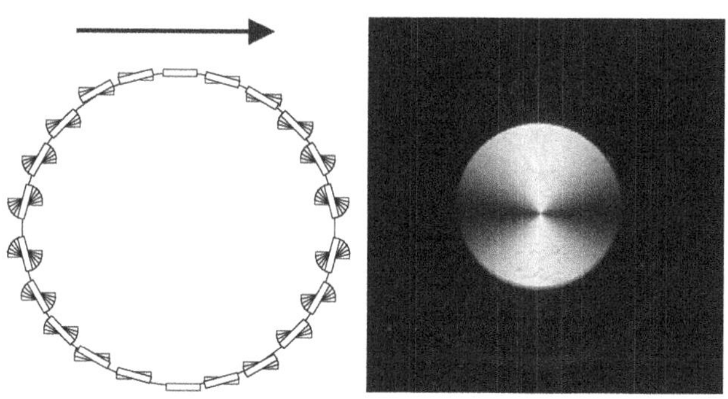

Figure 9.4 Structure of the director in the θ-cell. The arrow indicates the rubbing direction of one of the substrates. The other substrate is rubbed rotationally with a particular setup. Between crossed polarizers two black brushes become visible.

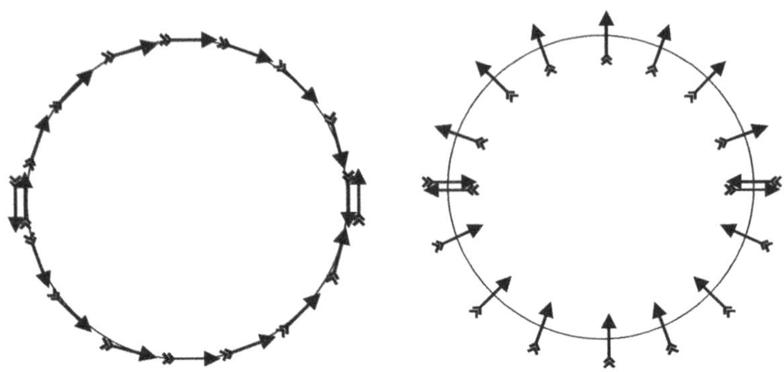

Figure 9.5 Azimuthal and radial polarizations. The arrows indicate the phase of the beams that has a mismatch at the disclination line. (*a*) Azimuthally polarized light is achieved for light incident parallel to the cell axis (rubbing direction). (*b*) Radially polarized light is achieved for light incident perpendicular to the cell axis.

along the diameter parallel to the cell axis arises. The defect line is caused by the different rotation sense in the two parts of the cell and, for a nonchiral liquid crystal mixture, is along the rubbing direction (cell axis).

The polarization rotation properties are discussed now for linear polarized light enters from the side with linear rubbing. Two different polarization states are created when linear polarized light enters parallel or perpendicular to the cell axis. Figure 9.5 visualizes the polarization states. Arrows are used to indicate the phase of the polarized light. The polarization converter is useful for producing a particular electric field distribution in the focal spot of high-numerical-aperture lenses (Stalder, 1996; Descrovi et al., 2004). The same concept of laterally modulating the polarization state can be used to depolarize light (Honma and Nose, 2004). An interesting alternative to producing radially polarized light is the use of a modified θ-cell. If a circularly rubbed surface substrate is combined with a homeotropically aligning substrate, a local hybrid texture is found (Lee et al., 1999). If such an element is filled with anisotropic absorbing liquid crystal mixtures with positive dichroic behavior, it forms a polarizer that transmits only radial polarized light. No defect line arises because a homogeneous texture is formed. Such an element is useful for the detection of the basic polarization properties of radiation. Due to the limited extinction ratio of liquid crystal dichroic mixtures, the contrast of such an element is rather limited.

9.3.2 Composed Films

Birefringent optical elements might be uniaxial or biaxial. However, due to the particular fabrication technology involved in thin polymer and liquid crystals with twisted structures on very thin layers, optical films can be created that do not have an optical axis in the classical sense. The classical definition states that the

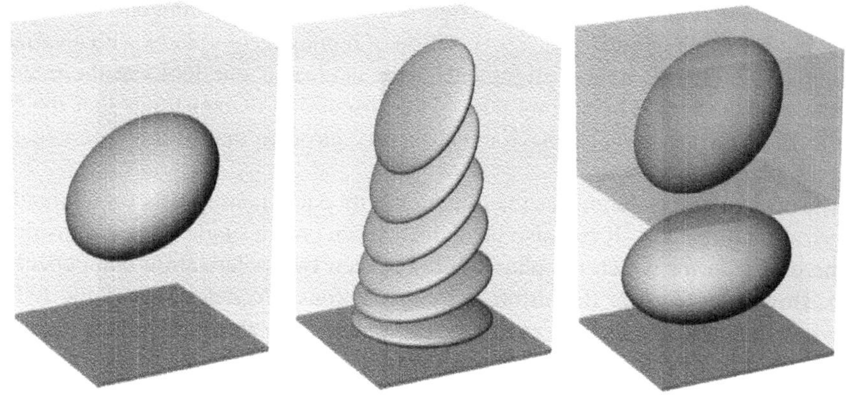

Figure 9.6 Three different thin optical films that show approximately the some optical properties: (*a*) A negative birefringent plate tilted 40° shows an optical axis; (*b*) a hybrid aligned liquid crystal film made from negative birefringent discotic polymer liquid crystal has no optical axis; and (*c*) the composed films from two negative uniaxial materials where both have a tilt angle can model a birefringent film without optical axis.

optical axis describes a direction in space where the propagation of electromagnetic waves is independent of the polarization of light. The light propagation in this direction is as it would be in an isotropic medium.

We will discuss here one particular problem that leads also to problems of definitions. We want to show the similarity between thin-film optical systems that are made of different anisotropic materials but show nearly identical optical properties. Figure 9.6 shows the different systems under investigation. One is a retarder film based on negatively birefringent material, the so-called Fuji film (Mori et al., 1997), and another is a composed system with two positive birefringent films superposed. A dispute is ongoing as to whether the optical systems described have an optical axis or not. Note that the definition of an optical axis is used here to describe a behavior of a nanostructured system with complicated internal structures. Each single element of the film has a well-defined optical anisotropy and the optical axis can be defined. The composed film does not show this property any longer. To prove this we present a comparison of the macroscopic optical behavior between a homogeneous uniaxial plate, a hybrid aligned discotic film (Fuji film), and a combination of two plates. The optical behavior of such birefringent films is principally characterized by the angular dependence of the retardation that can be produced or by the transmission between crossed polarizers when light passes through. This angular dependence, together with the negative birefringence, is of particular interest, because this property of the material permits the viewing angle of twisted nematic liquid crystal displays to be increased.

One important question to answer is whether this Fuji film can be regarded as a uniaxial plate. To answer this question, we consider two aspects. The first part demonstrates that the Fuji film is not, rigorously speaking, a homogeneous uniaxial plate, but must be regarded as a superposition of homogeneous plates with a variable tilt angle (also called hybrid alignment), as it is clear if one looks at the internal structure of the film. The second part of the discussion demonstrates that the Fuji hybrid film has an optical behavior that is very close to that of a single negative uniaxial plate.

The optical axis is defined as the direction with equal phase shifts for all incident polarization directions. It can also be defined when only a single plate is considered as the direction for which the phase shift between two polarization components is zero. However, the first definition is more rigorous. To determine if the hybrid film has an optical axis, we calculate the phase shift that the light experience when passing through the films for all the polarization directions and incidence angles. The incidence angle for which all the polarization directions have the same phase shift is considered as the optical axis of the film. For this we used a commercial simulation tool (Shintech LCD 2006), which is based on the Berreman 4 × 4 matrix calculation method (explained in Chapter 4). For modeling the Fuji film as described in the work of (Mori et al., 1997) we used the following parameters:

Retardation of the compound layer:	$R_{\text{th}} = \dfrac{n_z - (n_x + n_y)/2}{d} = -80$ nm
Mean angle:	40°
Axis orientation variation:	4° (planar) to 68° (hybrid) linear change
Material parameters assumed of the uniaxial negative birefringent material:	$\Delta n = -0.05$, $n_o = 1.55$, $n_e = 1.5$

A detailed analysis of the problem can be done by starting with simulations that show the angular dependence of the transmission between crossed polarizers for different angles of the film orientation. The polarizers and analyzer are rotated simultaneously and stay crossed. Figure 9.7 shows such simulations. It seems that at a certain position there is no change in intensity, which is an indication for an optical axis. In Figure 9.7 this position is found for an azimuthal angle of 0°. However, a detailed analysis shows that there is still retardation that changes at this position. To see this we show calculations for the incidence angles that have an azimuthal angle of 0°. So we only vary the polar angle, because this is the only angle where there is a certain chance of finding the optical axis. Figure 9.8 shows the transmission intensity of a Fuji film when it is placed between crossed polarizers, as a function of the polar incidence angle of the light. It can be viewed as a transversal cut of a conoscopic image. This transmission intensity is proportional to the phase shift between the two polarization components when they have passed through the film. The different curves represent the different orientations of the film with respect to the polarizers' orientations (in fact the different polarization directions). We see that there exists no incidence angle for which all curves have exactly the same transmission value. This means there is no propagation direction of the light for which all the polarization directions have the same phase shift. So, according

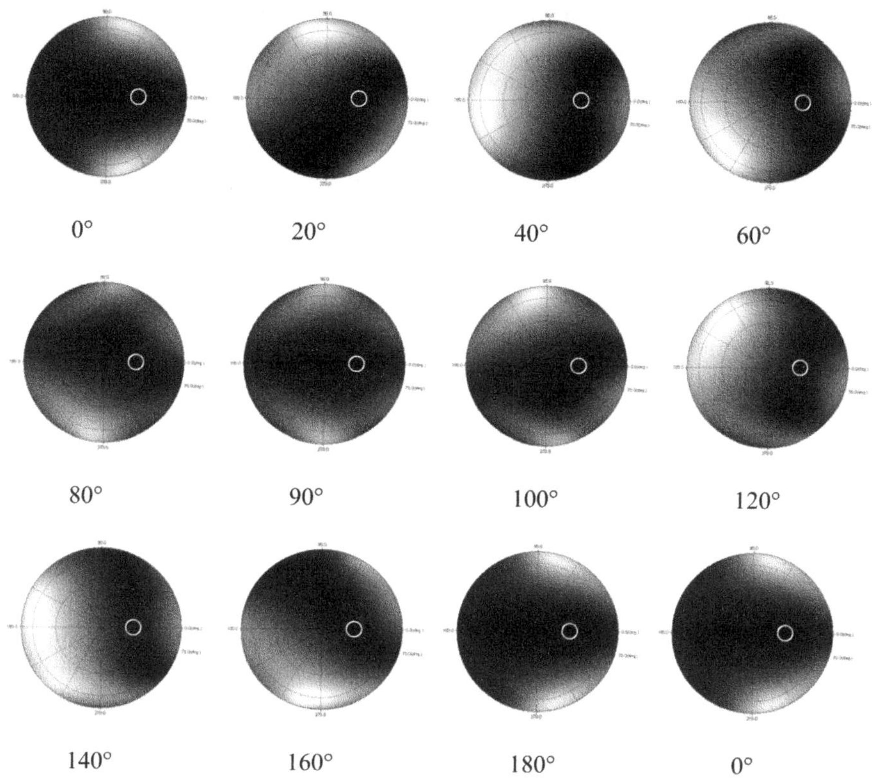

Figure 9.7 Angular dependence of the transmission between crossed polarizers for the Fuji film. The polarizers and analyzer are crossed and rotated simultaneously. At azimuth 0° and polar 30° there is a point that has nearly no change in intensity.

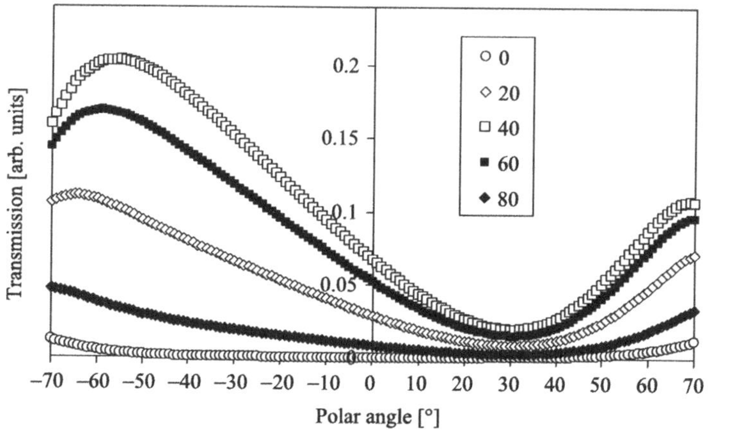

Figure 9.8 Transmission of the Fuji film between crossed polarizers at different orientations. At about 30° for the polar angle the transmission always has a minimum, but does not become zero. There is still transmission in this direction, which indicates birefringence.

to the definition given before, there is no optical axis! However, notice that for an incidence angle of 30° the transmitted intensities for the different polarization orientations are very close to each other. This shows that there is an "optical-axis-like behavior" at a tilt angle of approximately 30°. These small intensity fluctuations cannot be distinguished with conoscopic measurement methods because of the lack of contrast.

In order to show how close the hybrid (Fuji) and the homogeneous negative uniaxial films are, we have simulated a comparison of the angular dependence of the phase shift obtained with both configurations. We used the following simulations parameters:

Axis orientation:	40° obliquely oriented optical axis for a uniaxial element
Material parameters:	$\Delta n = -0.05$, $n_o = 1.55$, $n_e = 1.5$
Composed film	
Axis orientation:	azimuthal angle 0° and tilt angle 16°
	azimuthal angle 0° and tilt angle 48°
Birefringence:	$\Delta n = -0.05$, $n_o = 1.55$, $n_e = 1.5$
Film thickness:	$d_1 = 1.285$ μm, $d_2 = 1.285$ μm

We have simulated in Figure 9.9 three different configurations for an azimuthal angle of 0° (as in Fig. 9.8). The filled circle represents the hybrid Fuji film. The open squares represent the tilted uniaxial material with the parameters mentioned above, and the open triangles show results for a composed system of two negative uniaxial plates oriented at a tilt angle of 16° and 48° with the same azimuthal angle of 0°. As can be clearly seen, the angular dependence of the hybrid Fuji film is very similar to the curves with open squares (differences of 0 to 20% over

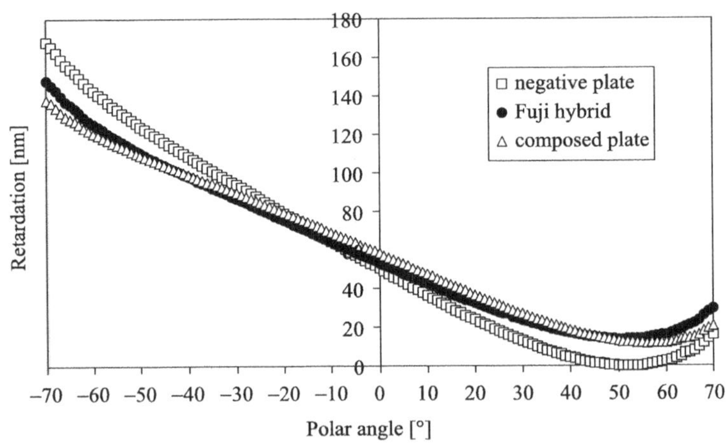

Figure 9.9 Retardation of the different thin films along the direction where the azimuthal angle is zero. Only the uniaxial plate has a well-defined optical axis, with retardation at about 50°. The composed films have no optical axis.

the whole angular range from $-70°$ to $70°$ incidence angle), representing a tilted negative uniaxial material. Both curves are asymmetric. We see also in Figure 9.9 that a combination of two uniaxial plates as presented by the open triangles can nearly perfectly match the hybrid Fuji film configuration. In addition, angular dependence simulations of conoscopic figures (not shown here) show that the two systems cannot be distinguished from each other. The overall optical behavior of the hybrid configuration present in the Fuji films can no longer be expressed in the classical terms of an optical axis. Strictly speaking, it has no optical axis, but it almost looks as if it has one. This slight deviation of the Fuji film from the perfect homogeneous uniaxial crystal is a second-order effect. It is not important for the correction of the viewing angle of a liquid crystal display. The more important point is the use of negative birefringence with a certain axis orientation to correct for the angular properties and retardation of the twisted nematic cell.

9.4 LIQUID CRYSTAL PHASE SHIFTERS

An often-used optical element in optical laboratories is the phase shifter. It is used to change the phase delay of a beam with respect to reference beam. A standard liquid crystal cell in planar texture and homogeneously switchable is actually nothing else than an element that allows to change the phase for one polarization component. The functional principle is based on the reorientation of the liquid crystal in an electric field due to its dielectric anisotropy. Different system can be realized (Blinov and Chrigrinov, 1994) and even twisted nematic cells were used for phase shift applications, but the most important ones is the planar nematic liquid crystal cell. We make a very short survey that gives the foundations of that what follows in the next chapters on modal control.

The basic effect is called electrical controlled birefringence and has already been explained in numerous textbooks (Blinov and Chrigrinov, 1994; Yeh and Gu, 1999). One considers a liquid crystal cell consisting of a homogeneously parallel aligned liquid crystal sandwiched between two substrates that have transparent electrodes. The texture is planar, as explained in Chapter 8. In the field off state the liquid crystal cell exhibits a phase retardation of

$$\Gamma = \frac{2\pi}{\lambda}(n_e - n_o)d, \tag{9.3}$$

where n_e and n_o are the principal refractive indices of the liquid crystal material and d the cell thickness. When an electric field is applied to the liquid crystal cell, the liquid crystal with positive dielectric anisotropy is reoriented and its tilt angle is changed towards the direction of the electric field. The director is rotated and an effective refractive index $n_e(\theta)$ has to be considered that depends on the angle of reorientation θ. This leads to a decrease of the phase retardation. The deformation in the liquid crystal cell is not uniform and one has to take the integral over the thickness to obtain the total retardation as the integral of all retardations, as was seen for the uniform hybrid texture in Section 8.3,

$$\Gamma = \frac{2\pi}{\lambda}\int_0^d (n_e(\theta) - n_o)\,dz, \tag{9.4}$$

where the integral is from $z = 0$ to $z = d$. The effective refractive index $n_{eff}(\theta)$ is given by Equation (2.61). The angle is now measured from the surface, which changes the equation to

$$n_{eff} = \frac{n_e n_o}{\sqrt{n_o^2 \cos^2\theta + n_e^2 \sin^2\theta}}, \tag{9.5}$$

where θ is the tilt angle of the director. It is important to repeat that $\theta(z)$ is a function of z and the tilt angle depends on the applied voltage. In the extreme case, by applying a strong electric field, the phase retardation tends to zero. If the anchoring of the liquid crystal at the surface substrates is strong planar, the retardation will never reach zero, which leads intrinsically to a maximum contrast ratio limitation. The important feature of this switching mode is the electrical distortion of the director. To see how the retardation as a function of the voltage develops, one has to know the dependence of the tilt angle $\theta(z)$ on the voltage. When an electric field is applied along the z-axis, the liquid crystal molecules are aligned towards the direction of the electric field, hence towards the z-direction. The distribution of the director tilt angle is a result of a balance between the elastic restoring forces and the electric force. The elastic restoring force is given by the elastic constants of the liquid crystal and the type of deformation and the boundary conditions, that is, the anchoring of the liquid crystal at the surface. For a planar liquid crystal cell the elastic energy density can be written as (Yeh and Gu, 1999)

$$u_{elastic} = \frac{1}{2}(K_1 \cos^2\theta + K_3 \sin^2\theta)\left(\frac{d\theta}{dz}\right)^2, \tag{9.6}$$

where θ is the tilt angle, K_1 the splay, and K_3 the bend elastic constants. The change of electrostatic energy due to the reorientation of the director can be written as

$$\Delta u_{electric} = \frac{1}{2\varepsilon_0}\frac{D_z^2}{\varepsilon_p \cos^2\theta + \varepsilon_s \sin^2\theta} - \frac{1}{2}\frac{D_z^2}{\varepsilon_s\varepsilon_0}, \tag{9.7}$$

where D_z is the z-component of the dielectric displacement field vector and should be held constant. The dielectric constants ε_p and ε_s are measured at low frequencies and in datasheets usually given at 1 kHz. They should not be confused with the dielectric constants at optical frequencies that determine the refractive index. Distortion occurs if the voltage is higher than a critical voltage:

$$V_{\text{threshold}} = E_{\text{threshold}} d = \pi \sqrt{\frac{K_1}{\Delta\varepsilon\varepsilon_0}}. \tag{9.8}$$

This transition behavior with threshold for liquid crystals is called the Fredericks transition. A numerical analysis can be carried out to obtain the director distribution $\theta(z)$. We show in Figure 9.10 the director profile for several applied voltages calculated using standard static theory. As a result of the high symmetry of the cell, the tilt angle is a symmetric function with respect to the midlayer $z = d/2$. Figure 9.11 shows measured data for the retardation of a 6 μm liquid crystal cell filled with the nematic mixture BL006 from Merck. Two things are important to note. The behavior of reorientation depends on the pretilt angle. The pretilt angle is the angle of the director at the surface substrate, which is caused by the alignment technology and is necessary for obtaining uniform alignment on large surface areas. The higher the pretilt the less sharp the threshold behavior. For thresholdless change of retardation, a high pretilt angle should be chosen. Strong anchoring of the liquid crystal director at the surface substrates is necessary to have a relaxation to the initial alignment state. Otherwise the liquid crystal cells will forget its designed alignment and the display becomes useless. The result of this condition is that the reorientation is never complete and leads to a thin layer with very strong deformation of the liquid crystal near the surface substrates. For characterization of such a layer the concept of coherence lengths can be used (de Gennes and Prost, 1993). Such a

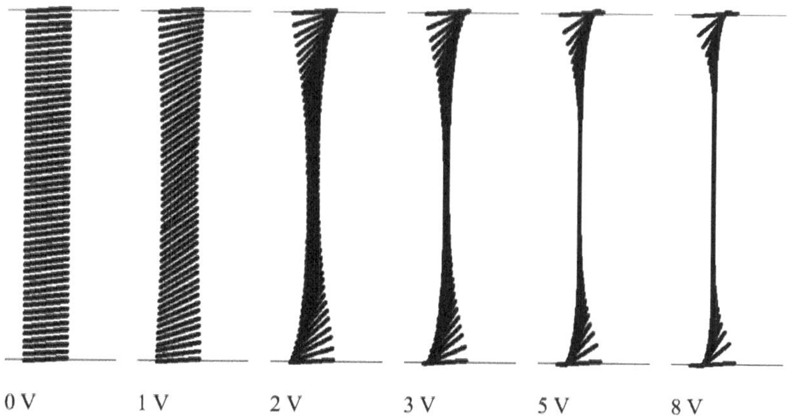

Figure 9.10 Simulated liquid crystal director configuration for different voltages in a liquid crystal planar cell (LCD Master Shintech, BL006, 6 μm, pretilt 5°).

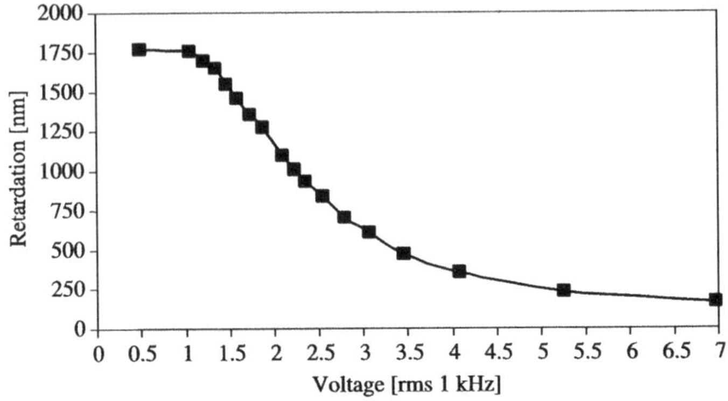

Figure 9.11 Measured retardation of a planar aligned liquid crystal cell. The phase shift becomes very small for high voltages (BL006, 6 μm, pretilt 5°).

coherence length gives a length scale with which to compare elastic distortion and deformations of the director fields by external forces. Optical devices that are surface sensitive, such as total internal reflection switches, will have particular conditions for operations. The thin layer prevents total internal reflection switches from being perfect (Soref and McMahon, 1980; Yang, 2003). On the other hand it may be used to study the surface anchoring effects of liquid crystals with varying substrates.

9.5 MODAL CONTROL ELEMENTS

The liquid crystal responds to the effective (rms) value of an applied alternating field. This fact allows the development of a new type of electrical addressing, called modal addressing, which is a continuous variation of the phase profile across a device. Both modal liquid crystal wavefront correctors and modal liquid crystal lenses (MLCLs) have been produced. Modal control elements are relatively simple to make and can produce a smooth change in deviation angle or focal length by control of the amplitude and frequency of the control voltages. However, these values need to be carefully selected in order to minimize phase distortions caused by both the nonideal distribution of the electric field across the elements and the nonlinear electro-optical response of the liquid crystal versus the applied field.

9.5.1 Prisms

Modal control can be used to design a prism of variable angle of deviation for polarized light. In the modal approach the lateral voltage gradient is neglected and the design is based on vertical switching only. All gradients in the plane of the device are assumed to be small. This holds for the voltage and for the refractive index gradients, but does not mean that there is no in-plane deformation. We will describe a design of a nonpixelated nematic liquid crystal deflector device (Riza and Khan, 2003; Khan and Riza, 2004).

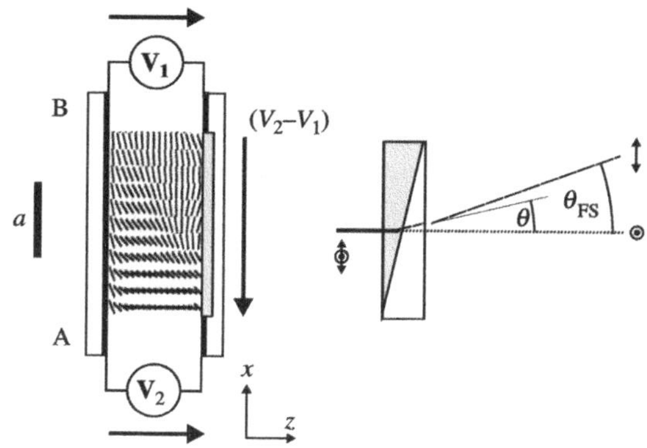

Figure 9.12 Principle of a liquid crystal deflector with modal control. A variable voltage drop is produced over the length of the element. The arrows indicate the direction of voltage drops. The free aperture diameter *a* is indicated on the left.

Figure 9.12 shows the principal construction of such a device, which is a sandwiched cell with transparent electrodes on each surface. Voltages V_1 and V_2 are applied along the vertical direction. The finite resistance indicated in gray leads to an in-plane voltage drop. As the basic design principle one can take the following procedure. The retardation voltage curve for a given liquid crystal slab thickness is calculated. Within this curve a linear region with reasonable phase shift is located. The phase shift has to be as high as possible because it will define the performance of the devices. This voltage range is chosen and the impedance of the device is selected to obtain the desired voltage drop over the surface area of the device. If the voltage drop is also linear, a linear phase profile will result. Such a device can be seen as a gradient index prism for beam deflection. If polarization insensitivity is needed, one uses two deflectors, each with orthogonal alignment orientation. The difficulty is now that the linear range of the retardation curve is very limited. To increase the performance of the device one needs to extent the voltage operation range to the nonlinear region, which complicates the design enormously. Let us discuss the idealized properties briefly to evaluate the basic parameters of the device.

The nematic liquid crystal material is sandwiched between an electrode (left in Fig. 9.12), which acts as the ground electrode, and a high-impedance control electrode (gray on the right in Fig. 9.12). The use of a high-resistance layer with two parallel electrical contacts was previously proposed in order to generate a smooth voltage ramp between these contacts to form a pixel-free deflector. Linear electrical contacts, which are used to apply the drive signal, are deposited at the edges of the control electrode. When a signal is applied to one of these linear electrical contacts, the voltage drops linearly across the high-impedance layer. The resulting electric field that is present across the nematic liquid crystal layer varies linearly across the aperture of the device, leading to an index modulation that varies across the aperture of the device in a near-linear fashion. Notice that this index modulation is along the nematic liquid crystal director and thus can be seen only by the component of the input polarized light that is also along the director that corresponds to the optical axis. With reference to Figure 9.12, a ray polarized along the director passing through the deflector device acquires a phase shift at the device position x, which can be expressed as

$$\Gamma(V, x) = \frac{2\pi}{\lambda} \int_0^d (n_e(\theta(V, x, z)) - n_o)\, dz, \tag{9.9}$$

where λ is the optical wavelength, d is the liquid crystal layer thickness, $n(\theta(V, x, z))$ is the electrically controlled refractive index the light sees, and V is the amplitude in volts. Looking at the two extreme ray positions at A and B, the beam deflection of the entire wavefront that is incident can be studied. Specifically, the A position ray suffers a phase shift given by

$$\Gamma(V_1) = \frac{2\pi}{\lambda} \int_0^d (n_{eff}(\theta(V_1, z)) - n_o)\, dz, \tag{9.10}$$

whereas the extreme ray at B suffers a phase shift given by

$$\Gamma(V_2) = \frac{2\pi}{\lambda} \int_0^d (n_{eff}(\theta(V_2, z)) - n_o)\, dz. \tag{9.11}$$

Hence the phase shift between the edges of the refracted plane wave exiting the nematic liquid crystal deflector is given by the difference, as

$$\Gamma(V_2) - \Gamma(V_1) = \frac{2\pi}{\lambda} \int_0^d (n_{eff}(\theta(V_2, z)) - n_{eff}(\theta(V_1, z)))\, dz, \tag{9.12}$$

where, between the A and B points, the incident plane wave acquires a linearly increasing phase shift if one chooses a voltage range with linear behavior for the effective refractive index. Otherwise a complicated retardation curve follows. In the linear regime the deviation of the beam is given by the ratio of the retardation and the free aperture over which this retardation is realized. The equation reads

$$\sin \Theta = \frac{\lambda}{2\pi} \frac{\Gamma(V_2) - \Gamma(V_1)}{a}. \tag{9.13}$$

To maximize the phase shift between A and B, one needs to use the whole retardation drop. If at maximum retardation (B in Fig. 9.12) the applied voltage is designed to be zero, the index seen by the incoming ray is essentially n_e, the liquid crystal material extraordinary index of refraction. Similarly, note that at the position of minimum retardation at location A, where the other device electrode is present, V can be increased until the retardation reaches n_o, where n_o is the ordinary refractive index. In this case the deflector is generating its maximum birefringence $\Delta n = n_e - n_o$ and hence also produces the largest phase shift between rays across the device aperture a. Note that the use of the whole retardation range usually does not give a linear drop of retardation over the device aperture and leads to deformed phase profiles that alters the device performance considerably. By evaluating Equation (9.13) the electrically controlled maximum deflector angle Θ can be written as

$$\sin \Theta_{\max} = \frac{\Delta n}{a} d. \tag{9.14}$$

A laser beam passing through the described thin deflector is refracted at the free space output of the deflector by an angle Θ, which depends on the average index n_{ave} of the liquid crystal material. The electrically set deflector angle Θ has therefore to be corrected to describe, with the index of the output media, the free space angle Θ_{FS}. This is done with Fresnel's formula for refraction. It leads to the expression

$$n_{\text{ave}} \sin \Theta = n \sin \Theta_{\text{FS}}. \tag{9.15}$$

For a parallel-rubbed liquid crystal cell and weak deformation, n_{ave} is given mainly as the extraordinary refractive index. To have high performance, that is, large deviation angles, material with a large birefringence should be incorporated into a thick cell with narrow electrode in-plane electrode distance a. The problem that then arises is the speed of the device. For thick liquid crystal layers that are used in devices with nonuniform equilibrium director configurations, switching times can be of the order of several seconds.

9.5.2 Lenses

Liquid crystal modal lenses are switchable lenses with a continuous phase variation across the lens (Loktev et al., 1998; Naumov et al., 1998a). The basic idea is to vary the electric field along the diameter of the lens. The different electric fields at different positions lead to different deformation strengths of the liquid crystal director. The principal idea is that the voltage gradient in plane can be neglected compared to the vertical one. Then the main reorientation force is the vertical to the substrate surfaces and the design has to make in that manner the retardation–voltage curve shown in Figure 9.11 can be used for element design. The voltage distribution in the plane of the element has to take into account the nonlinearity of this curve. In practice, the form of the retardation voltage function depends on the material parameters of the liquid crystal, the surface anchoring conditions, and also, slightly, on the thickness. A critical issue for such lenses is the minimization of aberrations.

The construction of a modal control liquid crystal lens is characterized by a high-resistance transparent electrode and a low-resistance control contact. The contact

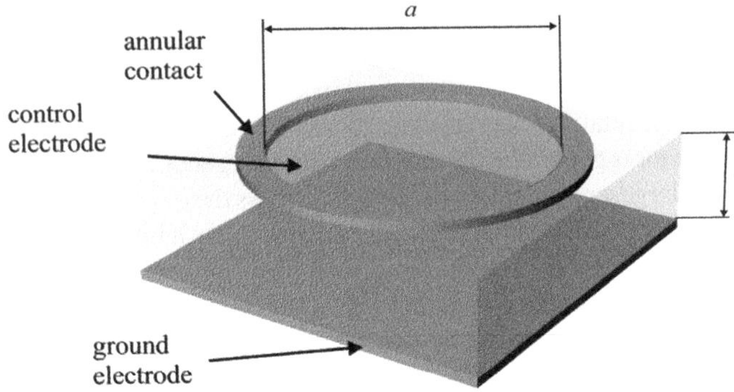

Figure 9.13 Electrode arrangement of a spherical modal control lens. The control electrode fills up all the active area defined by the diameter *a*. The liquid crystal layer is of thickness *d*. The ground electrode is used as the counter electrode to apply the switching voltage vertically. The diameter is usually several millimeters in size.

configuration determines the type of lens. For spherical lenses an annular contact is used and for cylindrical lenses two linear parallel contacts. Modal control liquid crystal lenses utilize the reactive parameters of the cell, that is, their electrical response to an AC field.

Figure 9.13 shows schematically a spherical modal control liquid crystal lens. The control voltage is applied to a low-resistance annular electrode around the active area of the lens. The voltage across the lens decreases radially towards the center of the lens, because of the potential divider that is formed by the high-resistance control electrode and the capacitance of the liquid crystal layer; that is, its impedance increases radially towards the lens center. This means that the voltage decreases radially towards the center. Conversely, the optical path length of the liquid crystal layer increases from the periphery to the aperture center. Note that the incident light must be linearly polarized along the liquid crystal's extraordinary axis. To operate with unpolarized light it is necessary to use two identical lenses with mutually orthogonal orientations or to apply the technique described by Love (1993) for reflection systems. The control voltage frequency defines the magnitude of the voltage across the liquid crystal, which in turn shapes the electro-optic response. The modal control liquid crystal lens focal length can be evaluated using the retardation difference between the edges and the center of the lens by making use of Fresnel's approximation,

$$f = \frac{\pi a^2}{\lambda} \frac{1}{\Gamma_{\text{center}} - \Gamma_{\text{edge}}}, \tag{9.16}$$

where Γ_{center} and Γ_{edge} are the retardations at the center and edges respectively, a is the radius of the active zone, and λ is the wavelength of incident light. The distribution of the rms voltage across the lens is described by a complicated function that depends on the nonlinear voltage–retardation dependency. In general, if a voltage of arbitrary magnitude and phase is applied to the cell, then the resulting phase distribution will be far from spherical, and the lens aberrations will be significant. However, for a certain relationship between the frequency of the applied voltage and the distributed impedance, the retardation distribution is close to a parabolic and good results are obtained for relatively weak lenses (Naumov et al., 1999; Vdovin et al., 2003).

Typical parameters for such modal lenses are a free aperture diameter of several millimeters, which results in a minimal focal length of approximately 0.5 m for 5 mm diameter. The numerical aperture is N.A. = 0.005 and therefore very low. Pretilted alignment with relatively high pretilt angles above 10° is used, which increases the minimal focal length due to a smaller initial retardation. The high-resistance electrode has a surface resistance of several $M\Omega/sq$. The pretilt and the lower high-resistance electrode allow higher liquid crystal switching speeds.

9.6 INTERFEROMETERS BASED ON POLARIZATION SPLITTING

Since the publishing of the book *Polarization Interferometers* (Françon and Mallick, 1971), only a few attempts have been made to take up new ideas and to develop applications. The biggest problem in using polarization interferometers is still that the materials are costly and sometimes not easy to handle. Up to now, no breakthrough has been made by using organic materials, mainly due to the lack of fabrication technology. In this chapter we would like to demonstrate two aspects: the usefulness of liquid crystal and polymers for interferometer applications and the advantages that can be drawn out of new optical configurations like twisted cells for the concrete design of elements. We discuss these points by describing an example of a spectrometer based on a Wollaston prism.

9.6.1 Microspectrometer

To obtain an impression of the complexity of design and fabrication when birefringent components are used we will discuss in the following chapter special examples. We start here with a polarization microspectrometer based on liquid crystal technology. The optical devices have already been described in several publications (Françon and Mallick, 1971; Padgett et al., 1994; Montarou and Gaylord, 2000). We discuss here possibilities for improving the performance when liquid crystal components are used.

A special arrangement of crystal slices allows the realization of static spectrometers with a large field of view, medium to good resolution, while being extremely robust. The operation principle is that of a Fourier transform spectrometer. Usually, a Fourier transform spectrometer operates by splitting an incoming beam into two parts, applying a differential phase shift, and then recombining the beams. The main advantages are the large optical throughput (known as the Jacquinot advantage), the multiplexing (known as the Fellgett advantage) and the wide spectral coverage. Moreover, if the phase shifts are varied spatially, there are no moving parts anymore, and the whole interferogram can be acquired on a detector array.

If polarization components are used, different propagation conditions for the extraordinary and ordinary rays forming the basis of the devices' polarizers are necessary. The intensity will be reduced by a factor of one-half due to absorption at the polarizer, and the response depends on the state of polarization of the incoming light. Although for grating spectrometers the spectrum is directly visible, the polarization spectrometer mainly works as a Fourier transform spectrometer. This demands the calculation of the spectra by a Fourier transform of the recorded signal. The resolution of the spectrometer is limited and given by the maximum retardation, which can be directly proportional to the thickness of the element. An angular dependence of the optical properties always exists and can limit the useful range of incident angles, the so-called field of view.

One possible design for a Fourier transform spectrometer based on polarization optics and amplitude division is shown in Figure 9.14. A Wollaston prism is placed between two linear polarizers. The amplitude division is realized by setting the entrance polarizers at 45° with respect to the optical axis of the first crystal wedge of the Wollaston prism. The uniaxial crystal splits the incoming linear polarization into two beams with orthogonal polarizations that propagate through the system subject to different retardations. The exit polarizer, which performs a projection, recombines these beams. A phase shift between these beams is created and if this phase shift is varied spatially, fringes appear. Due to the triangular shape of the crystal wedges and because the optical axes are perpendicular to each other, the phase retardation in the Wollaston prism is linear along wedge angle direction y. A linear phase shift appears between the two polarization directions with a maximum value that is given by the product of thickness of the Wollaston $2T$ and the birefringence Δn. Such systems have been widely studied for precise

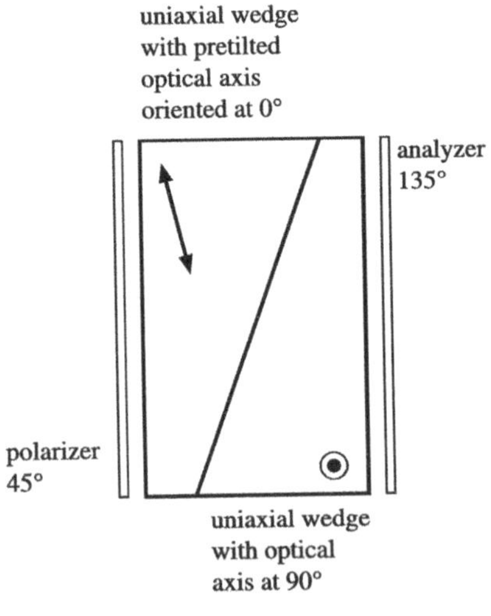

Figure 9.14 Fourier transform spectrometer based on a Wollaston prism. Two birefringent wedges are put together with different orientation of the optical axis. The thickness is $2T$.

wavelength determination and pollution measurements (Harvey et al., 1994; Courtial et al., 1997; Dunmore and Hanssen, 1998). The resolution (Padgett et al., 1994) is given by

$$\Delta\lambda = \frac{\lambda^2}{2T\Delta n} \tag{9.17}$$

One sees immediately that the resolution depends strongly on the wavelength, the thickness, and the birefringence. Special attention has to be given to the problems caused by the material, such as dispersion or temperature dependence of birefringence and absorption. Practical applications may need a resolution of about 1 nm at a wavelength of $\lambda = 800$ nm. To achieve this one calculates with Equation (9.17) a minimum retardation of $2T\Delta n \approx 650$ μm $\approx 800\,\lambda$. Looking for a miniaturized system, the birefringence should be high to limit the thickness of the system. A practical realization with quartz with a birefringence of $\Delta n = 0.009$ leads to a system that is about 72 mm thick that does not allow miniaturization. Crystals, such as calcite with a birefringence of $\Delta n = -0.17$, will reduce the thickness to a few millimeters. Liquid crystals and liquid crystal polymers can be used with birefringence as high as $\Delta n = 0.3$. This would allow a thickness of only 2 mm for a resolution of 1 nm at a wavelength of 800 nm.

A serious design problem is the imaging optic needed to image the spectrum onto the detector. This should be avoided in all cases to allow miniaturization. A problem encountered with this is the spatial coherence of the source. In fact, when an interferometer is illuminated by a spatially incoherent source, the interference fringes are localized in space (Chamberlain, 1979). The classical design with a conventional Wollaston prism shown in Figure 9.14 has the disadvantage that the plane of maximal contrast might be localized inside the prism. An additional lens would be necessary to image the interference pattern on the detector. Several approaches to pull out the plane of maximal contrast are used nowadays. A double Wollaston prism arrangement (Courtial et al., 1996; Prunet et al., 1999) or modified Wollaston prisms with a tilted optical axis (Patterson et al., 1996; Montarou and Gaylord, 1999) might be used. In the modified Wollaston prism, the optical axis is tilted with respect to the propagation direction and allows the extraction of the interferogram in an elegant manner. This is indicated in Figure 9.14. To calculate the exact position of the plane of maximal contrast, depending on parameters like birefringence and tilt angle, the light propagation in the interferometer has to be simulated. For this first design task one can use the fact that in an interferometer the position of maximum fringe contrast is found where the split rays are reunified. This allows the use of a standard polarization ray-tracing procedure to find the plane of interference. The coherence properties are not important in this particular case because no interference effects have to be searched. A sample simulation performed with commercial software (TRACE PRO Expert from Lambda Research Corporation (2004)) is shown in Figure 9.15. The refractive indices are set to $n_o = 1.5$ and $n_e = 1.7$. An angle of 45° is set for the optical axis of the first prism. The whole prism is 2 mm thick and 10×10 mm in area. An extended source with rays coming from a

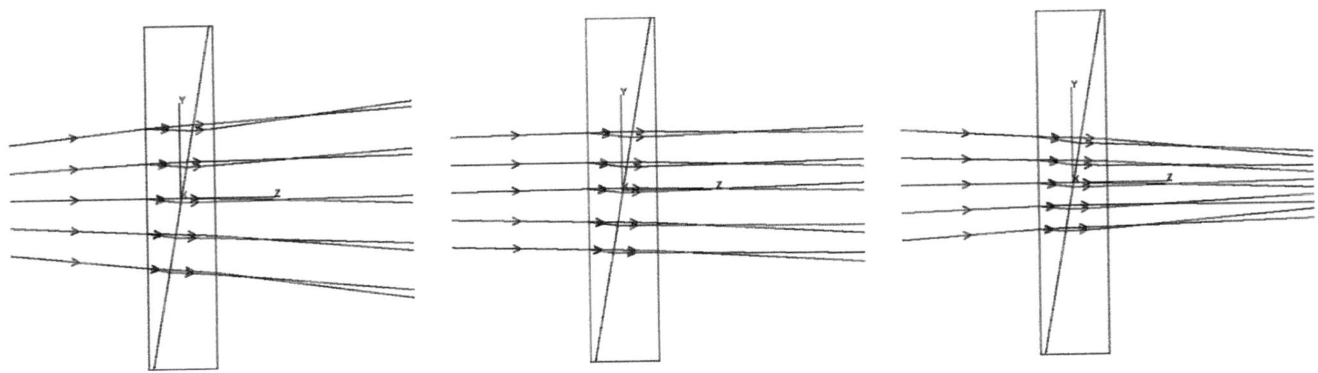

Figure 9.15 Incoherent ray tracing on Wollaston prisms. The crossing point of rays lies on a plane that is known as the plane of maximum contrast. The detector should be placed in this plane to obtain the best fringe contrast.

point 20 mm to the right of the prism is assumed. No particular polarization is set to the rays ray splitting is observed at the entrance of the prism. The interface between the two prisms redirects the rays and they are unified after the prism at different positions, forming a plane where the interference fringes have maximum contrast.

Changing the divergence of the source changes slightly the position where the rays intersect. This is illustrated for two different source divergences in Figure 9.15. If the source point is set at a very large distance to the left (−1600 mm), the light is collimated at the entrance and still focused behind the prism. Even when focused through the prism the plane of maximum contrast can be found. Such a behavior underlines the tolerance of such a configuration against extended sources (Montarou and Gaylord, 1999).

One important parameter for the presented system is the contrast of interference fringes, which cannot be simulated with simple ray tracing. Coherence properties of light, particularly the spatial coherence of the source, have to be included to study interference effects. This is made possible by using the Gaussian beam propagation methods, which provide coherent beam propagation and interference as described in Chapter 3.

Figure 9.16 shows sample simulations for the interference pattern if beams are combined incoherently and coherently for exactly the same geometry. Coherent ray tracing has to be used as it is available in advanced ray-tracing analysis programs (Breault Research Organization, 2005; Photon Engineering, 2005). In Figure 9.16*b*, an example simulated with ASAP (Breault Research, 2005) is presented for a completely coherent source. In this case the interference fringes have almost maximum contrast. The situation is ideal and interference happens all over the observation volume. For advanced analysis a spatially incoherent source can be defined with a set of point sources (Boer et al., 2004). From each point source, coherent rays have to be propagated. On the detector plane the intensities for each point source are summed up coherently and interference fringes appear. The spatially incoherent source is built up from point sources with slightly different wavelengths. If one wants to model a monochromatic source, the temporal coherence is fixed and the difference between the wavelengths for the source design should lie within the spectral width of the source. For laser light, the spectral width is extremely small and so the distance between sample wavelengths should also be small. The source is then a complex object that has, at every position in space, different wavelengths that have a value within the spectral width of the light. The distance between the sampling points together with the solid angle will define the spatial coherence (compare with Chapter 1). Both have to be set carefully to obtain reasonable results. That trick used is to distinguish in the simulation program the different sources and to prevent rays coming from different point sources from interfering. With such a model, interference fringes at every point behind the prism can be simulated for real sources and the contrast, the main criterion for electronic measurement, can be obtained.

An additional point to discuss for the design of a high-performance interferometer is the field of view. The system should have a wide field of view so as to

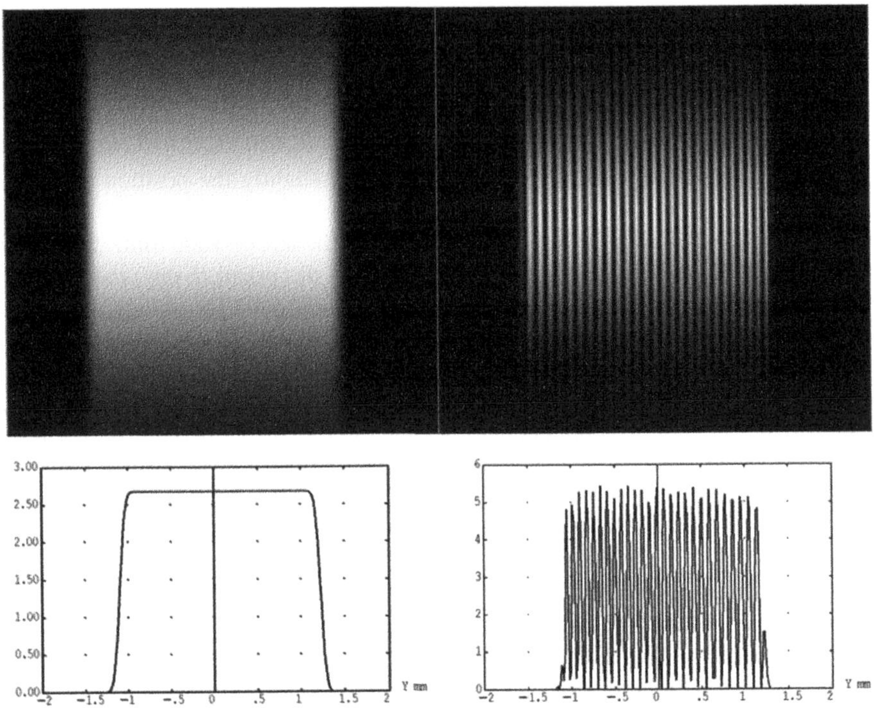

(*a*) (*b*)

Figure 9.16 Incoherent and coherent ray tracing for the analysis of interference effects of Wollaston prisms. (*a*) The incoherent ray tracing does not consider interference effects. (*b*) If coherent ray tracing interference fringes appear.

avoid the use of optical components at the entrance. The field of view is defined as the useful range of incident angles that will not reduce the contrast of the interferogram. A double Wollaston prism configuration is bulky, but allows the field of view to be increased. Using the advantage of liquid crystal technology to make twisted structures allows a nearly complete compensation and therefore a very high field of view (Boer et al., 2002). Unfortunately, this configuration does not allow an easy extraction of the plane of maximal contrast on a detector.

A more sophisticated configuration introduces a half wave plate between the two wedges of the Wollaston prism (Françon and Mallick, 1971; Courtial et al., 1996). Here the first wedge of the conventional Wollaston prism remains and is followed by a half wave plate that rotates both polarization components. Therefore, the second wedge has to be rotated with respect to the original Wollaston prism design. The field of view is increased without changing the plane of fringe localization when modified Wollaston prisms with tilted optical axes are used. Using a liquid crystal twist nematic cell as a polarization rotator avoids disturbing the dispersion effects of the wave plate. Figure 9.17 shows the related design. A twisted nematic cell is brought in between the two wedges and fulfils this tasks. It widens the field of

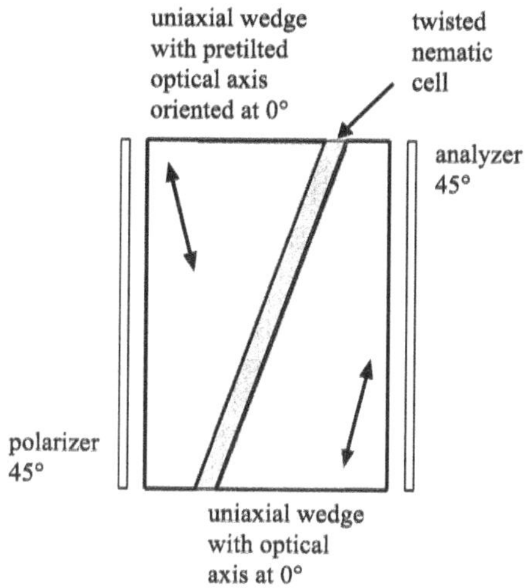

Figure 9.17 Modified configuration for automatic noise correction and contrast enhancement. The twisted nematic cell rotates the polarizations. The optical axes of the two Wollaston prisms have to be oriented parallel in order to obtain interference.

view and if it can be switched, a contrast enhancement can be obtained by measuring two interferograms, one with rotation of polarization and the other without (Boer et al., 2002). The liquid crystal components add here a remarkable feature that is not accessible otherwise. For such a complicated system including polarization rotation, simulation of optical performance is a challenge (Boer and Scharf, 2002). In particular, the twisted nematic configuration is not easily modeled in conventional ray tracing. Usually only the polarization rotation is assumed, which neglects angular dependence and the change of polarization state when propagated. Often such a simplification does not influence the system design to great degree and can be accepted.

9.7 BIREFRINGENT MICROLENSES

Liquid crystal lenses have long been investigated, and many different systems have been developed. There are macroscopic birefringent lenses (Sato, 1979), modal control lenses, refractive-like adaptive systems (Masuda et al., 1996; Gvozdarev and Newskaja, 1999; Scharf et al., 2000), microlenses immersed in liquid crystal hosts (Commander et al., 2000), and solid-state lenses. All have a limited range of variation of the focal length and usually the optical quality changes with the focal length. The most interesting feature is the electrically driven change of optical parameters such as focal length. However, there are additional applications for illumination and aberration control that need static birefringent micro-optical systems (Hain et al., 2001; Stapert et al., 2003). Here we will concentrate on static birefringent lenses, discuss their optical properties, and show the particularities of applications. The special case of birefringent microlenses will help the discussion by applying the results to a real application. Experimentally, there are several concepts needed to realize static microlenses, as discussed in Section 9.2. Two different types of birefringent microlens systems can be obtained easily: Replicated microlenses in isotropic materials can be immersed in a liquid crystal host, or the lens can be directly replicated in birefringent polymer material. Figure 9.18 shows the different concepts.

The easiest way to obtain birefringent microlenses of sufficiently good quality is to immerse microlenses fabricated with standard technology in a liquid crystal host. Lenses are usually replicated from master arrays (Nussbaum et al., 1997; Daly, 2001). For this, a conventional substrate with rubbed polyimide is brought close to the microlens array. Usually the distance is not much larger than the lens height and no additional spacer is needed to control the thickness of the liquid crystal layer. As an example, Figure 9.19 shows a replicated microlens array immersed in the liquid crystal mixture E7. The alignment directions of the substrates are parallel.

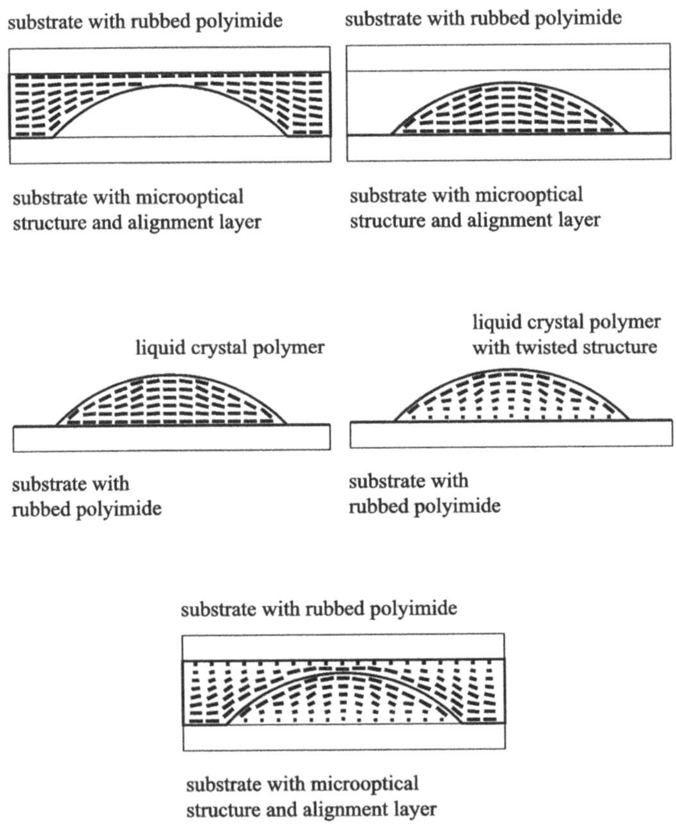

Figure 9.18 Some possible microlens arrays that can be fabricated by combinations of polymer technology. Immersion or direct replication can be used.

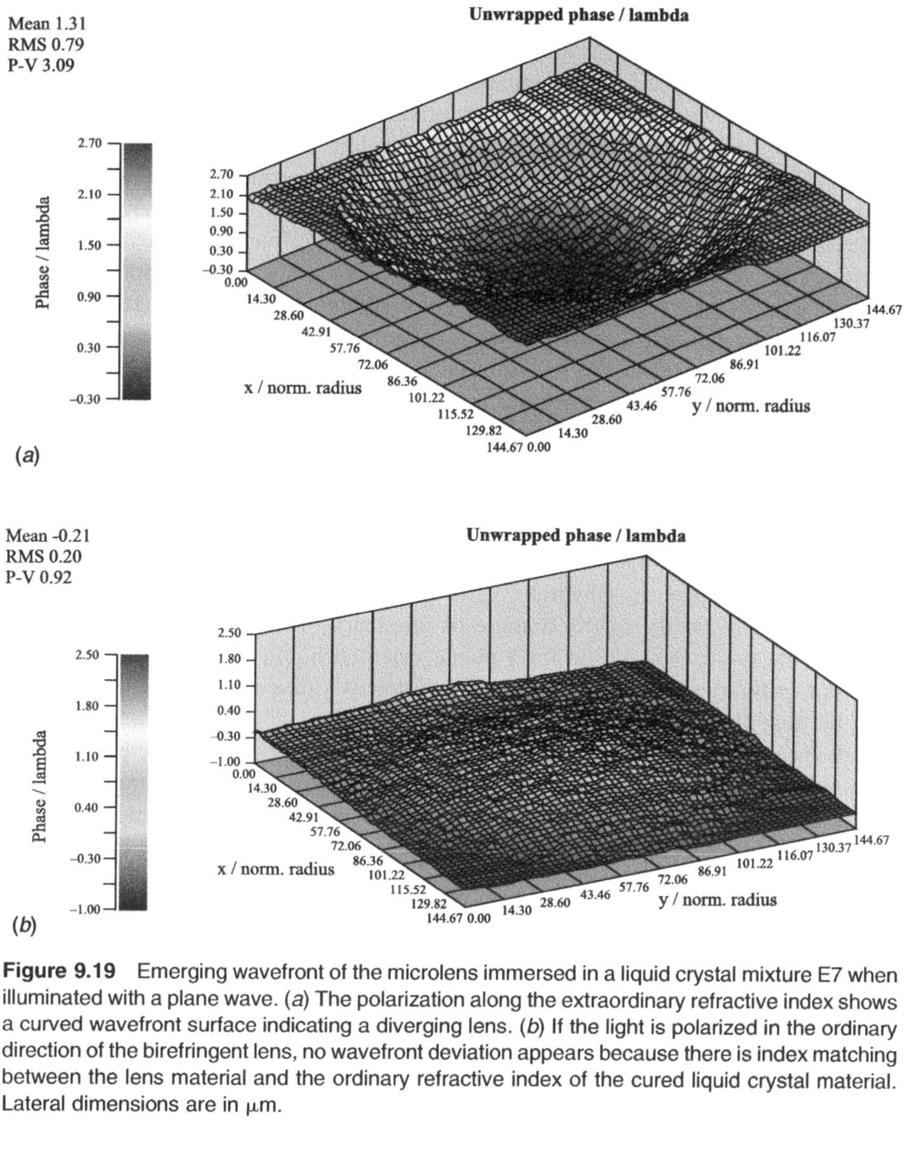

Figure 9.19 Emerging wavefront of the microlens immersed in a liquid crystal mixture E7 when illuminated with a plane wave. (*a*) The polarization along the extraordinary refractive index shows a curved wavefront surface indicating a diverging lens. (*b*) If the light is polarized in the ordinary direction of the birefringent lens, no wavefront deviation appears because there is index matching between the lens material and the ordinary refractive index of the cured liquid crystal material. Lateral dimensions are in μm.

This leads to index matching between the refractive index of the lens materials NOA 81 ($n = 1.56$) and the liquid crystal mixture E7 ($n_e = 1.75$ and $n_o = 1.532$). The lens function is obtained for one polarization only. The lens is diverging because the extraordinary refractive index of the immersion liquid crystal is larger than that of the lens material. The focal length of the system is $f = -2$ mm, which results in a numerical aperture of N.A. = 0.036. The wavefront emerging from such a lens is shown for two different polarizations of the incident light in Figure 9.19. The measurement was done at 633 nm and the total retardation is 1.7λ. If the polarization is along the ordinary direction, the wavefront is almost flat.

Liquid crystals in their liquid nematic state show strong temperature dependence of the birefringence. Generally, a wide temperature range is desired, especially for outdoor applications. For static birefringent microlens systems, one alternative is to use polymerizable liquid crystal mesogens. The structure can then be fixed with UV curing, for example. It is also interesting to have free-standing birefringent microlenses for a simple reason: For every optical design the refractive index is a key parameter. If one can no longer choose the refractive index in which the optical element is immersed, design freedom is restricted. For practical reasons, it is also desirable to have lenses that can be combined with conventional optical glue for assembling and index matching. Last, but not least, the numerical aperture is small for immersed systems, but can be much larger for free-standing lenses. It is possible to obtain defect-free structures over several centimeters square by replicating in birefringent mesogens. Lenses with a diameter of 145 μm can have focal lengths in air of about 400 μm.

Birefringent lens systems are known for their aberrations (Lesso et al., 2000). This is due to the fact that the birefringent axis orientation implies strong angular dependence (compare to Chapter 2). The extraordinary refractive index of the lens appears differently for each angle of incidence. Therefore, a strong coma is found as the typical aberration for a planar oriented liquid crystal polymer microlens. For ordinary polarized rays the system behaves like a conventional isotropic lens, because there is no angle dependence.

Advanced designs would include twist of the optical axis. To create birefringent elements with twist, the process is similar to the replication technique used for parallel alignment, with the difference that the master has a different alignment orientation with respect to the substrate. As an example, free-standing birefringent microlenses with a 90° twisted nematic structure are shown in Figure 9.20. The lenses have a diameter of 145 μm and a focal length in air of $f_{\text{ordinary}} = 0.47$ mm and $f_{\text{extraordinary}} = 0.37$ mm. These lenses were fabricated in the hope of obtaining an improved angular dependence of the optical properties. The assumption was that although the light is traveling through the lens and the polarization is rotated for both polarization directions, the angular dependence becomes less significant and consequently the asymmetry for the ordinary and extraordinary polarized ray will be reduced. In this context, it is instructive to study the interference fringes that are generated when the microlenses are illuminated by a plane wave. Although for a lens with planar texture the light in the extraordinary polarization shows aberrations, in the twisted system this is the case for the polarization of light that follows the ordinary refractive index. The extraordinary and ordinary polarization directions are easily identified by counting the fringes in the interferograms. Because of the higher refractive index for the extraordinary direction, more fringes appear for extraordinary rays and less for ordinary rays. A visual inspection leads to the assumption that the fringe deformation is less pronounced for the twisted system. We have not been able to prove this with the experimental equipment available (Scharf, 2005). Further investigation with diffraction limited microlenses adapted to the problem

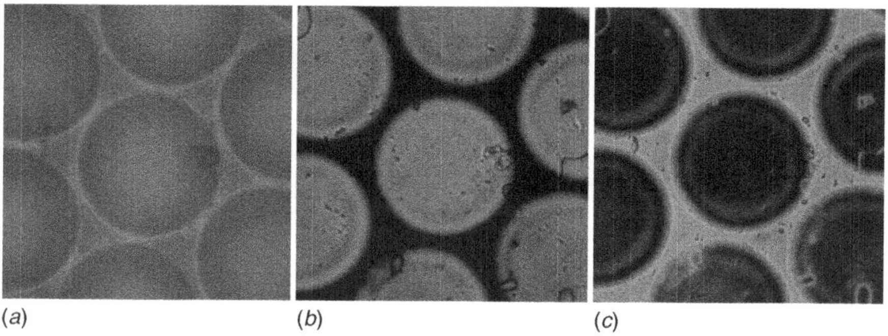

Figure 9.20 Replicated birefringent microlenses with a twisted nematic structure. The lenses are 145 μm in diameter and show the typical disclination lines when no chiral dopant is used. (*a*) Microphotograph without polarizers in reflection. (*b*) Between crossed polarizers in transmitted light and oriented along a polarizer, the lenses appear bright because of the waveguiding properties of the twist structure that rotated the polarization. (*c*) Seen between parallel polarizers the lenses become black. Interference colors at the lens rim indicate that the waveguide regime no longer holds for thin twisted birefringent optical systems.

of birefringent light propagation will show if this concept can lead to aberration corrected birefringent optical elements.

9.8 ELECTRICALLY SWITCHABLE MICROLENSES

Adaptive lenses are interesting for a lot of applications in machine vision and photonics. There are different possibilities for realizing such systems. A combination of a birefringent lens with a polarization rotator (Sato, 1979) leads to two distinct focal points. A microlens immersed in a switchable liquid crystal cell changes the focal dlength continuously for one linear polarization (Commander et al., 2000). Spatial inhomogeneous liquid crystal distributions have been studied intensively for microlens applications. For this there are two different configurations to distinguish, starting from different textures of the liquid crystal.

The microlens can be formed in a homeotropic texture (Gwozdarev, 1996; Scharf et al., 1999) or in a planer liquid crystal cell (Nose and Sato, 1989). Macroscopic liquid crystal lenses are often made with multi-electrode structures (de Jule and Riza, 1994) or spatially varying resistivity of the electrode structure (Loktev et al., 1998), and have already been reviewed in Section 9.5. We concentrate here on the concept of switchable microlenses with diameter less than 1 mm. Figure 9.21 presents the different concepts used in obtaining switchable microlenses. One can use liquid crystal materials to replace birefringent crystals and to form a bifocal system with liquid crystal polymers and a polarization rotator as shown in Figure 9.21*a*. In this case, the switching speed is that of the polarization rotator and is usually faster than 10 ms.

The quality of the lens is determined by the quality of the birefringent lens itself, which shows aberrations (Lesso et al., 2000). This can be tailored by using a differ-

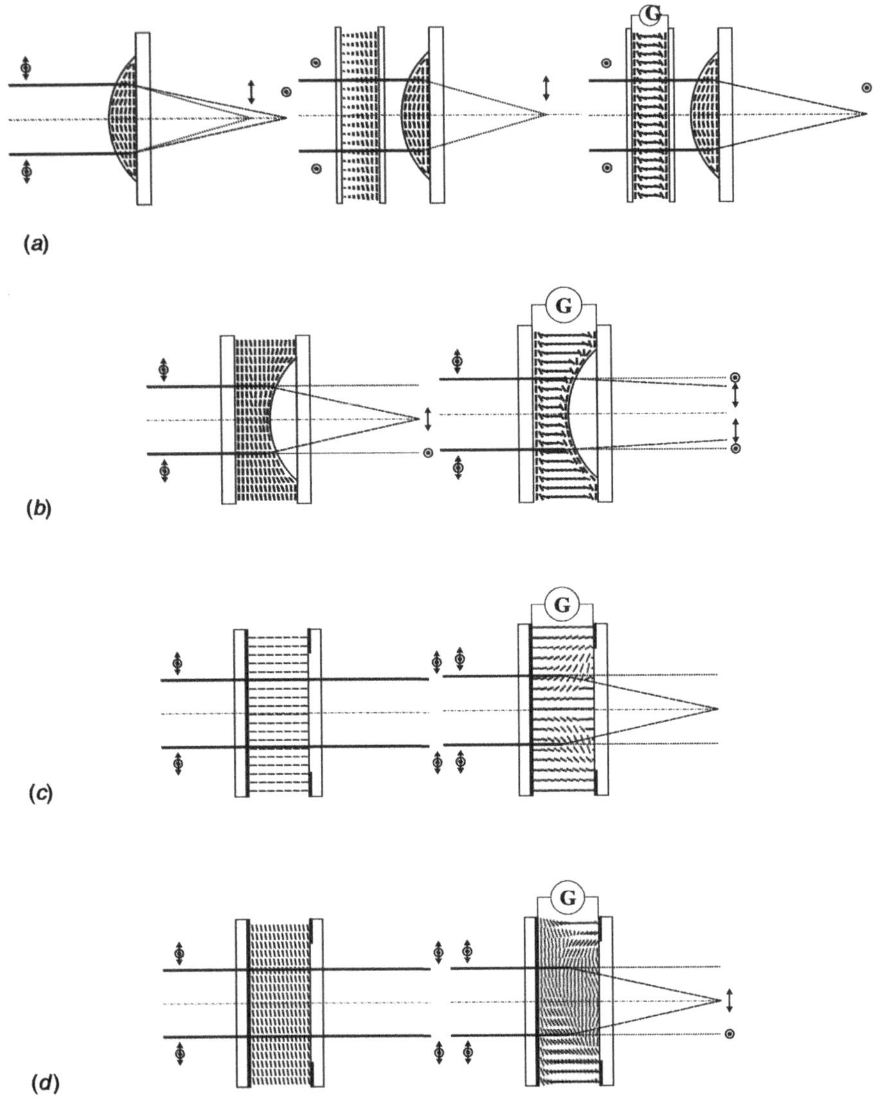

Figure 9.21 Switchable lens configurations based on birefringent materials. (*a*) A combination of passive birefringent lenses with polarization rotators allows switching between two focal points. (*b*) Immersing a lens in a switchable liquid crystal cell allows changing the surrounding refractive index of the lens. A continuously switchable lens system results. (*c*) In planar liquid crystal cells, spatially varying electric fields fringe at electrode edges to form a gradient index structure. That gradient index distribution can form a lens in homeotropic alignment. (*d*) For linear polarized light it is advantageous to use planar alignment and to deform the liquid crystal using fringing electric fields. The result is a microlens configuration with a small tilt because of the preferred direction of reorientation.

ent mechanism, for example by adapting the lens form to the aberrations or using immersions as shown in Figure 9.18, which might correct for certain aberrations. However, the system has only two focal points, which is sometimes a disadvantage. The liquid crystal offers the possibility of being deformed with spatially varying electric fields to form gradient index microlenses. The concept is different from the modal lens approach in the sense that a real three-dimensional deformation is now sought. The gradients in all directions are important. A continuous change of the focal length can be obtained by varying the applied voltage. A closer look at the systems shows the difficulties in the construction of high-quality liquid crystal microlenses based on this principle. Because of the anisotropy of the elastic and dielectric constants, it is difficult to obtain a circular distribution of the birefringence gradient for several voltages (Honma et al., 1997; Ye, 2001, 2003).

It is instructive here to discuss an example, because it will show general problems that appear for liquid crystal devices based on spatially inhomogeneous liquid crystal director distributions. We will have a closer look at a liquid crystal lens in a planar texture made by a structuring an electrode. A three-dimensional view of the system is shown in Figure 9.22. The geometry chosen here is the most simple. Only one electrode is structured, which prevents the alignment problems of substrates that appear when both substrates have structured electrodes (Masuda et al., 1991). The electrode pattern is in general prepared by photolithography. Holes of different diameters can be fabricated and the form is not limited to circles. The electrodes are usually transparent indium tin oxide (ITO), coated on to glass plates as substrates. One important aspect is the alignment of the liquid crystal. As mentioned above, homeotropic, hybrid, or planar alignment can be chosen, which might even be twisted. To obtain uniform structures it is convenient to use rubbed polyimides as the alignment layer. Such alignment layers produce a pretilt of the liquid crystal directors of 2–15° at the surface. Due to this a preferred direction for reorientation is given which leads to complicated director structures. However, the liquid crystal texture is well-defined, uniform, and strong planar anchoring conditions are assured. The cell thickness d is controlled with spacers, which are preferably not present in the active area. Plastic foils, spacer balls, or a microstructured

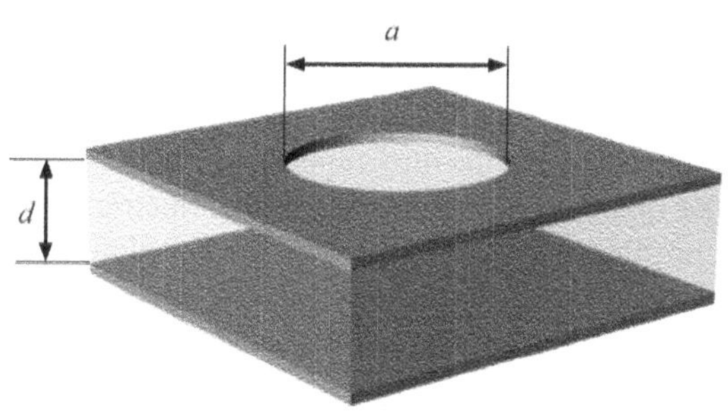

Figure 9.22 Electrode structure of a typical liquid crystal microlens system that is operated with fringing fields. The cover glass has a structured electrode and a hole diameter a. The bottom glass substrate has a complete counterelectrode. The thickness of the liquid crystal layer is d.

substrate serve to space the substrates. The thickness d varies between several tenths of a micron to several hundred microns. For such a liquid crystal lens one has many design parameters to consider: the geometry and form of the electrode hole, the diameter of the electrode opening a, and the thickness of the liquid crystal cell d. The pretilt angle and twist angle are additional parameters necessary to control the liquid crystal texture.

There are also the material properties of the liquid crystal, including the elastic properties and the electric properties (Nose et al., 1989; Masuda et al., 1996). There are three basic elastic constants and two dielectric constants. The latter might be altered by changing the driving frequency. To obtain high numerical apertures, materials with large birefringence are preferred. To obtain good optical quality for a system that uses this lens, one has to respect general design rules. We will consider the main tasks for obtaining a variable lens. We would like to create a circular symmetric lens system that allows the focal lengths to be changed.

For imaging, the lens should be diffraction limited for all focal lengths. The lens system should switch in a reasonable time. Designing an adaptive liquid crystal lens based on these considerations with the abovementioned parameter space that fulfils all requirements is a difficult task. In addition, it should work by changing only one or two control parameters: the electric field and/or the frequency of the driving voltage.

Let us have a look into the functional principle of such a lens system. We discuss the case for a planar aligned cell with a single patterned electrode as in Figure 9.22. At a certain moment the voltage is switched on and will lead to an electric field distribution in the liquid crystal cell. At time zero the electric field has a well-defined distribution that is already different for the two directions because of the dielectric anisotropy of the liquid crystal. When complete reorientation leads to equilibrium of the liquid crystal director distribution, the electric field distribution is different compared to its initial distribution. The equilibrium director configuration is determined by several facts. The electric field is the driving force that works against the elastic deformation of the liquid crystal. At the surfaces the director alignment is fixed due to the strong anchoring. The resulting director distribution contains all kinds of deformations: bend, twist, and splay. For each voltage, a specific equilibrium distribution of the director field is found. For illustration we show in Figure 9.23 simulations of the transmission obtained in a three-dimensional model with commercial software for liquid crystal director simulations (Shintech, 2006). Transmission distributions at different voltages are shown. The parameters of simulation are given in Table 9.1. The alignment of the liquid crystal is horizontal. A remarkable difference between the directions along and perpendicular to the alignment is seen and the intensity distributions are asymmetric. To show this in more detail, the director profiles along different sections are plotted in Figure 9.24 for a voltage of 9 V. The director distribution is symmetric perpendicular to the alignment and asymmetric parallel to the alignment. That leads to a tilt of the optical axis of the lens (Scharf et al., 2000).

This can be observed in measurements from a real system. Figure 9.25 shows a series of measured phase profiles for increasing voltages. The rubbing direction is horizontal. The system has a single patterned electrode structure, and is 50 μm

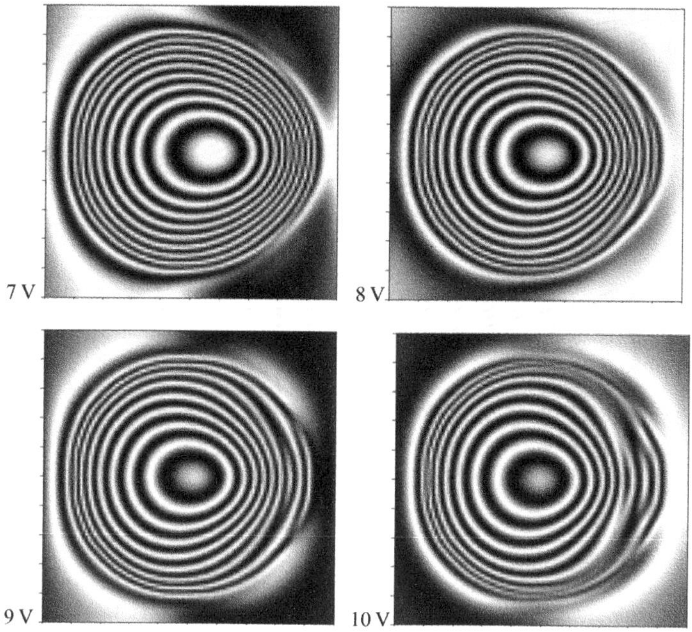

Figure 9.23 Transmsission of a liquid crystal lens between crossed polarizers at different voltages. The fringe pattern shows asymmetry due to the asymmetric reorientation of the planar aligned liquid crystal (parameters as given in Table 9.1).

thick and 140 μm wide. Lenses are arranged in a hexagonal array. The pitch of the lens array is 150 μm. The liquid crystal material is DF03 from Chisso Corp. Voltages (rms values) are varied between 4 and 7 V. Phases are plotted modulo of the wavelengths. The asymmetric deformation is easily seen. Vertically, the

Table 9.1 Parameters for the three-dimensional director field simulations.

Elastic constant K_1	15 pN
Elastic constant K_2	10 pN
Elastic constant K_3	20 pN
Pretilt angle (deg)	5
Rotational viscosity γ	0.01 Pas
Dielectric anisotropy $\Delta\varepsilon$	2
Dielectric constant ε_0	5.5
Internal simulation time	1500 ms
Refractive index n_0	1.5
Birefringence Δn	0.2246
Dielectric layer	0.4 μm, $\varepsilon = 2$
Cell thickness d	50 μm
Hole diameter	140 μm
Mesh size	$100 \times 100 \times 50$

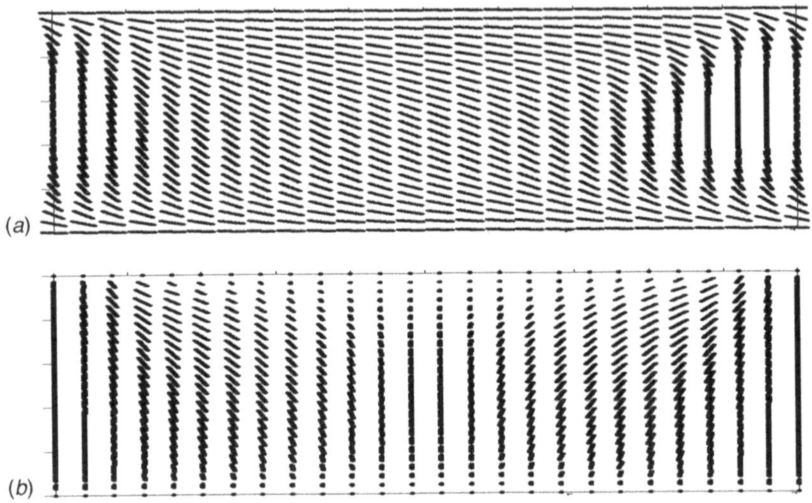

Figure 9.24 Cross-sections of the director fields for different directions in a liquid crystal lens. The electrode hole is on the top of the structure. (*a*) Section along the alignment direction. (*b*) Section perpendicular to the alignment direction. In the equilibrium the liquid crystal director field is asymmetrically deformed and shows all kinds of deformations: twist, tilt, and splay. The field size is 170 μm × 50 μm.

profile is symmetric and horizontally it is not. This is a consequence of the symmetry of the deformation profiles perpendicular and parallel to the alignment direction, as shown in the simulation results in Figure 9.24. For low voltages the center of the lens is at the right and moves to the left as the voltage is increased. The number of fringes increases from 4 to 5 V, and then decreases again when higher voltages are applied. Best performance is reached for a voltage at which the lens center is slightly off axis at about 5.5 V (not shown). We see that the focal lengths, and hence the number of fringes in the field, change only slightly in the operating voltage range. The optical properties are determined by the phase shift created over the cell thickness. At each position within the diameter of the lens, one has to sum all retardations. For strong electric fields, the fringing at the electrode edges is large, which leads to stronger deformation inside the electrode hole area. However, stronger electric fields lead also to much more complicated director deformations with twist. Twisted structures tend to turn linear polarized light of one polarization direction out of the plane. To prevent from this, the twisted texture should fulfil waveguide conditions. Under such condition the state of polarization changes less and is not lost for the functionality of the lens. For a fixed liquid crystal lens geometry, the numerical aperture is given by the maximum difference of retardation between the lens rim and the center. It becomes clear that a certain diameter-to-thickness ratio will lead to the most promising results. For a simple system the thickness should be about one-third of the diameter in order to work best (Nose et al., 1991). The adaptive lens has diffraction-limited working conditions only for a single voltage. Around that voltage the focal length changes only a few percent and has a plateau (Nose and Sato, 1989; Scharf et al.,

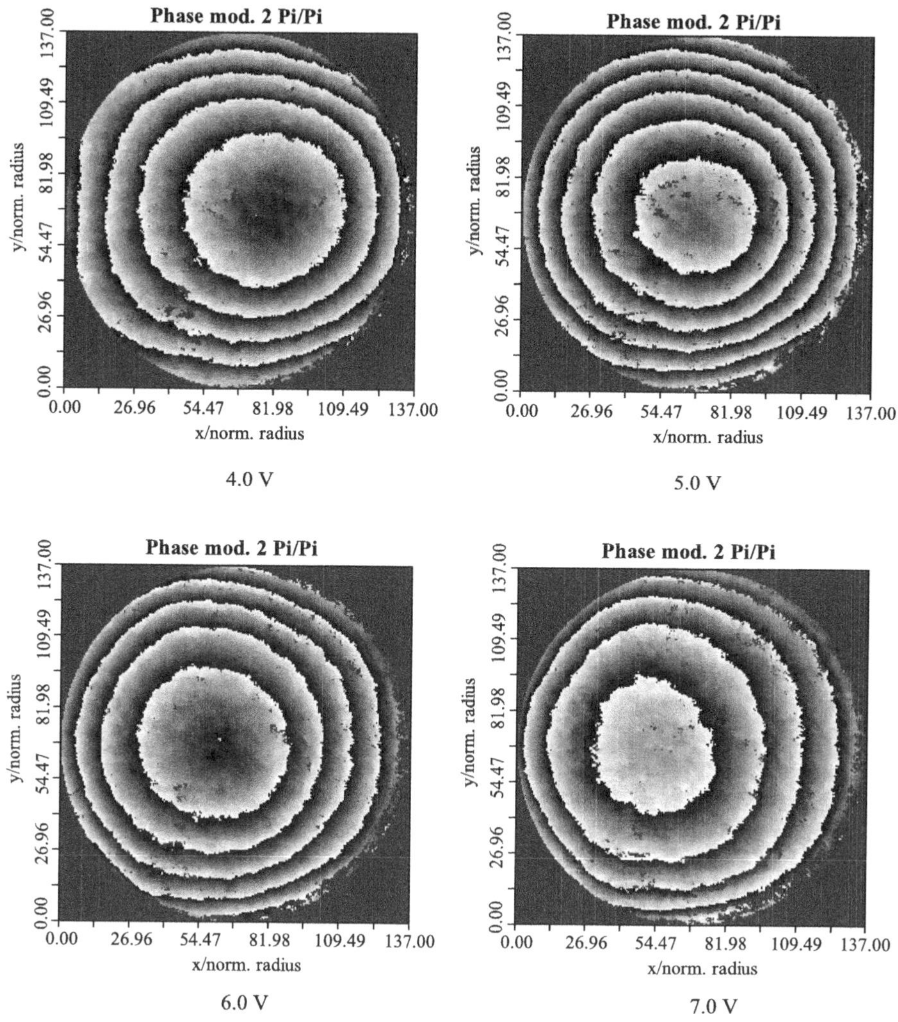

Figure 9.25 Interferometric measured phase profiles for a liquid crystal lens with a single patterned electrode arranged in a hexagonal array. Phase shifts for different voltages are shown modulo the wavelengths (thickness 50 μm, diameter 140 μm, liquid crystal mixture DF03, $\lambda = 633$ nm).

2000). For a remarkable change of focal lengths the quality worsens dramatically, as can be derived from Figure 9.25.

Optimization of a lens system based on this concept is difficult. Research has been carried out and a lot of particular approaches have been researched to develop design rules. All liquid crystal parameters are involved, which complicates the situation enormously. In addition, the system has the great disadvantage of being very slow. Several seconds are needed for liquid crystal layers of 50 μm to establish the equilibrium director distribution. Nevertheless, high-quality lenses have been obtained (Scharf et al.,

Table 9.2 Comparison of different liquid crystal lens concepts that allow the focal length to be changed.

Principle	Diameter [μm]	Thickness of Active Layer [μm]	Maximum N.A.	Optical Quality	Reference
Fringing field, homeotropic texture	90	22	0.15	Bad	Scharf et al., 1999
Fringing field, homeotropic texture	600	50	0.3		Gvozdarev and Newskaja, 1996
Fringing field, planar texture	750	50	0.06		Nose and Sato 1989
Fringing field, planar texture	290	25	0.08		Nose et al., 1992
Fringing field, planar texture	300	100	0.53	Good	Masuda et al., 1997
Fringing field, planar texture, elliptical electrode geometry	200–600	200	0.18	Good	Honma et al., 1999
Fringing field, planar texture	50–150	50	0.4	Diffraction limited	Scharf et al., 2000
Fringing field, planar texture	220	110	0.07	Good	Yanase et al., 2002
Hybrid structure, planar texture	450	20	0.05		Fan et al., 2005
Hybrid structure, planar texture	8000	300 + 300	0.1	Good	Wang, 2004
Fringing field, planar texture	60–120	50	0.07	Diffraction limited	Scharf et al., 2001
Spatial varying electric field, planar texture	4500	170	0.023	Good	Ye et al., 2004
Liquid crystal polymer composite	450	45	0.07		Ren et al., 2005b
Fringing field UV cured, planar texture	300–900	40–450	0.03	Good	Masuda et al., 1998a
Modal control planar texture	4000	25	0.01	Good	Naumov et al., 1998a
Immersed microlenses	150	27	0.15	Acceptable	Commander et al., 2000
Static birefringent, planar texture	150	50	0.2	Good	Scharf, 2005
Static birefringent, planar texture	1750	88	0.042	Acceptable	Stapert et al., 2003

2001; Honma et al., 2001; Wang et al., 2004). To obtain better control, additional electrodes are usually implemented. One of the new ideas is to use a multilevel electrode structure (Honma et al., 2001; Ye et al., 2004). However, even here, the complicated spatial director distribution makes an efficient optimization of the optical quality of liquid crystal microlenses difficult. The optics are remarkably influenced by the spatially varying twist angles of the director field. Three-dimensional simulations are necessary to explain reasonably the observed properties (Scharf et al., 2000; Yanase et al., 2001). It is important to note that the evaluation of quality has to be done for the lens as a whole. Analyzing sections of the birefringence along selected axes is not suitable for making judgements about all aberrations. Table 9.2 summarizes state-of-the-art microlens systems based on liquid crystal technology.

REFERENCES

Abramochkin, E.G., Vasiliev, A.A., Vashurin, P.V., Zhmurova, L.I., Ignatov, V.A., and Naumov, A.F. (1988) Controlled liquid crystal lens, *P. N. Lebedev Phys. Inst.* **194**, 18.

Asada, J. and Nishiwaki, S. (1994) Optical element converting linear polarization into circular tangential polarization, *Appl. Opt.* **34**, 4768–4771.

Bahadur, B. (Ed.) (1990–1992) Liquid Crystal – Application and Uses, Bd.1–3, World Scientific, Singapore.

Barberi, R., Giocondo, M., Sayko, G.V., and Zvedin, A.K. (1994) AFM experimental observation and fractal characterization of SiO coated plate for nematic bistable anchoring, *J. Phys. Cond. Mat.* **6**, A275–A278.

Berreman, D.W. (1973) Alignment of liquid crystals by grooved surfaces, *Mol. Cryst. Liq. Cryst.* **23**, 215–231.

Billings, B.H. (1975) Visual Fourier-transform spectroscopy with single crystal plate, *J. Opt. Soc. Am.* **65**, 817–824.

Blinov, L.M. and Chrigrinov, V.G. (1994) *Electrooptic Effects in Liquid Crystal Materials*, Springer, New York.

Boer, G., Scharf, T., and Dändliker, R. (2002a) Compact static Fourier transform spectrometer with large field of view based on liquid crystal technology, *Appl. Opt.* **41**, 1400–1407.

Boer, G. and Scharf, T. (2002b) Polarization ray tracing in twisted liquid crystal systems, *Mol. Cryst. Liq. Cryst.* **375**, 301–311.

Boer, G., Ruffieux, P., Scharf, T., Seitz, P., and Dändliker, R. (2004) Compact liquid crystal polymer Fourier transform spectrometer, *Appl. Opt.* **43**, 2201–2208.

Bomzon, Z., Kleiner, V., and Hasman, E. (2001) Formation of radially and azimuthally polarized light using space-variant subwavelength metal stripe gratings, *Appl. Phys. Lett.* **79**, 1587–1589.

Breault Research Organization (2005) *Advanced Systems Analysis Program ASAP*™, Tuscon, Arizona.

Brinkley, P.F., Kowel, S.T., and Chu, C. (1988) Liquid crystal adaptive lens: beam translation and field meshing, *Appl. Opt.* **27**, 4578.

Brosseau, C. (1998) *Fundamentals of Polarized Light*, Wiley, New York.

Bryan-Brown, G.P., Brown, C.V., Sage, I.C., and Hui, V.C. (1998) Voltage dependent anchoring of nematic liquid crystal on a grating surface, *Nature* **392**, 356–369.

Chamberlin, J. (1979) *The Principle of Interferometric Spectroscopy*, Wiley, Chichester.

Commander, L.G., Day, S.E., and Selviah, D.R. (2000) Variable focal length microlenses, *Opt. Comm.* **177**, 157–170.

Courtial, J., Patterson, B.A., Harvey, A.R., Sibbett, W., and Padgett, M.J. (1996) Design of a static Fourier-transform spectrometer with increased field of view, *Appl. Opt.* **35**, 6698–6702.

Courtial, J., Patterson, B.A., Hirst, W., Harvey, A.R., Duncan, A.J., Sibbett, W., and Padgett, M.J. (1997) Static Fourier-transform ultraviolet spectrometer for gas detection, *Appl. Opt.* **36**, 2813–2817.

Daly, D. (2001) *Microlens Arrays*, Taylor & Francis, New York.

Descrovi, E., Vaccaro, L., Nakagawa, W., Aeschimann, L., Staufer, U., and Herzig, H.P. (2004) Collection of transverse and longitudinal fields by means of aperturelss nanoprobes with different metal coating characteristics, *Appl. Phys. Lett.* **85**, 5340–5342.

Drzaic, P.S. (1995) *Liquid Crystal Dispersions*, World Scientific, Singapore.

Dunmore, F.J. and Hanssen, L.M. (1998) Miniature Fourier instrument for radiation thermometry, presented at the Fourier Transform Spectroscopy, 11th International Conference.

Fan, Y.H., Ren, H., Liang, X., Wang, H., and Wu, S.T. (2005) Liquid crystal microlens arrays with switchable positive and negative focal lengths, *J. Disp. Technol.* **1**, 151–156.

Françon, M. and Mallick, S. (1971) *Polarization Interferometers*, Wiley, New York.

Photon Engineering (2005) FRED simulation software.

de Gennes, P.G. and Prost, J. (1993) *The Physics of Liquid Crystals*, 2nd edn, Oxford University Press, Oxford.

Gwozdarev, A. and Newskaja, G.E. (1996) Homeotropic-aligned liquid crystal microlens properties, *SPIE* **2731**, 214–219.

Gvozdarev, A. and Newskaya, G. (1999) Optical properties of homogeneously and hybrid-aligned liquid crystal microlenses. *Mol. Cryst. Liq. Cryst.* **329**, 693–700.

Hain, M., Glockner, R., Bhattacharya, S., and Tschudi, T. (2001) Fast switching liquid crystal lenses for a dual focus digital versatile disc pickup, *Opt. Comm.* **188**, 291–299.

Harvey, R., Begbie, M., and Padgett, M.J. (1994) Stationary Fourier transform spectrometer for use as teaching tool, *Am. J. Phys.* **62**, 1033–1036.

Hecht, E. (1998) *Optics*, 3rd edn, Addison-Wesley, Reading.

Honma, M., Masuda, S., Nose, T., and Sato, S. (1997) Influence of the elastic constants on the optical properties of liquid crystal microlenses, *Jpn J. Appl. Phys.* **36**, 2765–2770.

Honma, M., Nose, T., and Sato, S. (1999) Optical properties of anamorphic liquid crystal microlenses and their application for laser diode collimation, *Jpn J. Appl. Phys.* **38**, 89–94.

Honma, M., Yamaguchi, R., and Sato, S. (2000) Application of a circularly homogeneously aligned liquid crystal cell to real time measurement of twist angles in twisted nematic lqiuid crystal cells, *Jpn J. Appl. Phys.* **39**, 2727–2731.

Honma, M., Nose, T., and Sato, S. (2001) Improved aberration properties of liquid crystal microlenses using stacked electrode structure, *Jpn J. Appl. Phys.* **40**, 1322–1327.

Honma, M. and Nose, T. (2004) Liquid crystal depolarizer consisting of randomly aligned hybrid orientation domains, *Appl. Opt.* **43**, 4667–4671.

Iizuka, K. (2002) *Elements of Photonics*, Vols 1 & 2, Wiley, New York.

Ishihara, S. (2005) How far has the molecular alignment of liquid crystals been elucidated? *J. Display Technol.* **1**, 30–40.

de Jule, M.C. and Riza, N.A. (1994) Three-terminal adaptive nematic liquid crystal lens device, *Opt. Lett.* **19**, 1013–1015.

Khan, S.A. and Riza, N.A. (2004) Demonstration of 3-dimensional wide angle laser beam scanner using liquid crystals, *Opt. Exp.*, **12**, 868–882.

Kim, S.R., Teixeira, A.I., Nealey, P.F., Wendt, A.E., and Abbott, N.L. (2002) Fabrication of polymeric substrates with well-defined nanometer-scale topography and tailored surface chemistry, *Adv. Mater.* **14**, 1468–1472.

Kornreich, P., Kowel, S.T., and Nouhi, A. (1984) Adaptive spherical lens, *Appl. Opt.* **23**, 2774–2777.

Lambda Research Corp. (2004) *TRACE PRO.*

Lee, J.H., Kim, H.R., and Lee, S.D. (1999) Polarization-insensitive wavelengths selection in an axially symmetric liquid crystal Fabry–Perot filter, *Appl. Phys. Lett.* **75**, 859–861.

Lesso, J.P., Duncam, A.J., Sibbett, W., and Padgett, M.J. (2000) Aberrations introduced by a lens made from a birefringnet material, *Appl. Opt.* **39**, 592–598.

Levy, U., Tsai, C.H., Pang, L., and Fainman, Y. (2004) Engineering space-variant inhomogeneous media for polarization control, *Opt. Lett.* **29**, 1718–1720.

Loktev, M. Yu., Naumov, A.F., Guralnik, I.R., and Vdovin, G. (1998) Liquid-crystal adaptive lenses with modal control, *Opt. Lett.* **23**, 992–994.

Love, G.D. (1993) Liquid-crystal phase modulator for unpolarized light, *Appl. Opt.* **32**, 2222–2223.

Love, G.D., Major, J.V., and Purvis, A. (1994) Liquid-crystal prisms for tip-tilt adaptive optics, *Opt. Lett.* **19**, 1170–1172.

Masuda, S., Sato, S., and Nose, T. (1991) Optical properties of a liquid crystal microlens with a symmetric electrode structure, *Jpn J. Appl. Phys.* **30**, L2110–L2112.

Masuda, S., Fujioka, S., Honma, M., and Sato, S. (1996) Dependence of optical properties on the device and material parameters in liquid crystal microlenses, *Jpn J. Appl. Phys.* **35**, 4668–4672.

Masuda, S., Takahashi, S., Nose, T., Sato, S., and Ito, H. (1997) Liquid-crystal microlens with a beam-steering function, *Appl. Opt.* **36**, 4772–4778.

Masuda, S., Nose, T., and Sato, S. (1998a) Optical properties of an UV-cured liquid-crystal microlens array, *Appl. Opt.* **37**, 2067–2073.

Masuda, S., Nose, T., and Sato, S. (1998b) Optical properties of a polymer-stabilized liquid crystal microlens. *Jpn J. Appl. Phys.* **37**, L1251–L1253.

Montarou, C. and Gaylord, T.K. (1999) Analysis and design of modified Wollaston prisms, *Appl. Opt.* **33**, 6604–6613.

Montarou, C. and Gaylord, T. (2000) Analysis and design of compact, static Fourier-transform spectrometers, *Appl. Opt.* **39**, 5762–5767.

Mori, H., Itoh, Y., Nishiura, Y., Nakamura, T., and Shinagawa, Y. (1996) Performance of a novel optical compensation film based on negative birefringence of discotic compound for wide viewing angle twisted nematic liquid crystal displays, *Jpn J. Appl. Phys.* **26**, 143–147.

Naumov, A.F. Loktev, M. Yu., Guralnik, I.R., and Vdovin, G.V. (1998a) Liquid crystal adaptive lenses with modal control, *Opt. Lett.* **23**, 992–994.

Naumov, A.F. and Vdovin, G.V. (1998b) Multichannel LC-based wavefront corrector with modal influence functions, *Opt. Lett.* **23**, 1550–1552.

Naumov, A.F., Love, G.D., Loktev, M.Y., and Vladimirov, F.L. (1999) Control optimization of spherical modal liquid crystal lenses. *Opt. Exp.* **4**, 344–352.

Niv, A., Biener, G., Kleiner, V., and Hasman, E. (2003) Formation of linearly polarized light with axial symmetry by use of space-variant subwavelength gratings, *Opt. Lett.* **28**, 510–512.

Nose, T. and Sato, S. (1989) A liquid crystal microlens obtained with a non-uniform electric field, *Liq. Cryst.* **5**, 1425–1433.

Nose, T., Sato, S., Yamaguchi, R., and Yanase, S. (1989) Relationship between lens properties and director orientation in a liquid crystal lens, *Liq. Cryst.* **5**, 1435–1442.

Nose, T., Masuda, S., and Sato, S. (1991) Optical properties of a microlens with symmetric electrode structure, *Jpn J. Appl. Phys.* **30**, L2110–L2112.

Nose, T., Masuda, S., and Sato, S. (1992) A liquid crystal microlens with hole-patterned electrodes on both sides, *Jpn J. Appl. Phys.* **31**, 1643–1646.

Nussbaum, P., Voelkel, R., and Herzig, H.P. (1997) Design, fabrication and testing of microlens arrays for sensors and microsystems, *Pure Appl. Opt.* **6**, 617–636.

Oswald, P. and Pieranski, P. (2002) *Les Cristaux Liquides* (in French), Gordon and Breach, Paris.

Padgett, M.J., Harvey, A.R., Duncan, A.J., and Sibbett, W. (1994) Single pulse fourier transform spectrometer having no moving parts, *Appl. Opt.* **33**, 6035–6040.

Patel, J.S. (1991) Polarization insensitive tunable liquid-crystal etalon filter, *Appl. Phys. Lett.* **59**, 1314–1316.

Patterson, B.A., Antoni, M., Courtial, J., Duncan, A.J., Sibbett, W., and Padgett, M.J. (1996) An ultra-compact static Fourier-transform spectrometer based on a single birefringent component, *Opt. Commun.* **130**, 1–6.

Photon Engineering LLC (2005) *Advanced optical engineering software program FRED*, Tuscon, Arizona.

V. Prakash, G., Kaczmarek, M., Dyadyusha, A., Baumberg, J.J., and D'Alessandro, G. (2005) Control of topological defects in microstructured liquid crystal cells, *Opt. Express* **13**, 2201–2209.

Prunet, S., Journet, B., and Fortunato, G. (1999) Exact calculation of the optical path difference and description of a new birefringent interferometer, *Opt. Eng.* **38**, 983–990.

Ren, H., Wu, J.R., Fan, Y.H., Lin, Y.H., and Wu, S.T. (2005a) Hermaphroditic liquid crystal microlens, *Opt. Lett.* **30**, 376–379.

Ren, H., Wu, J.R., Fan, Y.H., Lin, Y.H., and Wu, S.T. (2005b) Tunable-focus microlens array using nanosized polymer-dispersed liquid crystal droplets, *Opt. Comm.* **247**, 101–106.

Riza, N.A. and Khan, S.A. (2003) Programmable high speed polarization multiplexed optical scanner, *Opt. Lett.* **28**, 561–563.

Saleh, B.E.A. and Teich, M.C. (1991) *Fundamentals of Photonics*, Wiley, New York.

Sato, S. (1979) Liquid-crystal lens-cells with variable focal length, *Jpn J. Appl. Phys.* **18**, 1679–1684.

Scharf, T., Fontannaz, J., Bouvier, M., and Grupp, J. (1999) An adaptive microlens formed by homeotropic aligned liquid crystal with positive dielectric anisotropy, *Mol. Cryst. Liq. Cryst.* **331**, 2095–2103.

Scharf, T., Kipfer, P., Bouvier, M., and Grupp, J. (2000) Diffraction limited liquid crystal microlenses with planar alignment, *Jpn J. Appl. Phys.* **39**, 6629–6636.

Scharf, T., Cottier, K., and Dändliker, R. (2001) High quality adaptive liquid crystal microlenses, *Mol. Cryst. Liq. Cryst.* **366**, 2265–2272.

Scharf, T., Shlayen, A., Gernez, C., Basturk, N., and Grupp, J. (2004) Liquid crystal alignment on replicated nanostructured surfaces, *Mol. Cryst. Liq. Cryst.* **411**, 135–145.

Scharf, T. (2005) Static birefringent microlenses, *Opt. Laser Eng.* **43**, 317–327.

Shintech Inc. (2006) *LCD Master*, 90 Ogo-oku, Tabuse-Cho, Kumage-Gun, Yamaguchi-Ken, 742–1512, Japan.

Shurcliff, W.A. (1962) *Polarized Light*, Harvard University Press, Cambridge.

Soref, R.A. and McMahon, D.H. (1980) Total switching of unpolarized fiber light with a four-port electro-optical liquid-crystal device, *Opt. Lett.* **5**, 147–149.

Stalder, M. (1996) Active and passive components using liquid crystals, *SPIE* **2783**, 30–39.

Stalder, M. and Schadt, M. (1996a) Linear polarized light with axial symmetry generated by liquid crystal polarization converters, *Opt. Lett.* **21**, 1948–1950.

Stalder, M. and Schadt, M. (1996b) Polarization converters based on liquid crystal devices, *Mol. Cryst. Liq. Cryst.* **282**, 343–353.

Stapert, H.R., del Valle, S., Verstegen, E.J.K., van der Zande, B.M.I., Lub, J., and Stallinga, S. (2003) Photoreplicated anisotropic liquid-crystalline lenses for aberration control and dual-layer readout of optical discs, *Adv. Func. Mater.* **13**, 732–738.

Vdovin, G., Loktev, M., and Naumov, A. (2003) On the possibility of intraocular adaptive optics, *Opt. Exp.* **11**, 810–817.

Wang, B., Ye, M., and Sato, S. (2004) Lens of electrically controllable focal length made by a glass lens and a liquid crystal layers, *Appl. Opt.* **43**, 3420–3425.

Yanase, S., Ouchi, K., and Sato, S. (2001) Molecular orientation analysis of a design concept for optical properties of liquid crystal microlenses, *Jpn J. Appl. Phys.* **40**, 6514–6521.

Yanase, S., Ouchi, K., and Sato, S. (2002) Imaging properties of double-layered liquid crystal microlens, *Jpn J. Appl. Phys.* **41**, 3836–3840.

Yang, D.K. (2003) A simulation study of a liquid crystal optical switch based on total internal reflection, *J. Opt. A: P. Appl. Opt.* **5**, 402–408.

Ye, M. and Sato, S. (2001) Transient properties of a liquid crystal microlens, *Jpn J. Appl. Phys.* **40**, 6012–6016.

Ye, M. and Sato, S. (2003) Liquid crystal lenses with insulator layers for focusing light waves of arbitrary polarization, *Jpn J. Appl. Phys.* **42**, 6439–6440.

Ye, M., Wang, B., and Sato, S. (2004) Liquid crystal lens with focal length that varies in a wide range, *Appl. Opt.* **43**, 6407–6412.

Yeh, P. and Gu, C. (1999) *Optics of Liquid Crystal Displays*, Wiley, New York.

10

DIFFRACTIVE OPTICS WITH ANISOTROPIC MATERIALS

10.1 INTRODUCTION

In scalar diffraction theory, the diffractive optical element with a given phase profile is modeled as a thin element with a complex amplitude transmittance. The element retards the incident wavefront, and propagation of the new wavefront is modeled by appropriate scalar formulation, such as angular spectrum, Fresnel diffraction, or Fraunhofer diffraction (Goodman, 1996). Usually, diffractive optical elements are used in optical elements to split or deviate rays, to disperse rays, or to make uniform light fields. These are applications where the optical element is set in an isotropic environment. To be more specific, the light propagates from an isotropic material through the diffractive optical element and then again in the isotropic material. Because the polarization states decouple in isotropic materials, the light propagation can easily be modeled by taking onto account the polarization properties of the diffractive element only. One needs to know the polarization dependent complex amplitude transmittance function, which, for the simplest case, is a 2×2 matrix with complex elements.

Usually there are complex merit functions for a diffractive optical element. If for instance a beam steering device is required that divides a beam into beams of equal intensities, the one criteria of the quality might be the sum of these intensities and, in additional, the uniformity. In general, rather complicated design mechanisms have to be used to find a convenient design (Herzig, 1998). Although micro-optics is widely used, birefringent micro-optical elements are hard to find. The expense of fabrication in crystals prevents then from being widely used. For liquid crystal

Polarized Light in Liquid Crystals and Polymers. By Toralf Scharf

polymers a new way was developed recently to fabricate cost-effective elements (Broer et al., 1989; Broer, 1995). In the following sections we will discuss basic diffractive optical elements made of birefringent materials. In this category fall binary gratings and diffractive lenses, multilevel gratings, and fan-out elements.

Binary birefringent elements will first be considered. There are different kinds of systems that are of interest. Let us first present an overview of the main parameters that have to be respected when optical properties are studied. One important aspect is the fabrication technology of such devices. Using crystal materials, gratings can be fabricated by etching with high precision. If switchable systems made of liquid crystal material are considered, then "fringing" effects and defect structures have to be taken into account. This is particularly important when the smallest structure size of the element becomes comparable to the thickness of the element.

10.2 PRINCIPLES OF FOURIER OPTICS

We discuss here briefly light propagation in the far-field behind an optical element (Iizuka, 2002). Referring to Figure 10.1, the field distribution from a source is observed on a screen. The field distribution on the screen is called a diffraction pattern. We will demonstrate that the diffraction pattern can be elegantly expressed by the Fourier transform of the source. Let $E(x_i, y_i)$ represent the field at point P on the screen placed a distance z_i away from the source field $E(x_0, y_0)$. The distributed source $E(x_0, y_0)$ is considered as an ensemble of point sources. Each point source radiates a spherical wave. The field at the observation point P is comprised of contributions from an ensemble of fields radiated from all the point sources. The contribution of the point source located at (x_0, y_0) to point P at (x_i, y_i) is

$$\mathrm{d}E(x_i, y_i) = \frac{\mathrm{e}^{ikr}}{r} E(x_0, y_0)\mathrm{d}x_0\,\mathrm{d}y_0 \tag{10.1}$$

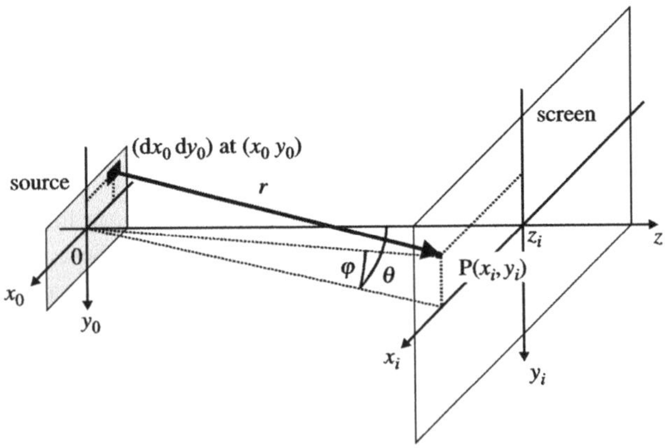

Figure 10.1 Field distribution from a source observed on a screen: general geometry of the diffraction problem.

where $E(x_0, y_0)$ is the magnitude of the point source located at (x_0, y_0) and r is the distance between (x_0, y_0) and (x_i, y_i). The distance r is expressed as

$$r = \sqrt{z_i^2 + (x_i - x_0)^2 + (y_i - y_0)^2} \tag{10.2}$$

The contribution of the spherical waves from all the point sources to $E(x_i, y_i)$ is

$$E(x_i, y_i) = K \iint \frac{e^{ikr}}{r} E(x_0, y_0) dx_0\, dy_0 \tag{10.3}$$

This equation is known as the Fresnel–Kirchhoff diffraction formula. The amplitude of the diffracted field is inversely proportional to its wavelength and is expressed as (Goodman, 1996)

$$K = \frac{1}{i\lambda} \tag{10.4}$$

If the point of observation is far enough away or in the vicinity of the z-axis (paraxial),

$$z_i^2 \gg (x_i - x_0)^2 + (y_i - y_0)^2 \tag{10.5}$$

then the distance r can be simplified by the binomial expansion as

$$r = z_i\left(1 + \frac{(x_i - x_0)^2 + (y_i - y_0)^2}{2z_i^2}\right) \tag{10.6}$$

which can be rewritten as

$$r \cong z_i + \frac{x_i^2 + y_i^2}{2z_i^2} - \frac{x_i x_0 + y_i y_0}{z_i} + \frac{x_0^2 + y_0^2}{2z_i^2} \tag{10.7}$$

The region z_i for which the approximate expression Equation (10.7) is valid is called the Fresnel region or the near-field region. As the distance is further increased in the z-direction, the last term in Equation (10.7) becomes negligible for the finite size of the source. This region of z_i is called the Fraunhofer region or far-field region. In this chapter we are concerned about the far-field. In the far-field region, the approximation for r is

$$r \cong z_i + \frac{x_i^2 + y_i^2}{2z_i^2} - \frac{x_i x_0 + y_i y_0}{z_i} \tag{10.8}$$

By substituting this approximation into the exponential term of the Fresnel–Kirchhoff diffraction formula, Equation (10.3), the field becomes

$$E(x_i, y_i) = \frac{1}{i\lambda z_i} e^{ik}(z_i + (x_i^2 + y_i^2)/2z_i) \iint_{-\infty}^{\infty} E(x_0, y_0) e^{-i2\pi(f_x x_0 + f_y y_0)} dx_0 \, dy_0$$

with

$$f_x = \frac{x_i}{\lambda z_i} \quad \text{and} \quad f_y = \frac{y_i}{\lambda z_i} \tag{10.9}$$

The integral is the two-dimensional Fourier transform of the field in the x, y domain into the f_x, f_y domain:

$$F\{g(x, y)\} = \iint_{-\infty}^{\infty} g(x, y) e^{-i2\pi(f_x x_0 + f_y y_0)} dx \, dy \tag{10.10}$$

Or in mathematical terms, the diffraction pattern is

$$\begin{aligned} E(x_i, y_i) = {} & \frac{1}{i\lambda z_i} e^{ik(z_i + (x_i^2 + y_i^2)/2z_i)} \\ & \times F\{E(x_0, y_0)\}_{f_x = x_i/\lambda z_i, f_y = y_i/\lambda z_i} \end{aligned} \tag{10.11}$$

where F denotes the Fourier transform. In short, the Fraunhofer diffraction pattern is the Fourier transform of the source field. Sometimes, the angular distribution rather than the planar distribution is desired. For this case, the azimuth angle φ and the elevation angle θ, measured with respect to the center of the source field, are approximated as $\sin\theta = x_i/z_i$ and $\sin\varphi = y_i/z_i$. Hence, the values in Equation (10.9) are

$$f_x = \frac{\sin\theta}{\lambda} \quad \text{and} \quad f_y = \frac{\sin\varphi}{\lambda} \tag{10.12}$$

The branch of optics that can be analyzed by means of the Fourier transform is categorized as Fourier optics. We will briefly explore the meaning of f_x and f_y in Equation (10.9). For simplicity, only the distribution in the (y_i, z_i) plane will be considered. Consider a typical phase and amplitude distribution of the field diffracted from an aperture source whose dimensions are much smaller than the distance to the screen. In the region far from the aperture, the phase distribution is more like that of a spherical wave. With the source placed at the origin, the phase front along the y-axis near $y_i = 0$ is always parallel to the y_i-axis. In the vicinity of this point, there is no variation in the phase of the field with respect to y. Hence, the field has zero spatial frequency at $y_i = 0$. (The variation of the field amplitude with y is normally much smaller than that of the phase.) As shown in Figure 10.2,

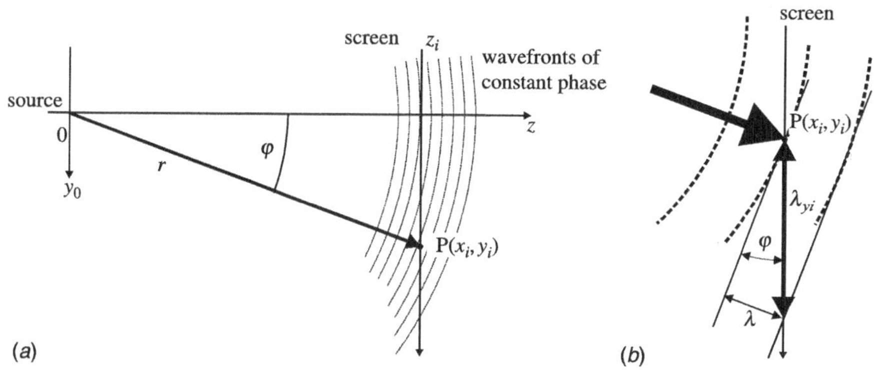

Figure 10.2 Field distribution from a source observed on a screen. (*a*) Distribution of the field in the $E(y_i, z_i)$ plane shown as a cut through Figure 10.1. (*b*) Detail at the observation point P(x_i, y_i) to illustrated the spatial frequency concept.

the change in the variation of phase increases as the point of observation P moves along the y_i-axis, such that, eventually, the wavelength λ_{yi} measured along the y_i-axis will approach the wavelength of free space. Mathematically, λ_{yi} at the observation point $P(y_i, z_i)$ is $\lambda_{yi} = \lambda/\sin\varphi$, as can be seen in Figure 10.2. The field located at P, therefore, has a spatial frequency of

$$f_{yi} = \frac{1}{\lambda_{yi}} = \frac{\sin\varphi}{\lambda} \cong \frac{y_i}{z_i\lambda} \tag{10.13}$$

The concept of spatial frequency is often used in optics to explain diffraction properties and the resolution of optical systems.

10.3 POLARIZATION PROPERTIES

To extend the abovementioned concept and include polarization properties we follow an approach that is a natural extension, based on a matrix concept (Moreno et al., 2004). It is based on the Jones matrix formalism and valid for fully polarized coherent monochromatic illumination. The approach is based on the Jones matrix introduced in Chapter 4 and describes, simultaneously, polarization and diffraction properties of the Fourier plane. Therefore the intensity and the local state of polarization in the Fourier plane are directly obtained. Let us consider a spatial variant thin polarization element. It can be described by a Jones matrix **J** with spatial dependence x, y as

$$\mathbf{J} = \begin{pmatrix} J_{11}(x, y) & J_{12}(x, y) \\ J_{21}(x, y) & J_{22}(x, y) \end{pmatrix} \tag{10.14}$$

When this element is illuminated with totally polarized light described by a Jones vector $\mathbf{V}_0$, the polarization state behind the diffractive polarization element is given by a Jones vector

$$\mathbf{V}_e(x,y) = \begin{pmatrix} V_e^s \\ V_e^p \end{pmatrix} = \begin{pmatrix} J_{11}(x,y) & J_{12}(x,y) \\ J_{21}(x,y) & J_{22}(x,y) \end{pmatrix} \begin{pmatrix} V_0^s \\ V_0^p \end{pmatrix} \tag{10.15}$$

where the subscripts s and p denote the two orthogonal polarizations in the horizontal and vertical direction, respectively. The Fourier diffraction pattern, which can be obtained through the Fraunhofer approximation as stated above, can be calculated for each component separately; thus

$$\mathbf{T}_e^{s,p}(f_x,f_y) = \frac{1}{i\lambda z_i} e^{ik(z_i+(x_i^2+y_i^2)/2z_i)} F\{\mathbf{V}_e^{s,p}(x_0,y_0)\}_{f_x=x_i/\lambda z_i,\, f_y=y_i/\lambda z_i} \tag{10.16}$$

As the Jones matrix and the Fourier transform are both linear operations it is possible to write the Jones vector in the Fourier plane in terms of the input light as

$$\mathbf{T}_e(f_x,f_y) = \begin{pmatrix} T_{11}(f_x,f_y) & T_{12}(f_x,f_y) \\ T_{21}(f_x,f_y) & T_{22}(f_x,f_y) \end{pmatrix} \begin{pmatrix} V_0^s \\ V_0^p \end{pmatrix} \tag{10.17}$$

The Jones matrix $\mathbf{T}(f_x, f_y)$ relates the spatially variant output Jones vector at the Fourier plane with the spatially uniform Jones vector at the input. This matrix is given by the Fourier transform of each of the components of the matrix and represents simultaneously the polarization and diffraction properties.

As mentioned above, when light propagates from an isotropic material through the anisotropic diffractive optical element and then again in the isotropic material, the polarization states decouple in the isotropic materials. Light propagation can easily be modeled by using Jones matrices in the isotropic media, even if one needs to calculate the properties of the elements with more sophisticated methods such as the 4×4 matrix method or FDTD approaches (see Chapters 4 and 5 for details). The difficulty sometimes lies in the calculation of transfer matrices between the birefringent element and the isotropic environment for different methods.

10.4 DIFFRACTION AT BINARY GRATINGS

In the discussion of liquid crystal gratings, two cases might be distinguished that can be treated with different optical design tools. Here we specifically speak about the treatment of the element itself. Systems exist that do not show a gradient index distribution of the refractive index and so can be handled with a classical thin element ray-tracing approach. This is the case for surface relief structures fabricated in crystals or polymer liquid crystals (Xu et al., 1995; Nieuborg et al., 1997) and for the refractive elements discussed in Chapter 9. In the same category fall the optical elements based on spatial light modulators (Holoeye Photonics, 2005). Filled systems, where the liquid crystals are incorporated into cells with microstructured surfaces, might also be considered for this category (Wang et al., 2000). The birefringent material is homogeneous in such cases. Only for very high spatial frequency, when the period is comparable to the wavelengths of light, do rigorous optical simulation methods have to be applied (Rokushima and Yamakita, 1984). On the other hand the configurations with spatial gradients are more often found when liquid crystals are used. Here complicated simulation tools are needed to achieve correct description of the optical properties. Two tasks are important: (1) to find the correct distribution of the director distribution of the liquid crystal with the gradient structure, and (2) to simulate the optical properties with rigorous methods. We want to continue the discussion with a basic description of binary gratings to show the limitations and possibilities of using birefringent devices to increase functionality of optical elements.

The Fraunhofer diffraction pattern of liquid crystal (NLC) binary gratings consists of a series of discrete spots corresponding to diffracted orders whose polarization states depend on the grating profile and on the type of the birefringence modulation involved Pellat-Finet (1995), Le Doucen (1998). We will discuss the different alignments and the interaction of polarized light with the grating. The gratings we are going to discuss here are assumed to be binary for the sake of simplicity. A period is then divided into two parts having different optical properties. For diffraction gratings, an amplitude and phase function are given for every point over the period. In the case of birefringent gratings, the Jones matrix can be used, which contains phases, amplitudes, and polarization information. A binary system can be described as having two Jones matrices ${}^{\mathrm{A}}\mathbf{J}$ and ${}^{\mathrm{B}}\mathbf{J}$. Each Jones matrix has four complex elements J_{ik}.

$$\mathbf{J} = \begin{pmatrix} J_{11} & J_{12} \\ J_{21} & J_{22} \end{pmatrix} \tag{10.18}$$

We want to focus our interest now on the diffraction pattern produced by such a grating. In the Fraunhofer approximation the Fourier transform of the electric field (complex) over the grating period describes the far-field properties of the grating. One can write for the transfer matrix $\mathbf{T}^{\mathrm{N}}$ of the Nth order of diffraction

$$\mathbf{T}^{\mathrm{N}} = \frac{1}{p}\int_{-p/2}^{p/2} \mathbf{J}(x)\mathrm{e}^{-\mathrm{i}\frac{2\pi}{p}xN}\,\mathrm{d}x \tag{10.19}$$

with the period p and the spatial coordinate x. The illustration in Figure 10.3 shows the principal geometry. Calculating the electric field of the diffraction pattern for the

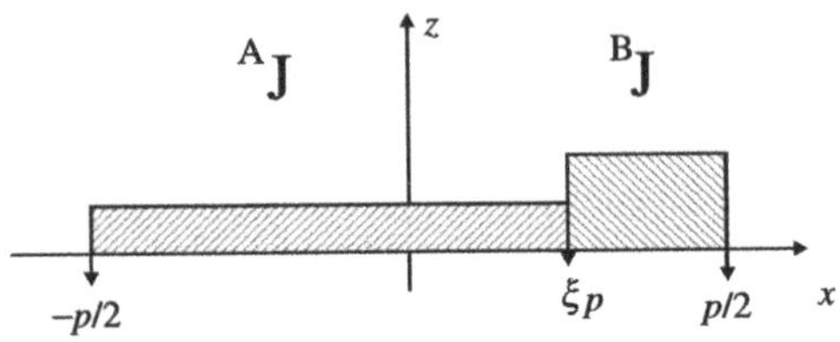

Figure 10.3 A binary grating composed of birefringent material can be described by two Jones matrices that are defined over the period p. The direction of propagation of light is z.

different orders N gives, by dividing the full period into parts, for the binary system

$$\mathbf{T}^{\mathrm{N}} = \frac{1}{p}\left[\int_{-p/2}^{\xi p} {}^{\mathrm{A}}\mathbf{J} \mathrm{e}^{-\mathrm{i}\frac{2\pi}{p}xN}\,\mathrm{d}x + \int_{\xi p}^{p/2} {}^{\mathrm{B}}\mathbf{J} \mathrm{e}^{-\mathrm{i}\frac{2\pi}{p}xN}\,\mathrm{d}x\right] \tag{10.20}$$

The constant ξ describes the repartition of the grating period between the two Jones matrices and can take values between $-\frac{1}{2}$ and $\frac{1}{2}$. Having in mind that the Jones matrix for each zone in the binary grating is constant, we find

$$\mathbf{T}^{\mathrm{N}} = \frac{1}{p}\left[{}^{\mathrm{A}}\mathbf{J}\int_{-p/2}^{\xi p} \mathrm{e}^{-\mathrm{i}\frac{2\pi}{p}xN}\mathrm{d}x + {}^{\mathrm{B}}\mathbf{J}\int_{\xi p}^{p/2} \mathrm{e}^{-\mathrm{i}\frac{2\pi}{p}xN}\mathrm{d}x\right] \tag{10.21}$$

and after integration

$$\mathbf{T}^{\mathrm{N}} = \frac{1}{p}\left[{}^{\mathrm{A}}\mathbf{J}\left|\frac{p\mathrm{i}}{2\pi N}\mathrm{e}^{-\mathrm{i}\frac{2\pi}{p}xN}\right|_{\frac{-p}{2}}^{\xi p} + {}^{\mathrm{B}}\mathbf{J}\left|\frac{p\mathrm{i}}{2\pi N}\mathrm{e}^{-\mathrm{i}\frac{2\pi}{p}xN}\right|_{\xi p}^{\frac{p}{2}}\right] \tag{10.22}$$

which leads to

$$\mathbf{T}^{\mathrm{N}} = \frac{i}{2\pi N}[{}^{\mathrm{A}}\mathbf{J}(\mathrm{e}^{-\mathrm{i}2\pi\xi N} - \mathrm{e}^{\mathrm{i}\pi N}) + {}^{\mathrm{B}}\mathbf{J}(\mathrm{e}^{-\mathrm{i}\pi N} - \mathrm{e}^{-\mathrm{i}2\pi\xi N})] \tag{10.23}$$

with $\mathrm{e}^{-\mathrm{i}\pi N} = \cos(-\pi N) + \mathrm{i}\sin(-\pi N) = \cos(\pi N) = \mathrm{e}^{\mathrm{i}\pi N}$ for natural $N = 0, \pm 1, \pm 2, \ldots$ and $\xi = (-\frac{1}{2}, \frac{1}{2})$ one finds

$$\mathbf{T}^{\mathrm{N}} = \frac{\mathrm{i}}{2\pi N}(\mathrm{e}^{i\pi N} - \mathrm{e}^{-\mathrm{i}2\pi\xi N})\,({}^{B}\mathbf{J} - {}^{A}\mathbf{J}) \tag{10.24}$$

Using the abbreviation

$$L(N,\xi) = \frac{\mathrm{i}}{2\pi N}(\mathrm{e}^{\mathrm{i}\pi N} - \mathrm{e}^{-\mathrm{i}2\pi\xi N}) \tag{10.25}$$

$$\begin{aligned} L(N,\xi) &= \frac{1}{2\mathrm{i}\pi N}(\mathrm{e}^{-\mathrm{i}\pi N 2\xi} - \mathrm{e}^{\mathrm{i}\pi N}) \\ &= \frac{\mathrm{e}^{-\mathrm{i}\pi N(\xi-\frac{1}{2})}}{2\mathrm{i}\pi N}\left(\mathrm{e}^{\frac{-\mathrm{i}\pi N(2\xi+1)}{2}} - \mathrm{e}^{\frac{\mathrm{i}\pi N(2\xi+1)}{2}}\right) \\ &= -\mathrm{e}^{-\mathrm{i}\pi N(\xi-\frac{1}{2})}\frac{\sin\left(\pi N\left(\xi+\frac{1}{2}\right)\right)}{\pi N} \end{aligned} \tag{10.26}$$

one can find a closed analytical expression for the transfer matrix $\mathbf{T}^{\mathrm{N}}$ of the Nth order as a function of the aspect ratio of the grating ξ, the Jones matrices, and the period

$$\mathbf{T}^{\mathrm{N}} = \mathrm{e}^{-\mathrm{i}\pi N(\xi-\frac{1}{2})}\frac{\sin\left(\pi N\left(\xi+\frac{1}{2}\right)\right)}{\pi N}({}^{\mathrm{A}}\mathbf{J} - {}^{\mathrm{B}}\mathbf{J}) \tag{10.27}$$

The diffraction efficiency is defined as the ratio between the incoming intensity and intensity in the diffracted order, and depends on the incoming polarization states. Let $\mathbf{V}$ be the input polarization state described as $\mathbf{V} = \begin{pmatrix} V_x \\ V_y \end{pmatrix}$ with the intensity $I_V = |V_x|^2 + |V_y|^2$ and $\mathbf{V}'$ being the polarization state at the output of the optical system with vector $\mathbf{V}' = \begin{pmatrix} V'_x \\ V'_y \end{pmatrix}$ and intensity $I_{V'} = |V'_x|^2 + |V'_y|^2$. With the Jones matrix elements of the difference Jones matrix $({}^{\mathrm{B}}\mathbf{J} - {}^{\mathrm{A}}\mathbf{J})$, V′ is found as

$$\mathbf{V}' = \begin{pmatrix} {}^{\mathrm{B}}J_{11} - {}^{\mathrm{A}}J_{11} & {}^{\mathrm{B}}J_{12} - {}^{\mathrm{A}}J_{12} \\ {}^{\mathrm{B}}J_{21} - {}^{\mathrm{A}}J_{21} & {}^{\mathrm{B}}J_{22} - {}^{\mathrm{A}}J_{22} \end{pmatrix}\begin{pmatrix} V_x \\ V_y \end{pmatrix} = {}^{\mathrm{B-A}}\mathbf{JV} \tag{10.28}$$

which leads directly to

$$\begin{aligned} \eta^N &= |L(N,\xi)|^2\left(|V'_x|^2 + |V'_y|^2\right) \\ \eta^N &= |L(N,\xi)|^2(|{}^{\mathrm{B-A}}J_{11}V_x + {}^{\mathrm{B-A}}J_{12}V_y|^2 + |{}^{\mathrm{B-A}}J_{21}V_x + {}^{\mathrm{B-A}}J_{22}V_y|^2) \end{aligned} \tag{10.29}$$

which becomes

$$\eta^N = \frac{\sin^2\left(\pi N\left(\xi + \frac{1}{2}\right)\right)}{(\pi N)^2}\left(|{}^{\mathrm{B-A}}J_{11}|^2 + |{}^{\mathrm{B-A}}J_{21}|^2\right)|V_x|^2$$
$$+ \left(|{}^{\mathrm{B-A}}J_{12}|^2 + |{}^{\mathrm{B-A}}J_{22}|^2\right)|V_y|^2$$
$$+ 2\mathrm{Re}\left(\left(\left({}^{\mathrm{B-A}}J_{12}{}^{\mathrm{B-A}}J_{11}^*\right) + \left({}^{\mathrm{B-A}}J_{21}^*{}^{\mathrm{B-A}}J_{22}\right)\right)V_y V_x^*\right) \tag{10.30}$$

Finally one finds that the polarization properties of the diffraction pattern are determined by an amplitude factor that represents the classical diffraction for binary gratings, and the difference of the two Jones matrices multiplied by the incoming polarization vector. All the interesting polarization properties could be derived by detailed discussion of this Jones matrix difference.

As an example we discuss the polarization properties of a grating where different domains rotate the polarization with different senses of rotation (Titus and Bos, 1997). The rotation of polarized light states is represented by the Jones matrix

$$\mathbf{J}^{\mathrm{rot}} = \begin{pmatrix} \cos\alpha & \sin\alpha \\ -\sin\alpha & \cos\alpha \end{pmatrix} \tag{10.31}$$

The difference Jones matrix is than given by

$$\begin{aligned} {}^{\mathrm{A}}\mathbf{J}(+\alpha) - {}^{\mathrm{B}}\mathbf{J}(-\alpha) &= \begin{pmatrix} \cos\alpha & \sin\alpha \\ -\sin\alpha & \cos\alpha \end{pmatrix} - \begin{pmatrix} \cos\alpha & -\sin\alpha \\ \sin\alpha & \cos\alpha \end{pmatrix} \\ &= \begin{pmatrix} 0 & 2\sin\alpha \\ -2\sin\alpha & 0 \end{pmatrix} = 2\sin\alpha\begin{pmatrix} 0 & 1 \\ -1 & 0 \end{pmatrix} \end{aligned} \tag{10.32}$$

With that the diffraction transfer matrix becomes

$$\mathbf{T}^{\mathrm{N}} = \mathrm{e}^{-\mathrm{i}\pi N\left(\xi - \frac{1}{2}\right)} \frac{\sin\left(\pi N\left(\xi + \frac{1}{2}\right)\right)}{\pi N} 2\sin\alpha\begin{pmatrix} 0 & 1 \\ -1 & 0 \end{pmatrix} \tag{10.33}$$

When we calculate the diffraction efficiencies that can be found with the Jones matrix difference in Equation (10.32), we obtain

$$\eta^N = \frac{\sin^2\left(\pi N\left(\xi + \frac{1}{2}\right)\right)}{(\pi N)^2} 4\sin\alpha\left(|V_x|^2 + |V_y|^2\right) \tag{10.34}$$

and with normalized intensity $I_V = |V_x|^2 + |V_y|^2 = 1$ at the input,

$$\eta^N = \frac{\sin^2\left(\pi N(\xi + \frac{1}{2})\right)}{(\pi N)^2} 4 \sin \alpha \tag{10.35}$$

Here we see that the element no longer has dependence on the state of polarization of incoming light. The system is polarization-independent. The diffraction efficiency is maximum for a rotation angle α of 90°.

Calculations like this allow the determination of the diffraction properties and show that in such cases polarization states and diffraction patterns are decoupled, in the sense that the diffraction pattern is due to the geometry and only the diffraction efficiencies are influenced by the polarization or more precisely by the local Jones matrix.

Polarization properties of birefringent phase gratings are discussed extensively in the literature (Galatalo, 1994; Mukohzaka, 1994; Pellar-Finet, 1995; Liu, 1997; Davis, 2001; Park, 2003). There is no principle difference in the discussion for systems realized with nematic or ferroelectric liquid crystals (Löfving, 1998; Brown, 2002, 2003, 2004) or application of amplitude and phase modulation (He, 1997). There are investigations done on reflective beam steering systems (Lu, 1998).

10.5 CONCEPTS AND FABRICATION

Diffractive optical elements are thin and their functionality is based on the periodicity of light waves. Thin elements can be fabricated with planar wafer technology, which is the standard for liquid crystal display fabrication. This is the reason why there are many more different concepts that exist for diffractive elements than for refractive optical components based on organic materials and liquid crystals. The most prominent example of a diffractive element made with liquid crystal and polymers is the liquid crystal display screen itself. In this case, the period is too large, and for use in diffractive optics, liquid crystal spatial light modulators with typical pixel sizes below 10 micron have found their way into several applications (Holoeye Photonics, 2005). Nearly all devices are electrically driven for tuning their optical properties. From the fabrication point of view there are differences, which we will summarize in the following. Let us state the different concepts. To form a diffractive element a spatially varying refractive index is necessary. This can be obtained with the following arrangements:

- Solid-state birefringent elements and immersed systems;
- Patterning the alignment of the liquid crystal at one or two substrate surfaces;
- Polymer dispersed gratings made by phase separation;
- Patterning the electrodes for binary or multilevel systems.

These approaches require different technologies, although all are planar and can be realized at the wafer level. In what follows we provide a brief introduction to the different subjects.

10.5.1 Solid-State Birefringent Elements and Immersed Systems

The principle of fabrication is comparable with the examples shown in Section 9.7. Surface microstructures can be immersed in anisotropic materials, birefringent optical elements can be combined with isotropic index-matching fluids, or anisotropic diffractive elements can be combined with birefringent materials (Liu, 1995; Jepsen and Gerritsen, 1996; Lester et al., 2005; Zhang et al., 2005). The difference with the examples presented in Chapter 9 is structure size. Although refractive elements should have a size much larger than the wavelengths of light, the diffractive optical components have dimensions that are comparable. For the visible region, the lateral structure size for refractive components should not be smaller than 30 μm. Diffractive components usually have a lateral dimension ranging from submicron to tens of microns (Hirabayashi et al., 1995; Strudwick and Lester, 1999). The change in dimensions also needs a different alignment technique for the liquid crystal. Although refractive elements are often aligned by rubbing or extra surface structuring, for diffractive structures the process can be easier because the structure itself can align the liquid crystal (Sakata and Nishimura, 2000). This is useful for gratings, but can cause complications for photonic crystal structures where multiple alignment states at the surface can be created due to the high symmetry of the structure. Figure 10.4 summarizes the different concepts. Fabrication is based on

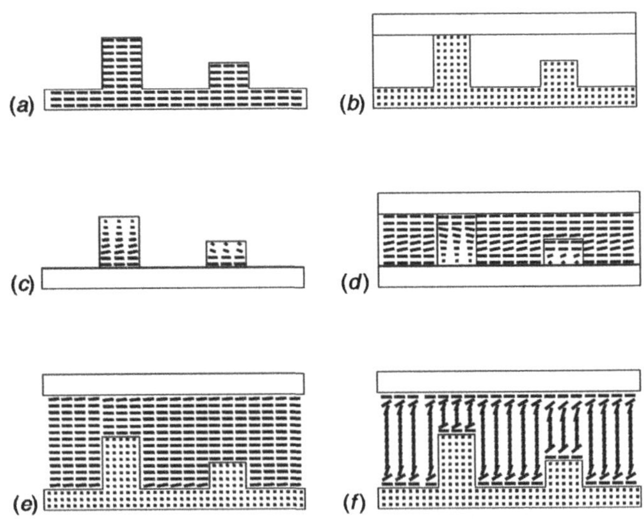

Figure 10.4 Diffractive optics with birefringent materials. The direction of the optical axis is indicated by the orientation of the small bars with the structure. Twisted and nontwisted structures are possible. Immersion of diffractive microstructures in liquid crystal leads to switchable components if electrodes are used too.

sandwiched cells with transparent electrodes that are complete over the whole surface area. Rubbing and nano- or microstructuring are used for aligning the liquid crystal. If homeotropic alignment is envisaged, polymers or silane treatments are often used to obtain an alignment of the director perpendicular to the surface. A few examples exist where birefringent elements are fabricated using photolithography and etching (Xu et al., 1995; Nieuborg et al., 1997). Twisted structures may only be fabricated in liquid crystal polymers and if one takes special care in the alignment techniques used.

10.5.2 Alignment Patterned Elements

Until now only homogeneous textures have been considered. A very elegant way to obtain grating structures is by modification of alignment on the substrate surfaces (Chen et al., 1995; Titus and Bos, 1997; Wen et al., 2002). The choice is between three main orientations. Figure 10.5 shows surfaces with different combinations of alignment patterning. To obtain such alignments different techniques have been suggested, among which alignment with polarized UV light is probably the most reliable although the materials choice in liquid crystals and alignment agent is still rather limited. With these techniques, homeotropic, planar, and tilted planar alignment can be obtained over reasonable surface areas and give all the design freedom needed (Schadt et al., 1992; Park et al., 2005). For small experimental test surfaces, atomic force microscope (AFM) pattering has become a tool of choice (Wen et al., 2002; Varghese et al., 2005). If larger areas are needed, photolithography, laser beam interference lithography, and replication techniques are used. However, not all techniques allow all kind of alignments. Patterning with an atomic force microscope gives mainly planar alignment, and the direction can be chosen. It imitates rubbing and is sometimes also called nanorubbing. Homeotropic domains are difficult to realize because on rubbed surfaces the liquid crystal prefers to lie flat on the surface. Double rubbing techniques, which protect one part of the surface and implement a different alignment on another part have also been used, but are delicate to handle. Rubbing strength has to be controlled carefully and masking has to be done with materials that will not pollute the rubbing cloth.

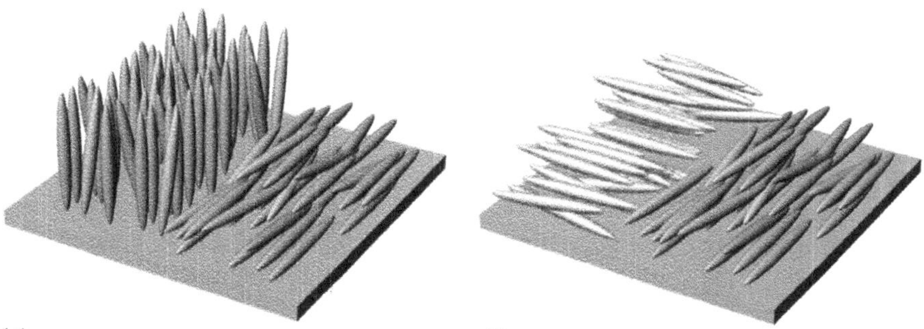

(*a*) (*b*)

Figure 10.5 Alignment patterning of liquid crystals. The planar texture can be combined with homeotropic texture as shown in (*a*) or with a different orientation of planar texture as shown in (*b*).

Patterned evaporation and ion beam surface treatments (Chaudhari et al., 1998, 2001) are further methods that can be used. A method used to generate a high-pretilt sample is chemical surface patterning. Here different chemicals are deposited on the surface, which leads to planar and homeotropic aligned domains (Park et al., 2005).

10.5.3 Switchable Diffractive Cholesteric Gratings

The cholesteric liquid crystals (cholesterics) are composed of molecules that are locally aligned preferentially in a particular direction represented by the director, which twists uniformly about an orthogonal direction. This results in a helical structure of a definite pitch. We consider a plane wavefront of linearly polarized light incident along a direction normal to the twist axis. When its electric vector is parallel to the twist axis, it emerges as a plane wavefront, whereas for the electric vector in the orthogonal state it emerges as a periodically corrugated wavefront. The latter case leads to diffraction of light as in a phase grating. A possibility for obtaining diffraction gratings is therefore to make use of the intrinsic periodicity of cholesterics (Subacius et al., 1997a; Wu et al., 2002). The grating itself is set by a cholesteric stripe pattern; no patterned electrodes are necessary. The grating can be formed due to instabilities when the liquid crystal is electrically driven (Kashnow and Bigelow, 1973) or the periodicity of the cholesteric texture can be responsible for diffraction. In the latter case the helical axis has to lie in the plane of the substrate surfaces. In such arrangements, it is possible to vary continuously a periodicity of the stripe texture by changing the applied voltage (Subacius et al., 1997b). This results in a deflection of a diffracted laser beam to a desired angle. Switching times of the order of 10 ms have been achieved. The grating direction can also be changed by applying a particular driving voltage (Fuh and Lin, 2002). When layer undulation of the cholesteric planar texture is used, two-dimensional diffraction patterns are found (Senyuk et al., 2005). The design parameters are limited to the intrinsic physical properties of the liquid crystal.

10.5.4 Polymer Dispersed Gratings

A mixture of polymer and liquid crystal can phase separate under certain conditions (Drzaic, 1995). This property can be used to make optical elements by photopolymerization (Sutherland et al., 1994; Duca et al., 1999; DeFilipo et al., 2001; Kossyrev et al., 2004; Caputo et al., 2005; Fan et al., 2005). With these methods, several high-resolution optical components have been fabricated. In particular, periodic structures in two and three dimensions have been successfully demonstrated (Escuti et al., 2003). Illumination with patterned UV illumination is a basic method for creating locally varying phase separation. The surface substrates do not need to be prestructured. Figure 10.6 shows the principal appearances of such devices after fabrication. Here the concept of mask illumination is used, which has low resolution. Holographic methods are equivalent if UV laser illumination

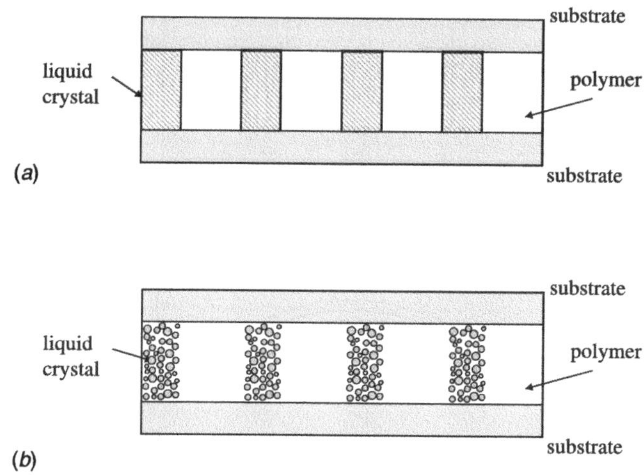

Figure 10.6 (*a*) Polymer dispersed grating with complete phase separation of polymer liquid crystal. (*b*) The phase separation can be performed with different processes that can lead also to mixed systems. The second example is a phase polymer dispersed liquid crystal polymer system.

is chosen (Bowley et al., 1999). Zones with polymer rich and liquid-crystal-rich composition formed have different refractive indices. Is is also possible to fill such gratings with liquid crystals to obtain tunability (Butler, 2000).

The variety of structure that can be fabricated is enormous. This is because of the fact that curing parameters can be chosen freely and there are many parameters that can be adjusted, such as mixture composition, temperature, exposure intensity, and dose. The phase separation can be complete (He, 1998b; Kim, 2004) or polymer dispersed composites can be made (Lee, 1998; Riza, 1998; Ren, 2003). The alignment at the borders between polymers and liquid crystal can be controlled with surfactants (Penterman et al., 2002). A drawback for these systems is the control of orientations and textures inside the confined volumes. This cannot be easily done. Another problem in dispersed systems might be the uniformity. Depending on the curing conditions and the stability of the process, different sizes of cavities can be found. These are statistically distributed in size and location. Rigorous optical simulations of such systems are difficult and not yet widely used for performance optimization.

10.5.5 Structured Electrodes

The most conventional method for realizing switchable diffractive elements is to pattern the electrodes in a planar liquid crystal cell (Murai, 1992; Kulick, 1995; Friedman, 1996; Klaus, 1997; Gu, 2005). This is based on conventional liquid crystal technology, including strong surface anchoring by rubbing. Several switching modes can be distinguished. With patterned electrodes one can switch

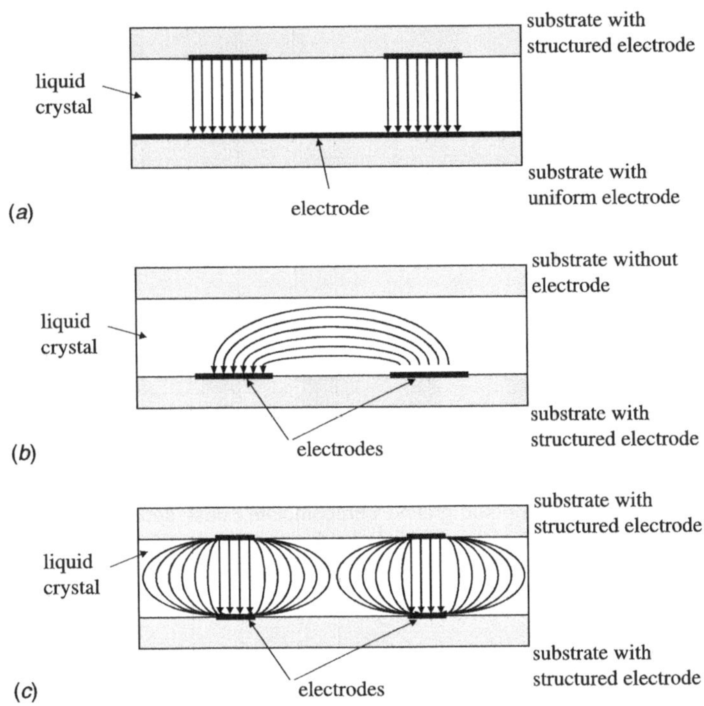

Figure 10.7 Switchable grating concepts. Three main types can be distinguished: (*a*) the vertically switched grating, (*b*) the horizontally or in-plane switched gratings, and (*c*) the fringing-field-based phase modulation.

vertically between the patterned electrode and the ground (Bouvier, 2000), vertically between patterned electrodes (O-he and Kondo, 1997; Okada et al., 1998; Fujieda, 2001), in-plane between different patterned electrodes, or one can use the fringing field emerging from thin electrodes for diffraction of light (Lindquist et al., 1994). All combinations of electrodes, patterned or continuous, can be imagined. Three different concepts are shown in Figure 10.7. The electrodes are structured by lithography and wet or dry etching. The resolution is limited and gratings with periods of a few micrometers can be realized at best. Normally the grating period is several tens of a micron for electrode structured gratings. The different grating structures will be discussed with regard to their optical properties and particularities in the next sections.

10.6 DIFFRACTIVE ELEMENTS DUE TO SURFACE MODIFICATIONS

10.6.1 Advanced Diffractive Optic Components: Fan-Outs

There are different ways in which liquid crystal switchable devices may be construced. All are based on periodically controlled orientation of the liquid crystal

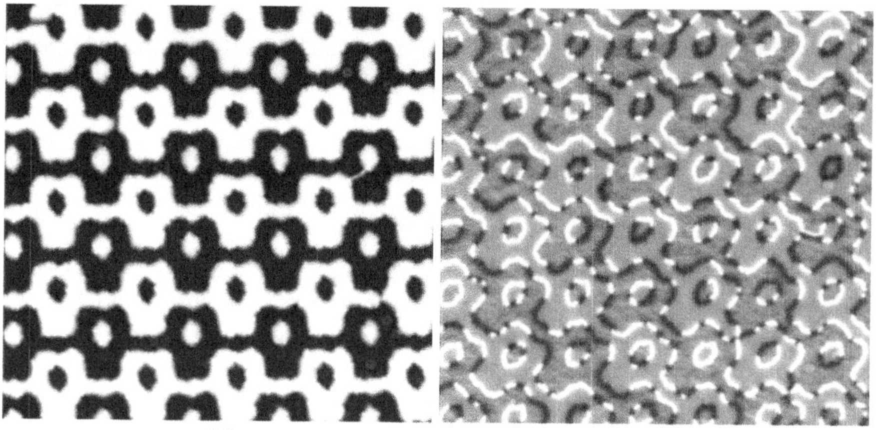

(a) (b)

Figure 10.8 Alignment patterned fan-out element fabricated with UV alignment. A thin liquid crystal polymer film is aligned and shows locally varying directions of the optical axis. The alignment of the polymer liquid crystal is hybrid. (a) The element is illuminated with parallel polarized light and observed in circular polarized light. (b) Between crossed polarizers and rotated at 45°, the domains show no contrast but the domain boundaries show distinct variations.

and the direction of the optical axis. If patterned electrodes are used, multiple processing steps are required to generate patterned electrodes, and the fringing field reduces the spatial resolution of the grating. In addition, the application is limited to structures with a connected electrode if standard cell geometry is applied. The diffraction pattern appears when the voltage is turned on. The device works in a normally-off mode. In the case of alignment patterning, single electrodes on each substrate can be used to electronically control the liquid crystal orientation. The optically controlled alignment process demonstrates high spatial and angular resolution of the local planar alignment of liquid crystals (Patel and Rastani, 1991; Schadt et al., 1992). By combining the high spatial and angular resolution of this process with the birefringent and electro-optical properties of the liquid crystal, binary phase holograms can be realized. Although for two patterned surfaces the alignment of the pattern causes difficulties, cells with one structured surface are easy to fabricate. The device works in a normally-on mode, because the pattern is always present and will be switched off by decreasing the retardation difference or the directional anisotropy of the different zones.

Figure 10.8 shows an example of a 4 × 4 fan-out element with patterned alignment. The element is fabricated with the technology introduced by ROLIC (Schadt et al., 1992). A liquid crystal polymer can be aligned with polymers in which the alignment direction and pretilt are generated by polarized UV light and the main propagation direction of illumination, respectively. It is a passive element that is fabricated in two steps. The alignment layer is spin-coated onto a substrate surface and exposed, through a mask, with polarized UV light to give a patterned alignment direction. A second exposure with perpendicular polarized UV light and without mask (float exposure) is done to address the nonexposed zones. The sample surface is alignment patterned and can be used in liquid crystal cells. In the example shown, a liquid crystal polymer is applied at a thickness of 1.1 μm. The alignment patterning orients the liquid crystal polymer and an element with locally varying alignment, and hence the direction of the fast axis is created. The alignment directions in our case are planar on the substrate surface. The other surface is an air–liquid crystal interface, which leads usually to homeotropic alignment. A closer look at Figure 10.8 reveals the details of such structures.

There are always defects at the boundaries between differently aligned zones. This will determine the resolution. The size of those defects is discussed in Chapter 8. There are different orientations of defects. One can observe in Figure 10.8*b* abrupt changes of appearance at the domain borders. To look at these problems in more detail we will, in the following, discuss the basic example of a binary alignment patterned grating. This will provide the necessary input for the design and fabrication of more complicated elements. Before the element can be assembled the design of the optical function has to be carried out. Different aspects are important. The resolution of the devices is limited due to surface defects between the domain boundaries. As a rule of thumb, one can say that this is about equal to the thickness of the liquid crystal layer. The design is carried out by standard means of diffractive optics design. Usually, methods such as simulated annealing and Fourier transform based algorithms are used to optimize the performance of binary optical elements (Herzig, 1998). A difficulty for high-performance devices is to include the fabrication errors in the design. Liquid crystal structures have an intrinsic limit of resolution and different models might be used to correct for this. Linear ramps (Yu et al., 2004) or a more complicated model using a kernel in the diffraction integral evaluation (Apter et al., 2004) have been proposed.

10.6.2 Liquid Crystal Gratings with Surface Defects

We will now discuss the case of alignment patterned gratings. The idea is to have only two different alignments on one substrate surface. The concepts can be expanded to the properties of more complicated situations with smooth alignment changes if necessary (Wen, 2002). There are two different situations, where only one substrate surface is alignment patterned and the countersubstrate is uniform, or where both substrates are patterned. In fabricating such elements the first variant is much less demanding, but the second offers more advanced features. Both systems have already been realized, for example, in the work of Chen et al. (1995) and Titus and Bos (1997).

We start by giving a brief overview of the systems that are of interest. Alignment patterning is based on different orientations of the liquid crystal director on the surface. There are two different extremes: homeotropic alignment and planar alignment. Planar alignment has the nematic director parallel to the surface and is in its homogeneous form additionally characterized by a direction, the alignment direction. Homeotropic alignment is rotationally symmetric with respect to the substrate surface normal. In a liquid crystal cell, several combinations of alignment patterning

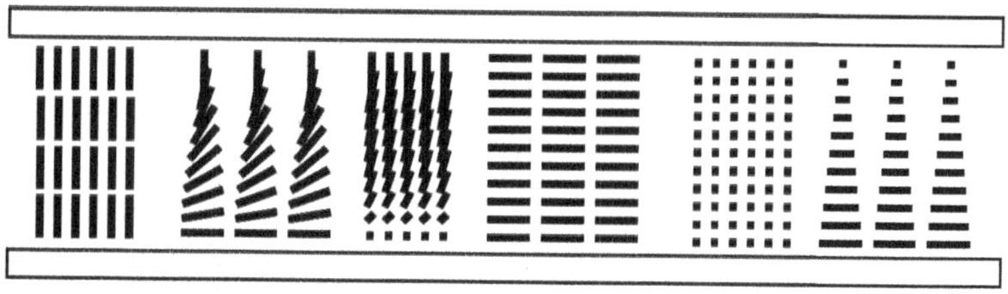

Figure 10.9 Birefringent gratings with alignment patterned surface where only homeotropic and planar alignment are used: (*a*) homeotropic alignment, (*b*) and (*c*) hybrid alignment with different orientations of the optical axis, (*d*) and (*e*) planar alignment with different orientations of the director, (*f*) twisted alignment where the twist sense can be left or right.

are possible. Figure 10.9 shows examples of combinations with planar and homeotropic alignment. We can see in Figure 10.9 that there are a multitude of textures in the liquid crystal cell if only cases at right angles are considered. In a diffraction device nearly all configurations can be realized with additional optimization of the orientational angles, depending on the purpose. Discussing all such cases is outside the scope of this chapter. What all theses textures have in common is that the alignment patterning produces defects at the surface. The type of defect and the elastic properties of the liquid crystal determine the performance of the device. Nearly all different defect types can be identified in a liquid crystal cell with only one alignment patterned surface. The discussion will therefore be limited to binary systems of the alignment patterns and to the case with only one structured surface.

Due to the defects at each alignment border, the device shows strong gradients in the refractive index. There are two major problems linked with the design and simulation of the optical properties of gradient index diffraction devices. First, the director profile has to be obtained, and secondly, the optical simulation becomes complicated, because the director field variations also cause a gradient in the refractive index that can no longer be neglected. The director profile can be found by various means, about which more details may be found in Chapter 8. An elegant way, however, is the conformal mapping technique, which has some fundamental limitations. Other techniques use finite difference calculations and allow adaptation to all kinds of liquid crystal parameters. Commercial programs are available that allow simulations in two and three dimensions (Shintech, 2005; Dimos, 2005).

Light propagation in anisotropic gradient index media has to be carried out using rigorous methods. We found earlier in Chapter 8 that the limit of spatial resolution of alignment patterned gratings is given by the thickness of the liquid crystal layer. Experimental results and simulations proved that a simple Jones matrix calculation is not sufficient to understand the optical properties of a spatial-resolution-limited grating because of strong gradients in the refractive index distribution. In order to investigate the propagation of light passing through a liquid crystalline structure of micrometer size, one uses rigorous FDTD methods, as presented in Chapter 5. This allows simulation of time-dependent electric and magnetic fields for a region that is two-dimensional and possibly anisotropic (Witzigmann et al., 1998; Titus et al., 2001; Scharf and Bohley, 2002). Perfectly matched layer (PML) and periodic boundary conditions are used to avoid disturbing reflection from the borders of the simulation box. The FDTD method, in contrast to the methods of geometrical and matrix optics, delivers results that include diffraction and scattering.

To illustrate this type of optical simulation we will discuss results for a model structure as shown in Figure 10.10. Planar alignment patterned surface substrates are combined with a homeotropic countersubstrate. The boundary conditions are periodic. Parameters for the liquid crystal mixture E7 were used and the simulation of the director profile was done using LCDMaster from Shintech (Shintech, 2005). The optical properties are simulated using the FDTD method, the results for which are shown in Figure 10.11. Snapshots of the electric field amplitude are presented for different light polarizations at the entrance. The simulation box is 3 μm high and 12 μm wide, with a discretization of 20 points per wavelength. The wavelength is $\lambda = 550$ nm. The liquid crystal layer has a thickness of 3 μm. The simulation box for light prorogation has periodic boundary conditions for the horizontal direction and is closed with perfectly matched layer conditions in the vertical direction. The spatial distribution of the birefringence causes particular intensity distributions at the output that cannot be calculated or correctly simulated with standard matrix methods. The structure represents a typical example for a polarization-independent grating. Because the domains have an optical axis orientation of 90° with respect to each other, the perpendicular linear polarizations will be influenced in the same way. The defect structure at the surface leads to all kinds of elastic deformations of the director structure. In particular, the twist deformation close to the alignment patterned surface leads to a situation where energy is transferred from one polarization

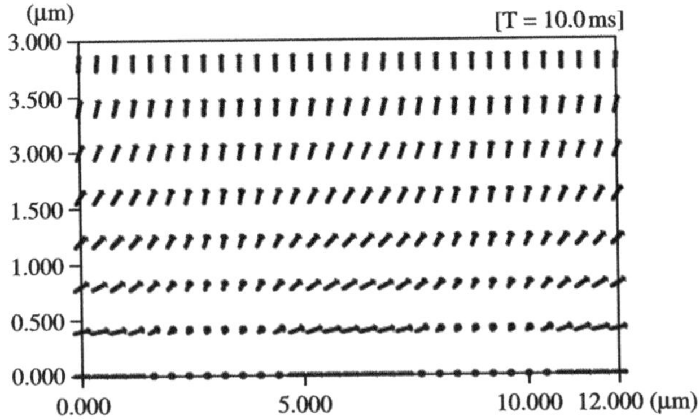

Figure 10.10 Surface alignment patterned grating configurations where one substrate gives homeotropic alignment and the other has planar domains of different orientations. The area of the patterning is 3 μm, the same as the thickness of the cell. Different configurations are possible depending on the defect structures at each border between the planar and homeotropic alignments.

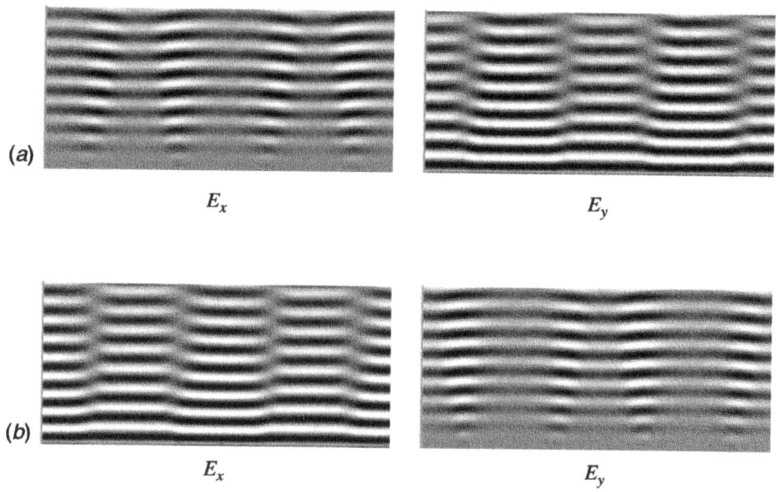

Figure 10.11 Intensity of light propagating through the asymmetric director field shown in Figure 10.10 ($2a = d = 3\,\mu\text{m}$, $\lambda = 550\,\text{nm}$) simulated with the FDTD anisotropic method (number of layers in steps of 27.5 nm, simulation box size $3\,\mu\text{m} \times 12\,\mu\text{m}$). (*a*) Incident polarization TM (E_y perpendicular to the plane of the paper); (*b*) incident polarization TE (E_x to the plane of the paper).

direction to another. This is visible in Figure 10.11 through the increase of contrast of the plotted phase profiles for the polarization of light that is not incident. In the example for TM incident polarization, the simulation viewgraph on the right in Figure 10.11 represents the incident electric field in E_y. At the exit plane of the liquid crystal layer, both polarizations are present due to the coupling caused by the twisted structure.

We see that the coupling efficiency is highest behind the defect position. This is understood when we take into account that the deformation is strongest there. The polarization independence is visible when simulations for TM and TE incident polarization are compared. Simulation gives results that are interchangeable with respect to the electric field components. The results for the E_y components when TM is incident are nearly identical to the E_x component when TE is incident. The same holds for the perpendicular fields. We can see this better by comparing the phase and amplitude profiles in Figures 10.12 and 10.13. The amplitude and phase profiles show variations in phase and amplitude for both electric field components at the output. In the case of TM incident polarization, the y-component of the electric field has a binary profile with a phase jump of less than π. The field component E_x has a pronounced amplitude dependence. The amplitude is zero at positions where the phase diverges. The phase profile is steplike, with a phase jump of π. In the case of TE incident polarization, the same curves are found but for interchanged electric field components. Now the E_y field component has the phase singularities at defect positions. We see that such a grating will be an efficient diffraction device but will not conserve the state of polarization.

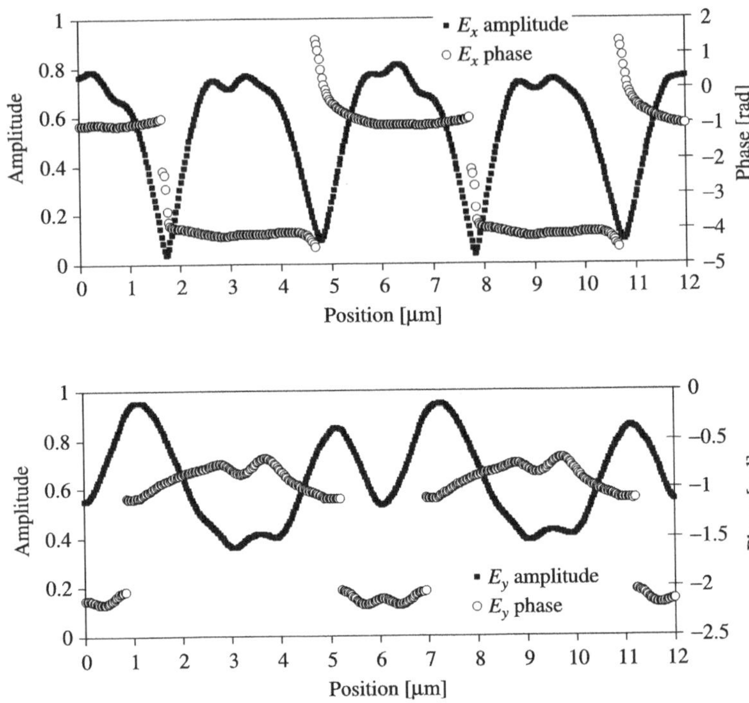

Figure 10.12 Simulated amplitude and phase profiles at the exit of the surface alignment patterned grating structure. The incident polarization is TM (E_y).

To calculate the state of polarization one identifies the amplitude and phase profile in Figures 10.12 and 10.13 with the components of the Jones vector and performs a Fourier transform as outlined in Sections 10.3 and 10.4. Note that the two simulations are usually not sufficient to determine all of the elements Jones matrix. The Jones matrix contains eight independent parameters. To calculate the complete Jones matrix, two additional simulations would be necessary.

The rigorous simulation includes anisotropy and gives the most reliable results. Other approximate methods have been studied to increase the speed of simulation or to develop analytical parameter dependencies. For structures that allow decoupling of polarization (see Section 2.5), an effective refractive index approach might be useful. The distribution of the effective refractive index with respect to a certain propagation direction can be calculated and used in an isotropic FDTD method or ray tracing. The results fail for high spatial resolution (Scharf and Bohley, 2002). The isotropic approach with the FDTD method allows the gradient to be taken into account and performance can be compared with advanced polarization ray-tracing methods (Kosmopoulos and Zenginoglou, 1987; Zenginoglou and Kosmopoulos, 1988, 1989). The advanced ray-tracing method was developed for several applications, but is not very well suited for twisted structures where polarizations are rotated. The shadow technique is of the same quality and can only be used

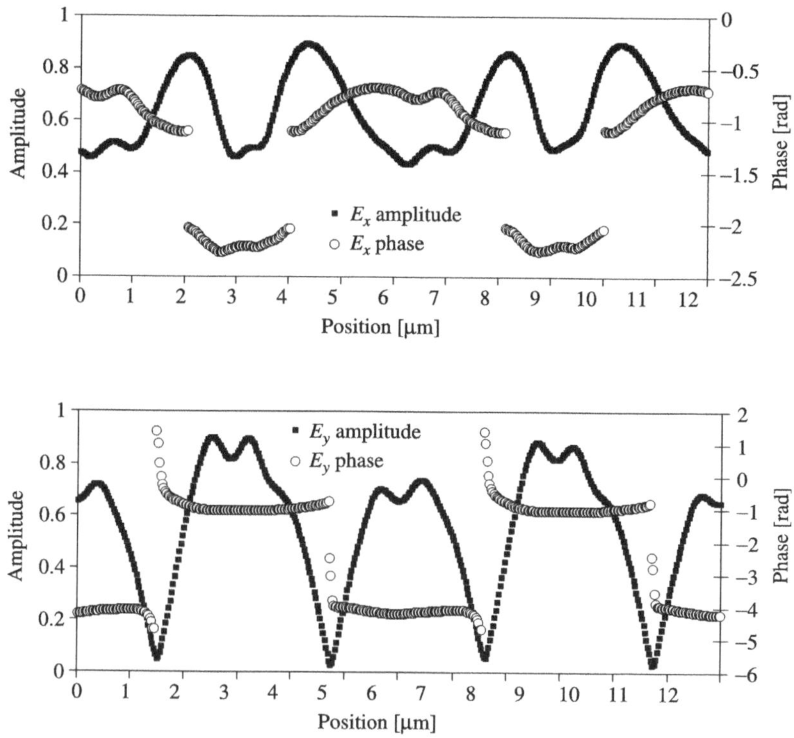

Figure 10.13 Simulated amplitude and phase profiles at the exit of the surface alignment patterned grating structure. The incident polarization is TE (E_x).

for large grating periods and weak gradients. A good rule of thumb is to study the crossing points of the rays. If the rays are crossing when propagating inside the structure, the ray-trace approximation should no longer be used. In thin films, matrix methods are often used that do not take gradient effects into account. In general, such matrix methods lead only to phase modulations and the amplitude modulations are underestimated. This results from the fact that the intensity modulation is only produced in the region of twist deformation and does not include deviation by gradients. The rigorous methods show amplitude variations due to the influence of the gradient in the birefringent distribution. The matrix method often gives symmetric distributions of amplitude and phases, but the gradient often causes asymmetries (Scharf and Bohley, 2002). Only the rigorous FDTD method including anisotropy shows the expected asymmetry of the intensity and phase, caused by the asymmetry of the structure.

The main criteria for characterizing such structures if they are arranged in gratings are the diffraction efficiencies. Diffraction efficiencies can be effectively calculated with the Fraunhofer approximation explained in Section 10.2, by Fourier transforming the complex amplitudes. Even for a structure size comparable with the thickness of the liquid crystal layer, the calculated efficiencies of the matrix method and the rigorous approach differ considerably. In general, the diffraction efficiencies calculated with the matrix methods are too large, and sometimes do not show the correct symmetry.

10.7 ELECTRICALLY SWITCHABLE GRATINGS

We have already learned about passive gratings in Sections 10.5 and 10.6. Most of these systems allow electrical actuation too. We will look at these possibilities and then discuss gratings that are made with structured electrodes. Although the first class of elements is usually switched by using continuous electrodes and works in a normally-on operation mode (no voltage equals device on), the second class develop their function only when voltage is applied. Switchable gratings with electrodes are often normally-off devices. Many prototypes have already been realized and studied. Others still wait for further, more intensive, study. Surface modification in particular, which leads to alignment patterned surfaces together with structured electrodes, still seems an interesting field of research. Here, in-plane and vertical electric fields might be considered. Recently, such concepts have led to switchable multistable devices (Kim et al., 2002). With advances in surface nanotechnology, such concepts will become technologically easily available. The advantage here is the combination of texture defects mediated by strong elastic deformations. With strong in-plane electric fields, one might increase switching performance, especially with regard to switching speed, and shorten the relaxation time.

10.7.1 Elements with Immersion

In Section 10.5 we have already mentioned some configurations for switchable gratings, fabricated by filling microstructured surfaces. One important aspect is the alignment of the liquid crystal inside the microstructure. In extreme cases the nematic director is locally oriented homeotropic or planar to the surface. Often the alignment is not known, and it is difficult to determine it. Figure 10.14 shows different configurations for a grating that has a homeotropic alignment countersubstrate. Homeotropic and planar alignment is assumed at the surfaces. The difference in the elastic distortion is remarkable and it is not easy to determine which alignment state is obtained in a real device. The situation becomes more and more difficult when the grating period becomes smaller. Inspection of Figure 10.14 makes it evident that the alignment at the surfaces determines the defect states of the textures. If the design principle of optical components is based on a classical optical design scheme, without taking into account the defect states, the element will not have good performance and the quality of the element will be determined by the defects (Strudwick and Lester, 1999; Lester et al., 2005). There are textures that are preferred for energetic reasons (Berreman, 1973). For instance, if both substrates are oriented planar and parallel to the grating, as seen in Figure 10.14*c*, no elastic deformation can be found. The texture can be defect free and is extremely reliable

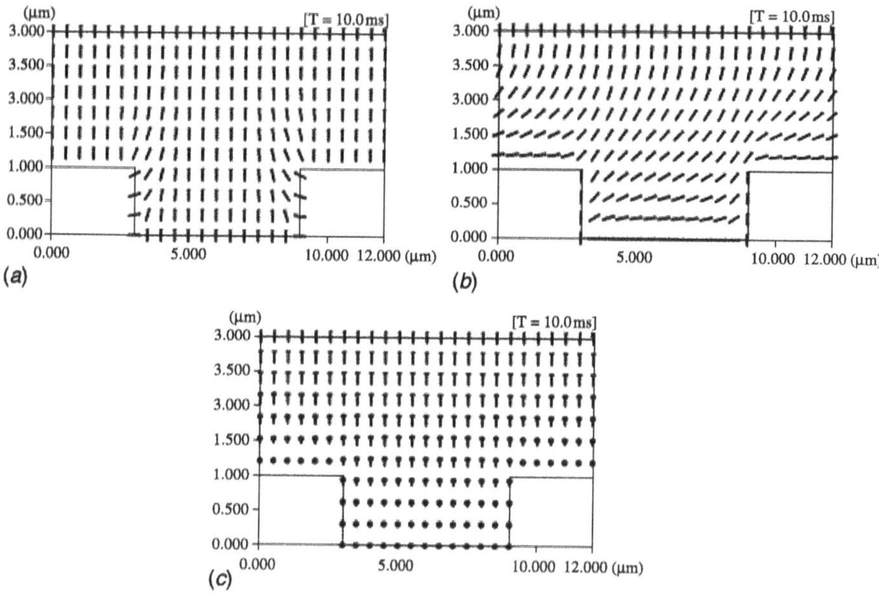

Figure 10.14 The alignment of the director influences the elastic deformation in filled grating structures. The squares at lower left and lower right indicate the microstructure of the surface. The three examples shown have (*a*) locally homeotropic alignment and (*b*) and (*c*) planar alignment.

in use for switchable systems. The alignment on the microstructure can be obtained by rubbing or by using incorporated nanostructures as indicated in Section 9.2. When such filled structures are switched, one has to take care of the electric field distribution, which is changed due to the protrusions. In extreme cases, a well-defined microstructure can produce electric field peaks that lead to defect states. (Wang et al., 2004; Lester et al., 2005). In general, voltages are moderate and the deformation is determined by the strong anchoring at the surface. Such switching is still in the elastic regime, allowing relaxation to the original state. If defects are involved that travel, relaxation shows a different mechanism, which leads to modified switching dynamics (Cheng and Thurston, 1981).

We want to give here an example of a grating with an ideal structure to show the influence of the protrusion on the deformation profile and finally on the performance of the grating. Consider a grating of period 12 μm with a height of 1.2 μm. The liquid crystal structure is considered to have planar alignment and the parameters for the liquid crystal mixture E7 are taken. Assuming that the protrusion has a refractive index of $n_p = 1.53$, one would get a retardation difference of $\Delta nd = (1.76 - 1.53)\ 1.2\ \mu\text{m} = 0.276$ nm, which corresponds to $\lambda/2$ for a wavelength of 550 nm. The theoretical switching threshold voltage for a uniform planar alignment is (Blinov and Chigrinov, 1996; Yeh, 1999)

$$V_{\text{th}} = \pi\sqrt{\frac{K}{\Delta\varepsilon\varepsilon_0}}. \tag{10.36}$$

With the dielectric anisotropy $\Delta\varepsilon = 13.8$ for E7 and assuming an average elastic constant $K = 11$ pN, one obtains

$$V_{\text{th}} = \pi\sqrt{\frac{11 \times 10^{-12}\,\text{Nm}}{13.8 \times 8.8542 \times 10^{-12}\,\text{F}}} = 0.942\,\text{V}.$$

In the off state, when the voltage is zero, the grating has maximum efficiency for TE polarization (normal to the plane). For TM polarization the index-matching condition is fulfilled and no diffraction effect is expected. The pretilt of the liquid crystal director is set to 5° on both substrates. Figure 10.15 shows a series of simulations for different voltages. Simulations of the director field were carried out with the Shintech LCDMaster (Shintech, 2005). Light propagation is simulated with FDTD methods to visualize the wavefronts inside the structure. We consider only the TE polarization.

Zero-voltage light propagation for the TE polarization corresponds to that found in isotropic binary gratings. There are effects at the edges of the binary grating that are well known from diffraction theory. If the voltage is increased above threshold, reorientation of the liquid crystal takes place and reduces the phase shift between adjacent zones. As can be seen by comparing parts (*a*), (*b*), and (*c*) of Figure 10.15, phase modulation is still present at 2 V and can only be suppressed at higher electric fields. However, at high fields, the fringing at the protrusions deforms the field lines and leads to unwanted tilts and twist deformation of the director. As a result, the quality of the diffraction element worsens. This is particularly true for the contrast between the on and off states, because the off state can no longer be switched completely off.

10.7.2 Alignment Patterned Gratings

It is not only filled structures that can be switched – the same is possible for the alignment patterned gratings. Different systems have been studied in work by Chen et al. (1995), Bos et al. (1995), Titus and Bos (1997), Yu et al. (2005). In common applications the electrodes are not structured. The most interesting elements are insensitive to the polarization, which can be achieved by a clever choice of the domains of the gratings. One possibility is to choose a perpendicular optical axis orientation, as discussed in Section 10.6. When such gratings are switched, the function is switched off. In other words, they work in the normally-on mode. Efficiencies are altered by losses due to defects at the surfaces. The function of the system can be switched complete off if high enough electric fields are used.

10.7.3 Switching and Fringing Fields

Switchable diffraction gratings based on liquid crystal have long been known. The First experiments were based around instabilities of the liquid crystal system itself (Kashnow and Bigelow, 1973; Blinov and Chigrinov, 1996). Since then, big steps have been made in technology, and spatial light modulators are nowadays used with feature size below 10 μm. They form remarkable diffraction gratings and are

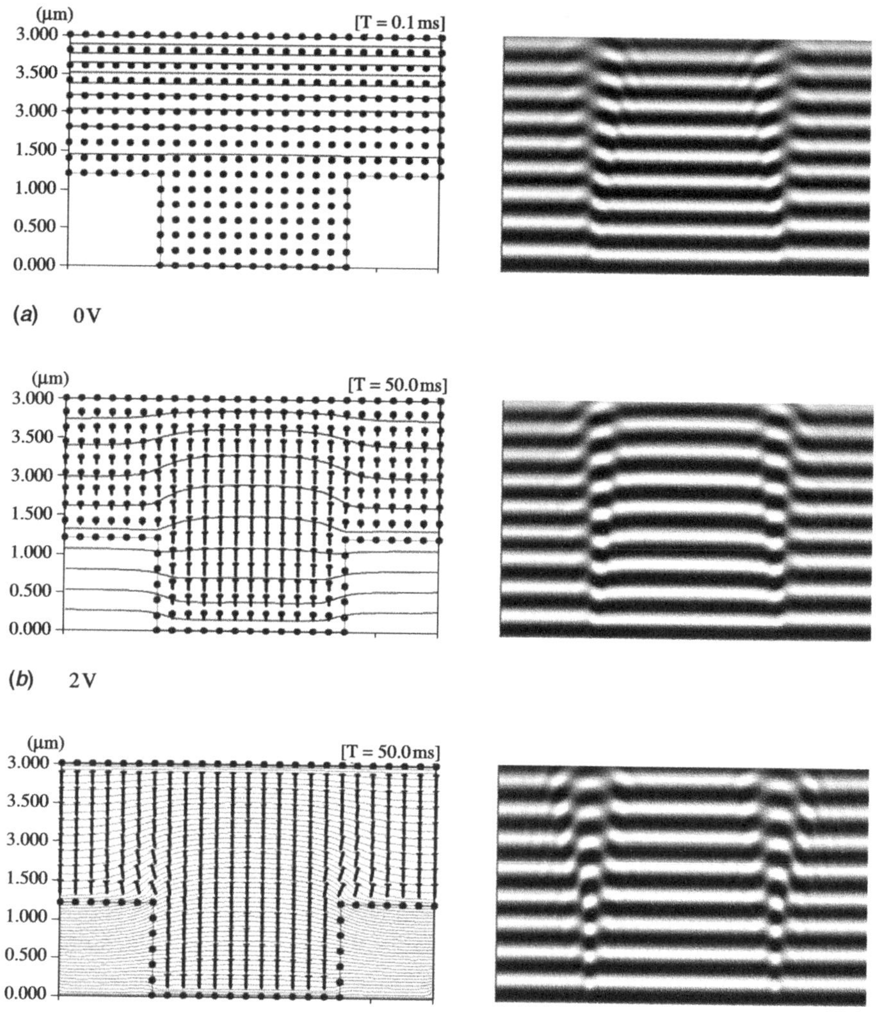

Figure 10.15 Switching of a filled grating and the propagation of TE polarized light through it for voltages (*a*) 0 V, (*b*) 2 V, and (*c*) 10 V. The electric field lines and the director distributions are shown. The light propagation is simulated for wavelengths of 550 nm. Plane wave illumination comes from below. The dimensions of the simulation box are 3 μm × 12 μm.

used as holographic systems for projection (Hällstig, 2004; Holoeye, 2005). The functional basis of these devices is the local reorientation of the liquid crystal in the electric field and, as already mentioned in Section 10.5, there are different directions of the field with respect to the cell geometry. The basic idea of the switchable grating is to start with a uniform liquid crystal texture and to modulate the phase or optical axis with electric fields created by structured electrodes. As we have seen in Chapter 10.5.5 there is a multitude of configurations and we cannot discuss all

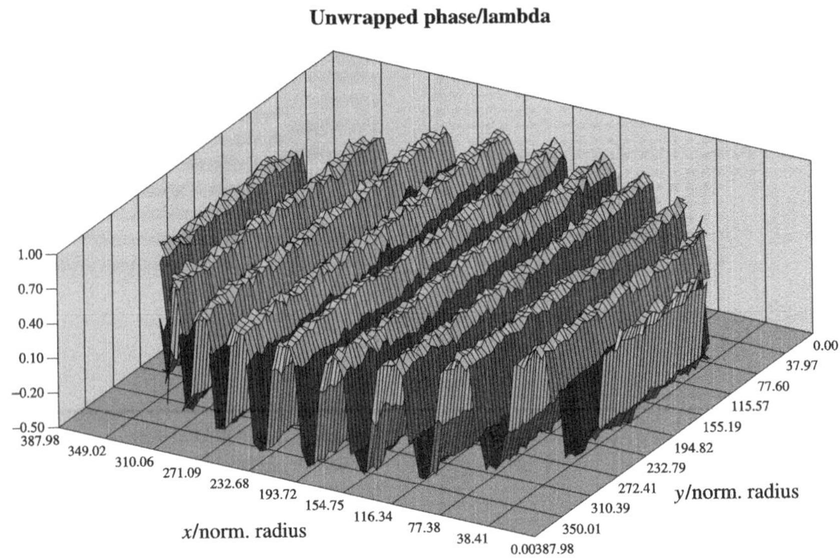

Figure 10.16 Snapshot of the phase profile at 1.8 V_{rms}. The imaged area is 360 μm × 390 μm. The unwrapped phase is shown and the phase scale ranges over 1.5λ. Measurements are carried out using a 633-nm He–Ne laser.

of them. Although switching can be mediated by vertical, planar, and fringing fields, the director can be aligned planar, homeotropic, hybrid, and twisted. More specialized literature may be consulted for details (Tanguay et al., 1984; MacManamon et al., 1993; Lindquist et al., 1994; Klaus et al., 1996; Stalder and Schadt, 1996; Bouvier and Scharf, 1997; Wolffer et al., 1998a; He, 1998; He and Sato, 1998; Sakata and Nishimura, 2000; Fujieda et al., 2001; Honma and Nose, 2004; Crawford et al., 2005). We shall discuss the basic principles of operation and the limitations by considering a real example. We use the simplest device that has planar alignment and a binary electrode structure. Structured electrodes are arranged only on one side and the other electrode is continuous. Liquid crystal alignment is carried out by rubbing, and the alignment direction is along the electrode grating. Figure 10.16 gives an impression of the quality of the phase profile modulation that can be obtained with standard liquid crystal technology. We show phase measurements for a 6-μm liquid crystal grating cell aligned along the grating direction. The grating period is 40 μm and the cell is filled with the liquid crystal mixture BL006 from Merck (Table 6.6). Such a system forms a pure phase grating and the switching is mainly vertical. Measurements were done with Mach–Zehnder interferometer that operates in polarized light. The area shown is 360 μm × 390 μm. One can see the phase profile is uniform, which indicates a defect-free elastic deformation. The phase jump at 1.8 V (rms 1 kHz) is approximately 1.5 λ.

An interesting inside into the functional principle is given when phase profiles for different voltages are compared. This can be seen in Figure 10.17. In the configuration under consideration, when the director is aligned along the electrodes,

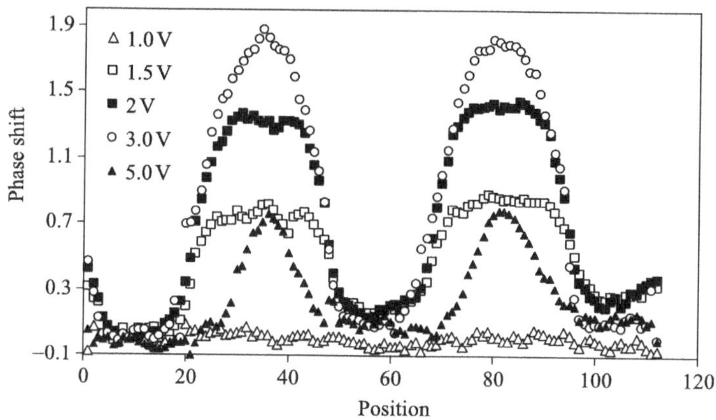

Figure 10.17 Phase profiles for different voltages for a liquid crystal grating. The period of the grating is 40 μm and the thickness of the cell is 6 μm.

the phase profiles are symmetric. This would not be the case for perpendicular alignment with respect to the electrodes (Bouvier, 2000). At 1 V the reorientation has not yet started. Low voltages at about 1.5 V lead to rather rectangular profiles. However, at 2 V, a smearing out of the phase profile is already visible. This is caused by the elastic deformation of the liquid crystal structure itself. Twist deformation occurs that has the smallest elastic constant. The width of this elastic deformation is about the cell thickness, 6 μm in our case.

The phase profile at higher voltages becomes deformed by the fringing field. This causes a deformed phase profile and a reduction of the effective phase shift as seen in the 5 V measurement. The fringing field has in-plane field and vertical field components. A separate discussion of vertical, in-plane, and fringing fields is not appropriate, because all effects appear at the same time. The fringing field at higher voltage extends also into the electrode gaps. There it causes reorientation of the liquid crystal director and reduces the effective phase shift. We can measure the effective phase step in our case. The results are plotted in Figure 10.18. The graph shows the measured phase shift for the grating device compared to a uniform planar-oriented liquid crystal cell of the same thickness. The retardation increases for both curves in the same way for small voltages up to 3 V_{rms}. Then the fringing field starts to considerably reorient the liquid crystal between the electrodes and the effective phase shift (retardation) for the grating device is decreased. In addition, the phase profile is altered.

Simulations of the director field for different voltages are presented in Figure 10.19. Thick lines represent the phase profile and thin lines are the electric field lines. The voltages are DC values and correspond to rms values in the measurements above. It is seen that the phase profile depths are small for 1.5 V, with a shape smoothed at the electrode edge due to elastic deformations. The phase profile is almost squared for 2.4 V, but the fringing field leads to wide deformations at the electrode edge. At this stage no remarkable in-plane orientation is visible. At 5 V, reorientation due to in-plane fields alters the director profile, and the phase profile shape has a modified aspect ratio. For the design of an efficient switchable grating

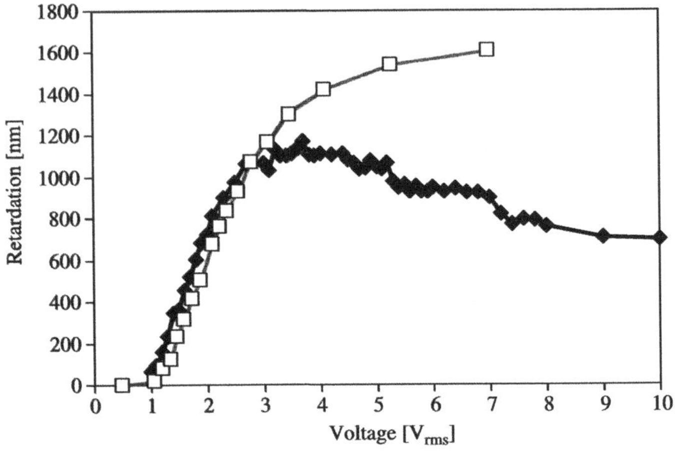

Figure 10.18 Effective retardation for a uniform planar liquid crystal cell (open squares) and a liquid crystal diffraction grating (filled squares). Due to fringing field effects between the electrodes of the diffraction grating, the retardation cannot reach its maximum (BL006, 6 μm, 40 μm grating period, $\lambda = 633$ nm).

one should consider a reduction of the electrode width by twice the liquid crystal cell thickness. To obtain high lateral resolution, the liquid crystal layer itself should be as thin as possible. This helps also to decrease the switching time, but sometimes causes technological difficulties. Liquid crystal cells with a thickness of less than 2 μm have already been demonstrated.

10.7.4 Multilevel Gratings with High Spatial Frequency

Nonmechanical beam steering devices are very attractive. Various schemes based on microlens arrays and phase arrays have recently been discussed. Reconfigurable phase array devices can be built with liquid crystals, which act as an electrically controlled blazed phase grating (Resler et al., 1996; Stalder and Schadt, 1996; Faris and He, 1999). The key idea is to use multilevel electrode gratings. Depending on the electrode number, the complexity of the electrical addressing schemes, related to the number of accessible steering angles, can be very high (Wolffer et al., 1998). A special case includes thick liquid crystal microprisms, which allow nearly continuous deviation angle with only two electrodes, but require higher driving voltages and exhibit only slow response times (Hirabayashi et al., 1995).

As an example we discuss a typical realization based on structured electrodes as shown in Figure 10.20. In fact, the switchable beam steering devices are high-resolution, one-dimensional phase modulators. The devices are liquid crystal cells, where the liquid crystal layer is sandwiched between two substrates formed by transparent electrodes on glass. One of the substrates has a structured conductive layer with an array of parallel electrodes. Applying different voltages to the electrodes will result in locally different nematic director deformation of the liquid crystal.

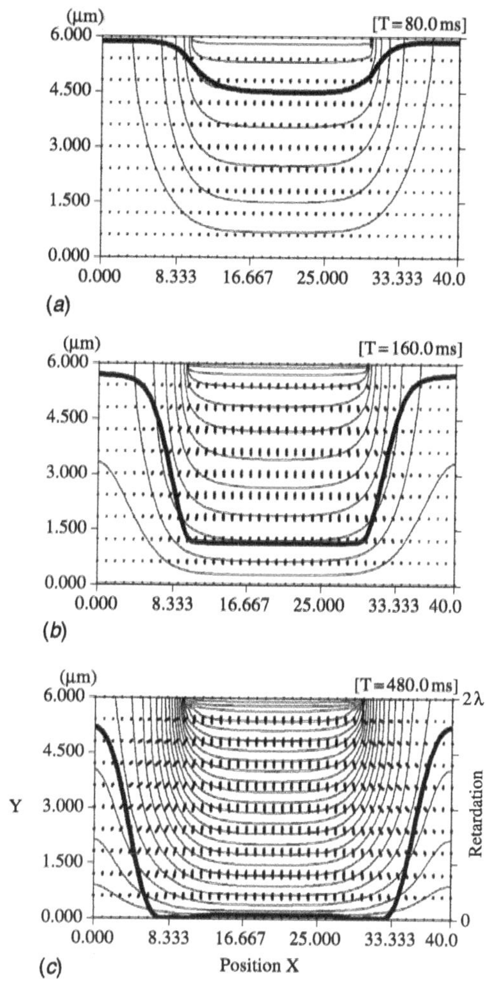

Figure 10.19 Simulated director profiles, electric field lines (thin lines) and retardation (thick lines) for three different voltages, (*a*) 1.5 V, (*b*) 2.4 V, and (*c*) 5 V, of nematic liquid crystal grating of 40 μm period. The parameters correspond to the measurements in Figure 10.17.

Because of the small distance between the electrodes, the transition of this variation between neighbouring electrodes becomes smooth. We would like to discuss particulars using as an example a high-resolution grating where the width and separation of the electrodes are 3 and 4 μm, respectively. The alignment is made by conventional rubbing techniques. Polyimide (PI 2545 from Nissan) is used as the alignment layer and is rubbed either parallel or perpendicular to the electrode grating. The other substrate has a continuous transparent ITO electrode and the polyimide layer is rubbed to produce planar antiparallel alignment. The devices have an effective thickness of 6 μm and are filled with liquid crystal BL006 from Merck. The resulting maximum modulation depth is equal to 2.6 times the

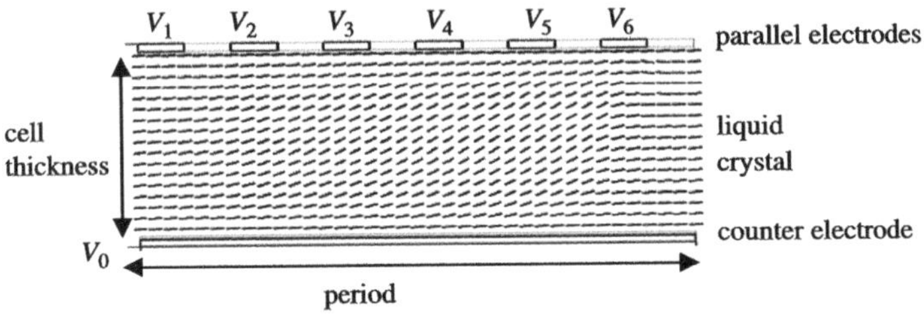

Figure 10.20 Schematic cross-section of the multilevel liquid crystal phase grating. One electrode is structured at high resolution. The counterelectrode is continuous.

wavelength. The electrode arrays have 192 electrodes, producing an active area of approximately 1.5 mm. The electrode array is externally grouped with a period of 12 electrodes. These 12 channels are then driven with AC signals of 1 kHz frequency and voltages in a range of 0–5 V_{rms} controlled by a computer. Experiments were performed with a He–Ne laser at 633 nm, and the diffraction pattern detected using a linear detector array with 512 pixels.

What can we expect from such a device? The main reason for fabricating such devices is its ability to modulate the steering angle and obtain high efficiency. In the present case and if all electrodes are used separately, the steering angle can be modulated from 0 to 6°. The diffraction efficiency changes, particularly when the number of electrodes gets smaller than 12. As an example of the beam steering capability of the device, diffraction patterns for 12 control electrodes are shown in Figure 10.21. The alignment is perpendicular to the electrode grating. We see in Figure 10.21 that the diffraction pattern for the large angle deviation peak shows unwanted intensity in other orders. This appears due to imperfections of the phase

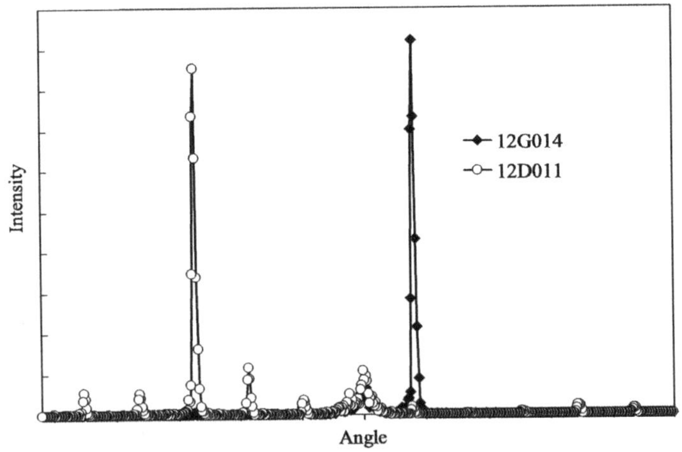

Figure 10.21 Diffraction patterns for different driving schemes with 12 control electrodes per period. The left and right beam steering capability is demonstrated. The liquid crystal is aligned perpendicular to the electrodes.

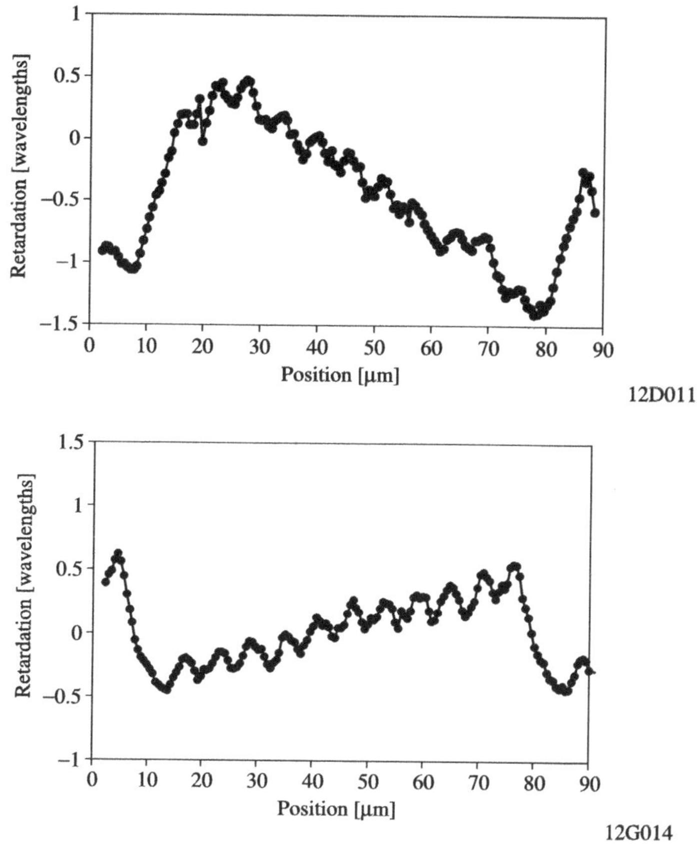

Figure 10.22 Measured phase profile for a high-resolution beam steering device with 3 μm electrode width and 3 μm interelectrode distance corresponding to the diffraction images of Figure 10.21. The liquid crystal is aligned perpendicular to the electrode grating as shown in Figure 10.20. Two different configurations are shown for steering at different directions.

profile. To switch into higher orders, the period of the grating can be reduced; this means, for instance, that six electrodes are grouped to form the new period for the second order. For the third order, three electrodes are taken as the new period. Another possibility is to increase the phase shift. This concept is used in Figure 10.21 to change the steering angle. The corresponding phase profiles are given in Figure 10.22. The phase profiles already show the typical problems of multi-electrode beam steering. For low voltages, ripples appear on the phase profiles because the reorientation is not complete. At higher voltages, the fringing between electrodes smooths the phase profile. Detailed investigations show that the diffraction efficiency can be very high and depends critically on the alignment of the liquid crystal with respect to the electrodes (Scharf et al., 2001). Parallel aligned cells show symmetric diffraction efficiency when the voltage profile is reversed; that is, the diffraction efficiency is approximately the same for both steering directions. On the other hand, perpendicular aligned devices show significant asymmetric behaviour. For a 12-level grating and for linear polarized light the efficiency in the first order for perpendicular alignment can reach over 60% and for parallel alignment it reaches only ~50%.

To understand better the limitations of the concept, it is worth discussing the electric field distribution. The liquid crystal deformation inside the cell depends on the applied electric fields and the orientation of the director with respect to the electrode grating. The vertical reorientation is used to cerate the phase profile, and in-plane electric fields comprise unwanted effects. Above the threshold for the Frederick's transition, vertical reorientation starts. The threshold voltage for the Frederick's transition of a planar aligned liquid crystal is given using Equation (10.36) and is calculated to be

$$V_{\mathrm{th}}^{\mathrm{Fredericks}} = \pi\sqrt{\frac{K}{\Delta\varepsilon\varepsilon_{\mathrm{o}}}} = 1.15\ \mathrm{V}, \tag{10.37}$$

where $K = 17.9$ pN and $\Delta\varepsilon = 17.3$ for BL006. In the case of planar alignment, the elastic constant only has contributions from bend and splay deformations. Let M be the number of separately controlled electrodes in a period, which in our case is 12. The first electrode, with V_1, might be grounded. To achieve a phase modulation, the voltage in the electrodes V_2 to V_M has to be higher than $V_{\mathrm{th}}^{\mathrm{Fredericks}}$. The strongest in-plane electric field appears between the electrodes M and 1 with a voltage V_M over the interdigital electrode distance. V_M is certainly larger than the threshold voltage $V_{\mathrm{th}}^{\mathrm{Fredericks}}$. These in-plane field effects become important for voltages above the threshold (Oh-e and Kondo, 1997), so that

$$V_{\mathrm{th}}^{\mathrm{in\text{-}plane}} = \pi\frac{p}{d}\sqrt{\frac{K}{\Delta\varepsilon\varepsilon_{\mathrm{o}}}}, \tag{10.38}$$

where p is the interelectrode distance, $d = 6\ \mu$m (cell thickness), K is an elastic constant, and $\Delta\varepsilon$ is the dielectric anisotropy.

To establish the effect of the in-plane electric field on the director profile, one has to distinguish two principal configurations: one where the molecular director is perpendicular to the electrode grating, and the other where the director is parallel to the electrode grating (Bouvier, 2000). If the liquid crystal director alignment is perpendicular to the electrode array, a tilt-only deformation appears. The deformation of the nematic directors is flat and always in the plane. Due to the in-plane electric field, the nematic directors in the areas between the electrodes are under an additional horizontal field component, which tends to avoid the tilt modulation. The transitions between the different electrodes are smoothed. The effect of the in-plane field is a reduction of the effective birefringence change for a given voltage. By changing the voltage levels, this effect could be compensated, and high diffraction efficiency is achieved if the voltage levels are optimized. If the alignment of the liquid crystal is parallel to the electrodes, the deformation of the liquid crystal director is much more complicated. The in-plane field is perpendicular to the director and that induces the twist deformation. Now, the elastic deformation is mainly twist and therefore the elastic constant K is approximately the twist elastic constant K_{22}. With our data, $K_{22} = 15$ pN, $\Delta\varepsilon = 17.3$, $d = 6$ μm, and $p = 3$ μm, we obtain, using Equation (10.38), $V_{\text{th}}^{\text{in-plane}} \approx 0.5$ V. The threshold for in-plane switching is smaller than for the planar homeotropic Frederick's transition. To minimize in-plane field switching, one has to increase the threshold $V_{\text{th}}^{\text{in-plane}}$, either by changing the geometry of the electrode arrangement (by changing p and d) or by searching materials with a high ratio of the elastic constants $K_{22}/K_{\text{splay-bend}}$. Twist deformation influences remarkably the state of polarization. As a consequence, the maximal diffraction efficiency will be reduced.

To clarify these points, further simulations of the director profile can be performed in two dimensions. An analysis of the phase profile gives the diffraction properties. To see the difference between parallel and perpendicular alignment we present measurements for two different configurations in Figure 10.23 and 10.24. The grating devices are in both cases controlled to produce binary phase modulation. Figure 10.23 shows the case for the perpendicular aligned grating with 3 μm electrode width and 3 μm interelectrode distance. The case of parallel alignment is demonstrated in an example with 4 μm electrode width and 4 μm interelectrode distance in Figure 10.24. For both examples, the resulting phase profiles are measured, for linear polarized light, with a Mach–Zehnder interferometer. The voltage levels are chosen to give a maximal phase shift of one wavelength. Valleys appear in the phase profile between adjacent electrodes. Parallel aligned gratings show much deeper valleys than perpendicular ones. Note that in each case the deformation of the nematic director field is determined by different elastic constants. For parallel aligned cells, the transition between two electrodes with different voltages is mainly determined by twist deformations. In this case,

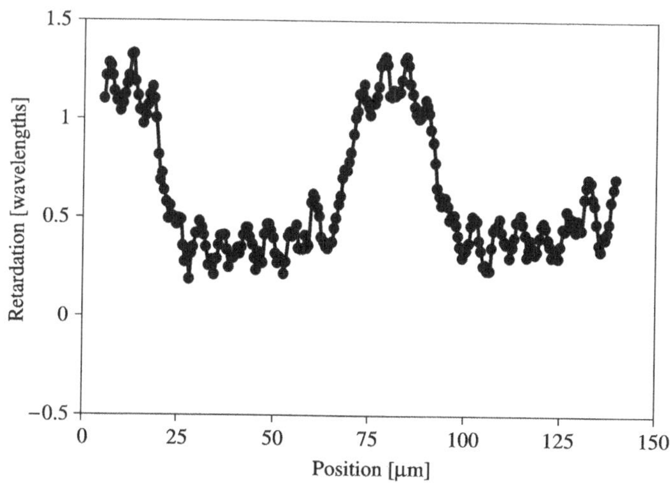

Figure 10.23 Phase profile of multi-electrode grating when driven in a binary mode. The grating has 3 μm electrodes and 3 μm interelectrode distance. The alignment is perpendicular to the electrodes.

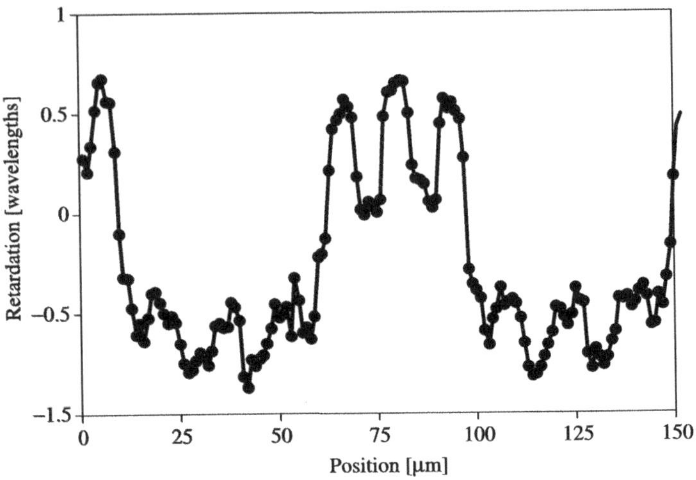

Figure 10.24 Phase profile of multi-electrode grating when driven in a binary mode and the alignment is parallel to the electrode grating. The grating has 4 μm electrodes and 4 μm interelectrode distance. Pronounced phase dips are seen in the switched zones.

the elastic constant with the lowest value is responsible for the deformation. A twist deformation of the director can change the state of polarization and the phase shift. For perpendicular alignment the bend and splay elastic constants are involved, and the elastic deformation energy becomes higher. The higher the elastic constant, the harder the deformation of the director profiles. This is why there is better smoothing of the phase profile for the perpendicular aligned grating.

The elastic deformation not only has consequences for the smoothness of the phase profile, but also for the fly-back length. The fly-back length Λ_{fb} is defined as part of the period Λ, which is needed to make the phase jump form π to 0. The simulations and the measurements show that in our case the fly-back length is about the thickness of the cell. This corresponds with results for binary gratings, where the resolution limit is given approximately by the thickness of the cell. A good model concept with which to judge the performance of a grating is infact the fly-back length. It is the lateral distance needed to reset the retardation to its initial value after one period. The fly-back length determines the maximum efficiency in the case of a perfect blazed grating profile, as in the case of perpendicular alignment. Within a simple model one finds for the efficiency η,

$$\eta = \eta_o \cdot \eta_{fb} = \eta_o \cdot \left(\frac{\Lambda - \Lambda_{fb}}{\Lambda} \right)^2, \tag{10.39}$$

where η_o is the initial efficiency for a perfect structure and η_{fb} is the reduction factor of the efficiency due to the fly-back length (Scharf et al., 2001). For a device with 12 electrodes, a period of $\Lambda = 72$ μm, and a thickness of $d = 6$ μm, we find a reduction factor of $\eta_{fb} = 0.84$. With this result it becomes clear that the maximum efficiency is determined by elastic deformations of the liquid crystal caused by the fly-back length. Although a short fly-back length demands a small cell thickness, the fine structure modulation becomes less pronounced for thicker cells. Therefore, the optimization process of the diffraction efficiency and switching performance becomes a complicated multiparameter problem. Recent efforts have been made to include all parameters in the design of high-resolution beam steering devices (Wolffer et al., 1998; Apter et al., 2004; Bahat-Treidel et al., 2005).

Following our argumentation above, the voltages required to obtain maximum diffraction efficiency in the case of perpendicular alignment are found to be higher than for a parallel aligned grating. This is because one has to compensate for the reduced tilt angle modulation due to the in-plane electric field effect. To maximize the diffraction efficiency, the voltage profile and the maximum phase shift have to be adapted. The asymmetry of the diffraction pattern for perpendicularly aligned gratings is a result of the pretilt angle and is also found in binary gratings. Because of the pretilt at the substrate surface, an initial direction of reorientation for the nematic director is given. This allows a uniform reorientation process without the appearance of domains of different tilt directions. However, the symmetry of the reorientation with respect to the electrode edges is no longer given. Thus, changing the beam steering direction by simply reversing the voltage profile does not give the same result for the perpendicularly aligned gratings. The main difference in efficiency between the two switching states is caused by the width of the fly-back length. The efficiency and performance of nematic liquid crystal diffractive gratings depend on the alignment direction of the nematic director with respect to the electrode grating. The main difference is that different elastic constants and distributions of the electric fields determine the deformation of the nematic director. For the perpendicular aligned cells the pretilt angle has a remarkable influence and depends on the sign of the selected ramp. Although perpendicular alignment shows slightly higher efficiency, the asymmetry for the reversed ramp is a major drawback for the electric driving if switching in both directions is needed.

The small distance between electrodes required for large diffraction angles results in strong interaction of the different electric fields. In-plane fields play an important role when discussing the performance of the grating. The fly-back length of the grating structure is given by the thickness of the liquid crystal cell and determines the maximum efficiency. Thus, with an increased number of electrodes, higher efficiency is obtained, but the electrical driving becomes more complex. In the liquid crystal grating with high spatial frequency of the electrodes, the director profile is provided by the equilibrium between the elastic and electrical forces. Both fields, the electric field and the director field, vary in space. This is the reason why it is very difficult to estimate the switching time. In general, switching times become very long for systems that have to be brought to an equilibrium director configuration. Usually one uses a special driving scheme. The behavior differs greatly when different diffraction orders are considered. Typical switching times are about 100 ms. The two different processes, switching on and switching off the grating, have approximately the same switching time. The switching times are long and cannot be reduced by using high driving voltages, because the blazed grating profile represents equilibrium between elastic and electric torque.

10.8 SWITCHABLE DIFFRACTIVE LENSES

Diffractive lenses are usually known as Fresnel lenses and can be made binary or multilevel. The easiest way to obtain a reconfigurable lens system is to use a spatial light modulator and to control the phase profiles of the pixels. This leads to lenses with low numerical aperture because of the low resolution of the modulator. If the pixel pitch of the modulator is $p_{\min}$, the maximum deviation angle for visible light $\Theta_{\max}$ is given by the grating equation as

$$\sin\Theta = \frac{\lambda}{p_{\min}}. \tag{10.40}$$

For a pixel pitch of 15 mm and a wavelength of $\lambda = 500$ nm, Equation (10.40) gives an angle of $\Theta_{\max} = 0.04°$. The numerical aperture is extremely limited. The maximum deflection angle is given by the resolution of the liquid crystal structure. To effectively realize diffractive optical elements, a phase jump of one wavelength should be realized. The retardation of conventional liquid crystal materials today is $\Delta n = 0.2$. For operation in the visible at $\lambda = 500$ nm, a minimum thickness of the devices of $d = \lambda/\Delta n = 0.5/0.2\ \mu\text{m} = 1\ \mu\text{m}$ is necessary. The lateral resolution of about the thickness of the liquid crystal layer is therefore 1 μm. We see that, using Equation (10.40), the maximum deflection angle is about 30°. To obtain such high N.A. values, two different concepts have been proposed: alignment patterning and fabrication of ring electrode structures.

10.8.1 Adaptive Lenses Using Liquid Crystal Displays

Pixelated liquid crystal displays have already been used extensively to make Fresnel lenses (Cottrell et al., 1990; Tam, 1992; Laude, 1998). If the liquid crystal display produces phase-only modulation, the Fresnel lenses can also encode nonuniform amplitude transmission filters. This makes it possible to encode special functions and correct for aberrations (Márquez et al., 2006). Optical elements are formed by local phase encoding and do not require special knowledge of polarization optics, despite the fact that the spatial light modulator is a polarization-sensitive device. To obtain better performance it is convenient to combine conventional lenses with a programmable diffractive Fresnel lens. The systems are polarization-dependent and can lead to increased functionality (Davis et al., 2004). The liquid crystal display affects only vertical polarization. Therefore the combination of the glass and the diffractive lens acts as a birefringent lens because vertical polarization is affected by both lenses, whereas horizontal polarization is affected only by the glass lens. Because this birefringent lens has different focal lengths for the two orthogonal polarizations, it produces two images with different locations and magnifications and with orthogonal states of polarization.

10.8.2 Alignment Patterned Lenses

Diffractive lenses with alignment patterning are binary systems and can be switched (Patel and Rastani, 1991). Their performance is limited by the defect structure and the resolution of the liquid crystal structure. Fabrication and operational principle are the same as for gratings (see Section 10.6 for details).

10.8.3 Polymer Dispersed Systems

Besides surface alignment patterning and electrode structuring, polymer-dispersed systems are proposed to realize switchable lenses. A tunable electronic lens using refractive index modulation in polymer-dispersed liquid crystal was obtained by a conventional fabrication technique due to phase separation of the polymer and the nematic liquid crystal from a mixture under UV irradiation (Fan et al., 2005). Under certain experimental conditions, a grin lens can be obtained with spatially varying refractive index (Ren et al., 2003a). The key element for fabricating the grin PDLC lens is a patterned photomask with circularly variable optical density in the radial direction. The liquid crystal monomer in the zones cured with a higher UV intensity has smaller liquid crystal droplets. Conversely, the zones with a weaker UV exposure have larger nanoscale droplets. When a uniform voltage is applied to the zone plate, the refractive indices of the zones with larger liquid crystal droplets are changed so that the beam diffraction efficiency and the focusing behaviors of the lens can be adjusted. As the nanoscale liquid crystal droplets in the polymer matrix are randomly distributed, the lens is polarization-independent and exhibits a fast switching time.

10.8.4 Binary Encoding and Fresnel Zone Plates

The simplest way to obtain high-N.A. switchable lenses is by implementation of an electrode pattern designed to form Fresnel zones. An example is shown in Figure 10.25. A typical lens consists of circular electrodes (Sun et al., 2000). Systems with linear electrode arrays have been considered too (Kowel et al.,

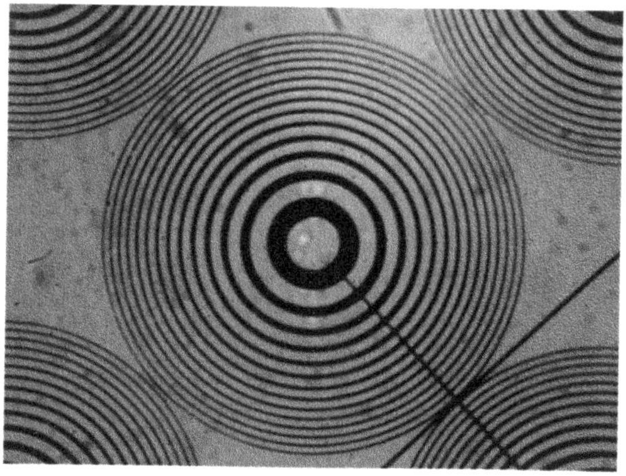

Figure 10.25 Part of a switchable Fresnel lens array. The lens diameter is 2 mm. The lens is seen between crossed polarizers at $\lambda = 633$ nm. Dark rings corresponds to the electrodes.

1984a; Riza and DeJule, 1994). We shall discuss a simple system, as shown in Figure 10.25, which has only one electrode structured surface. Such a binary system has a maximum efficiency of 41 percent for linear polarized light and leads to multiple focal points. To avoid multiple foci, a Gabor lens has been proposed (McOwan et al., 1993). In the example we discuss here, the electrode structure represents a binary diffractive lens structure as used in Fresnel zone plates (Hecht, 1998). The electrode areas are switched vertically and a phase profile can be produced. A typical realization has, for instance, 10 mm diameter and a design focal length of 100 mm. Measurements for such a system are shown in Figure 10.26 and Figure 10.27. The wavelength for characterization is 632.8 nm (He–Ne laser). The thickness is 6 μm and the cell was filled with the liquid crystal mixture ZLI 2461 from Merck ($\Delta n = 0.13$, $n_e = 1.622$, $n_o = 1.492$, $\Delta\varepsilon(1\ \text{kHz}) = 1$). The alignment is homogeneous planar antiparallel, implemented with standard polyimide rubbing. The measured diffraction efficiency for linear polarized light in Figure 10.26 corresponds to the intensity in the focal spot measured with a photodetector. For voltage below threshold, the lens already has a certain efficiency due to errors in the fabrication process. When threshold is reached, the efficiency increases and reaches a maximum at about 3.5 V. Here optimal working conditions are found for which the retardation between adjacent zones is half a wavelength. Further increase of the voltage leads to a decrease in efficiency. This is caused by unfavorable phase shift and additionally by the influence of fringing fields. The fringing fields cause in-plane reorientation of the nematic director and worsen the performance of the device in the same way as for high-resolution gratings (see Sections 10.6 and 10.7). The lens has the particularity that the grating period varies over the radius. In the center, large uniform electrode areas are present and at the edges the grating has high spatial period. In the case of a lens with 10 mm diameter and 100 mm focal length, the grating period at the lens rim is 12 μm. An additional problem if a planar aligned liquid crystal structure is used in combination with circular electrodes is the varying orientation of the liquid crystal director with respect to the electrodes (Sato et al., 1989). For low lateral resolution areas this is not important. However, for

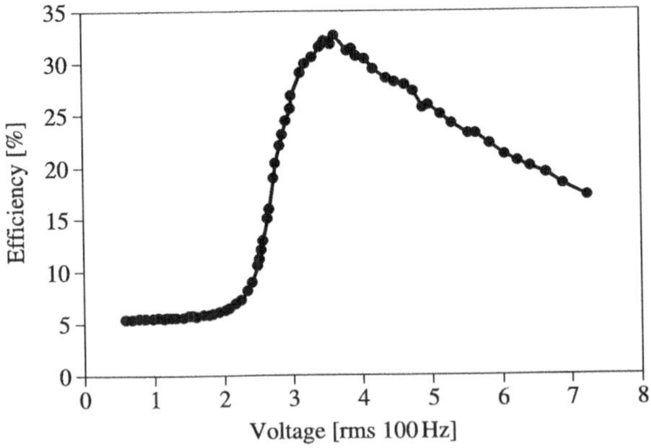

Figure 10.26 Efficiency of a binary switchable Fresnel lens as a function of the voltage.

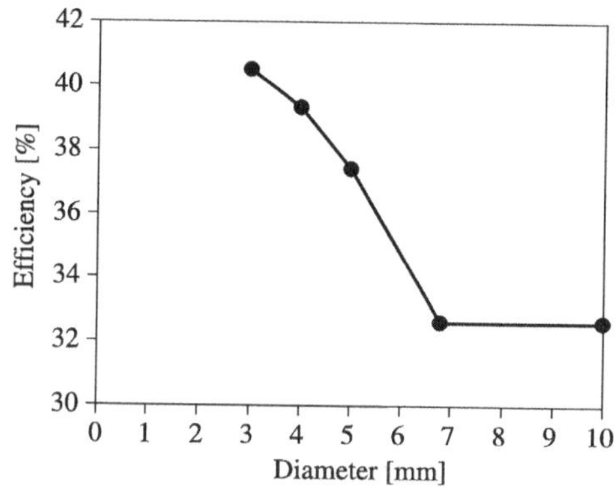

Figure 10.27 Efficiency as a function of the clear apertures of the lens. The limited resolution of the liquid crystal lens system becomes evident for high-spatial-frequency areas of the lens system. At larger diameter the lens performance decreases drastically (applied voltage 3.5 V_{rms}).

the highest resolution at the outer rim it alters the performance of the lens. At a given voltage, fringing properties along and perpendicular to the alignment direction are different, as seen in Chapter 8. To obtain an impression of such effects on the device performance, we show in Figure 10.27 measurement of the efficiency as a function of the diameter. For small diameters of the lens and low numerical apertures, the lens has nearly theoretical diffraction efficiency. The efficiency decreases with increasing diameter to approximately 31%. The decrease is caused by the intrinsic resolution limit given by the fringing field. This can be corrected, reducing the electrode area to about the thickness of the liquid crystal layer. If in-plane reorientation of the liquid crystal director occurs, the phase profiled for the design polarization is altered. To explain all the effects in detail, simulations are necessary that include all effects. To date this has not been done, although multilevel lenses based on such concepts attract more and more attention for variable focus glass applications.

REFERENCES

Apter, B., Efron, U., and Bahat-Treidel, E. (2004) On the fringing-field effect in liquid crystal beam-steering devices, *Appl. Opt.* **43**, 11–19.

Bahat-Treidel, E., Apter, B., and Efron, U. (2005) Experimental study of phase-step broadening by fringing fields in a three-electrode liquid crystal cell, *Appl. Opt.* **44**, 2989–2995.

Berreman, D.W. (1973) Optics in smoothly varying anisotropic planar structures: Application to liquid crystal twist cells, *J. Opt. Soc. Am. A* **63**(11), 1374–1380.

Blinov, L.M. and Chigrinov, V.G. (1996) *Electrooptic Effects in Liquid Crystal Materials*, Springer, New York.

Bos, P.J., Chen, J., Doane, J.W., Smith, B., Holton, C., and Glenn, W. (1995) An optically active diffraction grating device for a high efficiency light valve, *SID Digest*, **1995**, 601–604.

Bouvier, M. and Scharf, T. (2000) Analysis of nematic-liquid-crystal binary gratings with high spatial frequency, *Opt. Eng.* **39**, 2129–2137.

Bouvier, M. (2000) Diffractive liquid crystals optical elements, Ph.D. thesis, University of Neuchâtel.

Bowley, C.C., Yuan, H., and Crawford, G.P. (1999) Morphology of holographically-formed polymer dispersed liquid crystals (H-PDLC), *Mol. Cryst. Liq. Cryst. Technol. A* **331**, 2069–2076.

Broer, D.J., Hikmet, R.A.M., and Challa, G. (1989) In-situ photopolymerization of oriented liquid crystalline acrylates, 4. Influence of lateral methyl substituent on monomer oriented polymer network properties of a mesogenic diacrylate, *Makromol. Chem.* **190**, 3201–3215.

Broer, D.J. (1995) Creation of supramolecular thin film architectures with liquid-crystalline networks, *Mol. Cryst. Liq. Cryst.* **261**, 513–523.

Brown, C.V., Kriezis, E.E., and Elston, S.J. (2002) Optical diffraction from a liquid crystal phase grating, *J. Appl. Phys.* **91**, 3495–3500.

Brown, C.V. and Kriezis, E.E. (2003) Calculation of the efficiency of polarization insensitive surface stabilized ferroelectric liquid crystal diffraction gratings, *Appl. Opt.* **42**, 2257–2263.

Brown, C.V. and Kriezis, E.E. (2004) Surface-stabilized ferroelectric liquid crystal diffraction gratings with micrometer-scale pitch, *Appl. Opt.* **43**, 5287–5294.

Butler, J. and Malcuit, M.S. (2000) Diffraction properties of highly birefringent liquid crystal composite gratings, *Opt. Lett.* **25**, 420–422.

Caputo, R., De Sio, L., Veltri, A., and Umeton, C. (2005) Observation of two-wave coupling during the formation of POLICRYPS diffraction gratings, *Opt. Lett.* **30**, 1840–1842.

Chaudhari, P., Lacey, J.A., Lien, S.C.A., and Speidell, J.L. (1998) Atomic beam alignment of liquid crystals, *Jpn J. Appl. Phys.* **37**, L55–L56.

Chaudhari, P., Lacey, J., Doyle, J., Galligan, E., Lien, S.C.A., Callegari, A., Hougham, G., Lang, N.D., Andry, P.S., John, R., Yang, K.H., Lu, M., Cai, C., Speidell, J., Purushothaman, S., Ritsko, J., Samant, M., Stöhr, J., Nakagawa, Y., Katoh, Y., Saitoh, Y., Sakai, K., Satoh, H., Odahara, S., Nakano, H., Nakagaki, J., and Shiota, Y. (2001) Atomic-beam alignment of inorganic materials for liquid-crystal displays, *Nature* **411**, 56–59.

Chen, J., Bos, P.J., Vithana, H., and Johnson, D.L. (1995) An electro-optical controlled liquid crystal diffraction grating, *Appl. Phys. Lett.* **67**, 2588–2590.

Cheng, J. and Thurston, R.N. (1981) The propagation of disclinations in bistable switching, *J. Appl. Phys.* **52**, 2766–2775.

Cottrell, D.M., Davis, J.A., Hedman, T.R., and Lilly, R.A. (1990) Multiple imaging phase-encoded optical elements written as programmable spatial light modulators, *Appl. Opt.* **29**, 2505–2909.

Crawford, G.P., Eakin, J.N., Radcliffe, M.D., Callan-Jones, A., and Pelcovits, R.A. (2005) Liquid-crystal diffraction gratings using polarization holography alignment techniques, *J. Appl. Phys.* **98**, 123102.

Davis, J.A., Adachi, J., Fernandez-Pousa, C.R., and Moreno, I. (2001) Polarization beam splitters using polarization diffraction gratings, *Opt. Lett.*, **26**, 587–589.

Davis, J.A. and Evans, G.H. (2004) Polarizing binary diffraction grating beam splitter, *Opt. Lett.* **29**, 1443–1445.

Davis, J.A., Evans, G.H., Crabtree, K., and Moreno, I. (2004) Programmable birefringent lenses with a liquid-crystal display, *Appl. Opt.* **43**, 6235–6241.

De Filpo, G., Nicoletta, F.P., Macchione, M., Cupelli, D., and Chidichimo, G. (2001) Switchable gratings from polymerized nematic emulsions, *Adv. Funct. Mater.* **11**, 457–460.

DIMOS (2005) *DIMOS 2.0*, Autronic Melchers GmbH, Greschbachstr. 29, D-76229 Karlsruhe, Germany, available at http://www.autronic-melchers.com/.

Drzaic, P.S. (1995) *Liquid Crystal Dispersions*, World Scientific, Singapore.

Duca, D., Sukhov, A.V., and Umeton, C. (1999) Detailed experimental investigations on recording of switchable diffraction gratings in polymer dispersed liquid crystal films by UV laser curing, *Liq. Cryst.* **26**, 931–937.

Escuti, M.J., Qi, J., and Crawford, G.P. (2003) Tunable face-centered-cubic photonic crystal formed in holographic polymer dispersed liquid crystals, *Opt. Lett.* **28**, 522–524.

Fan, Y., Ren, H., and Wu, S.T. (2005) Electrically switchable Fresnel lens using a polymer-separated composite film, *Opt. Exp.* **13**, 4142–4147.

Faris, S.M. and He, Z. (1999) *Beam Steering Device for Large-Area Displays*, SID International Symposium, San José, CA, 1094–1097.

Friedman, L.J., Hobbs, D.S., Lieberman, S., Corkum, D.L., Nguyen, H.Q., Resler, D.P., Sharp, R.C., and Dorschner, T.A. (1996) Spatially resolved phase imaging of a programmable liquid crystal grating, *Appl. Opt.* **35**, 6236–6240.

Fuh, A.Y.G., Lin, C.H., and Huang, C.Y. (2002) Dynamic pattern formation and beam-steering characteristics of cholesteric gratings, *Jpn J. Appl. Phys.* **41**, 211–218.

Fujieda, I. (2001) Liquid crystal phase grating based on in plane switching, *Appl. Opt.* **40**, 6252–6259.

Fujieda, I., Mikami, O., and Ozawa, A. (2003) Active optical interconnect based on liquid-crystal grating, *Appl. Opt.* **42**, 1520–1525.

Galatola, P., Oldano, C., and Kumar, P.B.S. (1994) Symmetry properties of anisotropic dielectric gratings, *J. Opt. Soc. Am. A* **11**, 1332–1341.

Goodman, J.W. (1996) *Introduction to Fourier Optics*, 2nd International edn, McGraw-Hill, New York.

Gu, L., Chen, X., Jiang, Y., Liu, J., and Chen, R.T. (2005) High-spatial-frequency liquid crystal phase gratings with double-sided striped electrodes, *SPIE* **5741**, 180–189.

Hällstig, E. (1994) *Nematic Liquid Crystal Spatial Light Modulators for Laser Beam Steering*, Acta Universitatis Upsaliensis, Uppsala.

He, Z., Ling, Z., Nose, T., and Sato, S. (1997) Liquid crystal display devices with slit patterned electrode structures, *Jpn J. Appl. Phys.* **36**, 2128–2133.

He, W.T., Nose, T., and Sato, S. (1998b) Novel liquid crystal grating with a relief structure by simple UV irradiation process, *Jpn J. Appl. Phys.* **37**, 4066–4069.

He, W.T., Nose, T., and Sato, S. (1998c) A double layer polymer/liquid crystal grating with polarization direction independent diffraction properties, *Jpn J. Appl. Phys.* **37**, L795–L797.

He, Z., Nose, T., and Sato, S. (1998a) Polarization properties of an amplitude nematic liquid crystal grating, *Opt. Eng.* **37**(11), 2885–2898.

He, Z. and Sato, S. (1998) Polarization properties of inversely twisted nematic liquid crystal gratings, *Appl. Opt.* **37**, 6755–6763.

Hecht, E. (1998) *Optics*, 3rd edn, Addison Wesley, Reading.

Herzig, H.-P. (Ed.) (1998) *Micro-Optics*, Taylor and Francis, London.

Hirabayashi, K., Yamamoto, T., and Yamaguchi, M. (1995) Free-space optical interconnections with liquid-crystal microprism arrays, *Appl. Opt.* **34**, 2571–2580.

Holoeye Photonics AG (2005) Albert-Einstein-Str. 14, 12489 Berlin-Adlershof, Germany, www.holoeye.com.

Honma, M. and Nose, T. (2004) Liquid crystal blazed grating with azimuthally distributed liquid-crystal directors, *Appl. Opt.* **43**, 5193–5197.

Iizuka, K. (2002) *Elements of Photonics*, Vols 1 & 2, Wiley, New York.

Jepsen, M. and Gerritsen, H.J. (1996) Liquid crystal filled gratings with high diffraction efficiency, *Opt. Lett.* **21**, 1081–1083.

Kashnow, R.A. and Bigelow, J.E. (1973) Diffraction from a liquid crystal phase grating, *Appl. Opt.* **12**, 2302–2304.

Kim, J.H., Yoneya, M., and Yokoyama, H. (2002) Tri-stable nematic liquid-crystal device using AFM-fabricated micro-patterned surface alignment, *Nature* **420**, 159–162.

Kim, S.H. and Chien, L.C. (2004) Liquid crystal blazed gratings formed in a single step though photo-induced localization of polymer into a prismatic structure, *Opt. Exp.* **12**, 1238–1242.

Klaus, W., Shinomori, S., Ide, M., Tsuchiya, M., and Kamiya, T. (1996) Efficient beam deflector with a blazed liquid crystal phase grating, Presented at Optical Computing, Sendai, 1996, 28–29.

Klaus, W., Ide, M., Morokawa, S., Tsuchiya, M., and Kamiya, T. (1997) Angle-independent beam steering using a liquid crystal grating with multi-resistive electrodes, *Opt. Comm.* **138**, 151–157.

Kosmopoulos, J.A. and Zenginoglou, H.M. (1987) Geometrical optics approach to the obliquely illuminated nematic liquid crystal grating: numerical results, *Appl. Opt.* **26**, 1714–1721.

Kossyrev, P., Sousa, M.E., and Crawford, G. (2004) One and two dimensionally polymer networks in liquid crystals for switchable diffractive optical applications, *Adv. Funct. Mat.* **14**, 1227–1232.

Kowel, S.T., Cleverly, D.S., and Kornreich, P.G. (1984a) Focusing by electrical modulation of refraction in a liquid crystal cell, *Appl. Opt.* **23**, 278–289.

Kulick, J.H., Jarem, J.M., Lindquist, R.G., Kowel, S.T., Friends, M.W., and Leslie, T.M. (1995) Electrostatic and diffraction analysis of a liquid crystal device utilizing fringing fields: applications to three dimensional displays, *Appl. Opt.* **34**, 1901–1922.

Laude, V. (1998) Twisted nematic liquid crystal pixelated active lens, *Opt. Comm.* **153**, 134–152.

Le Doucen, M. and Pellet-Finet, P. (1998) Polarization properties and diffraction efficiencies of binary anisotropic gratings: general study and experiments on ferreoelectric liquid crystals, *Opt. Comm.* **151**, 321–330.

Lee, S.N., Chien, L.C., and Sprunt, S. (1998) Polymer-stabilized diffraction gratings from cholesteric liquid crystals, *Appl. Phys. Lett.* **72**, 885–887.

Lester, G.A., Coulston, S.J., and Strudwick, A.M. (2005) Defect suppression in liquid crystal filled polymer devices, *J. Opt. A: Pure Appl. Opt.* **7**, 290–295.

Lindquist, R.G., Kulick, J.H., Nordin, G.P., Jarem, J.M., Kowel, S.T.J., and Friends, M. (1994) High-resolution liquid crystal phase grating formed by fringing fields from interdigitated electrodes, *Opt. Lett.* **19**, 670–672.

Liu, S. and Chen, Y. (1995) Multilevel binary phase grating polarization device with a birefringent substrate, *Opt. Lett.* **20**, 1832–1834.

Liu, S.Q., Chen, Y.S., Wu, N.J., and Ignatiev, A. (1997) Realization of a polarizer employing the combined effects of birefringence and diffraction, *Opt. Lett.* **22**, 1518–1520.

Löfving, B., and Hard, S. (1998) Beam steering with two ferroelectric liquid crystal spatial light modulators, *Opt. Lett.* **23**, 1541–1543.

Lu, M. and Yang, K. (1998) Nematic liquid crystal phase gratings for reflective spatial light modulators, *Jpn. J. Appl. Phys.* **37**, L587–L590.

MacManamon, P.F., Watson, E.A., Dorschner, T.A., and Barnes, L.J. (1993) Applications look at the use of liquid crystal writable gratings for steering passive radiation, *Opt. Eng.* **32**, 2657–2664.

Márquez, A., Iemmi, C., Campos, J., and Yzuel, M.J. (2006) Achromatic diffractive lens written onto a liquid crystal display, *Opt. Lett.* **31**, 392–394.

McOwan, P.W., Gordan, M.S., and Hossack, W.J. (1993) A switchable liquid crystal Gabor lens, *Opt. Comm.* **103**, 189–193.

Moreno, I., Yzuel, M.J., Campos, J., and Vargas, A. (2004) Jones matrix approach for polarization Fourier optics, *J. Mod. Optics* **51**, 2031–2038.

Mukohzaka, N., Yoshida, N., Toyoda, H., Kobayashi, Y., and Hara, T. (1994) Diffraction efficiency analysis of a parallel-aligned nematic-liquid-crystal spatial light modulator, *Appl. Opt.* **33**, 2804–2811.

Murai, H., Gotoh, T., Suzuki, M., Hasegawa, E., and Mizoguchi, K. (1992) Electro-optical properties for liquid crystal phase gratings, *SPIE* **1665**, 230–239.

Nieuborg, N., Kirk, A., Morlion, B., Thienpont, H., and Veretennicoff, I. (1997) Polarization-selective diffractive optical elements with an index-matching gap material, *Appl. Optics* **36**, 4681–4685.

Oh-e, M. and Kondo, K. (1997) The in-plane switching of homogeneously aligned liquid crystals, *Liq. Cryst.* **22**, 379–390.

Okada, H., Bos, P.J., and Onnagawa, H. (1998) In-plane liquid crystal beam steering devices with a beam separation structure, *Jpn J. Appl. Phys.* **37**, 2576–2580.

Park, J.H., Yu, C.J., Kim, J., Chung, S.Y., and Lee, S.D. (2003) Concept of a liquid crystal polarization beamsplitter based on binary phase gratings, *Appl. Phys. Lett.* **83**, 1918–1920.

Park, S., Padeste, C., Schift, H., Gobrecht, J., and Scharf, T. (2005) Chemical nanopatterns via nanoimprint lithography for simultaneous control over azimuthal and polar alignment of liquid crystals, *Adv. Mater.* **17**, 1398–1401.

Patel, J.S. and Rastani, K. (1991) Electrically controlled polarization-independent liquid crystal Fresnel lens array, *Opt. Lett.* **16**, 532–534.

Pellat-Finet, P. and Le Doucen, M. (1995) Polarization properties of birefringence gratings, *Optik* **100**, 159–166.

Penterman, R., Klink, S.I., de Koning, H., Nisato, G., and Broer, D.J. (2002) Single-substrate liquid-crystal displays by photo-enforced stratification, *Nature* **417**, 55–58.

Pérennè, F. and Wu, Z.Y. (1997) Resolution and response time dependence of ferroelectric liquid crystal optically addressed spatial light modulators on grating profiles, *Appl. Opt.* **36**, 3825–3834.

Ren, H., Fan, Y.H., and Wu, S.T. (2003a) Tunable Fresnel lens using nanoscale polymer-dispersed liquid crystals, *Appl. Phys. Lett.* **83**, 1515–1517.

Ren, H., Fan, Y.H., and Wu, S.T. (2003b) Prism grating using polymer-stabilized nematic liquid crystal, *Appl. Phys. Lett.* **82**, 3168–3170.

Resler, D.P., Hobbs, D.S., Sharp, R.C., Friedman, L.J., and Dorschner, T.A. (1996) High-efficiency liquid-crystal optical phased-array beam steering, *Opt. Lett.* **21**, 689–691.

Riza, N.A. and DeJule, M.C. (1994) Three-terminal adaptive nematic liquid-crystal lens device, *Opt. Lett.* **19**, 1013–1015.

Riza, N. and Madamopoulos, N. (1998) Photonic delay line using electrically switched gratings in polymer dispersed liquid crystals, *Opt. Eng.* **37**, 3061–3065.

Rokushima, K. and Yamakita, J. (1983) Analysis of anisotropic dielectric gratings, *J. Soc. Am.* **73**, 901–908.

Sakata, H. and Nishimura, M. (2000) Switchable zero-order diffraction filters using fine-pitch phase gratings filled with liquid crystals, *Jpn J. Appl. Phys.* **39**, 1516–1521.

Sato, S., Nose, T. Yamaguchi, R., and Yanase, S. (1989) Relationship between lens properties and director orientation in a liquid crystal lens, *Liq. Cryst.* **5**, 1435–1442.

Schadt, M., Schmitt, K., Kozinkov, V., and Chigrinov, V. (1992) Surface induced parallel alignment by linearly polymerized photopolymers, *Jpn J. Appl. Phys.* **31**, 2155–2164.

Scharf, T., and Bohley, C. (2002) Light propagation through alignment patterned liquid crystal gratings, *Mol. Cryst. Liq. Cryst.* **375**, 491–500.

Scharf, T., Bouvier, M., and Dändliker, R. (2001) Multilevel nematic phase gratings, *SPIE* **4418**, 31–37.

Senyuk, B.I., Smalyukh, I.I., and Lavrentovich, O.D. (2005) Switchable two-dimensional gratings based on field-induced layer undulations in cholesteric liquid crystals, *Opt. Lett.* **30**, 349–351.

Shintech (2005) *LCDMaster*, Shintch Inc., 90 Ogo-oku, Tabuse-Cho, Kumage-Gun, Yamaguchi-Ken, 742–1512, Japan, available at http://www.shintech.jp/.

Stalder, M. and Schadt, M. (1996) *Beam steering devices based on switchable liquid crystal blazed gratings*, Presented at the Eurodisplay '96, Birmingham, UK.

Strudwick, A.M. and Lester, G.A. (1999) Electrically controlled phase grating for instrumentation applications, *Electron. Lett.* **35**, 1374–1376.

Subacius, D., Bos, P.J., and Lavrentovich, O.D. (1997) Switchable diffractive cholesteric gratings, *Appl. Phys. Lett.* **71**, 1350–1352.

Subacius, D., Shiyanovskii, S.V., Bos, Ph., and Lavrentovich, O.D. (1997b) Cholesteric gratings with field-controlled period, *Appl. Phys. Lett.* **71**, 3323–3325.

Sun, Y., Kowel, S.T., and Nordin, G.P. (2000) Circular geometry liquid crystal adaptive lens with Fresnel lens phase profile, *Proc. SPIE* **3955**, 120–131.

Sutherland, R.L., Tondiglia, V.P., Natarajan, L.V., Bunning, T.J., and Adams, W.W. (1994) Electrically switchable volume gratings in polymer-dispersed liquid crystals, *Appl. Phys. Lett.* **64**, 1074–1076.

Tam, E.C. (1992) Smart electro-optical zoom lens, *Opt. Lett.* **17**, 369–371.

Tanguay, A.R., Jr., Chavel, P., Strand, T.C., Wu, C.S. (1984) Polarization properties of the variable-grating-mode liquid crystal device, *Opt. Lett.* **9**, 174–176.

Titus, C.M. and Bos, P.J. (1997) Efficient, polarization-independent, reflective liquid crystal phase grating, *Appl. Phys. Lett.* **71**, 2239–2241.

Titus, C.M., Kelly, J.R., Gartland, E.C., Shinyanovski, S.V., Anderson, J.A., and Bos, P.J. (2001) Asymmetric behavior of liquid-crystal diffraction gratings, *Opt. Lett.* **26**, 1188–1190.

Varghese, S., Crawford, G.P., Bastiaansen, C.W.M., de Boer, D.K.G., and Broer, D.J. (2005) High pretilt four-domain twisted nematic liquid crystal display by microrubbing: Process, characterization, and optical simulation, *J. Appl. Phys.* **97**, 053101.

Wang, X., Wilson, D., Muller, R., Maker, P., and Psaltis, D. (2000) Liquid-crystal blazed-grating beam deflector, *Appl. Opt.* **39**, 6545–6555.

Wang, B., Wang, X., and Bos, P.J. (2004) Finite-difference time-domain calculations of a liquid-crystal-based switchable Bragg grating, *J. Opt. Soc. Am. A* **21**, 1067–1072.

Wen, B., Petschek, R.G., and Rosenblatt, C. (2002) Nematic liquid-crystal polarization gratings by modification of surface alignment, *Appl. Opt.* **41**, 1246–1250.

Witzigmann, B., Regli, P., and Fichtner, W. (1998) Rigorous electromagnetic simulation of liquid crystal displays, *J. Opt. Soc. Am. A* **15**, 753–757.

Wolffer, N., Vinouze, B., and Gravey, P. (1998) Holographic switching between single mode fibres based on electrically addressed nematic liquid crystal gratings with high deflection accuracy, *Opt. Comm.* **160**, 42–46.

Wu, J.J., Wu, Y.S., Chen, F.C., and Chen, S.H. (2002) Formation of phase grating in planar aligned cholesteric liquid crystal film, *Jpn J. Appl. Phys.* **41**, L1318–L1320.

Xu, F., Ford, J., and Fainman, Y. (1995) Polarization-selective computer generated holograms: design, fabrication and applications, *Appl. Opt.* **34**, 256–266.

Yeh, P. and Gu, C. (1999) *Optics of Liquid Crystal Displays*, Wiley, New York.

Yu, C.-J., Park, J.-H., Kim, J., Jung, M.-S., and Lee, S.-D. (2004) Design of binary diffraction gratings of liquid crystals in linear graded phase model, *Appl. Opt.* **43**, 1783–1788.

Yu, C.J., Kim, D.W., Kim, J., and Lee, S.D. (2005) Polarization gratings based on a photo-aligned liquid crystal in an oppositely twisted binary configuration, *Opt. Lett.* **30**, 1995–1997.

Zenginoglou, H.M. and Kosmopoulos, J.A. (1988) Geometrical optics approach to the obliquely illuminated nematic liquid crystal diffraction grating, *Appl. Opt.* **27**, 3898–3901.

Zenginoglou, H.M. and Kosmopoulos, J.A. (1989) Geometrical optics approach to the nematic liquid crystal grating: leading term formulas, *Appl. Opt.* **28**, 3516–3519.

Zhang, Y., Wang, B., Bos, P.J., Colegrove, J., and Chung, D.B. (2005) High-efficiency, liquid-crystal-based, controllable diffraction grating, *J. Opt. Soc. Am. A* **22**, 2510–2515.

11

BRAGG DIFFRACTION

11.1 REFLECTION BY MULTILAYER STRUCTURES

The propagation of electromagnetic radiation in periodic media exhibits many interesting features. Diffraction of light from periodic strain variations accompanying a sound wave and forbidden bands in photonic crystals are two examples. The phenomena are widely used in filtering and mirror design. The general properties of periodic layered media have been widely discussed in the literature (Born and Wolf, 1980; Heavens, 1991; Yariv and Yeh, 2003). Usually, matrix methods are used to calculate the reflection and transmission of such systems. The theory has a strong formal analogy to the quantum theory of electrons in crystals and one can make use of the concept of Bloch waves, forbidden bands, and evanescent waves. The simplest form of period media is the stratified periodic layered structure. The optical properties of such structures can be treated with the theory given in Chapter 4, based on matrix methods. We do not want to repeat here the analytical approach, as it is very well described in the existing literature. However, for a multilayer structure, there exists a reflection band. The reflected intensity depends on the number of layers and the refractive index difference of the layer materials. Which wavelengths will be reflected depends on the period and refractive index. The width of the reflection band is found by using the period and the reflective index difference of the materials. At oblique angles of incidence, the reflection band shifts to shorter wavelengths. If several stacks with different periods are combined, (chirp) broadband reflection can be obtained. Useful for practical applications are analytical equations for calculating reflected intensities as a function of such given parameters as layer thickness, number of layers, and refractive indices. We will develop these formulas here based on well-established concepts. More complicated layer structures are searched by numerical simulations. Examples are given throughout this chapter.

11.1.1 Coupled Wave Description of Bragg Diffraction

An interesting approximate approach for solving problems with disturbances is the coupled mode theory developed originally for phase holograms (Dändliker, 1997, Yariv, 1989). The concept of coupled waves is used to describe the propagation of optical waves in a dielectric medium, which can be characterized by any kind of perturbation of the dielectric polarization with respect to a basic situation with known solutions. Often, weak perturbations that is small difference between refractive indices of successive media, is used as approximation. An approximation that does not lead to particular restrictions in the case of organic materials as their index of refraction vary mainly between 1.3 and 2. The coupled mode theory considers variations of the refractive index as perturbations that lead to coupling between the unperturbed normal mode of the structure. For Bragg diffraction periodic perturbations are interesting. We assume that the stratified structure is homogeneous in x and y. The dielectric tensor as a function of z can be expressed as

$$\varepsilon(z) = \varepsilon_{\mathrm{u}} + \delta\varepsilon(z), \tag{11.1}$$

where ε_{u} is the unperturbed part of the dielectric tensor and $\delta\varepsilon$ is the periodic perturbation in the z-direction. It is the only periodically varying part of the dielectric tensor. We assume that the electric field can be expressed as a linear sum of the normal modes and can be written as

$$E = \sum_k A_k(z) E_k(x, y) \mathrm{e}^{i(\omega t - \beta_k z)} \tag{11.2}$$

with an amplitude of the normal modes $E_{\mathrm{m}}(x, y)$ and an propagation constant β_m. The normal modes are solutions of the wave equation

$$\left(\frac{\partial^2}{\partial x^2} + \frac{\partial^2}{\partial y^2} + \omega^2 \varepsilon_0 \varepsilon_u \mu_0 - \beta_k^2\right) E_k = 0 \tag{11.3}$$

that describes the wave propagation of the homogeneous medium with index ε_{u} and without perturbation. Let us consider the propagation of an unperturbed mode $E_1 \mathrm{e}^{\mathrm{i}(\omega t - \beta_1 z)}$ in the perturbed medium described by the dielectric tensor Eq. (11.1). The presence of the dielectric perturbation gives rise to a new perturbation polarization term ΔP

$$\Delta P = \varepsilon_0 \, \delta\varepsilon \, E_1 \mathrm{e}^{i(\omega t - \beta_1 z)} \tag{11.4}$$

This polarization wave can feed energy into some other mode $E_2 e^{i(\omega t - \beta_2 z)}$ when we say that the dielectric perturbation couples between modes E_1 and E_2. The normal modes form a complete set and fulfill the equation (Yariv, 2003)

$$\int E_k^* E_l \, dx \, dy = W_k \delta_{kl} \tag{11.5}$$

where δ_{kl} is the Dirac delta function. Starting from the Maxwell equations (Chapter 2) for nonmagnetic materials ($\mu = 1$) the wave equation, including perturbation, can be written as

$$(\nabla^2 + \omega^2 \mu_0 \varepsilon_0 \varepsilon_u + \omega^2 \mu_0 \varepsilon_0 \delta\varepsilon (z)) E(x, y, z) = 0 \tag{11.6}$$

One substitutes the electric field E with the normal mode description in Equation (11.2) in the wave equation and using Equation (11.3) one gets

$$\sum_k \left(\frac{d^2 A_k(z)}{dz^2} - 2i\beta_k \frac{dA_k(z)}{dz} \right) E_k(x, y) e^{-i\beta_k z} = -\omega^2 \varepsilon_0 \mu_0 \sum_k A_k(z) \delta\varepsilon E_k(x, y) e^{-i\beta_k z} \tag{11.7}$$

One assumes that the dielectric perturbation is "weak", so that the variations of the mode amplitudes are small along z and satisfy the condition

$$\left| \frac{d^2 A_k(z)}{dz^2} \right| \ll \left| \beta_k \frac{dA_k(z)}{dz} \right|. \tag{11.8}$$

Thus, neglecting the second derivative in Eq. (11.7) leads to

$$\sum_k 2i\beta_k \frac{dA_k(z)}{dz} E_k(x, y) e^{-i\beta_k z} = \omega^2 \varepsilon_0 \mu_0 \sum_k A_k(z) \delta\varepsilon E_k(x, y) e^{-i\beta_k z}. \tag{11.9}$$

The orthogonality relation, Equation (11.5), can be used to eliminate the electric field. One take the scalar product with E_l^* and integrates over the area. The result is

$$\frac{dA_k(z)}{dz} = \frac{\omega^2 \varepsilon_0 \mu_0}{2i\beta_k W_k} \sum_l A_l(z) e^{i(\beta_k - \beta_l)z} \int E_k^*(x, y) \delta\varepsilon E_l(x, y) dx \, dy \tag{11.10}$$

The dielectric perturbation $\delta\varepsilon$ is assumed as a periodic function in z and can be expanded into a Fourier series as

$$\delta\varepsilon = \sum_{m \neq 0} \varepsilon_m e^{-im2\pi z/p}, \tag{11.11}$$

where the summation is taken over all m except $m = 0$. This is because of the definition of $\delta\varepsilon$ in Equation (11.1) where the constant part ε_u is already separated. The period of the structure is p. Substituting $\delta\varepsilon$ in Equation (11.10) leads to

$$\frac{\mathrm{d}A_k(z)}{\mathrm{d}z} = \frac{\omega^2\varepsilon_0\mu_0}{2\mathrm{i}\beta_k W_k}\sum_l\sum_{m\neq 0} A_l(z)\mathrm{e}^{\mathrm{i}(\beta_k-\beta_l)z}\mathrm{e}^{-\mathrm{i}m2\pi z/p}\int E_y^*(x,y)\varepsilon_m E_l(x,y)\mathrm{d}x\,\mathrm{d}y \quad (11.12)$$

The coupling coefficient can be defined as

$$C_{kl}^{(m)}(z) = \frac{\omega^2\varepsilon_0\mu_0}{2\beta_k W_k}\int E_k{}^*\varepsilon_m E_l\,\mathrm{d}x\,\mathrm{d}y \quad (11.13)$$

and reflects the magnitude of coupling between the kth and lth mode due to the mth Fourier component of the dielectric perturbation $\delta\varepsilon$. Equation (11.12) can be together with Equation (11.13) written in the form

$$\frac{\mathrm{d}A_k(z)}{\mathrm{d}z} = -\mathrm{i}\sum_l\sum_{m\neq 0} C_{kl}^{(m)}A_l(z)\mathrm{e}^{\mathrm{i}(\beta_k-\beta_l-2m\pi/p)z} \quad (11.14)$$

Equation (11.14) represents a set of coupled linear differential equations. In principle, an infinity number of modes are involved but in practice, only a few modes are used for calculation. In particular, the case where only two modes are strongly coupled is of importance. This is the case when resonant coupling appears, which takes place when the condition

$$\beta_k - \beta_l - \frac{2\pi m}{p} = 0 \quad (11.15)$$

is fulfilled for some integer m. Equation (11.15) is known as the Bragg condition because of its analogy to x-ray diffraction by crystals.

Let us discuss the case in Figure 11.1 where a plane wave propagates in the y-z plane. Let the incident wave be represented by a plane wave with spatial propagation factor $\exp(-\mathrm{i}k_y y - \mathrm{i}\beta z)$. In the case of Bragg reflection, the incident wave will by strongly coupled with a reflected wave with a spatial propagation factor $\exp(-\mathrm{i}k_y y + \mathrm{i}\beta z)$. The constant β is the component of the wavevector perpendicular to the relevant crystal planes. It follows that the periodicity p needs to satisfy

$$\beta - (-\beta) = 2\beta = m\frac{2\pi}{p}. \quad (11.16)$$

The propagation constant β is the component of k in the z-direction. From Figure 11.1 we see that with angle of incidence θ, the wavevector becomes $\beta = k\cos\theta$, which leads directly to

$$2p\cos\theta = m\lambda, \qquad m = 1, 2, 3, \ldots. \quad (11.17)$$

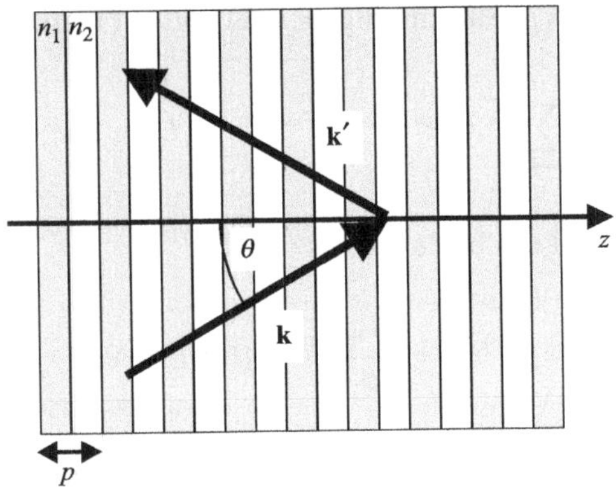

Figure 11.1 Bragg reflection geometry for stratified media. The multilayer system is composed of layers with refractive indices n_1 and n_2. The structure is periodic in z with a period p.

This condition is necessary, but not sufficient. The intensity depends very much on the Fourier expansion coefficients of the dielectric constant.

11.1.2 Calculation of Reflection Coefficients

If one considers the reflection coefficients of a Bragg mirror, the calculation can be carried out by using only two modes E_1 and E_2 and treating contradirectional coupling. One considers only coupling of an incident mode with a propagating mode of opposite direction of propagation. The mode propagating in the opposite direction is the reflection we are looking for. We start to write down the coupled mode equations for two coefficients A_1 and A_2:

$$\frac{\mathrm{d}A_1(z)}{\mathrm{d}z} = -\mathrm{i}C_{12}^{(m)}A_2(z)\mathrm{e}^{\mathrm{i}\Delta\beta z} \quad \text{and} \quad \frac{\mathrm{d}A_2(z)}{\mathrm{d}z} = -\mathrm{i}C_{21}^{(-m)}A_1(z)\mathrm{e}^{-\mathrm{i}\Delta\beta z} \tag{11.18}$$

where

$$\Delta\beta = \beta_1 - \beta_2 - \frac{2\pi m}{\mathrm{p}}. \tag{11.19}$$

In our case, when ε is only a function of z, the normal modes are plane waves and the Fourier coefficient ε_m in Eq. (11.11) becomes constant.

One can extend the concept of coupling coefficients to take the vectorial nature of the light into account. Doing so the coupling coefficients can be expressed as

$$C_{12}^{(m)}(z) = \frac{\omega^2\varepsilon_0\mu_0}{2\beta_1}\mathbf{E}_1^*\varepsilon_m\mathbf{E}_2 \quad \text{and} \quad C_{21}^{(-m)}(z) = \frac{\omega^2\varepsilon_0\mu_0}{2\beta_2}\mathbf{E}_2^*\varepsilon_{-m}\mathbf{E}_1 \tag{11.20}$$

The vectors $\mathbf{E}_1$ and $\mathbf{E}_2$ are unit vectors that represent polarization vectors of the plane wave. For unit vectors, W_k in Equation (11.5) is equal to one. One can identify them as the Jones vectors. The coupling coefficient depends on the polarization of the modes and on the tensor properties of the Fourier expansion coefficients of the dielectric constant ε_m that now is a tensor. The sign of the coupling coefficients is very important and depend on the direction of propagation. When the modes are propagating in opposite directions with $\beta_1 > 0$ and $\beta_2 < 0$ the coupling coefficients have different signs. If only two modes are taken into account and contradirectional coupling the coefficients $C_{12}^{(m)}$ and $C_{21}^{(-m)}$ fulfill the condition

$$C_{12}^{(m)} = -C_{21}^{(-m)*} = \kappa \tag{11.21}$$

This is a consequence of Equation (11.13) provided that the dielectric tensor is hermitian. Insert Equation (11.21) into Equation (11.18) the coupled wave equation become

$$\frac{\mathrm{d}A_1(z)}{\mathrm{d}z} = -\mathrm{i}\kappa A_2(z)\mathrm{e}^{\mathrm{i}\Delta\beta z} \quad \text{and} \quad \frac{\mathrm{d}A_2(z)}{\mathrm{d}z} = \mathrm{i}\kappa^* A_1(z)\mathrm{e}^{-\mathrm{i}\Delta\beta z} \tag{11.22}$$

The complex amplitudes A_1 and A_2 are the amplitudes of the normalized modes. Therefore $|A_1|^2$ and $|A_2|^2$ represents the power flow in modes 1 and 2, respectively. The net power flow in positive z-direction is the difference $|A_1|^2 - |A_2|^2$. The conservation of energy requires that

$$\frac{\mathrm{d}}{\mathrm{d}z}\left(|A_1|^2 - |A_2|^2\right) = 0. \tag{11.23}$$

To obtain the reflectivity of the structure one has to solve the coupled set of equations with respect to the boundary conditions. At position $z = 0$ and $z = L$, one usually assumes certain values for A_1 and A_2. To calculate the reflectivity we assume $A_1(0) = 1$ at $z = 0$ and $A_2(L) = 0$ at $z = L$. The distance L describes the coupling length and in general is equal to the thickness of the sample. The general solution is obtained by integration from 0 to z and is given as (Yariv and Yeh, 2003)

$$A_1(z) = \mathrm{e}^{\mathrm{i}(\Delta\beta/2)z}\frac{s\cosh s(L-z) + \mathrm{i}(\Delta\beta/2)\sinh s(L-z)}{s\cosh sL + \mathrm{i}(\Delta\beta/2)\sinh sL}$$

and

$$A_2(z) = \mathrm{e}^{-\mathrm{i}(\Delta\beta/2)z}\frac{-\mathrm{i}\kappa^*\sinh s(L-z)}{s\cosh sL + \mathrm{i}(\Delta\beta/2)\sinh sL} \tag{11.24}$$

where $s^2 = \kappa\kappa^* - (\Delta\beta/2)^2$. The power exchange between the modes decreases with increasing $\Delta\beta$. A complete exchange of power is only possible for $\Delta\beta = 0$. The reflectivity R of a Bragg reflector is defined as the ratio of the energy in the incident mode A_1 to that in the reflected mode A_2 at the entrance of the reflector ($z = 0$), and can be written as

$$R = \frac{|A_2(0)|^2}{|A_1(0)|^2}. \tag{11.25}$$

With the amplitude values calculated in Equation (11.24) this becomes

$$R = \frac{|\kappa|^2 \sinh^2 sL}{s^2 \cosh^2 sL + (\Delta\beta/2)^2 \sinh^2 sL}. \tag{11.26}$$

Maximum reflectivity occurs at $\Delta\beta = 0$, which leads to

$$R_{\max} = \tanh^2 |\kappa| L. \tag{11.27}$$

The spectrum of such reflectors consists of a series of peaks and a series of sidelobes. The bandwidth of the main peak is given approximately by

$$\Delta\beta = 4|\kappa|, \tag{11.28}$$

because at $\Delta\beta = \pm 2|\kappa|$ the parameter $s^2 = |\kappa|^2 - (\Delta\beta/2)^2$ becomes zero. We see from Equation (11.27) that high reflectance requires that $|\kappa|L > 1$. The coupling coefficient κ has to be determined through the Fourier transform of the dielectric tensor. If one assumes a Bragg reflector composed of isotropic media with refractive indices n_1 and n_2, with the layer normal in the z-direction, one can write

$$\varepsilon(z) = \begin{cases} n_2^2, & 0 < z < p/2 \\ n_1^2, & p/2 < z < p. \end{cases} \tag{11.29}$$

Here we assume equal layer thicknesses of half the period. The period is p and the condition

$$\varepsilon(z) = \varepsilon(z + p) \tag{11.30}$$

is fulfilled. The dielectric constant can be broken down into a constant value and a varying function in z. For that, a periodic square wave function $f(z)$ is introduced Starting with Equation 11.35:

$$f(z) = \begin{cases} 1, & 0 < z < p/2 \\ -1, & p/2 < z < p. \end{cases} \tag{11.31a}$$

$$f(z) = \sum_{m \neq 0} \frac{i(1 - \cos m\pi)}{m\pi} e^{-im2\pi z/p} \tag{11.31b}$$

where Equation 11.31b gives the Fourier transform of 11.31a.
One can write for $\varepsilon(z)$

$$\varepsilon(z) = \frac{n_1^2 + n_2^2}{2} + f(z)\frac{n_2^2 - n_1^2}{2} = \bar{n}^2 + f(z)\frac{n_2^2 - n_1^2}{2}. \tag{11.32}$$

The normal modes are plane waves and have a wavenumber k that is given by the geometric average of the refractive indices as

$$k = \left(\frac{\omega}{c_0}\right)^2 \bar{n} = \frac{2\pi}{\lambda}\bar{n} \tag{11.33}$$

These plane waves are TE and TM waves according to their polarization state. As both dielectric constants are scalars, polarization mode coupling does not exist. Only waves of the same polarization can couple. In addition, this is only possible in contradirectional coupling because of the phase-matching condition of Equation (11.19), which cannot be fulfilled for codirectional coupling. For TE and TM waves, different coupling constants are found, which are caused by the difference in projection of the wavevector onto the z-direction. In Figure 11.1 the electric field components of the TE polarization are perpendicular to the plane of the paper, and those of the TM polarization lie in the plane of the paper.

Let θ be the angle between the wavevector $\mathbf{k}$ and the z-axis and $\mathbf{k}'$ the wavevector of the reflected wave. The coupling constants in Equation (11.20) become (Yariv, Yeh, 2003)

$$\kappa = \begin{cases} \mathrm{i}\dfrac{(1-\cos m\pi)}{2m\lambda\cos\theta}\dfrac{\sqrt{2}(n_2^2-n_1^2)}{\sqrt{n_2^2+n_1^2}}, & \text{TE waves} \\ \mathrm{i}\dfrac{(1-\cos m\pi)}{2m\lambda\cos\theta}\dfrac{\sqrt{2}(n_2^2-n_1^2)}{\sqrt{n_2^2+n_1^2}}\cos 2\theta, & \text{TM waves.} \end{cases} \tag{11.34}$$

The difference between polarizations is a directional factor $\cos 2\theta$, which is the cosine of the angle between the polarization vectors of TM waves. The important property of the phase mismatch, $\Delta\beta$, is given by

$$\Delta\beta = 2k\cos\theta - m\frac{2\pi}{p} = \frac{4\pi}{\lambda}\bar{n}\cos\theta - m\frac{2\pi}{p}. \tag{11.35}$$

For normal incidence $\theta = 0$, there is no difference for TE and TM polarization. We can set, additionally, $m = 1$ and obtain the coupling constants

$$\kappa = \frac{i}{\lambda}\frac{\sqrt{2}(n_2^2-n_1^2)}{\sqrt{n_2^2+n_1^2}}. \tag{11.36}$$

With this, the maximum reflectivity can be calculated:

$$R_{\max} = \tanh^2\left(\left|\frac{\sqrt{2}(n_2^2-n_1^2)}{\sqrt{n_2^2+n_1^2}}\right|\frac{L}{\lambda}\right). \tag{11.37}$$

This therefore depends on the lengths of the reflector L and the refractive indices. Let us calculate the reflection coefficient of a Bragg reflector for a typical set of values. Figure 11.2 shows the maximum reflectivity for $n_1 = 1.5$ and different values of n_2. The maximum reflectivity increases for thicker reflectors. The larger the refractive index difference the higher the reflectivity.

Next we are interested in the shape of the spectrum of the reflection curve as a function of the wavelengths, as given by Equation (11.26). Figure 11.3 shows

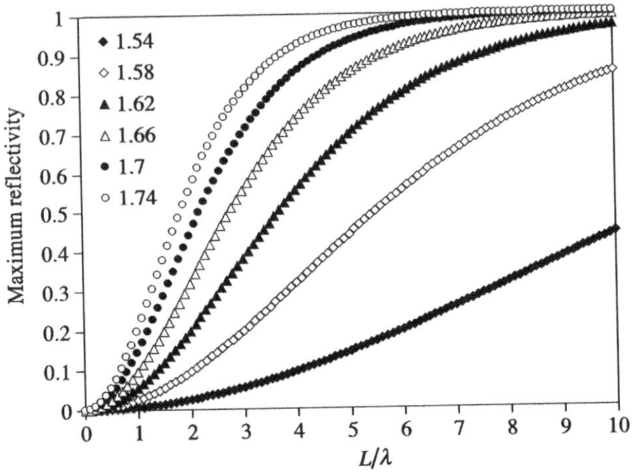

Figure 11.2 Maximum reflectivities of a Bragg reflector at normal incidence as a function of the ratio of wavelength L to wavelength λ. The value of n_1 is fixed at $n_1 = 1.5$ and n_2 is varied.

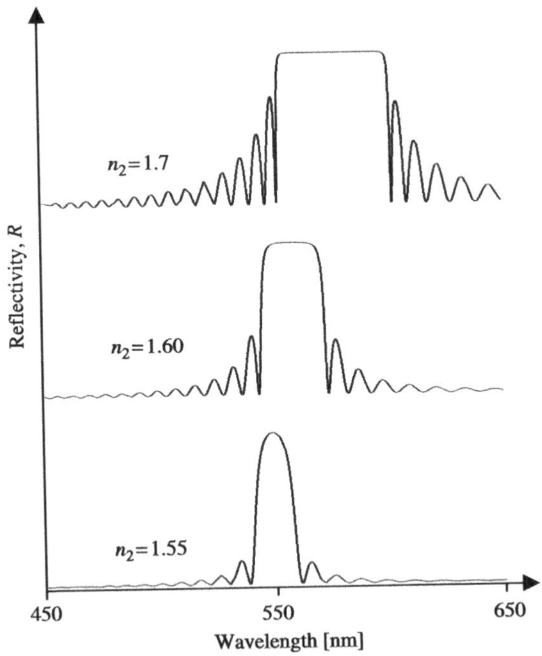

Figure 11.3 Reflectivity curves for different values of n_2. The reflectivity reaches 1 for each curve. Calculations are made for the parameters $n_1 = 1.5$, $p = 0.18\ \mu m$, and $L = 10\ \mu m$.

reflection spectra for $n_1 = 1.5$ and different values of n_2. The coupling length was set to $L = 10$ μm. The maximum reflectivity is obtained at $\Delta\beta = 0$. For normal incidence $\theta = 0$ and with $m = 1$, the wavelength of maximum reflectivity in Equation (11.35) becomes

$$\lambda_{\max} = 2p\bar{n} \qquad \text{with} \qquad \bar{n} = \sqrt{\frac{n_1^2 + n_2^2}{2}}. \tag{11.38}$$

To calculate the width of the reflection band we use the equation $\Delta\beta = \pm 2|\kappa|$ for the left and the right band edge. This leads to

$$\frac{2\pi}{\lambda_1}\bar{n} - \frac{\pi}{p} = \frac{1}{\lambda_1}\frac{(n_2^2 - n_1^2)}{\bar{n}} \qquad \text{and} \qquad \frac{2\pi}{\lambda_2}\bar{n} - \frac{\pi}{p} = -\frac{1}{\lambda_2}\frac{(n_2^2 - n_1^2)}{\bar{n}}. \tag{11.39}$$

Solving for λ_1 and λ_2 and taking the difference, one obtains

$$\Delta\lambda = 2\frac{p}{\pi}\frac{(n_2^2 - n_1^2)}{\bar{n}} = p(n_2 - n_1)\frac{4}{\pi}\frac{(n_2 + n_1)}{2}\frac{1}{\bar{n}} \approx p(n_2 - n_1). \tag{11.40}$$

For small differences of the refractive index, one can set the geometric average equal to the mean value. The bandwidth of the reflection band is approximately proportional to the product of the refractive index difference and the period of the grating. For polymer multilayer structures, one achieves refractive index differences of 0.1. To arrive at observable effects with reflectivities larger than 80%, one has to have coupling lengths of about 10 periods.

11.2 POLYMER FILMS

Multilayer Bragg reflectors and corrugated gratings are widely used in optical components and optoelectronics (interference filters, antireflection coatings, vertical cavity surface emitting laser (VCSEL), distributed Bragg reflectors (DBR), distributed feedback lasers (DFB)). Usually such devices are fabricated using inorganic oxides, mainly due to the large refractive index contrast and well-known process (sputtering, vapor deposition) giving excellent control of the deposited thicknesses. Polymers are generally less used because of their low refractive index contrast and therefore the need for a larger number of layers for an equivalent reflection efficiency. The deposited thickness is more difficult control, depending on solution dilution, the speed parameters of the spin-coater, humidity, and surface cleanliness. The recent development of organic light emitting diodes (OLED) and other organic optical structures illustrates the trend towards all-organic optical components, mainly as a result of economic objectives. The reduction of costs of optical components is achieved by using new processes of fabrication such as replication and spin-coating instead of costly e-beam writing and molecular deposition under vaccum. In this regard, it seems important to develop basic optical functionalities to be integrated into devices. Of particular interest are multilayer Bragg structures because of their optical properties. Such structures can be fabricated by spin-coating as a stack of thin layers on a glass substrate (Alvarez et al., 2003).

We shall show an example of fabrication to give an idea of the problems encountered in polymer multilayer fabrication. Two different low-cost polymer materials in solution are used: poly(N-vinylcarbazole) (PVK) in toluene and poly(vinyl alcohol) (PVA) in water. These two polymers were chosen for their optical and mechanical properties. An additional important aspect is the solubility of the polymers in solvents. The two different polymers used to form thin layers must be soluble to spin-coat homogeneous layers. They should be soluble in two different solvents and the mixtures have to deposit in succession, each on the other. They should therefore be immiscible. To obtain thin layers, solubility is an issue. It can be selected by properly choosing the molecular weight of the polymers. Polymers PVA and PVK have a refractive index difference of about 0.15 in the visible and permit multilayer coatings with interlayer problems. A Bragg multilayer was designed to have a reflection band around 620 nm at normal incidence, and was composed of five pairs of layers. The PVA and PVK layers are about 100 nm and 90 nm thick, respectively. The obtained devices are stable and robust. The following parameters were are used for the design of the optical multilayer structures:

PVA refractive index $n_{\mathrm{PVA}} = 1.56$,
PVK refractive index $n_{\mathrm{PVK}} = 1.72$ (mean value at wavelength 500 nm),
PVA thickness = 100 nm (quarter-wave plate for 620 nm),
PVK thickness = 90 nm (quarter-wave plate for 620 nm).

Figure 11.4 shows the measurement of typical reflection intensities obtained from such a multilayer structure. Reflected intensity is shown as a function of the incidence angle for wavelengths from 400 to 750 nm. Measurements are done in

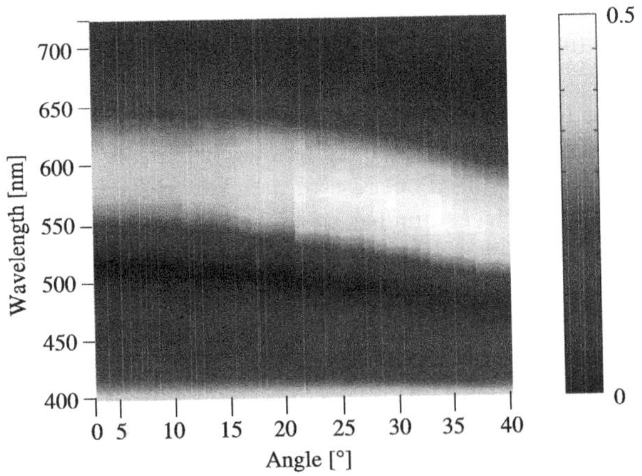

Figure 11.4 Reflection measurement for Bragg structure made of 10 layers on a glass substrate, PVA thickness = 100 nm, PVK thickness = 90 nm (PVA first layer).

specular reflection. A well-pronounced reflection band is visible that shifts to shorter wavelengths for increasing angle of incidence. The shift to shorter wavelengths is a typical phenomenon for multilayer Bragg reflectors.

11.3 GIANT POLARIZATION OPTICS

Multilayer mirrors that maintain or increase their reflectivity with increasing incidence angle can be constructed using polymers that exhibit large birefringence in their indices of refraction. Multilayer polymeric interference mirrors were demonstrated in the late 1960s (Alfrey et al., 1969). The most important feature of these multilayer interference stacks is the index variation in the thickness direction. Such films can yield optical results that are difficult or impossible to achieve with conventional multilayer optical designs. The materials and processes necessary to fabricate such films are amenable to large-scale manufacturing. (Weber et al., 2000). The fabrication process is based on standard polymer foil fabrication. Thin films are produced by stretching and rolling. Stretching a polymer film induces a preferred orientation and can induce birefringence depending on the base material.

Imagine a multilayer stack of different polymers that act distinctly differently on stretching. One has a large induced birefringence that the other does not. If such a film is stretched, a multilayer film with isotropic and anisotropic thin films forms. Figure 11.5 shows a multilayer stack composed of two different materials. One is isotropic and the other is anisotropic. The ellipse and balls at the right indicate the index ellipsoid for each layer. As an example of the strengths of the concept, imagine a situation where the ordinary refractive index is index-matched to the

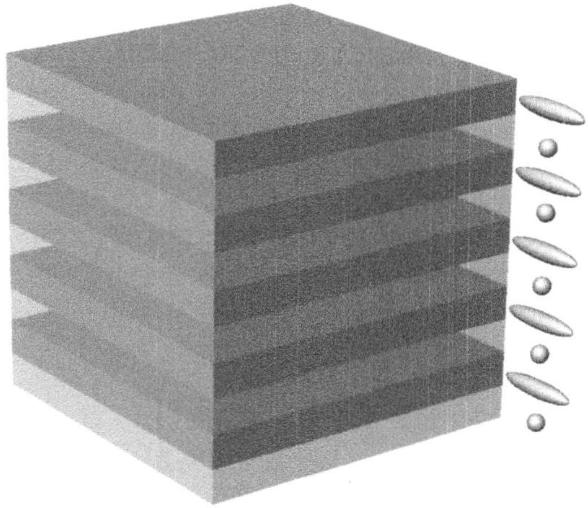

Figure 11.5 Principal structure of a Bragg reflector system with birefringent films. Such films can act as polarizers with high efficiency and low loss if index matching and Bragg reflection are combined.

refractive index of the isotropic layer. In such a situation, the stack would not influence propagation of the polarization of the light that is parallel to the ordinary index of refraction. Light with this polarization experiences an isotropic material. On the contrary, the polarization state parallel to the extraordinary refractive index of the anisotropic layer experiences a multilayer film. If additionally this multilayer film is a Bragg mirror, one polarization shall be reflected while the other is transmitted. A multilayer design that allows polarizers to be made for the whole visible spectrum can be obtained by apodization or chirping (see Section 11.6). A key limitation of conventional multilayer mirrors is Brewster's law, which predicts the decrease of reflection for p-polarized light at material interfaces with increasing incidence angle. Specifically, Brewster's law states that there is an angle of incidence (Brewster's angle) for which the reflectivity for p-polarized light vanishes at a material interface. As a result, a multilayer interference mirror that is designed to have a 1% loss for reflection of p-polarized light (99% reflectivity) at normal incidence can have many times that loss at high incidence angles. Using highly birefringent polymers, one can design multilayer mirrors that maintain or increase their reflectivity with increasing incidence angle. The key to obtaining such effects is birefringence in the thin films. The inventors (Weber et al., 2000) of these structures called them "giant birefringent." A central feature of giant birefringent optics is improved control of the reflectivity of p-polarized light.

11.4 REFLECTION BY CHOLESTERIC LIQUID CRYSTALS

In the cholesteric phase of a liquid crystal, the averaged molecular direction changes linearly from plane to plane so that the molecular structure forms a helix. As a result, we obtain some unique and interesting optical properties. The structure can act as a Bragg reflector and shows strong optical rotatory power. The optically important parameters of the cholesteric liquid crystal are the period of the cholesteric, called the pitch p, and the principal refractive indices n_o (ordinary) and n_e (extraordinary). A homogeneous texture can be seen as formed by planes with equal director orientation. Between these planes, the molecular director makes a full turn of 2π. If selective-reflection optical applications in the visible are sought, the values for the pitch are between 300 nm and 500 nm.

We are mainly interested in finding the reflection properties. For further treatment, we assume a positive uniaxial liquid crystal material with $n_o < n_e$. A uniaxial liquid crystal has in its principal local coordinate system the dielectric tensor

$$\boldsymbol{\varepsilon}(z) = \begin{pmatrix} n_e^2 & 0 & 0 \\ 0 & n_o^2 & 0 \\ 0 & 0 & n_o^2 \end{pmatrix}. \tag{11.41}$$

For the cholesteric liquid crystal layer it is assumed that the optical axis of the molecule remains in the x–y-plane but changes its orientation depending on the position in z by the pitch p. This assumption can be made for most cholesteric liquid crystals.

In the laboratory coordinate system the dielectric tensor for rotation of the uniaxial local tensor around z can be rewritten by using Equation (4.16). With $\theta = 0$ and $\varphi = 2\pi z/p$, it becomes, in the laboratory coordinate system,

$$\boldsymbol{\varepsilon}(z) = \begin{pmatrix} \bar{\varepsilon} + \Delta\varepsilon\cos\dfrac{4\pi z}{p} & \Delta\varepsilon\sin\dfrac{4\pi z}{p} & 0 \\ \Delta\varepsilon\sin\dfrac{4\pi z}{p} & \bar{\varepsilon} - \Delta\varepsilon\cos\dfrac{4\pi z}{p} & 0 \\ 0 & 0 & n_o^2 \end{pmatrix}, \qquad (11.42)$$

where $\bar{\varepsilon} = (n_e^2 + n_o^2)/2$ and $\Delta\varepsilon = (n_e^2 - n_o^2)/2$, and $\boldsymbol{\varepsilon}$ is a periodic function of z with a period of $p/2$. Light propagation of plane waves along the helical axis z can be studied, leading to two analytic solutions with elliptical polarizations (de Vries, 1951). Furthermore it can be shown that in the frequency gap $n_o p < \lambda < n_e p$, circularly polarized light is almost fully reflected if it has the handedness of the cholesteric liquid crystal helix, and is transmitted if it has the opposite handedness. Using the two known analytical solutions as modes in the coupled wave analysis, the reflectivity can be found with coupled mode analysis (Yeh and Gu, 1999).

For illustration, we will study the case of normal incidence. Light travels along z and we ignore the z-component of the electric field. The dielectric tensor can be written as a two-component system

$$\boldsymbol{\varepsilon}(z) = \bar{\varepsilon}\begin{pmatrix} 1 & 0 \\ 0 & 1 \end{pmatrix} + \Delta\varepsilon \begin{pmatrix} \cos\dfrac{4\pi z}{p} & \sin\dfrac{4\pi z}{p} \\ \sin\dfrac{4\pi z}{p} & -\cos\dfrac{4\pi z}{p} \end{pmatrix} = \boldsymbol{\varepsilon}_u + \boldsymbol{\delta\varepsilon} \qquad (11.43)$$

where ($\bar{\varepsilon} = (n_e^2 + n_o^2)/2$ and $\Delta\varepsilon = (n_e^2 - n_o^2)/2$). The second term in Equation (11.43) represents the perturbation. The coupled wave theory starts to establish a set of unperturbed solutions. In analogy to the calculation above, one can directly identify the propagation constant for the unperturbed solution. The propagation constant is determined by the first part of the tensor in Equation (11.43) $\bar{\varepsilon}$ and is given by

$$k = \frac{\omega}{c_0}\sqrt{\bar{\varepsilon}} = \frac{2\pi}{\lambda}\sqrt{\frac{n_e^2 + n_o^2}{2}}. \qquad (11.44)$$

The wavenumber k is equivalent to the propagation constant β in Chapter 1.1. The second term is the dielectric perturbation, which can be further decomposed into

$$\boldsymbol{\delta\varepsilon}(z) = \Delta\varepsilon \begin{pmatrix} \cos\dfrac{4\pi z}{p} & \sin\dfrac{4\pi z}{p} \\ \sin\dfrac{4\pi z}{p} & -\cos\dfrac{4\pi z}{p} \end{pmatrix} = \Delta\varepsilon\left(\frac{1}{2}\begin{pmatrix} 1 & -\mathrm{i} \\ -\mathrm{i} & -1 \end{pmatrix} e^{\mathrm{i}\frac{4\pi z}{p}} + \frac{1}{2}\begin{pmatrix} 1 & \mathrm{i} \\ \mathrm{i} & -1 \end{pmatrix} e^{-\mathrm{i}\frac{4\pi z}{p}}\right).$$

(11.45)

Consider now an incident and reflecting plane wave as $\mathbf{E_i}e^{i(\omega t-kz)}$ and $\mathbf{E_r}e^{i(\omega t+kz)}$. The perturbation $\boldsymbol{\delta\varepsilon}$ that leads to coupling is given as (see Section 11.1)

$$\Delta\mathbf{P} = \varepsilon_0 \boldsymbol{\delta\varepsilon}(z)\mathbf{E_i}e^{i(\omega t-kz)}. \tag{11.46}$$

The structure of $\boldsymbol{\delta\varepsilon}$ will determine how effectively the incident wave radiates into the reflected wave. Compared to calculations in Section 11.1, we have to respect the polarization of light because of the tensorial properties of the dielectric constant. For normal incidence, there are two conditions to fulfil:

$$2k = \frac{4\pi}{p} \qquad \text{and} \qquad \mathbf{E_r^*}\boldsymbol{\delta\varepsilon}(z)\mathbf{E_i} \neq 0. \tag{11.47}$$

The first condition is the Bragg condition, as we know from Section 11.1. The second condition describes the coupling. If the coupling is not present, no energy will be transferred from one to the other mode and no reflection occurs. Because $\boldsymbol{\delta\varepsilon}(z)$ has a certain sense of rotation, the coupling is different for left-handed and right-handed polarized light.

To illustrate this we calculate the perturbation term $\Delta\mathbf{P}$ in Equation (11.46) for different polarizations. The electric field vectors $\mathbf{E}$ are identified with Jones vectors. We use the notation for the different states of polarization from Section 4.6. All states of polarization can be seen to be composed of two orthogonal polarizations. For convenience, we break down the incident electric field vector $\mathbf{E}_i$ as

$$\mathbf{E_i}e^{i(\omega t-kz)} = \left(V_x\begin{pmatrix}1\\0\end{pmatrix} + V_y\begin{pmatrix}0\\1\end{pmatrix}\right)e^{i(\omega t-kz)} \tag{11.48}$$

with V_x and V_y complex. To describe linear polarization, V_x and V_y would be rational. The perturbation $\Delta\mathbf{P}$ in Equation (11.46) becomes

$$\begin{aligned}\boldsymbol{\delta\varepsilon}(z)\mathbf{E}_i e^{i(\omega t-kz)} &= \Delta\varepsilon\left(\frac{1}{2}\begin{pmatrix}1 & -i\\ -i & -1\end{pmatrix}e^{i\frac{4\pi z}{p}} + \frac{1}{2}\begin{pmatrix}1 & i\\ i & -1\end{pmatrix}e^{-i\frac{4\pi z}{p}}\right)\\ &\quad\times\left(V_x\begin{pmatrix}1\\0\end{pmatrix} + V_y\begin{pmatrix}0\\1\end{pmatrix}\right)e^{i(\omega t-kz)}.\end{aligned} \tag{11.49}$$

One finds, after multiplication,

$$\begin{aligned}\boldsymbol{\delta\varepsilon}(z)\mathbf{E}_i e^{i(\omega t-kz)} &= \frac{\Delta\varepsilon}{2}\left(\left(V_x\begin{pmatrix}1\\-i\end{pmatrix} - iV_y\begin{pmatrix}1\\-i\end{pmatrix}\right)e^{i\frac{4\pi z}{p}-ikz}\right.\\ &\quad\left.+\left(V_x\begin{pmatrix}1\\i\end{pmatrix} + iV_y\begin{pmatrix}1\\i\end{pmatrix}\right)e^{-i\frac{4\pi z}{p}-ikz}\right)e^{i\omega t}.\end{aligned} \tag{11.50}$$

If, additionally, we take into account the Bragg condition of Equation (11.47), we end up with the perturbation term that is of interest. We find

$$\boldsymbol{\delta\varepsilon}(z)\mathbf{E}_i e^{i(\omega t-kz)} = \frac{\Delta\varepsilon}{2}\begin{pmatrix}1\\-i\end{pmatrix}(V_x - iV_y)e^{i(\omega t+kz)}. \tag{11.51}$$

We see that the propagation on the right-hand side in Equation (11.51) is now in the $-z$ direction, which is the desired reflection we are looking for. The perturbation leads to coupling in a particular polarization state. The perturbation term becomes zero if the components of the incident polarized light in Equation (11.48) fulfil the condition $(V_x - iV_y) = 0$. This is fulfilled when for instance $V_x = 1$ and $V_y = -i$, which describes right handed circular polarized light. Note that in Equation (11.48) we fixed the propagation direction in the $+z$ direction, so the coupling is zero for incident right handed polarized light propagating in the $+z$ direction. For left handed circular polarized light with $V_x = 1$ and $V_y = i$ and propagating along the $+z$ direction coupling becomes maximum. For coupled mode analysis, it seems preferential to take circular polarized light as the unperturbed modes. We write the electric field as

$$\mathbf{E} = A_1(z)\frac{1}{\sqrt{2}}\begin{pmatrix}1\\i\end{pmatrix}e^{i(\omega t-kz)} + A_2(z)\frac{1}{\sqrt{2}}\begin{pmatrix}1\\-i\end{pmatrix}e^{i(\omega t+kz)} = A_1(z)\mathbf{E}_1 + A_2(z)\mathbf{E}_2$$

$$\text{with } \mathbf{E}_1 = \frac{1}{\sqrt{2}}\begin{pmatrix}1\\i\end{pmatrix} \quad \text{and} \quad \mathbf{E}_2 = \frac{1}{\sqrt{2}}\begin{pmatrix}1\\-i\end{pmatrix}. \tag{11.52}$$

This corresponds to Equation (11.2) when one identifies the normal modes with the unit polarization vectors $\mathbf{E}_1$ and $\mathbf{E}_2$. The propagation constants β_1 and β_2 in Equation (11.2) are $\beta_1 = \mathrm{k}$ and $\beta_2 = -\mathrm{k}$ with k defined in Equation (11.44). Using Equations (11.21) and (11.22) the coupled wave equations for the amplitude factor A_1 and A_2 can be written as

$$\frac{dA_1(z)}{dz} = -i\kappa A_2(z)e^{i\Delta kz} \quad \text{and} \quad \frac{dA_2(z)}{dz} = i\kappa^* A_1(z)e^{-i\Delta kz} \tag{11.53}$$

The Bragg condition in Equation (11.19) becomes

$$\Delta k = 2k - \frac{4\pi}{p} \tag{11.54}$$

where $m = 2$. That can be seen when the matrix expression for the refractive index in Equation (11.45) is compared with the definition of the Fourier series in Equation (11.11). The non-vanishing part of the coupling coefficient κ is calculated after Equation (11.20) as

$$\kappa = \frac{\omega^2\varepsilon_0\mu_0}{2k}\mathbf{E}_1^*\frac{\Delta\varepsilon}{2}\begin{pmatrix}1 & i\\ i & -1\end{pmatrix}\mathbf{E}_2 = \frac{\pi}{\lambda}\frac{\Delta\varepsilon}{\sqrt{\bar{\varepsilon}}} \tag{11.55}$$

Remember that $\bar{\boldsymbol{\varepsilon}} = (n_e^2 + n_o^2)/2$ and $\Delta\varepsilon = (n_e^2 - n_o^2)/2$ and the wavevector k is again Equation (11.44). Equations (11.53) are formally identical to Equation (11.22) and will therefore have the same solutions. The only differences are the

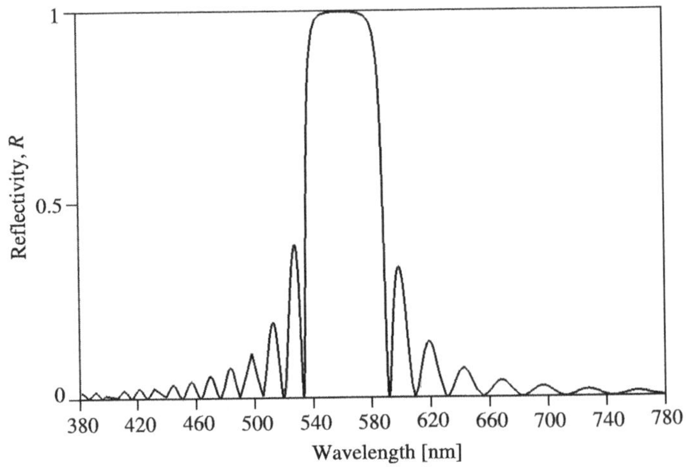

Figure 11.6 Calculated reflection spectra at normal incidence for a cholesteric liquid crystal reflector. Incident light has the same handedness as the cholesteric liquid crystal. The parameters used are $n_e = 1.7$, $n_o = 1.5$, $p = 300$ nm, and $L = 5$ μm.

different forms of the Bragg condition and the value of the propagation constant κ. Using $s^2 = \kappa^2 - (\Delta k/2)^2$ in Equation (11.26), the reflectivity is

$$R = \frac{|\kappa|^2 \sinh^2 sL}{s^2 \cosh^2 sL + (\Delta k/2)^2 \sinh^2 sL}. \tag{11.56}$$

The calculated reflectivity for a limited wavelength range is shown in Figure 11.6. The maximum reflectivity is given for $\Delta k = 0$ at the wavelengths

$$\lambda_{\max} = p\bar{n}. \tag{11.57}$$

Note that, contrary to the isotropic multilayers discussed in Section 11.1, the wavelengths of maximum reflection are found as the product of the period and the average refractive index. The difference between Equation (11.57) and Equation (11.38) is a factor 2. Therefore the period of the cholesteric liquid crystal, the pitch p, for applications in the visible is between 180 and 400 nm. The width of the reflection band is found as

$$\Delta\lambda = p(n_e - n_o). \tag{11.58}$$

The analytical solution for the reflectance exists only for normal incidence. Numerical solutions are offered for oblique incidence using 4×4 matrix methods. The matrix method presented in Chapter 4 allows the calculation of light propagation in a cholesteric material with spatially varying optical parameters, that is, varying pitch or refraction indices of the liquid crystal material along the helical axis. We will show in the next sections why this can be useful.

11.5 COLOR PROPERTIES OF CHOLESTERIC BRAGG REFLECTORS

Because of their selective reflection and bistability properties, cholesteric liquid crystals are also of interest for reflective display applications. Reflective displays are suitable for portable information systems because of their low power consumption (Wu and Yang, 2001). Using three cholesteric liquid crystal cells providing different colors, a controllable display based on an additive color mixture is feasible. In order to verify the available color space of this display type, it is important to determine chromaticity and luminance of the three basic colors red, green, and blue, produced by the cholesteric liquid crystal. All following reflected intensities are calculated for circularly polarized light of the handedness of the cholesteric liquid crystal material. The determination of chromaticity and luminance is based on the assumption of an identical light source and the color-matching functions of the CIE 1931 Standard Colorimetric System. To learn more about color calculation please refer to Chapter 7. The chromaticity is defined by the values of x, y corresponding to this standard. The spectrum of the reflected light depends on material properties of the cholesteric liquid crystal (ordinary and extraordinary refraction indices n_o and n_e, thickness d, and pitch p). The typical shape of a corresponding reflection spectrum is rectangular-like with so-called sidelobes (Fig. 11.6). For color calculation, one can model the reflectivity without considering the sidelobes. The reflectivity switches on at the wavelength $\lambda_1 = pn_o$ and shuts down at $\lambda_2 = pn_e$.

By assumption of a rectangle shape of the reflection spectrum one can compute the luminance and the chromaticity in a simple manner. In order to obtain the material properties of cholesteric liquid crystals with desired chromaticity and luminance, one has to solve the inverse problem. For that purpose the chromaticity and luminance of rectangular reflection spectra with the borders $\lambda_1 = pn_o$ and $\lambda_2 = pn_e$ is studied. For applications in the visible and for materials with refractive indices $n_o = 1.5$ and n_e between 1.55 and 1.8, the period p can vary between 250 and 500 nm. The material parameters were sorted by the corresponding resulting chromaticity and entered in a table of extraordinary refraction index versus chromaticity (Fig. 11.7) and in a table of pitch versus chromaticity (Fig. 11.8). In the table of luminance versus chromaticity (Fig. 11.9) one can read the luminance of the light that is reflected by a cholesteric liquid crystal with the material parameters p and n_e. The tables in Figures 11.7–11.9 are used as follows. Let us define first a color we want to realize with color coordinates x, y. To realize this color with cholesteric liquid crystal reflection we search the material parameters. In Figure 11.7 we find the extraordinary refractive index and in Figure 11.8 the pitch is given. For green light at 550 nm with $x = 0.275$ and $y = 0.7$, we obtain, for example, $n_e = 1.55$ and $p = 361$ nm. The luminance for such a structure is given in Figure 11.9 as 0.16.

To provide an additional example we plot in Figure 11.10 the spectra for three different colors with the same luminance. It shows the reflection spectra of single cholesteric liquid crystals for a blue, green, and red reflecting cell. Each of these reflectance spectra has a luminance of 20%. The product of the birefringence and the pitch determines the width of the reflection spectra. The refractive indices

y (×10^{-3}) \ x (×10^{-3})	25	50	75	100	125	150	175	200	225	250	275	300	325	350	375	400	425	450	475	500	525	550	575	600	625	650	675	700
800			1.55	1.55	1.55																							
775		1.55	1.57	1.58	1.57	1.55	1.55																					
750		1.57	1.60	1.62	1.62	1.62	1.58	1.55																				
725	1.56	1.58	1.60	1.64	1.65	1.66	1.66	1.62	1.56	1.55																		
700	1.55	1.60	1.62	1.65	1.67	1.68	1.69	1.68	1.66	1.55	1.55																	
675	1.55	1.60	1.64	1.67	1.69	1.71	1.72	1.71	1.71	1.70	1.55	1.55																
650	1.56	1.62	1.64	1.67	1.71	1.72	1.74	1.74	1.74	1.74	1.71	1.59	1.55															
625	1.56	1.62	1.67	1.69	1.71	1.74	1.75	1.77	1.77	1.78	1.77	1.75	1.62	1.55														
600	1.58	1.62	1.67	1.70	1.73	1.75	1.77	1.79	1.80	1.80	1.80	1.80	1.77	1.65	1.55													
575	1.57	1.63	1.68	1.70	1.73	1.77	1.79	1.80						1.80	1.65	1.55												
550	1.58	1.63	1.68	1.70	1.74	1.77	1.79									1.69	1.55											
525	1.57	1.63	1.68	1.71	1.76	1.80	1.80										1.77	1.55										
500	1.56	1.64	1.69	1.72	1.77	1.80												1.76	1.55									
475	1.56	1.63	1.68	1.72	1.77	1.80														1.55								
450	1.55	1.63	1.69	1.73	1.77																1.55							
425	1.56	1.63	1.68	1.73	1.78																	1.55						
400	1.55	1.62	1.68	1.74	1.78																		1.55					
375	1.56	1.62	1.67	1.74	1.80																			1.56				
350	1.55	1.61	1.68	1.75	1.80																				1.55			
325		1.57	1.67	1.74	1.80																					1.56		
300		1.55	1.66	1.74																							1.55	
275		1.56	1.65	1.73																								1.56
250		1.55	1.63	1.73																								
225		1.57	1.61	1.72																								
200			1.56	1.69																								
175			1.55	1.66	1.80																							
150			1.55	1.64	1.79																							
125				1.58	1.77																							
100				1.55	1.71																							
75				1.55	1.64																							
50					1.55																							
25					1.55	1.65																						

Chromaticity

Extraordinary refractive index, n_e

Figure 11.7 Chromaticity of reflection from an idealized cholesteric Bragg reflector. The parameter is the extraordinary refraction index. ((x, y) color coordinates in the CIE 1931 system.)

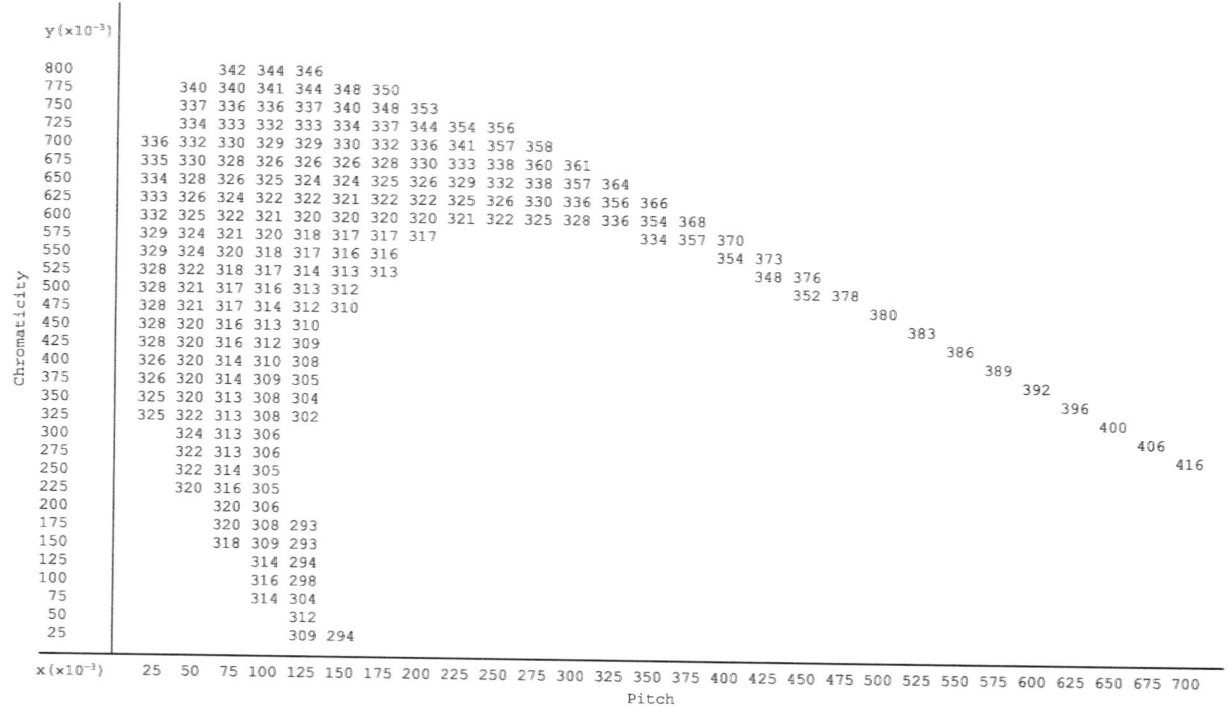

Figure 11.8 Chromaticity of reflection from an idealized cholesteric Bragg reflector. The parameter is the pitch of the cholesteric liquid crystal. ((x, y) color coordinates in the CIE 1931 system.)

Chromaticity y (×10⁻³) versus x (×10⁻³); cell values: Luminance

y (×10⁻³) \ x (×10⁻³)	25	50	75	100	125	150	175	200	225	250	275	300	325	350
800			0.10	0.12	0.12									
775		0.11	0.14	0.18	0.18	0.13	0.13							
750		0.12	0.20	0.25	0.28	0.30	0.21	0.16						
725	0.10	0.14	0.19	0.28	0.31	0.36	0.39	0.34	0.18	0.14				
700	0.08	0.16	0.22	0.29	0.35	0.40	0.43	0.45	0.44	0.16	0.16			
675	0.07	0.15	0.24	0.32	0.37	0.43	0.46	0.49	0.52	0.53	0.17	0.17		
650	0.08	0.18	0.23	0.29	0.38	0.44	0.49	0.52	0.57	0.59	0.58	0.29	0.17	
625	0.08	0.18	0.27	0.33	0.36	0.46	0.50	0.55	0.60	0.63	0.65	0.64	0.38	0.17
600	0.10	0.17	0.26	0.33	0.41	0.46	0.52	0.57	0.62	0.65	0.68	0.70	0.70	0.47
575	0.08	0.17	0.26	0.31	0.39	0.47	0.52	0.58						0.76
550	0.08	0.16	0.26	0.30	0.37	0.47	0.51							
525	0.07	0.16	0.25	0.30	0.41	0.49	0.51							
500	0.06	0.16	0.25	0.30	0.42	0.49								
475	0.05	0.13	0.22	0.30	0.40	0.48								
450	0.04	0.14	0.22	0.29	0.38									
425	0.05	0.12	0.20	0.29	0.38									
400	0.04	0.11	0.20	0.29	0.37									
375	0.04	0.10	0.17	0.27	0.39									
350	0.04	0.09	0.17	0.28	0.37									
325		0.05	0.16	0.24	0.35									
300		0.03	0.13	0.23										
275		0.04	0.12	0.21										
250		0.03	0.09	0.20										
225		0.04	0.07	0.17										
200			0.03	0.13										
175			0.03	0.10	0.20									
150			0.02	0.08	0.19									
125				0.03	0.15									
100				0.02	0.09									
75				0.02	0.05									
50					0.02									
25					0.01	0.03								

y (×10⁻³) \ x (×10⁻³)	375	400	425	450	475	500	525	550	575	600	625	650	675	700
600	0.17													
575	0.46	0.18												
550		0.57	0.18											
525			0.73	0.18										
500				0.71	0.15									
475						0.15								
450							0.14							
425								0.13						
400									0.14					
375										0.14				
350											0.11			
325												0.11		
300													0.08	
275														0.05

Figure 11.9 Luminance of cholesteric Bragg reflector calculated with the parameters from Figures 11.7 and 11.8. ((x, y) color coordinates in the CIE 1931 system.)

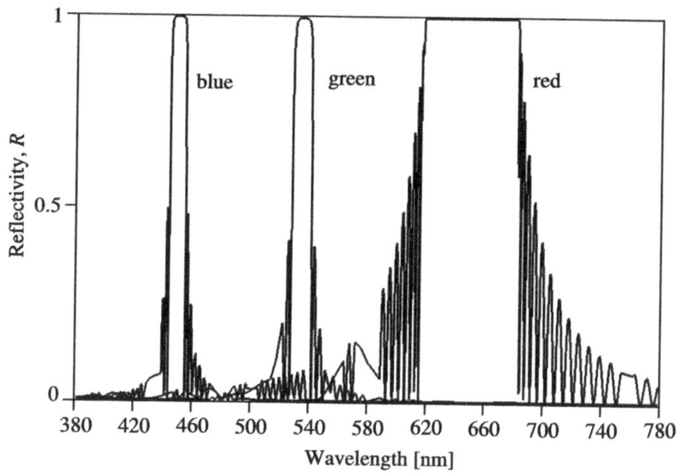

Figure 11.10 Reflection spectra for three cholesteric liquid crystal reflectors: blue, $p = 294$ nm, $n_o = 1.5$, $n_e = 1.55$; green, $p = 350$ nm, $n_o = 1.5$, $n_e = 1.55$; red, $p = 396$ nm, $n_o = 1.5$, $n_e = 1.77$.

times the pitch gives the wavelength positions of the reflection band. To summarize, for the design of cholesteric liquid crystal displays of maximal luminance and best colors, it is not sufficient to vary the pitch, but it is also necessary to take into account different refractive indices. For instance, a maximal luminance of 0.76 in circularly polarized light can be reached with a pitch of 343 nm, that is, for a gold color, for refraction indices of $n_o = 1.5$ and $n_e = 1.8$. The red and blue corner regions are represented only by reflection spectra delivering very low luminance below 0.05.

11.6 APODIZATION OF CHOLESTERIC BRAGG FILTERS

Apodization of Bragg reflectors is used for shaping the form of the reflection spectra. Several applications need to have large band reflection. The easiest way to achieve this is to combine several Bragg mirrors with overlapping reflection bands. This can be done by fabricating a multilayer stack or by finding technological ways to make it in an integral fabrication process (Chelix, 2005). The design can be made by using the formulas developped in Section 11.4. We have to remember that the reflection band is limited by the values $\lambda_1 = n_o p$ and $\lambda_2 = n_e p$. For a thick multilayer structure, where the effectivity can by seen as maximum, one can tune the pitch within a single material. The reflection band is then limited by $\lambda_1 = n_o p_{\min}$ and $\lambda_2 = n_e p_{\max}$, which allows very large reflection bands to be made. The question arises whether the form of the reflection curve can also be designed. The selective reflection property of the cholesteric liquid crystal makes it appropriate to design stopband filters. This is an optical device that reflects, at best, 100% over a given spectral band and transmits 100% on each side of the reflected band. Such filters have been realized for many years by using layers consisting of quarter-wave stacks of alternating low and high index layers. This arrangement provides a reflection band for a preferred wavelength region with sidelobes around the reflecting region as seen in Section 11.1. In order to suppress these sidelobs, the refractive index value can be modulated (or "apodized"). These sidelobe reduced interference filters are called "rugate" filters (Southwell, 1989; Abu-Safia et al., 1993). Following the idea of apodization, designs can be found for a cholesteric Bragg filter with reduced sidelobes by modulating the extraordinary refraction index along the helical axis of the cholesteric liquid crystal. The sidelobes of the reflected intensity are mostly suppressed. One can compare various types of apodization functions, such as trapezoidal and exponential (Bohley, 2004).

11.7 REFLECTION BY DISPERSED CHOLESTERIC LIQUID CRYSTALS

The cholesteric liquid crystal display can be made multistable (Wu and Yang, 2001). This is based on the fact that domains formed in multistable structure do not relax back, but need a "reset" of the texture with external electric fields. For reasons of stability in the mentioned display devices, mixtures of liquid crystals and polymers are often but not necessarily used. If a small amount of polymer is dispersed in the liquid crystal, the system will be divided into domains, which are bound by thin polymer borders (Drzaic, 1995). A model structure is shown schematically in Figure 11.11. The cholesteric texture liquid crystal display assumes different

Figure 11.11 Domain structure. A simple model of a multidomain polymer-dispersed liquid crystal layer, also called "gel." The liquid crystal slab can be seen as divided into several domains of different size. A typical domain volume dimension is 1 μm^3. Each domain can be characterized by a helical orientation and it is assumed that within the domain the structure is homogeneous.

Figure 11.12 The three main states of helix orientation: planar, focal conic, and homeotropic. The transition between these states can be simulated by changing two parameters describing the texture: the average helix orientation and the standard deviation of helix orientation distribution.

stable states. We consider three states, which appear in the following order if an electric field is applied: planar state (reflection), focal conic state (forward scattering), and uniformly lying state (transmission) (Wu and Yang, 2001). The different states are illustrated schematically in Figure 11.12. In the case of the planar texture, the liquid crystal molecules are parallel to the surface of the cell and the incident light is selectively reflected by the system. In the focal conic texture, the orientation of the helical axes is assumed to be randomly distributed, but centered perpendicular to the substrate surfaces. In the case of the uniformly lying state, the helix axes lie parallel to the substrate and no reflection occurs. In the focal conic state the reflection depends on the distribution of the helical axis orientation and the internal structure of the multidomain texture.

We will now describe a model to simulate reflection properties such as intensity, chromaticity, and depolarization of light by these multidomain structured cholesteric liquid crystals. The investigations will be made using a statistical approach that is based on the 4×4 matrix method. We consider an inhomogeneous random material with the help of a quasi-three-dimensional ray-trace method. The simulation combines incoherent and coherent light propagation. The outline of the process is as follows. Small areas of the devices are simulated as a column, and light is propagated coherently through it. Stokes vectors of the reflected light are calculated that contain intensity values. Intensities for different columns are added and form a Stokes vector for the whole structure. This intensity addition is a completely incoherent process. Taking into account the domain structure of the focal conic textures of a cholesteric liquid crystal, one has to consider, in addition to the properties of the liquid crystal material, parameters such as number and thickness of domains, the orientations of the helices inside the domains, and the refractive index of the polymer that forms the borders of the domains.

John and colleagues investigated the influence of pitch variations and helical axis orientation distribution on a cholesteric system by averaging the reflection spectra, neglecting the internal structure of the multidomain texture (John et al., 1995). We will call this approach the single-domain model. Depolarization properties are underestimated because the reflection of a single-domain liquid crystal Bragg structure always produces the same polarization state. An advanced model can be applied that includes variations of the helical axis orientation and different domain sizes inside the liquid crystal cell. This model is called the multidomain model. It is set up as follows. One simulates the propagation of light through successive domains in a column while the orientation of the helix inside the domains is assumed to have a certain distribution. Each domain is modeled as a stratified medium and the Jones matrices of reflection and transmission are calculated for each domain with the 4×4 Berreman method (see Chapter 4). The inclination of the domain borders with respect to the propagation direction is included by taking into account an angle of incidence. The Jones matrices of reflection and transmission for a column consisting of different domains are then calculated by multiplication of the corresponding individual matrices. The reflection spectra of the columns are calculated as an averaged sum of a number of reflection spectra of randomly generated liquid crystal multidomain sample structures (columns in Fig. 11.13). The helix orientations and domain sizes are distributed statistically normal. As a measure for the distribution one can take the variance and the corresponding standard deviation (SD) of the helix orientation. Testing this model one obtains the predicted development of the reflection spectra by changing parameters like domain size, angle of incidence, and birefringence (Bohley et al., 2001). To simulate the optical properties for different states of the multistable system, the standard deviation of the orientation of the helical axis can be used as a relevant parameter. This gives the possibility to associate the applied voltages with a helical axis distribution and the internal domain structure in the liquid crystal cell (Bohley et al., 2001).

An interesting point of discussion is the dependence of the color coordinates on the variance of the helical axis (Bohley and Scharf, 2003). Figure 11.14 contains results of simulations and experimentally obtained color coordinates (Davis et al., 1998 (●) and Hashimoto et al., 1998 (◉)). For blue and green colors the

(*a*) (*b*) (*c*)

Figure 11.13 (*a*) To model the cholesteric domain structure, a sample volume, a column, is chosen that contains several domains. (*b*) The single-domain model treats light propagation as going through various samples with different but uniform helix orientations. (*c*) The multidomain model simulates propagation of light through different examples, which are divided into different domains of different helical orientations.

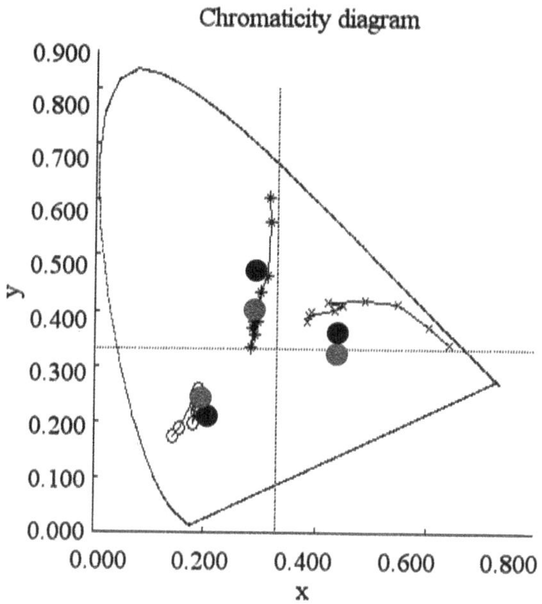

Figure 11.14 Chromaticities of three cholesteric liquid crystals ($n_o = 1.55$ $n_e = 1.8$ as above; pitches (p) 290 nm, blue, o; 331 nm, green, $*$; 400 nm, red, $\times$; linear polarized light at normal incidence). The chromaticities go from the outer to the inner space with increasing variance of the helix axis orientation.

simulation with the multidomain model corresponds well with experimentally findings. The calculated color coordinates for the red color, however, do not fit. This might be due to additives like absortive dyes in the red liquid crystal mixture or to different liquid crystal parameters. In practice, most reflective displays are designed as matrices of controllable basic color cells that lie in a common plane (Yeh and Gu, 1999; Wu and Yang, 2001). If the basic cells (blue, green, and red) consist of cholesteric liquid crystal material, a stacked design can be used. Three basic color cells are arranged one behind the other. This is possible because the circularly polarized light of same handedness as the cholesteric liquid crystal fully transmits light outside the reflection band. Using the multidomain method, one can investigate the state and degree of polarization of the light that is reflected and transmitted by stacked cholesteric liquid crystal systems. Considering the question of different combinations of the handedness of three stacked cholesteric liquid crystal cells (red–green–blue) one has to compare two principal arrangements: First, the three cells have the same handedness (+++); secondly, the two outer cells have the same handedness and the inner cell has the opposite handedness (+ − +). The spectra of the light passing through these two arrangements are different, the intensities of the + − + constellation being higher, as observed in experiments (Bohley and Scharf, 2002). The reason is that single cholesteric liquid crystal cells reflect the part of the incident light that is circularly polarized with the same handedness as the cholesteric liquid crystal director helix. The parts that are circularly polarized with the opposite handedness are transmitted for all wavelengths. Therefore, the $+ - +$ constellation has a better power budget. This is obvious in the perfectly aligned single-domain cholesteric liquid crystal display. For multidomain cholesteric liquid crystal systems the advantage is not so clear, because of the depolarization effects that are present (Section 11.8). In the case of a perfect planar texture, it is noticed that differences in the reflection spectra mainly appear in the regions between the edges of the reflection bands of the single cells. For a larger standard deviation of the helix orientation, the spectra differ more. Surprisingly, in simulations, the intensity for the $+ - +$ configuration, is about twice as high as the $+++$ configuration, which was not expected at all (Bohley and Scharf, 2002). This effect appears not only for the "white" state (all cells in full reflection), but also for "color" states, where the single basic cells are only partially switched.

11.8 DEPOLARIZATION EFFECTS BY POLYMER DISPERSED CHOLESTERIC LIQUID CRYSTALS

The light reflected by polymer-dispersed cholesteric liquid crystal displays in a multidomain configuration can be seen as being composed of many plane light waves of different intensities and polarization states. In order to investigate the polarization properties of the completely reflected light, one can use the concept of the Stokes column. See Chapter 1, which gives a brief introduction. To calculate the optical properties of a cholesteric liquid crystal texture, the Berreman matrix method is convenient. If the optical properties of anisotropic stratified media are simulated with the Berreman matrix, one can find Jones vectors to describe the reflected and transmitted intensities in the surrounding isotropic material. These Jones vectors can be transformed into Stokes columns using Equation (1.10). A number of incident plane waves results in a number of Stokes columns. Adding the individual contributions of the corresponding Stokes elements describes the observation with incoherent light. The result is a Stokes column that describes the texture, including the statistics of domain distribution. In detail, the Berreman method applied to the polymer-dispersed cholesteric liquid crystal layer delivers the reflected light in the form of n Jones vectors. The result for each texture column is

$$\begin{pmatrix} E_{xi} \\ E_{yi} \end{pmatrix} = \begin{pmatrix} E_{x0i} \\ E_{y0i} \cos \Delta_i \end{pmatrix}, \qquad i = 1, \ldots, n. \tag{11.59}$$

They are different, because of the propagation through different columns of the domain-structured layer. These Jones vectors can be obtained for instance by means of the multidomain method described in Section 11.7. Applying the principle of incoherent superposition of individual Stokes vectors, one can now write

$$\mathbf{S} = \begin{pmatrix} S_0 \\ S_1 \\ S_2 \\ S_3 \end{pmatrix} = \begin{pmatrix} \sum_{i=1}^{n} E_{x0i}^2 + E_{y0i}^2 \\ \sum_{i=1}^{n} E_{x0i}^2 - E_{y0i}^2 \\ \sum_{i=1}^{n} 2E_{x0i}E_{y0i} \cos \Delta_i \\ \sum_{i=1}^{n} 2E_{x0i}E_{y0i} \sin \Delta_i \end{pmatrix}. \tag{11.60}$$

This is the reflected light in Stokes column representation. The degree of polarization can now be calculated with Equation (1.15),

$$P = \frac{\sqrt{S_1^2 + S_2^2 + S_3^2}}{S_0}. \tag{11.61}$$

In order to represent the polarization of the nth contributions, we use the Poincaré sphere. The modulus of the Jones vector corresponds to the intensity of the individual contributions S_{0i}. An elegant method is to represent the intensity of each contribution by the distance of the points from the center of the Poincaré sphere. In that way, natural light would be represented by a uniform distribution of states on and inside the Poincaré sphere. Light of single linear polarization would be represented by one point on the equator. Figure 11.15 shows a series of Poincaré spheres with a radius of 1 in stereo representation. To see the stereoscopic effect one has to look at the images with relaxed eyes. Each sphere contains 50 examples of reflection. The simulated structure is a 10-domain-structured cholesteric liquid crystal with a thickness of 10 μm. The refraction indices are $n_o = 1.516$ and $n_e = 1.744$, with a pitch of 331 nm. Linear polarized light with an intensity of 2 is incident. The state of polarization of the incident light is represented by the point marked with + seen at the right side of the equator of the spheres. Increasing the standard deviation of the helix orientation causes a larger distribution of the polarization states and a shift towards the center, because of the decreasing intensity of the overall reflection. A decreasing degree of polarization is seen when moving from the planar state to the focal conic state. Such behavior has also been found experimentally (Khan et al., 2001). Using Equation (11.61), one is able to look at the development of the degree of polarization for different helical axis orientation distributions in a multidomain structured liquid crystal (Bohley and Scharf, 2002). For larger standard deviation of the helical axis orientation, the degree of polarizations becomes smaller because the state of completely disordered helical orientations is approached. If one compares the single-domain model and a multidomain model, one finds that a multidomain model gives a smaller degree of polarization and is therefore in better agreement with experimental results (Khan et al., 2001).

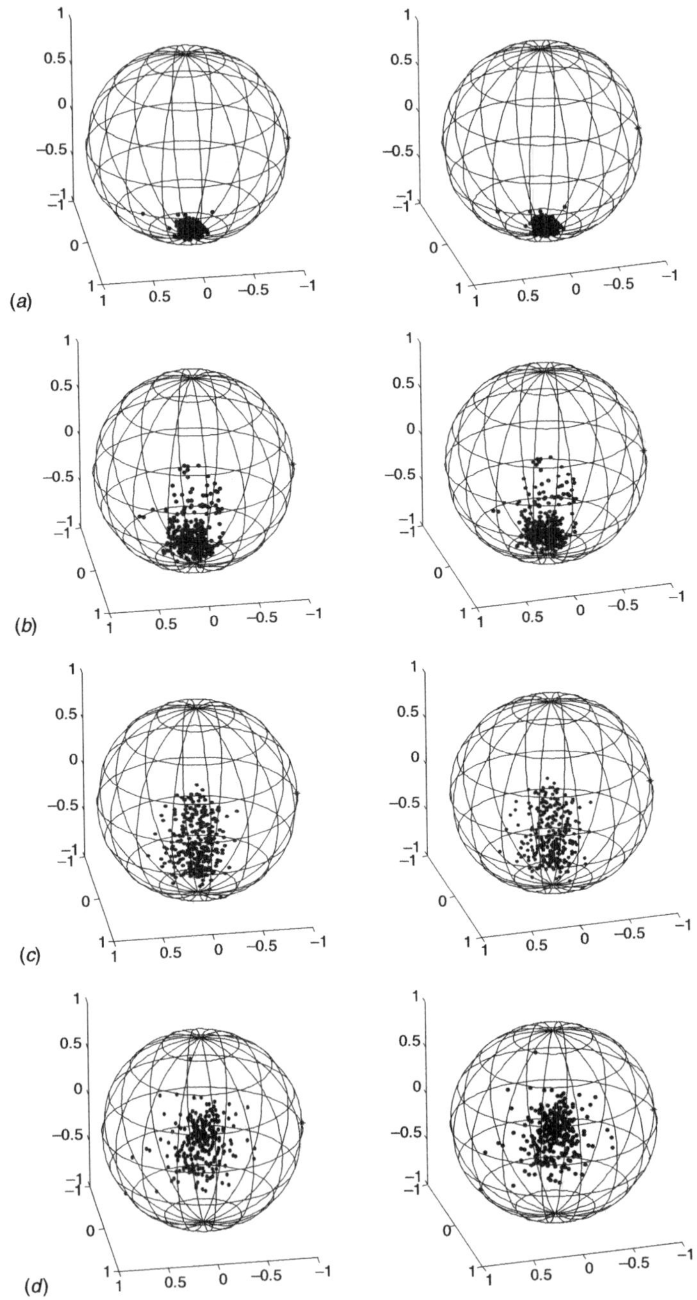

Figure 11.15 Poincaré sphere representation in stereoscopic view with intensity-weighted polarizations of light, which is reflected from a polymer-dispersed cholesteric liquid crystal. Calculation was carried out with the multidomain method for linear polarized light at normal incidence (wavelength 520 nm) and different standard deviations for the helix orientation: (*a*) 0°, (*b*) 20°, (*c*) 40°, and (*d*) 60°. (Courtesy of Dr Christian Bohley.)

11.9 DEFECT STRUCTURES IN CHOLESTERIC BRAGG REFLECTORS

With the ability to fabricate high-quality cholesteric mirrors in polymer liquid crystal the use of such elements in photonic devices has increased rapidly. Most evident is the replacement of multilayer mirrors by cholesteric Bragg structures. If two well-designed Bragg reflectors are brought together, they create a cavity with specific transmission and reflection properties. With such a concept, Fabry–Perot etalons of a special type have been realized (Stockley et al., 1999). We consider now defects in chiral structures that serve the same purpose: to create small transmission or reflection bands. The basic properties of defect modes in cholesteric liquid crystals are reviewed in the work of Kopp et al. (2003a). The cholesteric Bragg reflector can be seen as a one-dimensional photonic crystal structure. The important role of defects in photonic crystals has been understood from the very beginning of the investigation of photonic crystals. However, it arose originally from the design of classical thin film filters (Heavens, 1991). In photonic crystal structures, defects can serve for example as perfect lossless waveguides or as long-lived laser cavities (Yablonovitch et al., 1991; Joannopoulos et al., 1995). Many types of defects have been studied. They can be produced by removing or adding material or by altering the refractive index of one or a number of layers. Consider a multilayer stack that forms a Bragg reflector. Introducing a quarter-wavelength space in the middle of such a sample produces a defect line transmission in the middle of the reflection band. Such a defect is widely used to produce all dielectric interference filters (Heavens, 1991) and high-Q laser cavities in VCSELs (Yokoyama and Ujihara, 1995). Such a defect can also be produced in helical structures. Possible defect structures are shown schematically in Figure 11.16. One possibility is to add an isotropic layer in the middle of a cholesteric liquid crystal Bragg reflector (Yang et al., 1999). A second is a chiral twist defect that can be created by rotating one part of the sample about its helical axis (Kopp and Genack, 2002a,b; Schmidtke, 2003). Modifying the chiral twist angle from 0 to 180° tunes the defect frequency from the low- to the high-frequency band edge. A 90° twist produces a photonic defect with a frequency at the center of the stop band. The defect has also been examined in a macroscopic model of a chiral structure constructed from overhead transparency (Kopp, 2003b).

To simulate the optics of such structures, matrix methods are used. We will present a few examples to show the dependence on basic parameters and provide a simulation of the transmission. As a first example we consider the effect of an isotropic layer on the reflection band (Yang et al., 1999). The results are given in Figure 11.17. All calculations are made with the 4×4 Berreman matrix method and for normal incidence. Figure 11.17 shows the reflection for linear polarized light. A cholesteric stack with the following parameters is assumed: pitch $p = 300$ nm, $n_e = 1.77$, $n_o = 1.5$, thickness of the cholesteric slab is 6 μm. These parameters give a reflection band of 80 nm width, centered at 490 nm. The isotropic material is set in the middle of the structure so that it is sandwiched between cholesteric slabs of 3 μm thickness on each side. A very sharp peak of zero transmittance

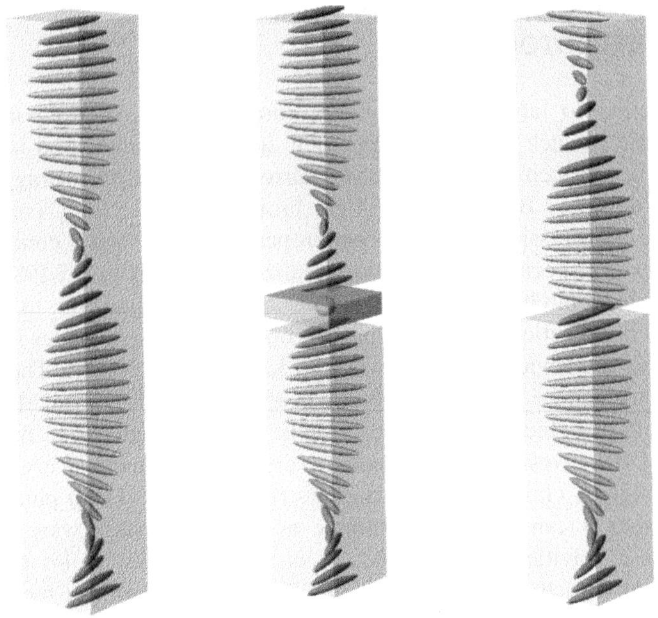

(*a*) (*b*) (*c*)

Figure 11.16 Schematic of defects in periodic cholesteric structures. (*a*) perfect cholesteric liquid crystal structure; (*b*) isotropic layer sandwiched between cholesteric Bragg reflectors; (*c*) chiral twist defect produced in periodic cholesteric structure by twisting the top of the sample relative to the bottom (90°).

appears. The position of the defect line transmission can be tuned by changing the thickness of the isotropic layer. Figure 11.17*b* shows the peak positions for different thicknesses varying from 0.38 μm to 0.42 μm. In the range shown, peak position changes linearly with thickness. Such a defect corresponds to the filter design concept for isotropic thin films. For thicker isotropic layers a multitude of defect states are created (Kopp et al., 2003a).

An additional possibility for creating transmission and reflection peaks for cholesteric structures are orientational defects. The perfect helix is interrupted and set together again with a discontinuity in the twist angle. We consider now the results of matrix calculations for the chiral twist defect with different twist angles. Figure 11.18 shows simulated transmission spectra. The parameters are the same as for the example above. The cholesteric slab is divided into two parts of 3 μm thickness. The upper part is rotated by a twist angle. The chiral twist defect introduces a single localized mode into the bandgap with a polarization of the same handedness as the structure. In contrast, an additional spacing introduced into isotropic periodic structures creates a degenerate pair of modes. As a result, the polarization of the chiral twist defect mode is independent of the polarization of the exciting mode, whereas the polarization of the wave inside a one-dimensional isotropic system always matches that of the exciting radiation. The localized chiral defect

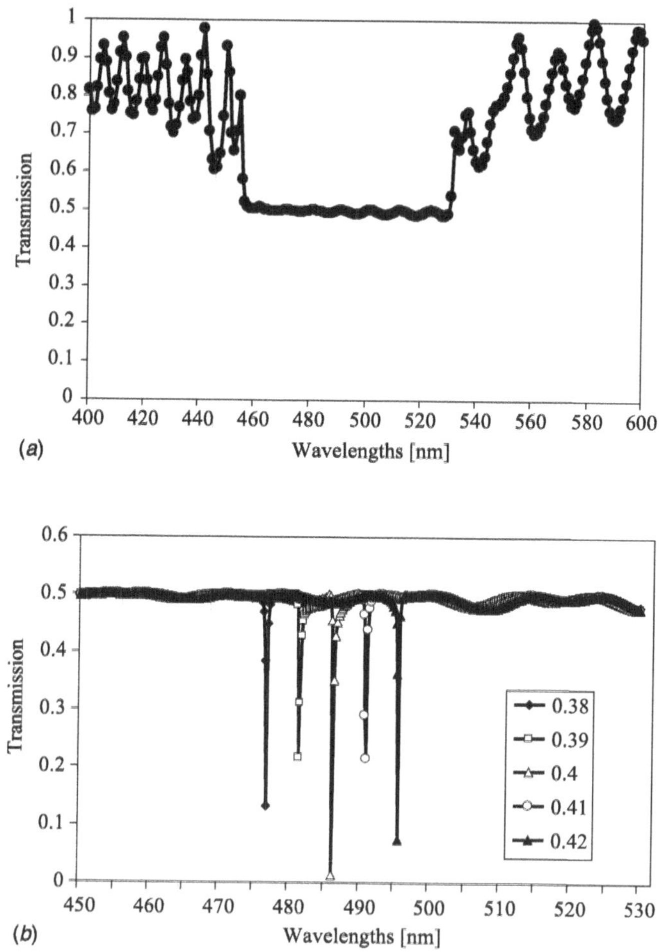

Figure 11.17 (*a*) Transmission of the 6 μm layer of cholesteric liquid crystal with $n_o = 1.5$, $n_e = 1.77$, and $p = 300$. (*b*) Transmission spectra in the reflection band if an isotropic layer is put in the middle of the structure. The thickness of the isotropic film is varied between 0.38 μm and 0.42 μm with $n_{isotropic} = 1.5$. The peak of zero transmission can be moved within the band by changing the thickness of the isotropic layer.

mode gives rise to a crossover in the nature of propagation (Kopp et al., 2003a,b). Below a crossover thickness, the localized mode is excited most efficiently by a wave with the same handedness as the structure and exhibits an exponential spatial distribution of the energy density inside the sample and a peak in transmission at the defect frequency. Above the crossover, the defect mode is only effectively excited by the oppositely polarized wave and a resonant peak appears in reflection. Although the polarization is transformed within the sample, the polarization of the transmitted and reflected waves is the same as that of the incident wave. Detailed descriptions of polarization states of defect modes are still a matter of

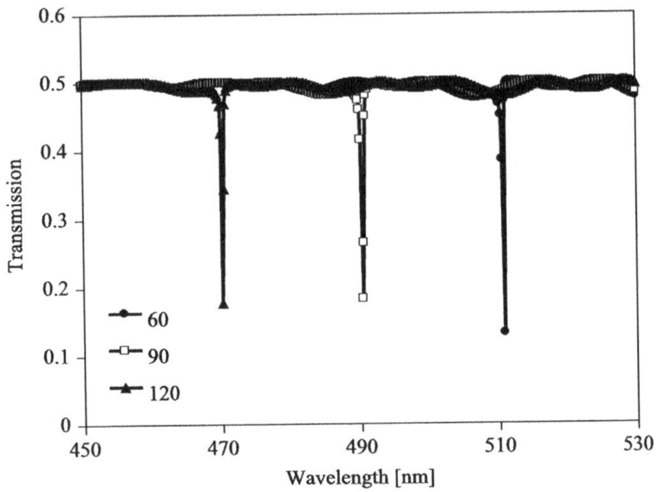

Figure 11.18 Localized modes on a cholesteric liquid crystal slab with orientational defects. At the middle of the 6 μm layer is a discontinuity of the twist angle with different values ($n_o = 1.5$, $n_e = 1.77$, $p = 300$ nm, thickness $d_1 = 3$ μm, $d_2 = 3$ μm).

intensive research (Wang and Lakhtakia, 2005). Combinations of cholesteric mirrors and switchable elements lead to interesting device concepts (Song et al., 2004). Adaptive systems can be made by utilizing a variable-phase plate between two cholesteric mirrors. A very modern application is the design of optical cavities for lasing (Kopp, 2003a; Song et al., 2006).

11.10 STRUCTURED CHOLESTERIC BRAGG FILTERS

The combination of cholesteric Bragg reflection with surface relief grating has not attracted so much attention yet. The systems are complicated and interference effects are found that are combinations of the photonic-crystal-like behavior of the cholesteric reflector and the diffraction effect of the surface relief grating. This leads to effects similar to those known from butterfly wings (Berthier, 2003). Cholesteric reflection is used by small animals to produce a colored appearance (Neville and Cavaney, 1969). Fabrication can be done using replication techniques (Broer, 1995; Scharf, 2005) that allow to create surface relief structure on a cholesteric layer. Figure 11.19 shows a scanning electron microscope micrograph of such a structure. The diffraction of the grating leads to a modified angular dependence of reflection properties. The dispersion properties can be reasonably modified. Figure 11.20 shows conoscopic images under different polarizer orientations. On the left, the polarizers are crossed. Transmission is small and only light within the Bragg reflection band is seen. For comparison, a grating made from isotropic material with comparable refractive index does not show any transmittance in this case. If the polarizers are turned to be parallel to the analyzer, diffraction of light

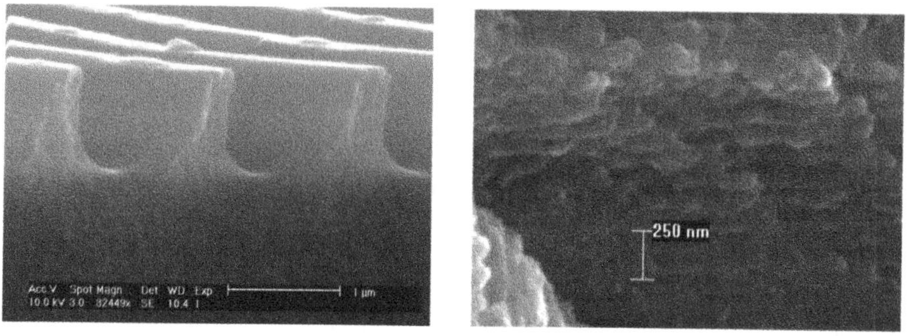

Figure 11.19 Replicated diffraction grating in cholesteric liquid crystal polymer (polysiloxane from Wacker Chemie Germany SLM 90520 Wacker). The pitch of the measured structure is approximately 250 nm. The Bragg reflection is in the blue for normal incidence. The grating has a period of 1 μm.

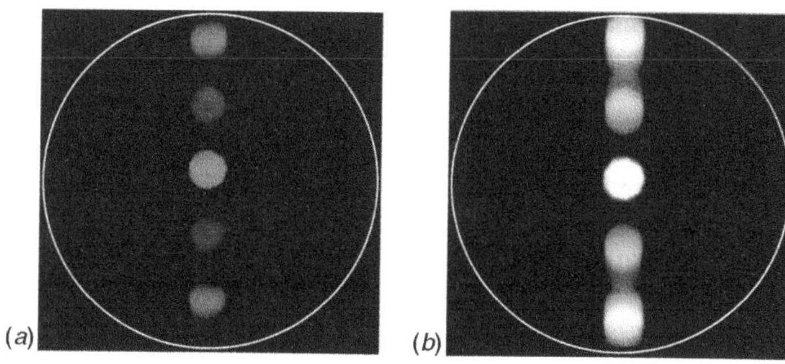

Figure 11.20 White light conoscopic diffraction images in transmission for the structure composed of a grating on a cholesteric liquid crystal Bragg reflector. (*a*) The liquid crystal structure under crossed polarizers. (*b*) Polarizers oriented at 45° to each other. In the conoscopic image, diffraction orders become visible as spots positioned at different positions. (Polysiloxane from Wacker Chemie Germany SLM 90520 Wacker.)

is seen with modifications of the transmission. This is shown in Figure 11.20*b*. The situation reflects the coupling between the grating transmission and the light propagation in the cholesteric liquid crystal.

11.11 PLANE WAVE APPROACH TO THE OPTICS OF BLUE PHASES

The optics of blue phase has been investigated using crystallographic methods (Jerome and Pieronski, 1989), many-wave optics methods (Belyakov and Dmitrienko, 1989), and Fourier methods based on simplified assumptions about the dielectric tensor properties (Hornreich and Shtrikman, 1993). As a fast and reliable method to study the optics of liquid crystalline blue phases, one can use the 4 × 4 Berreman matrix method. For the special case of the blue phase structures, a 4 × 4 matrix method was already been suggested by Berreman (1984).

11.11.1 Matrix Method for Anisotropic Crystals

The 4 × 4 Berreman method is based on the assumption that light propagation can be described as reflection (or transmission) by a stratified medium, as described Chapter 4. Before one starts to simulate optical properties, one has to model the director structure of the blue phase structure. A geometrical model has been used that implements double twist cylinders and fills up space between them with isotropic material (Bohley, 2004). The director structure is very complicated and not easy to visualize. A model unit cell for a blue phase II crystal is shown in Figure 11.21*a*. In order to find the Bragg reflection of the crystal for a certain direction, one divides the crystal into parallel and equidistant planes that are perpendicular to the propagation direction. This is illustrated in Figure 11.21*b*. The important point for light propagation simulations is that the dielectric tensors in each of these parallel planes are now averaged. The resultant index ellipsoid is usually biaxial. The average is made arithmetically, which is justified if the wavelength of the incident light is much larger than the period of the anisotropic crystal. One obtains for a certain propagation direction of light a stack of dielectric tensors along the propagation direction through the crystal. In what follows, this procedure will be applied to a blue phase crystal. Figure 11.22*a* shows the typical appearance of director planes. Two examples are shown for a blue phase II type crystal. In Figure 11.22*b*, the corresponding averaged biaxial index-ellipsoids are illustrated. Layers represent lattice planes when the unit cell is cut at different heights. The calculated principal dielectric susceptibilities are different for the principal directions. It is instructive to see the succession of dielectric tensor components on traveling through the layer structure.

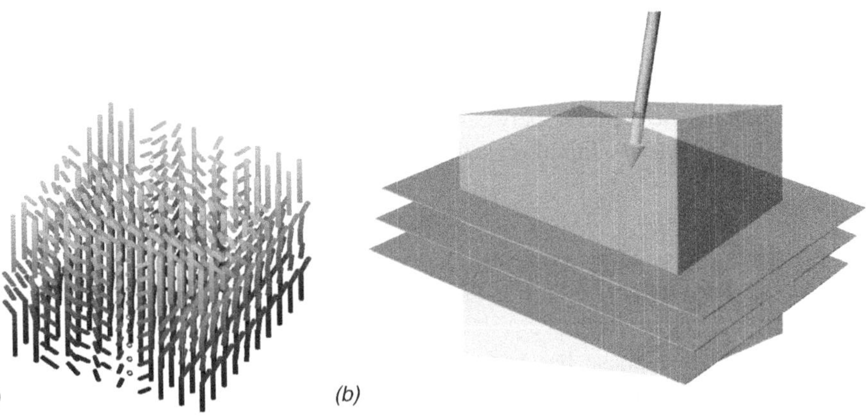

(a) *(b)*

Figure 11.21 Unit cell of a blue phase II model structure, and cuts perpendicular to the incident plane wave. The planes are normal to the propagation direction and spaced by *h*. The incident plane wave is symbolized by the arrow.

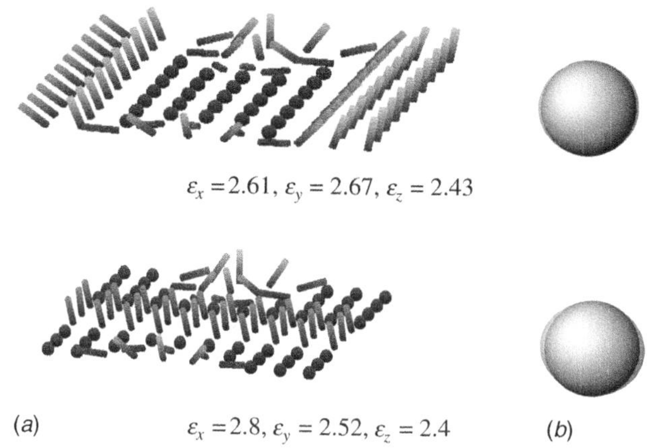

Figure 11.22 Examples of two different cuts through the blue phase II director field (uniaxial elements) and the corresponding (biaxial) index ellipsoids, obtained by averaging the dielectric tensors of the [100] lattice planes.

Figure 11.23 shows the biaxial index-ellipsoid components in more detail. The dielectric tensors may be observed passing through a blue phase II lattice in the [100] orientation at a depth of 840 nm. This corresponds to 3.5 lattice periods. The eigenvalues of the averaged dielectric tensors are ε_x, ε_y and ε_z. In the simulation, the ordinary refractive index n_o was set to 1.5 and the extraordinary index n_e to 1.8. The eigenvalues of the averaged dielectric tensor are between 2.25 and 3.24. Passing through the cubic crystal at normal incidence, one obtains averaged dielectric tensors with the same period as the spatial period of the lattice (here 240 nm). In

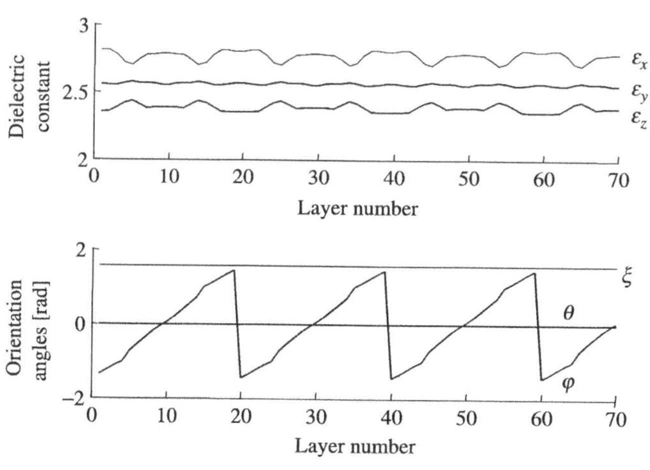

Figure 11.23 Principal values of the biaxial index ellipsoid and orientation angles when travelled though a blue phase II model structure at normal incidence (θ-angle with the x–y plane, φ-angle with the x–z plane, ξ-rotation angle around the direction specifed by θ and φ). The incidence angle is $\varphi = 0°$, and $\theta = 90°$.

our example this period corresponds to 20 layers of thickness 12 nm. The orientation angles of the biaxial index ellipsoid change within the period. It is interesting to observe that the semiaxes of the biaxial index ellipsoid vary only a little around the mean value. The mean values n_x, n_y, and n_z are distinctly different. One can introduce orientational angles to describe the local orientation of the index ellipsoid. The definition used here introduces the angles θ, φ, and ξ. With the help of these angles, the local coordinate system of the dielectric tensor can be transformed in the laboratory frame system. The angle θ represents tilt angle, φ is twist, and ξ represents a rotation around an axis parallel to the largest refractive index (Bohley, 2004). The point here is that the angles ξ and θ remain constant and only φ increases linearly. This means that the biaxial index ellipsoid keeps approximately the same form, but changes its orientation by twisting when the light traverses the blue phase crystal in the [100] direction. This behavior is reminiscent of the helical phase of a cholesteric nematic liquid crystal from which the structure is composed.

The situation is more complicated for oblique incidence (Bohley, 2004). For a certain direction of incidence one can calculate a stack of dielectric tensors. This stratified medium is then used to calculate the optical properties of the lattice using the 4×4 matrix method of Berreman. This method can be applied to anisotropic crystals. The outputs are Jones matrices of reflection and transmission for a chosen wavelength and angle of incidence. Note that this method is a plane wave approach and neglects lateral scattering in the medium.

11.11.2 Reflection Spectra

Reflection spectra of blue phase layer structures are possible, including the polarization properties. The parameters chosen for the sample simulations are a blue phase II in [100] orientation with a thickness of 10 μm, $n_o = 1.5$, and $n_e = 1.8$. The space between the double-twist cylinders as considered as filled with an isotropic space with the average refraction index $n = 1.61$, fulfiling the relation $n^2 = (n_e^2 + 2n_o^2)/3$ (Blinov and Chrigrinov, 1994). Figure 11.24 shows the calculated reflection spectra for different polarizations. The light is at normal incidence in Figure 11.24a and at oblique incidence at 20° in Figure 11.24b. For all types of polarization we obtain a selective reflection, which is strongest for the circularly polarized light of the handedness of the chiral nematic material. As for the helical phase of the chiral nematic liquid crystal material, the reflection spectrum is shifted to shorter wavelengths for oblique incidence.

11.11.3 Interference Diffraction Patterns

In order to investigate the optical behavior of a blue phase II layer, one can evaluate the Jones matrices in reflection and transmission for every direction of incidence and for different wavelengths. This is similar to the observation of conoscopic figures in the microscope (see Chapter 7 for this observation method). The intensities can be mapped into a plane and one obtains intensity and phase distributions, which are called "diffraction patterns." Figure 11.25 shows simulated diffraction patterns for

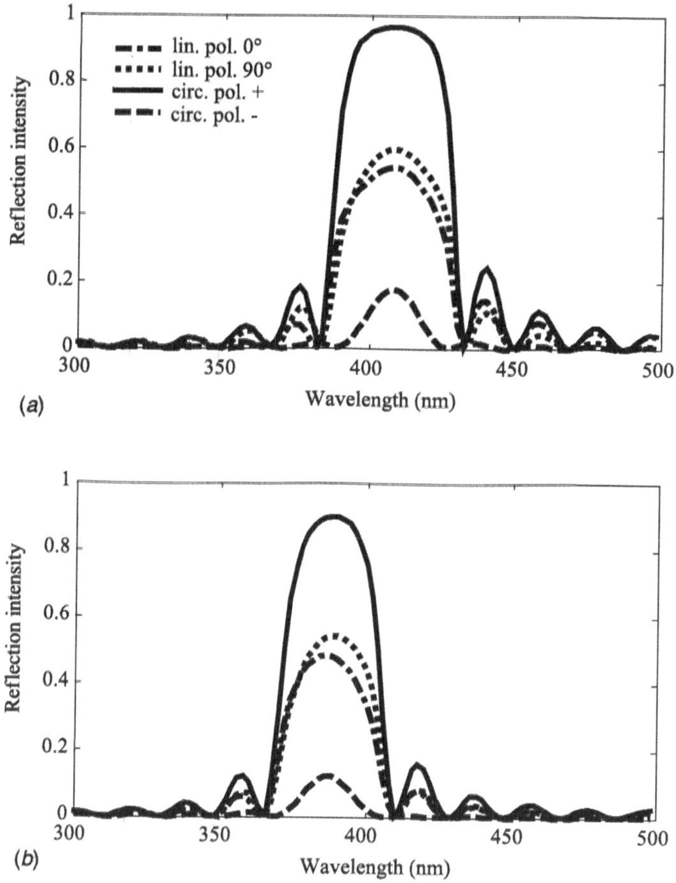

Figure 11.24 Reflection spectra of a blue phase II at normal (*a*) and oblique incidence at an angle of 20° (*b*) for different polarizations of the incident light. Strong selective reflection is observed for the circular polarization of the same handedness as the chiral material.

a blue phase II crystal in [100] and [110] orientations for the same material parameters as before. The incident light is right-handed circularly polarized. Figure 11.25 shows the so-called Kossel lines in four-fold symmetry for the [100] orientation and two-fold symmetry, as is expected for the [110] orientation. The simulations are carried out for different wavelengths and the "opening" of the Kossel lines for decreasing wavelengths is nicely seen in the simulations. The center of the Kossel lines corresponds to crystallographic axis (Oswald and Pieranski, 2000). The matrix method allows the investigation of the intensity behavior of Kossel lines at that crossing points (Bohley and Scharf, 2003). The polarization properties of the reflected light on the Kossel lines are additional substantial information that can be obtained from such simulations. By analyzing the ellipticity of the Bragg reflected light on the Kossel line, it is possible to compare simulated and measured results. Flack and colleagues measured Mueller matrices of Bragg

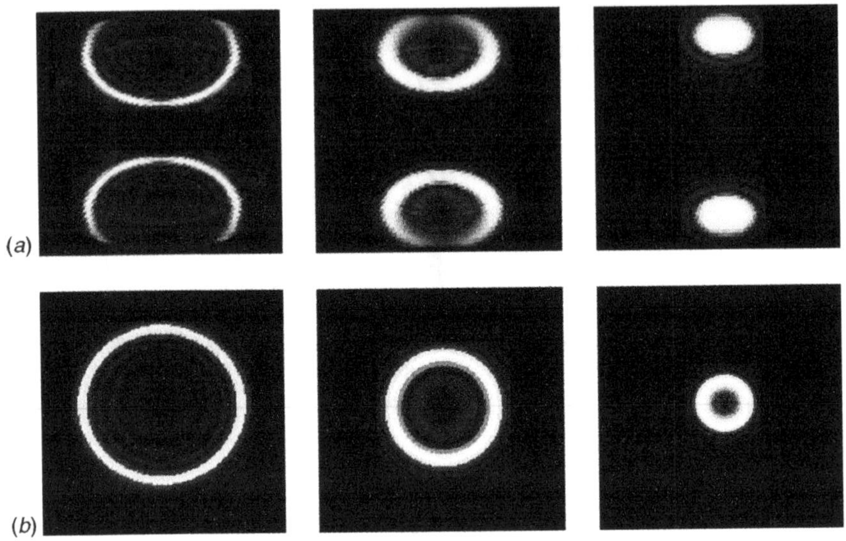

Figure 11.25 Simulated reflection diffraction patterns of blue phase II in (*a*) [110] and (*b*) [100] orientation for different wavelengths (left to right: 424 nm, 570 nm, and 652 nm). The structure is cubic, with a cell side length of 200 nm, $n_o = 1.5$, $n_e = 1.8$. Interference patterns are composed of reflection intensities calculated for directions up to an incident angle of $\theta = 70°$.

reflected light for different types of blue phases, with linear polarized incoming light and oblique incidence at $\theta = 45°$ (Flack et al., 1982). From measurements at single points they concluded that the Mueller matrices of the blue phase I and blue phase II phases are those of a homogeneous elliptical polarizer with an ellipticity of approximately 0.7 (where the ellipticity is defined as the amplitude ratio of the axes of the polarization ellipse). This means that the blue phase crystal reflects a specific elliptical polarization of the ellipticity 0.7. When the reflection behavior along a whole Kossel line was simulated, the same principal behavior was observed, but other values for the ellipticity, depending on the rectascension φ of the direction of the incident light, were found (Bohley, 2004). The simulation allows the calculation of the corresponding normalized Mueller matrices M for different angles of orientation. The measurement corresponds to an average of the main orientations at $\varphi = 0°$, 45°, and 90°. The simulated averaged Mueller matrices have the form

$$\mathrm{M}_{\mathrm{averaged}} = \begin{pmatrix} 1 & 0.072 & 0.093 & 0.921 \\ 0.072 & 0.014 & 0.001 & 0.108 \\ -0.093 & -0.001 & -0.026 & -0.068 \\ 0.922 & 0.108 & 0.068 & 0.96 \end{pmatrix}. \tag{11.62}$$

Such matrices were calculated from the Jones matrices obtained by adding Stokes columns as described in Section 11.8 for the cholesteric domain structure. Such an average matrix $\mathrm{M}_{\mathrm{average}}$ describes the Mueller matrix of the plane wave, which

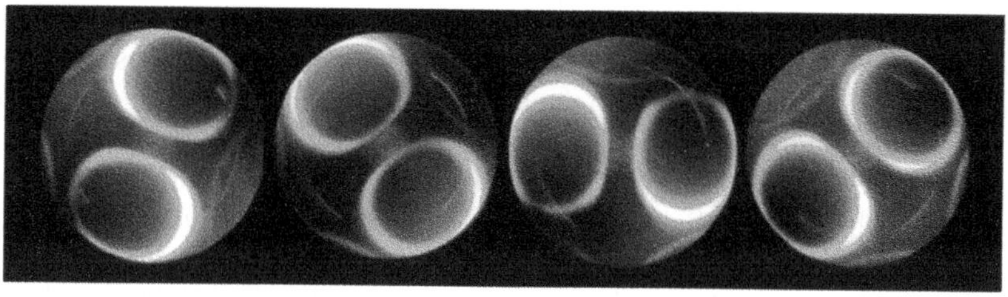

Figure 11.26 Photograph of diffraction patterns between crossed polarizers. The wavelength is 546 nm. The blue phase was obtained by a mixture of 55% chiral dopant CB15 in the liquid crystal mixture E48 (all from Merck).

is reflected by a polycrystalline specimen allowing all azimuthal orientations of the crystal platelets.

The reflected natural (nonpolarized) light from the Mueller matrix in Equation (11.62) has a Stokes column like a reflecting elliptical polarizer, but the major axis of the corresponding polarization ellipse changes slightly with the rectascension angle. This is the normalized matrix of an elliptical polarizer (Goldstein, 2003) and it is in good agreement with the single point measurements of Flack et al. (1982) for polycrystalline specimens. To prove the simulation concept it is useful to compare the simulation results with experiments. The reflection behavior of the blue phase II can be investigated with the conoscopic setup of the polarization microscope (Chapter 7). Diffraction patterns of the blue phase crystals can be obtained, which show the Kossel lines corresponding to the Bragg reflection caused by the crystalline structure of the blue phase II. For the observation of diffraction patterns and their polarization properties, high-quality large crystals are required. Therefore, for the following figures industrially fabricated test cells with rubbed polyimide coating were used. The typical thickness of the liquid crystal layer is 6 μm. The surface substrates were 300 μm thick, allowing the use of microscope objectives with high numerical aperture. The diffraction patterns in Figure 11.26 are digital photographs, taken in the focal plane of the polarization microscope objective (100 × 1.3 Oil Zeiss) at 546 nm with an interference filter that has full-width half-maximum of 10 nm. Between crossed polarizers, the Kossel lines for single crystals of the blue phase are not uniform in intensity. As an example, a platelet with a [100] crystal orientation is shown in Figure 11.26 at different orientations. One obtains a diffraction pattern with two-fold symmetry. When the blue phase crystal is rotated between crossed polarizers, the intensity of the Kossel lines varies depending on the rotational position. Corresponding diffraction patterns can be simulated and show good agreement (Bohley, 2003).

11.11.4 Dye Doped Liquid Crystal Blue Phase

In some recent publications fluorescent-dye-doped liquid crystals are used to establish functionalities like fluorescence in chiral nematic liquid crystal hosts (Schmidtke and Stille, 2003b). Absorptive dyes are mainly used to achieve special functionalities. The Heilmeier display, for instance, use a dichroic dyed Fréedericksz cell, which makes use of its anisotropic absorption feature (Wu and Yang, 2001).

Furthermore, for some liquid crystal devices, the need for external polarizing filters can be reduced or eliminated. The light propagation in chiral structures with dyes is changed significantly. This leads to the assumption that incorporation of absorptive dyes allows for tuning the optical reflection and transmission properties of liquid crystalline blue phases. Liquid crystals that contain suitable dichroic (anisotropic) dyes can absorb polarized light. Depending on the dye material used, one can have absorption parallel or perpendicular to the director orientation. The established guest–host interaction between the anisotropic dye molecule and the liquid crystal causes an alignment of the dye. This additionally provides the possibility of modulating the light propagation. The structure of the blue phase is very fragile and small changes in the composition of the mixture can be enough to destroy the three-dimensional self-organized nanostructure. To stabilize the phases and use it in optical applications, several concepts are promising. One of these is the photopolymerization of the blue phase and another is polymerization by gel formation and stabilization (Kikuchi et al., 2002). Other concepts are based on direct molecular interaction to extend the phase range (Coles and Pivnenko, 2005).

REFERENCES

Abu-Safia, H., Al-Sharif, A., and Abu Aljarayesh, I. (1993) Rugate filter sidelobe suppression using half-apodization, *Appl. Opt.* **32**, 4831–4835.

Alfrey, T., Gurnee, E.F., and Schrenk, W.J. (1969) Physical optics of iridescent multilayer plastic films, *Polym. Eng. Sci.* **9**, 400–404.

Alvarez, A.L., Tito, J., Vaello, M.B., Velasquez, P., Mallavia, R., Sanchez-Lopez, M.M., and Fernandez de Avila, S. (2003) B3Polymeric multilayers for integration into photonic devices, *Thin Solid Films* **433**, 277–280.

Belyakov, V.A. and Dmitrienko, V.E. (1989) Many wave optics of blue phases, *Liq. Cryst.* **5**, 839–836.

Berreman, D.W. and Scheffer, T.J. (1970) *Phys. Rev. Lett.* **25**, 577–581.

Berreman, D.W. (1984) In *Liquid Crystals and Ordered Fluids*, Vol. 4, Griffen, A.C., and Johnson, J.F. (Eds), Plenum Press, New York.

Berthier, S. (2003) *Iridescences – les Couleurs Physiques des Insectes*, Springer, Paris.

Blinov, L.M. and Chrigrinov, V.G. (1994) *Electrooptic Effects in Liquid Crystal Materials*, Springer, New York.

Bohley, C., Scharf, T., Klappert, R., and Grupp, J. (2001) Reflection of multi-domain structured cholesteric liquid crystals, *SPIE* **4463**, 181–187.

Bohley, C. and Scharf, T. (2002) Depolarization effects of light reflected by domain structured cholesteric liquid crystals, *Opt. Comm.* **214**, 193–198.

Bohley, C. and Scharf, T. (2003) Blue phases as photonic crystals, *SPIE* **5184**, 202–208.

Bohley, C. (2004) PhD thesis, University of Neuchâtel, Neuchâtel.

Born, M. and Wolf, E. (1980) *Principles of Optics*, 6th edn, Pergamon, New York.

Broer, D.J. (1995) Creation of supramolecular thin film architectures with liquid-crystalline networks, *Mol. Cryst. Liq. Cryst.* **261**, 513–523.

Chelix (2005) *CLC One-Component Narrowband CLC Polarizers, CLC One-Component Broadband CLC Polarizers*, Chelix Technologies Corp. 520 Mercury Drive Sunnyvale, CA 94085, USA.

Coles, H.J. and Pivnenko, M.N. (2005) Liquid crystal blue phases with a wide temperature range, *Nature* **436**, 997–1000.

Dändliker, R. (1997) Coupled waves: A powerful concept in modern optics. *Proc. SPIE* **3190**, 279–288.

Davis, D., Kahn, A., Huang, X.Y., Doane, J.W., and Jones, C. (1998) Eight color high-resolution reflective cholesteric LCDs, *SID Digest* **31**(2).

de Vries, H. (1951) Rotary power and other optical properties of certain liquid crystals, *Acta Cryst.* **4**, 219–226.

Flack, J.H., Crooker, P.P., and Svoboda, R.C. (1982) Measurements of the Mueller matrices of blue-phase structures, *Phys. Rev. A* **26**, 723–726.

Goldstein, D. (2003) *Polarized Light*, Marcel Dekker, New York.

Hashimoto, K. Okada, M., Nishiguchi, K., Masazumi, N., Yamakawa, E., and Taniguchi, T. (1998) Reflective color display using cholesteric liquid crystals, *SID Digest* **31**(1).

Heavens, O.S. (1991) *Optical Properties of Thin Films*, Dover, New York.

Hornreich, R.M. and Shtrikman, S. (1993) Photonic bands in simple and body-centered-cubic cholesteric blue phases, *Phys. Rev. E* **47**, 2067–2072.

Jerome, B. and Pieranski, P. (1989) Kossel diagrams of blue phases, *Liq. Cryst.* **5**, 799–812.

Joannopoulos, J.D., Meade, R.D., and Winn, J.N. (1995) *Photonic Crystals: Molding the Flow of Light*, Princeton University Press, Princeton.

St. John, W.D., Fritz, W.J., Lu, Z.J., and Yang, D.-K. (1995) Optical properties of bistable cholesteric reflective displays, *Phys. Rev. E* **51**, 1191–1198.

Khan, A., Huang, X.-Y., Armbruster, R., Nicholson, F., Miller, N., Wall, B., and Doane, J.W. (2001) *SID 01 Digest* **27**.

Kikuchi, H., Yokota, M., Hisakado, Y., Yang, H., and Kajiyama, T. (2002) *Nature Mater.* **1**, 64.

Kopp, V.I. and Genack, A.Z. (2002a) US Patent no. 6 396 859.

Kopp, V.I. and Genack, A.Z. (2002b) *Phys. Rev. Lett.* **89**, 033901.

Kopp, V.I., Zhang, Z.-Q., and Genack, A.Z. (2003a) Lasing in chiral photonic structures, *Prog. Quant. Electron.*, **27**, 369–416.

Kopp, V.I., Bose, R., and Genack, A.Z. (2003b) Transmission through chiral twist defects in anisotropic periodic structures, *Opt. Lett.* **28**, 349–351.

Neville, A.C. and Cavaney, S. (1969) Scarabaeid beetle exocuticle as an optical analogue of cholesteric liquid crystals, *Biol. Rev.* **44**, 531–562.

Oswald, P. and Pieranski, J.P. (2000) *Les Cristaux Liquides*, Gordon and Breach, Paris.

Scharf, T. (2005) Static birefringent microlenses, *Opt. Laser Eng.* **43**, 317–327.

Schmidtke, J. and Stille, W. (2003) Fluorescence of a dye-doped cholesteric liquid crystal film in the region of the stop band: theory and experiment, *Eur. Phys. J. B* **31**, 179–194.

Schmidtke, J., Stille, W., and Finkelmann, H. (2003) Defect mode emission of a dye doped cholesteric polymer network, *Phys. Rev. Lett.* **90**, 083902.

Song, M.H., Park, B., Shin, K.C., Ohta, T., Tsunoda, Y., Hoshi, H., Takanashi, Y., Ishikawa, K., Watanabe, J., Nishimura, S., Toyooka, T., Zhu, Z., Swager, T., and Takezoe, H. (2004)

Effect of phase retardation on defect mode lasing in polymeric cholesteric liquid crystals, *Adv. Mater.* **16**, 779–783.

Song, M.H., Ha, N.Y., Amemiya, K., Park, B., Takanishi, Y., Ishikawa, K., Wu, J.W., Nishimura, S., Toyooka, T., and Takezoe, H. (2006) Defect mode lasing with lowered threshold in a three-layered hetero-cholesteric liquid crystal structure, *Adv. Mater.* **18**, 193–197.

Southwell, W. (1989) Using apodization functions to reduce sidelobes in rugate filters, *Appl. Opt.* **28**, 5091–5094.

Stockley, J.E., Sharp, G.D., and Johnson, K.M. (1999) Fabry–Perot etalon with polymer cholesteric liquid-crystal mirrors, *Opt. Lett.* **24**, 55–57.

Wang, F. and Lakhtakia, A. (2005) Defect modes in multisection helical photonic crystals, *Opt. Exp.* **13**, 7319–7335.

Weber, M.F., Stover, C.A., Gilbert, L.R., Nevitt, T.J., and Ouderkirk, A.J. (2000) Giant polarization optics, *Science* **287**, 2451–2456.

Wu, S.-T., and Yang, D.-K. (2001) *Reflective Liquid Crystal Displays*, Wiley, New York.

Yablonovitch, E., Gmitter, T.J., Meade, R.D., Rappe, A.M., Brommer, K.D., and Joannopoulos, J.D. (1991) *Phys. Rev. Lett.* **67**, 3380–3385.

Yang, Y.C., Kee, C.S., Kim, J.E., Park, H.Y., Lee, J.C., and Jeon, Y.J. (1999) Photonic defect modes of cholesteric liquid crystals, *Phys. Rev. E* **60**, 6852–6854.

Yariv, A., and Yeh, P. (2003) *Optical Waves in Crystals*, Wiley, New York.

Yeh, P. and Gu, C. (1999) *Optics of Liquid Crystal Displays*, Wiley, New York.

Yokoyama, H. and Ujihara, K. (Eds) (1995) *Spontaneous Emission and Laser Oscillation in Microcavities*, CRC Press, Boca Raton, FL.

INDEX

Polarized Light in Liquid Crystals and Polymers. By Toralf Scharf

Printed and bound by CPI Group (UK) Ltd, Croydon, CR0 4YY
16/09/2026
14963443-0003